The Mesenteric Organ in Health and Disease

Eli D. Ehrenpreis · John C. Alverdy ·
Steven D. Wexner
Editors

The Mesenteric Organ in Health and Disease

Editors
Eli D. Ehrenpreis
Department of Medicine
Advocate Lutheran General Hospital
Park Ridge, IL, USA

John C. Alverdy
Department of Surgery
University of Chicago Medicine
Chicago, IL, USA

Steven D. Wexner
Department of Colon and Rectal Surgery
Cleveland Clinic Florida
Weston, FL, USA

ISBN 978-3-030-71965-4 ISBN 978-3-030-71963-0 (eBook)
https://doi.org/10.1007/978-3-030-71963-0

This Springer imprint is published by the registered company Springer Nature Switzerland AG
The registered company address is: Gewerbestrasse 11, 6330 Cham, Switzerland

This book is dedicated with love to my wife Ana and my children Benjamin, Jamie, and Joseph; all creative scholars dedicated to making the world a better place.

EDE

All surgeons have the unique privilege of the opportunity to help patients by improving the quality of and/or prolonging their lives. Academic surgeons are also blessed by being able to share our knowledge and skills with others allowing the multifold multiplication of their practices to countless other patients who we will never meet. The desire to teach can be fulfilled locally by training residents, fellows, and students and globally through research and education. I have been incredibly fortunate during the last 33 years to have been able to avail myself of teaching trainees at Cleveland Clinic Florida. The common denominator among these endeavors has been the commitment of my philanthropic friends who have supported my research and educational endeavors at Cleveland Clinic Florida since 1988. Although none of these magnanimous altruistic friends are surgeons, their gifts mean that by sharing my vision, they have also made a global impact on the health of countless patients, perhaps in every country. Accordingly, it is with immense pleasure and eternal gratitude that I express my deepest appreciation to my friends by dedicating this textbook to them:

Nick Caporella
Ellen Cates-Smith and Sandford Smith
Aliko Dangote
Dr. Danny Dosoretz and Chely Dosoretz
Itchko Ezratti and Diana Morrisson
Theo and Connie Folz
Bernard and Juliette Klepach
Bill Livek
Bernie Marcus
Earl and Nancy Stewart
Herbert and Doreen Wigwe.

SDW

Foreword

The conventional model of abdominal anatomy described multiple and separate mesenteries. Recent advancements in our understanding of the mesenteric organ demonstrate that there is one mesentery and that all abdominal digestive organs are in direct connection with the single mesenteric organ. These advances prompted a re-appraisal of human mesenteric embryology that in turn led to the discovery that abdominal digestive organs develop within and remain in close association with the mesenteric frame. This has major implications for human biology in general. For example, it means the abdomen is anatomically organized into two principle compartments; the mesenteric and non-mesenteric compartment (i.e., the mesenteric model). The mesenteric model of human abdominal anatomy overcomes problems associated with the peritoneal-based model and in so doing reconciles embryological, anatomical, surgical, and radiological approaches to the abdomen as a group.

The centrally positioned mesenteric continuum is of far greater importance in human biology than was previously thought. It follows that it is of considerable importance to pathological derangements of human biology, i.e., diseases. The following is *the* foundation text on mesenteric diseases and is thus timely. Written by established expert scientists and clinicians it is carefully designed to provide information in an accessible format but also to offer a platform for use for reference purposes. The editors and assembled authors should be congratulated on compiling the first reference text on mesenteric diseases. It is an invaluable resource that will become part of teaching curricula in medical, radiological, pathological, and surgical training programs. Most importantly, this will translate into substantial clinical benefits for patients with abdominal and systematic diseases.

Limerick, Ireland JC Coffey

Preface

Almost two decades ago, a fifty-one-year-old male who was a local politician came to see me for a cluster of concerning symptoms. The patient complained of pain across his lower abdomen that was worse with activity. Furthermore, although a large man, he had a poor appetite and had lost a significant amount of weight in the prior two months. He was alarmed by another symptom: awakening with sweats that soaked his pajamas and bedsheets. A physical examination revealed an obvious abdominal mass that was firm and tender to the touch. The mass enveloped a substantial portion of the right lumbar, umbilical, and hypogastric regions of the abdomen. Anticipating a diagnosis of sarcoma, and a dismal prognosis, I ordered imaging of the abdomen with computerized tomography. As expected, a large abdominal mass was found. However, a subsequent biopsy demonstrated chronic inflammation and fat necrosis consistent with the diagnosis of mesenteric panniculitis. After finding virtually no clinical resources to assist in managing this condition, I embarked on a journey to try to treat my new patient and to offer innovative approaches to the disease. With the help of an enthusiastic group of medical residents and gastroenterology fellows, my group began publishing a variety of observational studies based on a gathering number of patients with mesenteric panniculitis that were coming to the office. These included a new method to score the severity of symptoms for future pharmaceutical studies called the Mesenteric Panniculitis Symptom Assessment Score (MPSAS). Understanding nuances of the clinical care of patients with mesenteric panniculitis followed, allowing for the solidification of diagnostic and therapeutic approaches of mesenteric panniculitis. Many questions arose when caring for patients with a rare disease that had fewer than 300 hundred cases described in the medical literature. For example, what happens if patients fail or are intolerant to the few initial therapies that are recommended? What is the synergy that occurs between the physical and psychological components of a benign disease that nobody knows how to treat? What future risks can be anticipated and what is the appropriate follow up for patients with mesenteric panniculitis? One can anticipate similar difficulties when caring for patients with other rare or poorly understood medical conditions. Our group ultimately developed several new potential treatments for mesenteric panniculitis including thalidomide, enteric coated budesonide (unpublished), and low dose naltrexone using the MPSAS as a guide to drug response. Because many patients that are first diagnosed with "mesenteric panniculitis" or "misty mesentery" based solely on CT findings without confirmatory biopsies, we also explored the relationship between mesenteric panniculitis and malignancy, an important piece of the mesenteric panniculitis puzzle.

The mesentery, generally viewed as a fragmented group of connective tissues that functioned to attached the gastrointestinal organs to the abdominal cavity, has undergone anatomic description for centuries. Of interest, a drawing of the mesentery by Leonardo Da Vinci from about 1508 that is owned by the Royal Collection Trust shows his deep understanding of its gross anatomy, most likely developed from its careful dissection away from adjacent abdominal organs. In Da Vinci's view, the mesocolon and mesorectum are shown in contiguity suggesting a single mesenteric structure whose point of origin appears to be at the mesenteric root near the small intestine. This view was contrary to the more modern anatomists who described a fragmented, unorganized version of the mesentery. After a detailed review of historic anatomic descriptions as well as gross and microscopic examination of cadaveric and surgical resections, J Calvin Coffey, a colorectal surgeon from Limerick Ireland, and his group published a paper in the Lancet Gastroenterology and Hepatology in 2016 that re-examined the relationship of the mesentery and its surrounding structures. Their proposal that the mesentery is a distinct organ in the body first raised a great deal of interest in the scientific and lay press and subsequently resulted in the acceptance of this concept in the scientific world. In fact, the mesentery now appears on the list of 79 organs of the human body on Wikipedia and is classified as a gastrointestinal organ. Aside from the curiosity that the finding of a "new organ in the body" creates, this concept allows for an expanded understanding of the role that the mesentery may play in the pathophysiology of diseases affecting other organ systems.

Most medical diseases of the mesentery have not received the degree of attention as the mesenteric vascular disorders, including ischemic enteropathy and ischemic colitis. Ischemic enteropathy is among the deadliest medical condition affecting the mesentery and has been the focus of important work and medical breakthroughs in the past few decades. Other medical diseases of the mesentery are generally viewed in the miscellaneous category of gut disorders and have remained relegated as such. In the surgical literature, the mesentery has garnered more attention. Inclusion of mesenteric resection in the management of rectal cancer has resulted in significant improvement in prognosis in these patients. More recent work has also suggested that removal of portions of the mesentery with intestinal resections may enhance the outcomes of these surgeries in patients with Crohn's disease. Since the time that my patient walked through the office door, a greater understanding of mesenteric involvement in local and systemic disease processes has occurred. The single organ model of the mesentery is a potential catalyst toward understanding the role of the mesentery in the maintenance of the homeostasis in healthy individuals.

These fresh developments in the field of mesenteric disorders and my prior experience with mesenteric panniculitis directed me to think about creating a new book specifically focused on the mesentery. To fully actualize the process, it was vital to engage surgical colleagues to write and edit major portions of the work.

I was fortunate to receive enthusiastic interest in the book from Dr. John Alverdy, a talented surgeon and Professor of Surgery at the University of Chicago, whose life work has focused on the effects of bacterial translocation from the gastrointestinal tract, the source of which occurs at the interface of the mesentery and the gut. Dr. Steven Wexner, Chairman of the Department of Colorectal Surgery at Cleveland Clinic Florida, is an innovator in the field of colorectal surgery who has also become an eager partner in this endeavor. His work on laparoscopic mesorectal resections, mesenteric lengthening techniques, and advanced laparoscopic surgeries for Crohn's disease has made him an ideal coauthor and coeditor for the project.

In putting together this first book devoted specifically to diseases of the mesentery, we had several important challenges. Our goal was to create an easy to read and clinically relevant book. However, since we wished to bring the reader up to speed on translational aspects of normal and abnormal mesenteric function, a detailed review of mesenteric anatomy and physiology was also required. To accomplish these goals, we have enlisted a number of additional authors and co-authors to create the chapters of this book. All authors were instructed to focus on clinical features of their subjects. Illustrative examples and succinct tables are included throughout to assist in the understanding of material that is presented. We have been very fortunate to have enlisted Dr. Abraham Dachman, Professor of Radiology from the University of Chicago who created a library of radiographic images of mesenteric diseases for use in the book.

The structure of the book consists of introductory chapters, followed by a review of the anatomy, embryology, and physiology of the mesentery. After an overview of relevant medical procedures, the role of the mesentery in systemic medical disorders is covered in several chapters. The book is completed with chapters for specific medical and surgical diseases of the mesentery.

As authors and editors, we respectfully submit our book as the first endeavor to assemble what is known about the mesentery in a single volume. We hope in turn that our efforts will enhance the appreciation of the mesentery in normal and disease states and that this will stimulate further developments that benefit patient care.

Park Ridge, IL, USA Eli D. Ehrenpreis

Contents

Part VIII Surgery for Individual Conditions

Part IX Future

Contributors

John C. Alverdy Department of Surgery, University of Chicago, Chicago, IL, USA

John Anagnostakos Department of Surgery, Sinai Hospital of Baltimore, Baltimore, MD, USA

Karl Andersen Department of Medicine, Advocate Lutheran General Hospital, Park Ridge, IL, USA

Shoaib Ashfaq School of Medicine and Department of Surgery, University of Limerick Hospital Group, Raheen, Dooradoyle, Limerick, Ireland

Charles Broy Department of Gastroenterology, Edward Hines Jr VA Hospital, Hines, IL, USA

Sarah Burroughs Department of Medicine, Advocate Lutheran General Hospital, Park Ridge, IL, USA

Kevin G. Byrnes School of Medicine and Department of Surgery, University of Limerick Hospital Group, Raheen, Dooradoyle, Limerick, Ireland

Mariane G. M. Camargo Digestive Disease and Surgery Institute, Department of Colorectal Surgery, Cleveland Clinic, Cleveland, OH, USA

Adib Chaus Department of Internal Medicine – Cardiology, Advocate Lutheran General Hospital, Park Ridge, IL, USA

J. Calvin Coffey School of Medicine and Department of Surgery, University of Limerick Hospital Group, Raheen, Dooradoyle, Limerick, Ireland

Tara M. Connelly School of Medicine and Department of Surgery, University of Limerick Hospital Group, Raheen, Dooradoyle, Limerick, Ireland

Kristen T. Crowell Department of Surgery, Division of Colon and Rectal Surgery, Beth Israel Deaconess Medical Center, Boston, MA, USA

Abraham H. Dachman Department of Radiology, The University of Chicago, Chicago, IL, USA

Colum P. Dunne School of Medicine and Centre for Interventions in Infection, Inflammation and Immunity, University of Limerick, Castleroy, Limerick, Ireland

Suzanne S. Dunne School of Medicine and Centre for Interventions in Infection, Inflammation and Immunity, University of Limerick, Castleroy, Limerick, Ireland

Eli D. Ehrenpreis Department of Medicine, Advocate Lutheran General Hospital, Park Ridge, IL, USA

Sameh Hany Emile Department of Colorectal Surgery, Mansoura University Hospitals, Mansoura, Dakahlia, Egypt

Christina A. Fleming School of Medicine and Department of Surgery, University of Limerick Hospital Group, Raheen, Dooradoyle, Limerick, Ireland

Sara Gaines Department of Surgery, University of Chicago, Chicago, IL, USA

Siddhartha Guru Department of Medicine, Advocate Lutheran General Hospital, Park Ridge, IL, USA

Saad Hashmi Department of Internal Medicine, Advocate Lutheran General Hospital, Park Ridge, IL, USA

Paul T. Hernandez Department of Surgery, Division of Colon and Rectal Surgery, University of Pennsylvania, Perelman School of Medicine, Philadelphia, PA, USA

Ryan T. Hoff Department of Medicine, Advocate Lutheran General Hospital, Park Ridge, IL, USA

Richard A. Jacobson Department of Surgery, Rush University Medical Center, Chicago, IL, USA

J. Joshua Smith Department of Colorectal Surgery, Memorial Sloan Kettering Cancer Center, New York, NY, USA

Deborah S. Keller Division of Colon and Rectal Surgery, NewYork-Presbyterian, Columbia University Medical Center, New York, NY, USA

Robert C. Keskey Department of Surgery, University of Chicago, Chicago, IL, USA

Hermann Kessler Digestive Disease and Surgery Institute, Department of Colorectal Surgery, Cleveland Clinic, Cleveland, OH, USA

Ahmed Khattab Department of Internal Medicine, Division of Gastroenterology, Advocate Lutheran General Hospital, Park Ridge, IL, USA

Miranda G. Kiernan School of Medicine and Centre for Interventions in Infection, Inflammation and Immunity, University of Limerick, Castleroy, Limerick, Ireland
School of Medicine and Department of Surgery, University of Limerick Hospital Group, Raheen, Dooradoyle, Limerick, Ireland

David H. Kruchko Department of Medicine, Advocate Lutheran General Hospital, Park Ridge, IL, USA

Chloe Lee Department of Internal Medicine, Advocate Lutheran General Hospital, Park Ridge, IL, USA

Craig A. Messick Department of Surgical Oncology, The University of Texas MD Anderson Cancer Center, Houston, TX, USA

Assad Munis Department of Medicine, Advocate Lutheran General Hospital, Park Ridge, IL, USA

Vincent Obias Department of Colorectal Surgery, George Washington University Medical Faculty Associates, Washington, DC, US

Peter O'Leary School of Medicine and Department of Surgery, University of Limerick Hospital Group, Raheen, Dooradoyle, Limerick, Ireland

Emily Papazian Department of Surgery, University of Chicago, Chicago, IL, USA

Salvatore Parascandola Department of Colorectal Surgery, George Washington University Medical Faculty Associates, Washington, DC, US

Amir Patel Department of Gastroenterology and Hepatology, Medical College of Wisconsin/Froedtert Hospital, Milwaukee, WI, USA

Jeffrey Prochot Department of Internal Medicine, Advocate Lutheran General Hospital, Park Ridge, IL, USA

Cindy G. Pulido Department of Medicine, Advocate Lutheran General Hospital, Park Ridge, IL, USA

Felipe Quezada-Díaz Department of Colorectal Surgery, Memorial Sloan Kettering Cancer Center, New York, NY, USA

Joseph V. Sakran Division of Acute Care Surgery and Adult Trauma Services, Department of Surgery, Johns Hopkins University, Baltimore, MD, USA

Dorsa Samsami Department of Internal Medicine, Advocate Lutheran General Hospital, Park Ridge, IL, USA

Nicole M. Saur Department of Surgery, Division of Colon and Rectal Surgery, University of Pennsylvania, Perelman School of Medicine, Philadelphia, PA, USA

Rishabh Sehgal Department of Colorectal Surgery, University Hospital Limerick, Limerick, Ireland

Natasha Shah Department of Gastroenterology, Advocate Lutheran General Hospital, Park Ridge, IL, USA

Sherief Shawki Department of Colon and Rectal Surgery, Cleveland Clinic, Cleveland, OH, USA

Khaja M. Siraj Department of Internal Medicine, NorthShore Medical Group, Evanston, IL, USA

Scott Sorensen The University of Chicago, Chicago, IL, USA

Winson Jianhong Tan Department of Colorectal Surgery, Memorial Sloan Kettering Cancer Center, New York, NY, USA

Nyi Nyi Tun Department of Internal Medicine, Miami Valley Hospital, Dayton, OH, USA

Jasper B. van Praag Department of Surgery, University Medical Center Groningen, Groningen, The Netherlands

Dara Walsh School of Medicine and Department of Surgery, University of Limerick Hospital Group, Raheen, Dooradoyle, Limerick, Ireland

Steven D. Wexner Department of Colon and Rectal Surgery, Cleveland Clinic Florida, Weston, IL, USA

Ashley J. Williamson Department of Surgery, University of Chicago, Chicago, IL, USA

Joshua H. Wolf Department of Surgery, LifeBridge Health, Baltimore, MD, USA

Evan G. Wong Division of Acute Care Surgery and Adult Trauma Services, Department of Surgery, Johns Hopkins University, Baltimore, MD, USA

Thomas Q. Xu Department of Surgery, Rush University Medical Center, Chicago, IL, USA

Part I
General

1 Redefining the Mesentery as an Organ

J. Calvin Coffey and Peter O'Leary

Definition of the Mesentery

The mesentery is defined as the collection of tissues that maintains all abdominal digestive organs in position and in continuity with other systems. This definition differs considerably from the classic definition which describes the mesentery as a double fold of peritoneum that maintains the intestine in position.

Once the shape of the mesentery was clarified it became apparent that it was continuous, substantive, and important from a functional perspective. It was also noted that it could be readily differentiated from structures with which it was continuous. Further studies demonstrated that mesenteric continuity applied from the oesophago-gastric to anorectal junction. In turn, it was observed that all abdominal digestive organs are continuous with the mesentery.

These findings indicated that the mesentery has an essential function. In the embryo, it serves as an incubator for digestive organs, providing a platform on which they can develop. In the adult it also has an essential function in maintaining the organization of the digestive system as well as its connectivity with all other systems.

These observations led our group to suggest that it be designated an organ in itself.

Historic Development of the Understanding of the Mesentery

As our current understanding of the mesentery differs considerably from that depicted in the classic model, it is important to describe how our current understanding of the mesentery developed. Renaissance anatomists visually depicted a

J. C. Coffey (✉) · P. O'Leary
School of Medicine and Department of Surgery, University of Limerick Hospital Group, Raheen, Dooradoyle, Limerick, Ireland
e-mail: calvin.coffey@ul.ie

E. D. Ehrenpreis et al. (eds.), *The Mesenteric Organ in Health and Disease*,
https://doi.org/10.1007/978-3-030-71963-0_1

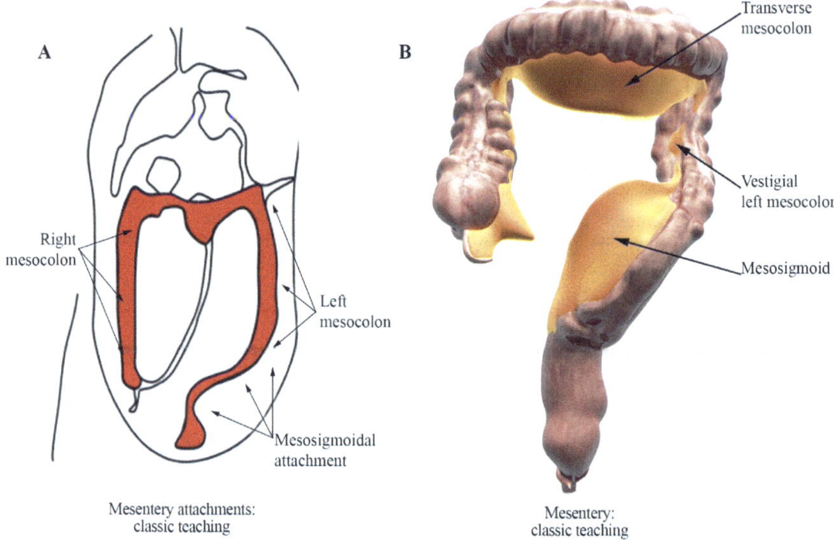

Fig. 1.1 Classic model depicting multiple mesenteries and a complex peritoneal landscape

continuous mesentery but they did not textually describe it as such. In general, textual descriptions of the mesentery were remarkably limited. In 1858 Henry Gray described the "root" of the small intestinal mesentery as attaching along a particular trajectory across the posterior abdominal wall. He described multiple mesenteries, as opposed to one continuous mesentery (Fig. 1.1). This concept was supported by observations of numerous surgeons including Sir Frederick Treves.

The Gray model was indoctrinated in reference texts for a number of reasons, until recently when it was noted that the mesentery is in fact continuous and that all abdominal digestive organs are continuous with it. The last point is particularly important as it means that our understanding of how all digestive organs are connected to the rest of the body, must change.

It is now recognized that the mesentery is centrally positioned within the abdominal digestive system, linking all digestive organs, and connecting these collectively to the posterior abdominal wall (Fig. 1.2). Moreover, the mesentery, and digestive organs, determine and explain the peritoneal landscape. As the latter determines the shape of the peritoneal cavity, it follows the anatomy of the peritoneal cavity is primarily determined and explained by the mesentery and associated digestive system organs (Fig. 1.3).

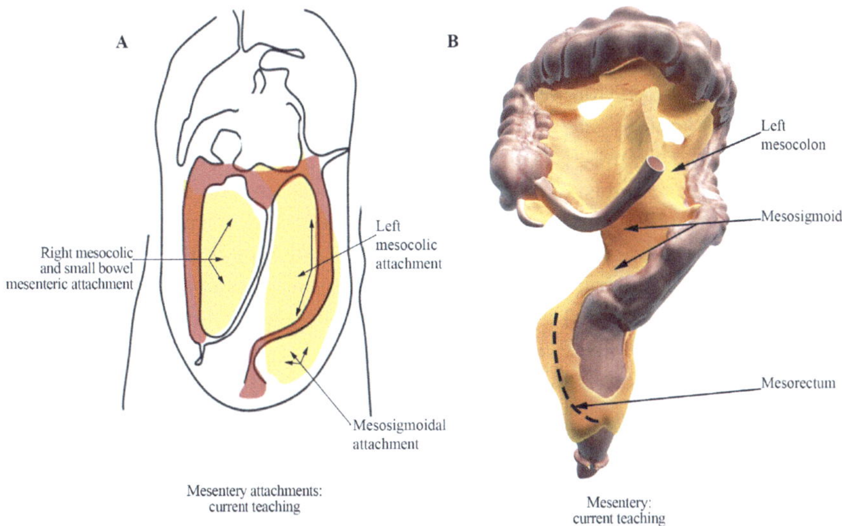

Fig. 1.2 Current model depicting a single continuous mesentery. Mesenteric continuity has major implications for our understanding of the anatomical organization of the abdomen

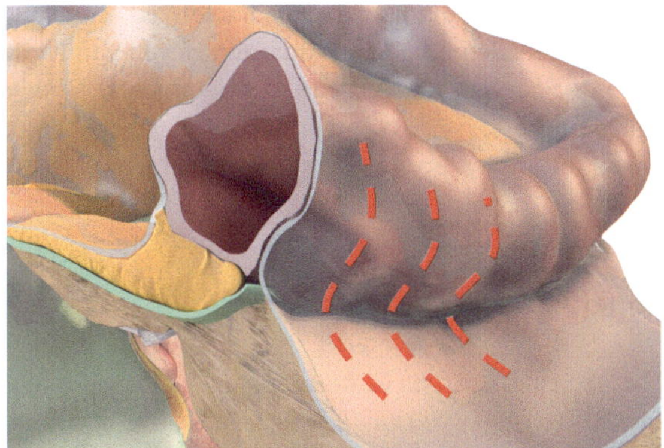

Fig. 1.3 3D digital sculpture demonstrating the peritoneal reflection and the position of the peritoneum in the mesenteric-based model of abdominal anatomy

The Science of the Mesentery

The proposal that the mesentery is an organ was surprising in that many held that all human anatomy was well established. The proposal generated considerable media attention and argument. Whether the mesentery is an organ or not is a point for discussion, and as with any discussion there are for and against sides.

Arguing against the concept that it is an organ, one commentator argued that whilst the mesentery is not an organ, it does occupy the same "hierarchical level" as an organ. The argument was based on multiple points but chief amongst these was the concept the mesentery does not serve an essential and unique function. Others commentated that more information was required in order to make a definitive decision on the matter.More recently, anatomists have moved towards arguing that anatomical terminology such as "organ" requires updating, and that classification systems need updating.

Apart from these points, there appears to be little else in formal literature arguing against the concept of the mesenteric organ. Indeed, many scientists are using the definition as an opportunity to further characterize and explore the mesentery. It is not possible to reference all articles of relevance as these are continuing to emerge in numerous languages and across multiple scientific and clinical disciplines. In many respects this book, reflects that movement.

If we are to consider the mesentery as an organ, then we must challenge it in the same way we would any organ. In keeping with this, the current text summarizes diseases of the mesentery and is the first reference text that explains how these may be diagnosed and treated. Yi Li et al. recently published a comprehensive review of lymphatics of the mesenteric organ. Britta Siegmund's group described the importance of the mesenteric organ in the context of intestinal and systemic immunity. Rivera et al. elegantly summarized the role of mesenteric events in multiple diseases, in multi-organ dysfunction syndrome and the systemic inflammatory response. Argikar et al. characterized the pharamacobiology of the mesenteric organ demonstrating several opportunities for further scientific exploration. Several authors have characterized mesenteric-organ inputs into hepatic, pancreatic, splenic and intestinal development. These are the subject of a focused issue of Seminars in Cell and Developmental Biology. The title of the issue, currently in press, is "Mesenteric Organogenesis." Indeed, several authors have even linked the mesentery to correlates in numerous traditional medicines.

Conclusion

The mesenteric organ is being given increased attention in the forthcoming reference text *Gray's Surgical Anatomy*, in the upcoming edition of *Gray's Anatomy* and in reference textbooks on human embryology (including the most recent edition of Langmans Embryology). These are important developments that signify increasing acceptance of the importance of the mesentery in scientific and clinical practice.

It is imperative that we now exploit developments in order to enhance our ability to diagnose and treat human disease. In that regard, there is no more fitting place to start, than with diseases of the mesentery itself.

Suggested Reading

1. Coffey JC, Sehgal R, Culligan K, Dunne C, McGrath D, Lawes N, Walsh D. Terminology and nomenclature in colonic surgery: universal application of a rule-based approach derived from updates on mesenteric anatomy. Tech Coloproctol. 2014;18:789–94.
2. Sehgal R, Coffey JC. Standardization of the nomenclature based on contemporary mesocolic anatomy is paramount prior to performing a complete mesocolic excision. Int J Colorectal Dis. 2014;29:543–4.
3. Coffey JC, O'Leary DP. The mesentery: structure, function, and role in disease. Lancet Gastroenterol Hepatol. 2016;1(3):238–47.
4. Culligan K, Walsh S, Dunne C, Walsh M, Ryan S, Quondamatteo F, Dockery P, Coffey JC. The mesocolon: a histological and electron microscopic characterization of the mesenteric attachment of the colon prior to and after surgical mobilization. Ann Surg. 2014;260 (6):1048–56.
5. Hikspoors JPJM, Kruepunga N, Mommen GMC, Peeters JPWU, Hülsman CJM, Eleonore Köhler S, Lamers WH. The development of the dorsal mesentery in human embryos and fetuses. Semin Cell Dev Biol. 2018.
6. Byrnes KG, Walsh D, Lewton-Brain P, McDermott K, Coffey JC. Anatomy of the mesentery: historical development and recent advances. Semin Cell Dev Biol. 2018. pii: S1084–9521(18) 30204–0.
7. Kumar A, Kishan V, Jacob TG, Kant K, Faiq MA. Evidence of continuity of mesentery from duodenum to rectum from human cadaveric dissection—a video vignette. Colorectal Dis. 2017;19(12):1119–20.
8. Coffey JC, Dillon M, Sehgal R, Dockery P, Quondamatteo F, Walsh D, Walsh L. Mesenteric-based surgery exploits gastrointestinal, peritoneal, mesenteric and fascial continuity from duodenojejunal flexure to the anorectal junction—a review. Dig Surg. 2015;32(4):291–300.
9. Coffey JC, Culligan K, Walsh LG, Sehgal R, Dunne C, McGrath D, Walsh D, Moore M, taunton M, Scanlon T, Dewhurst C, Kenny B, O'Riordan C, O'Brien JM, Quondamatteo F, Dockery P. An appraisal of the computed axial tomographic appearance of the human mesentery based on mesenteric contiguity from the duodenojejunal flexure to the mesorectal level. Eur Radiol. 2016;26(3):714–21.
10. Coffey JC, O'leary DP. Defining the mesentery as an organ and what this means for understanding its roles in digestive disorders. Expert Rev Gastroenterol Hepatol. 2017;11 (8):703–5.
11. Byrnes KG, McDermott K, Coffey JC. Development of mesenteric tissues. Semin Cell Dev Biol. 2018. pii: S1084–9521(18)30234–9.
12. Gray, H. Anatomy, descriptive and surgical. New York;1858.
13. Treves F. Lectures on the anatomy of the intestinal canal and peritoneum in man. Br Med J. 1885;1:470–4.
14. Culligan K, Coffey JC, Kiran RP, Kalady M, Lavery IC, Remzi FH. The mesocolon: a prospective observational study. Colorectal Dis. 2012;14(4):421–8; discussion 428–30.
15. Neumann PE. Another new organ! is this a golden age of discovery in anatomy? Clin Anat. 2018;31(5):648–9. https://doi.org/10.1002/ca.23184 Epub 2018 May 25.
16. Kumar A, Ghosh SK, Faiq MA, Deshmukh VR, Kumari C, Pareek V.: A brief review of recent discoveries in human anatomy. QJM. 2018.
17. Neumann PE. Organ or not? Prolegomenon to organology. Clin Anat. 2017;30(3):288–9.
18. Argikar AA, Argikar UA. The mesentery: an ADME perspective on a 'new' organ. Drug Metab Rev. 2018;50(3):398–405.
19. Li Y, Ge Y, Gong J, Zhu W, Cao L, Guo Z, Gu L, Li J. Mesenteric lymphatic vessel density is associated with disease behavior and postoperative recurrence in Crohn's disease. J Gastrointest Surg. 2018;22(12):2125–32.

20. Siegmund B. Mesenteric fat in Crohn's disease: the hot spot of inflammation? Gut. 2012;61 (1):3–5.
21. Rivera ED, Coffey JC, Walsh D, Ehrenpreis ED. The mesentery, systemic inflammation, and Crohn's disease. Inflamm Bowel Dis. 2019;25(2):226–34.
22. Byrnes KG, McDermott KW, Coffey JC. Mesenteric organogenesis. Semin Cell Dev Biol. 2018.
23. Gong CZ, Liu W. Convergence of medicines: west meets east in newly-discovered organs and functions. Chin J Integr Med. 2018.

Embryology of the Mesentery

2

Kevin G. Byrnes and J. Calvin Coffey

Introduction

Past appraisals of embryonic development of the mesentery aimed to reconcile observations with the concept of multiple mesenteries. As it is now accepted that there is one, single and continuous mesentery, this prompts a re-appraisal of mesenteric embryology. That process is currently underway, and as such, it is somewhat premature to provide a detailed description of mesenteric embryology.

It is, however, possible to mention some points that hold, irrespective of recent advancements in our understanding of anatomy. Previously, the mesentery was regarded as an inert scaffold that structurally supported digestive organ development, but with little direct contributions to the latter. Clarification of mesenteric structure permitted systematical investigation of the mesentery and with this came the realization that the mesentery is far more active in digestive organ and related organogenesis, than previously thought. This led to the development of the field of mesenteric organogenesis, the study of organ development related to the mesentery.

Embryological Development of the Mesentery and Related Organs

The Ventral and Dorsal Mesentery

The mesentery develops early during development, as a membrane that spans the developing coelom which is the forerunner of the peritoneal cavity (Fig. 2.1). The following discussion will focus on the upper region of the abdominal mesentery,

K. G. Byrnes · J. C. Coffey (✉)
School of Medicine and Department of Surgery, University of Limerick Hospital Group, Raheen, Dooradoyle, Limerick, Ireland
e-mail: calvin.coffey@ul.ie

E. D. Ehrenpreis et al. (eds.), *The Mesenteric Organ in Health and Disease*,
https://doi.org/10.1007/978-3-030-71963-0_2

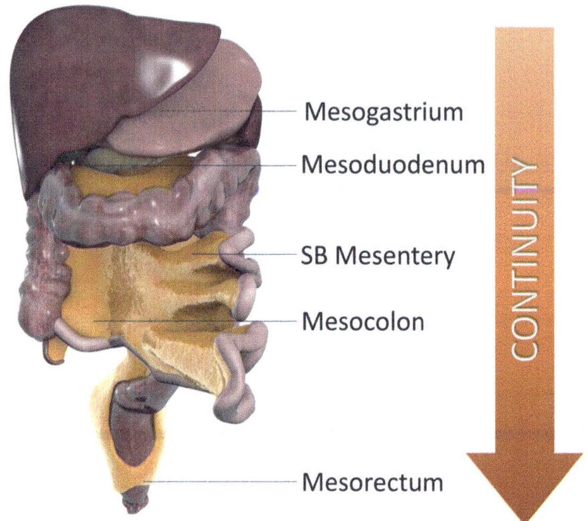

Fig. 2.1 Mesenteric continuity extending throughout abdominal digestive system

which extends from anterior to the posterior wall of the coelom, It is arbitrarily divided into two regions, based on the presence of endoderm (from which the intestine will derive). The region of the mesentery anterior to the developing intestine is the ventral mesentery, and posterior is the dorsal mesentery. The ventral mesentery has a lower limit extending from the developing diaphragm to the umbilicus. The dorsal mesentery extends beyond this to the region of the anorectal junction.

Development of the Upper Region of Mesentery and Associated Organs

All abdominal digestive organs develop in and remain continuous with, the mesentery (Fig. 2.2). In the upper region, the liver develops in the ventral mesentery and quickly outgrows the boundaries of this. At full term and in the adult, the ventral mesentery in which the liver developed is retained as a mesenteric brace, the front region of which corresponds to the falciform ligament and the back region corresponds to the ventral mesentery. Both falciform ligament and ventral mesentery converge and meet at the oesophagogastric mesenteric pedicle (see chapter on *Anatomy*).

A region of the upper mesentery invaginates to give rise to a mesenteric bubble or diverticulum that is oriented to the left. The spleen develops in the left-most corner of this bubble of mesentery. The stomach will develop as a region of developing endoderm, in the anterior leaf of the bubble. The mesentery between the stomach and spleen will extend inferiorly into the coelom, and give rise to the greater omentum.

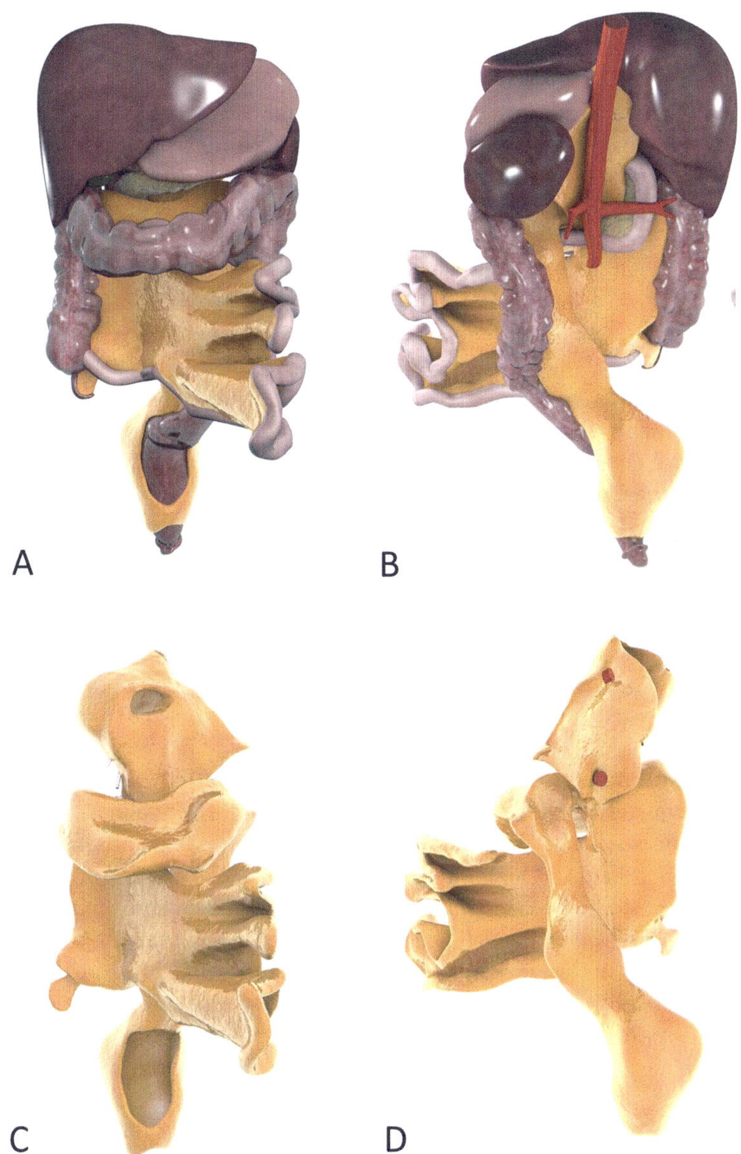

Fig. 2.2 Abdominal digestive organs centered on a mesenteric frame. **a** Anterior view of mesentery with contiguous abdominal digestive organs in situ **b** Posterior view of mesentery with contiguous abdominal digestive organs in situ **c** Anterior view of mesenteric frame extending from mesogastrium to mesorectum **d** Posterior view of mesenteric frame extending from mesogastrium to mesorectum

Development of the Lower Region of Mesentery and Associated Organs

At present the precise morphological changes the mesentery undergoes during development, are unknown. Data is emerging, following the application of a number of techniques, pointing to several processes, but these remain to be validated. For example, our group has examined individual stages of embryological development, reconstructed the shape of the mesentery at each stage (in the context of mesenteric continuity), then merged each stage in animation to depict the processes involved. A second approach involved reverse engineering the development of the mesentery, working from the adult shape, in reverse, back to the stage at which the mesentery was a sheet like structure spanning the coelom. The shape changes identified using both these techniques are complex, due to the presence of the ventral mesentery in the upper region, and evolving relationships with digestive organs as these develop in tandem. Although these findings are preliminary several statements can be made about mesenteric development.

Firstly, most development of the mid and lower intestine-mesenteric complex occurs, as mentioned, at the periphery of the mesentery. The developing gut tube elongates faster than the mesentery causing the mesentery and gut tube to form a coil. A fixed set of coils arises early in development, and these are the platform on which the remainder of development occurs. At this point in development, the duodenal region forms a coil to the right of the midline. The developing hindgut has already lateralised to the left of the midline. Between left and right lateralized regions, the gut tube has a hairpin-like conformation and the mesentery occupies the space between limbs of the hairpin.

The intestine of the proximal limb of hairpin elongates substantially, generating multiple additional coils. These events are not replicated in the distal limb of hairpin. Instead, the distal limb of developing gut tube and mesentery, elongate in tandem. As a result of in-tandem development, they take up a position first in the upper region of the developing abdominal cavity, and thereafter in the right flank of the cavity. The last change is important is at causes the highly coiled region of intestine and mesentery (the forerunner of the ileum and jejunum) to take up a position in the central region of the abdomen.

Development of the Peritoneum and Peritoneal Cavity

Most mesenteric organogenesis occurs at the periphery of the mesentery. By full-term, the mesentery and digestive organs have acquired their final conformation and topographical organization in the coelom. In tandem with this, mechanisms of mesenteric attachment (see chapter on *Anatomy*) develop and assist in maintaining the mesentery and abdominal digestive system in position. The peritoneal reflection is a major mechanism by which the mesentery is held in position (Fig. 2.3). However, the reflection appears to continue to develop after birth. As the peritoneal landscape determines the shape of the peritoneal cavity, it follows that the

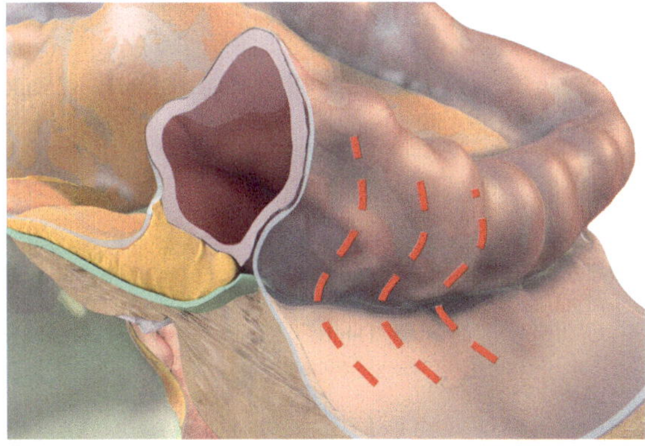

Fig. 2.3 Topography of lateral peritoneal reflection following completion of mesenteric development. Red broken line indicates reflection

peritoneal cavity develops both before and after birth. In fact, changes in the peritoneal landscape occur throughout life, under the influence of several natural and artificial (i.e. surgical interruption) factors.

Migration Across the Mesentery

Several cell types use the mesentery as a platform across which to migrate. These include cells that migrate from the neural crest to the developing intestine where they will contribute to the enteric nervous system. Primordial germ cells migrate from the developing hindgut, across the mesentery to the genital ridges where they will ultimately give rise to adult germ cell.

Regenerative Capacity of the Mesentery and Bioengineering of Abdominal Digestive Organs

The field of mesenteric organogenesis received a considerable boost with the identification of mesenteric continuity. Prior to this development, multiple groups had been working separately on the embryological development of digestive system organs in humans and other species. For example, the sea cucumber and related species expel its gut tube under the influence of certain stimuli. This is followed by the regeneration of a new gut tube at the mesenteric margin once occupied by the previous intestine.

With clarification of mesenteric shape, it became apparent that previously divergent fields shared a common element in mesenteric inputs. These and other events have prompted renewed interest and focus on recapitulating the mesenteric milieu, in bioengineering of liver, pancreatic and intestinal tissue. It is likely that advances in this area of human embryology could assist in addressing the current deficit in digestive organs available for transplantation.

Suggested Reading

1. Coffey JC, O'Leary DP. The mesentery: structure, function, and role in disease. Lancet Gastroenterol Hepatol. 2016;1:238–47. https://doi.org/10.1016/S2468-1253(16)30026-7.
2. Coffey JC, Dockery P. Colorectal cancer: surgery for colorectal cancer—standardization required. Nat Rev Gastroenterol Hepatol. 2016;13:256–7. https://doi.org/10.1038/nrgastro.2016.40.
3. Byrnes KG, Walsh D, Dockery P, McDermott K, Coffey JC. Anatomy of the mesentery: current understanding and mechanisms of attachment. Semin Cell Dev Biol. 2018. https://doi.org/10.1016/j.semcdb.2018.10.004.
4. Culligan K, Walsh S, Dunne C, Walsh M, Ryan S, Quondamatteo F, Dockery P, Coffey JC. The mesocolon. Ann Surg. 2014;260:1048–56. https://doi.org/10.1097/SLA.0000000000000323.
5. Byrnes KG, Walsh D, Lewton-Brain P, McDermott K, Coffey JC. Anatomy of the mesentery: historical development and recent advances. Semin Cell Dev Biol. 2018. https://doi.org/10.1016/j.semcdb.2018.10.003.
6. Sadler TW. Langman's medical embryology. 2018.
7. Yang L, Li L-C, Wang X, Wang W-H, Wang Y-C, Xu C-R. The contributions of mesoderm-derived cells in liver development. Semin Cell Dev Biol 2018. https://doi.org/10.1016/j.semcdb.2018.09.003.
8. Byrnes K, Walsh D, Lowery A, Lamers WH, Mcdermott KW, Coffey JC. Reconstructing the embryological development of the mesentery. Ir J Med Sci. 2017;186:S330–S330.
9. Byrnes KG, Walsh D, Ullah MF, Hashmi O, Westby D, Garritt A, Mirapeix R, Lamers W, Wu Y, Zhang SX. AB065. 229. On the growth and form of the mesentery, Mesentery and Peritoneum. 2 (2018).
10. Savin T, Kurpios NA, Shyer AE, Florescu P, Liang H, Mahadevan L, Tabin CJ. On the growth and form of the gut. Nature. 2011;476:57–62. https://doi.org/10.1038/nature10277.
11. Kastelein AW, Vos LMC, de Jong KH, van Baal JOAM, Nieuwland R, van Noorden CJF, Roovers J-PWR, Lok CAR. Embryology, anatomy, physiology and pathophysiology of the peritoneum and the peritoneal vasculature. Semin Cell Dev Biol. 2018. https://doi.org/10.1016/j.semcdb.2018.09.007.
12. Byrnes KG, McDermott K, Coffey JC. Development of mesenteric tissues. Semin Cell Dev Biol. 2018. https://doi.org/10.1016/j.semcdb.2018.10.005.
13. Nishiyama C, Uesaka T, Manabe T, Yonekura Y, Nagasawa T, Newgreen DF, Young HM, Enomoto H. Trans-mesenteric neural crest cells are the principal source of the colonic enteric nervous system. Nat Neurosci. 2012;15:1211–8. https://doi.org/10.1038/nn.3184.
14. Hen G, Sela-Donenfeld D, 'A narrow bridge home': the dorsal mesentery in primordial germ cell migration. Semin Cell Dev Biol (2018). https://doi.org/10.1016/j.semcdb.2018.08.010.
15. García-Arrarás JE, Bello SA, Malavez S. The mesentery as the epicenter for intestinal regeneration. Semin Cell Dev Biol (2018). https://doi.org/10.1016/j.semcdb.2018.09.001.
16. Byrnes KG, McDermott KW, Coffey JC. Mesenteric organogenesis. Semin Cell Dev Biol. 2018. https://doi.org/10.1016/j.semcdb.2018.10.006.

General Anatomy of the Mesentery

3

Christina A. Fleming, Dara Walsh, and J. Calvin Coffey

Introduction

The mesentery is a continuous structure with which all abdominal digestive organs are continuous (i.e. from liver to rectum). This description differs markedly from descriptions currently used by many reference textbooks of anatomy, when describing the anatomy of the abdomen in general terms. These texts describe the classic model of abdominal anatomy which held that there were multiple mesenteries, and that individual digestive system organs (with exception the small intestine, transverse colon and sigmoid) attached directly to the posterior abdominal wall. The peritoneal landscape that stemmed from the older model was complex and unpredictable (Fig. 3.1).

According to the current model, the mesentery is continuous and all abdominal digestive organs first attach directly to the mesentery. In turn the mesentery collectively connects them to the posterior abdominal wall. In keeping with this, their vascular, lymphatic and biliary circuitry are housed within the mesentery. In addition, the resultant peritoneal landscape is entirely explained by the anatomy of the underlying mesentery and digestive system (Fig. 3.2).

Given the above, it is first important to describe the mesentery, the abdominal component of the digestive system, and the peritoneum. As the mesentery collectively holds the entirety of the abdominal digestive system in position, the mechanisms by which it does so must be described. This is essential as visceral surgery almost invariably requires a disruption of these anatomical mechanisms.

C. A. Fleming · D. Walsh · J. C. Coffey (✉)
School of Medicine and Department of Surgery, University of Limerick Hospital Group, Raheen, Dooradoyle, Limerick, Ireland
e-mail: Calvin.coffey@ul.ie

E. D. Ehrenpreis et al. (eds.), *The Mesenteric Organ in Health and Disease*,
https://doi.org/10.1007/978-3-030-71963-0_3

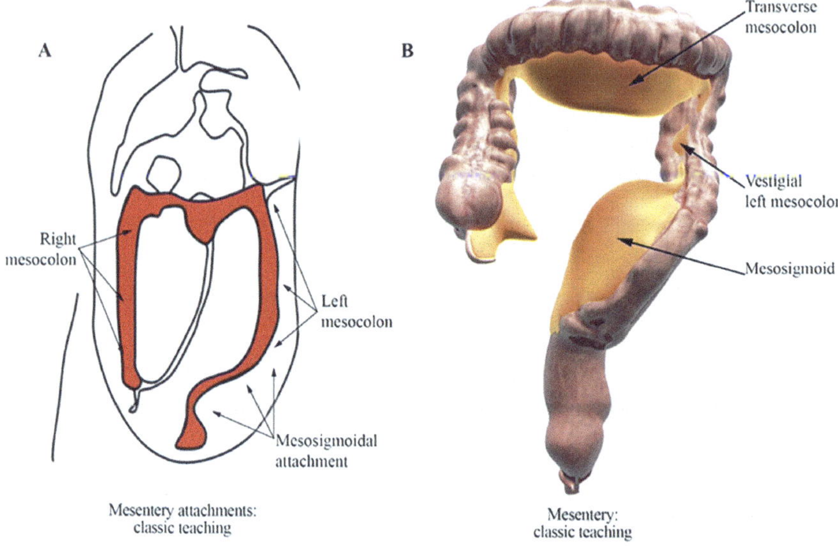

Fig. 3.1 Schematic illustration of the peritoneal model of abdominal anatomy

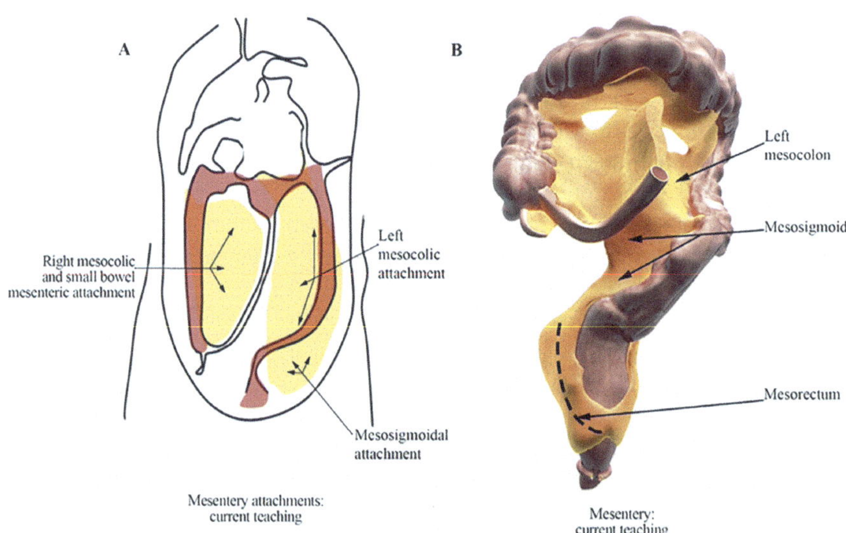

Fig. 3.2 Schematic model of the mesenteric model of abdominal anatomy

A detailed description of different mesenteric regions is beyond the scope of this chapter. The reader is referred to ***Mesenteric Principles of Gastrointestinal Surgery: Basic and Applied Principles***, for a detailed description of the mesentery distal to the duodenojejunal flexure.

The Mesentery

The mesentery is continuous from oesophagogastric to anorectal junction. The posterior surface of the mesentery is held flush against the posterior abdominal wall and a fascial layer (Toldt's fascia) is interposed between both. The continuous surface of the mesentery is interrupted only where the superior mesenteric artery, the coeliac trunk and the inferior mesenteric artery enter it from posteriorly. A further interruption of this anatomical landscape occurs where hepatic veins enter the inferior vena cava.

The mesentery does not have a corresponding anterior surface as it fans out to bridge the space between the posterior abdominal wall, and all abdominal digestive organs. The latter (the abdominal digestive organs) are located at the periphery of the mesentery. The digestive organs and mesentery collectively make up the anatomical abdominal digestive system. These anatomical relationships arose early during embryological development when the mesentery provided an incubator-like support for developing organs, contributing cells and connective tissue to these. Given these contributions, digestive organs remain continuous with different regions of the mesentery, throughout, adult life.

The following description relates to the anatomy of the *in situ* mesentery. This is an important point as once the mesentery has been detached and disconnected, its shape bears minimal resemblance to the shape it had in situ. The mesentery is divisible into upper, central and lower regions. The upper and lower regions are joined by a central region located at the level of the pancreas. At this region, the mesoduodenal region of mesentery detaches from the posterior abdominal wall, coils posterior to the superior mesenteric artery, then fans out as the small intestinal region of mesentery.

It is best to interpret the upper region of mesentery as a continuous structure interrupted by individual organs such as the stomach, spleen and liver (Fig. 3.3). This pattern of interruption enables one arbitrarily subdivide the upper region into the dorsal mesogastrium, greater omentum, ventral mesogastrium and falciform ligament. A review of the embryological development of the mesentery and these organs helps generate an understanding of the shape of the mesentery in the adult human.

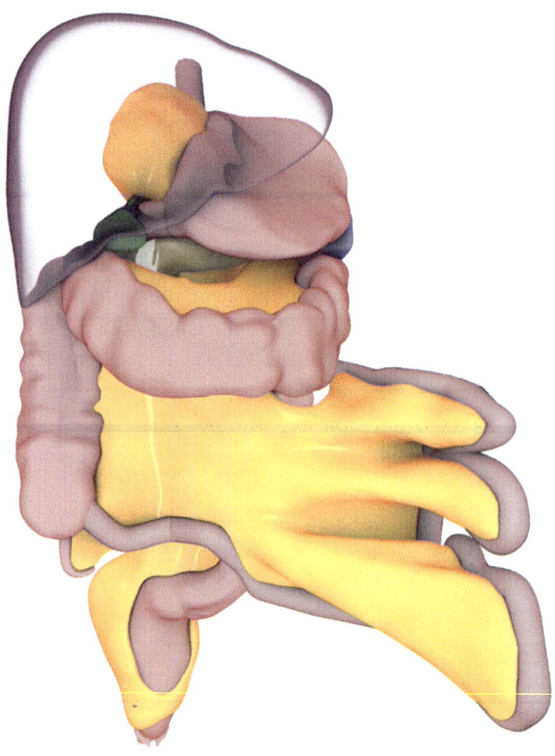

Fig. 3.3 Schematic illustration demonstrating the relationship between abdominal digestive organs and the mesentery

The Dorsal Mesogatsrium and Mesoduodenum

The dorsal mesogastrium is a region of upper mesentery attached to the posterior abdominal wall, over a broad area. It detaches from the posterior abdominal wall laterally, in the left upper quadrant and continues anteriorly as the greater omentum (Fig. 3.4). The spleen is contiguous with the region where the dorsal mesogastrium detaches. If one were to look at the posterior surface of the dorsal mesogastrium, and follow it distally, one would see that it tapers at the level of the duodenum as the mesoduodenum. Next, if you were to rotate the abdominal digestive system so the duodenum and pancreas are in view, you would see the mesoduodenum narrows and detaches from the posterior abdominal wall at the posterior of the superior mesenteric artery.

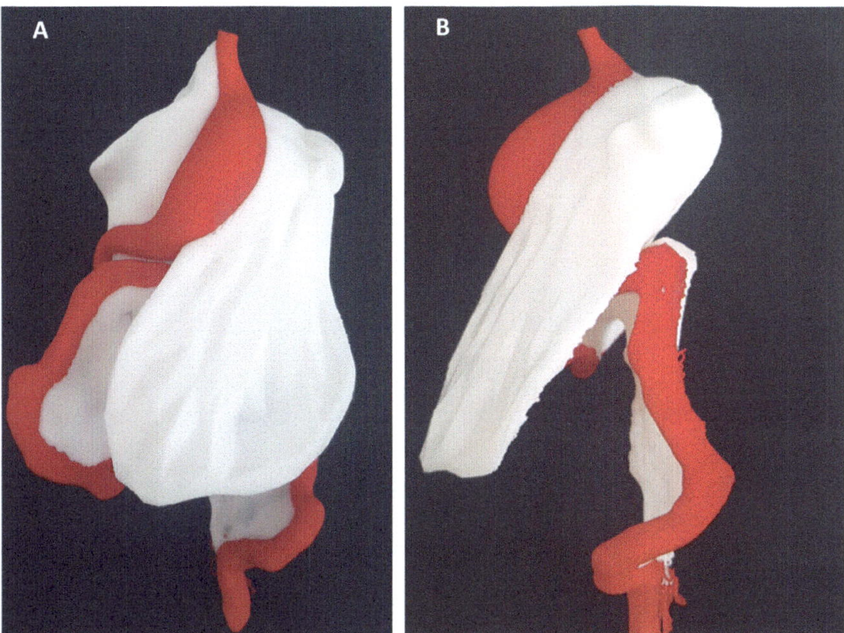

Fig. 3.4 Images of 3D printed intestine (red) and regions of mesentery (white). **a** front on view of the entire abdominal digestive system (except the liver). **b** side view (from left) of same model, demonstrating the shape of the dorsal mesogastrium as this comes around anteriorly to continue as the greater omentum

The Greater Omentum

The greater omentum is interrupted by the stomach which is positioned between it and the ventral mesogastrium (previously called the lesser omentum). The ventral mesogastrium is a flange of mesentery that extends from the lesser curvature of the stomach to the liver and was the ventral region of mesentery in the embryo. At the proximal end of the ventral mesogastrium the mesentery narrows and converges with the dorsal mesogastrium. This occurs close to the oesophagogastric junction, to generate a mesenteric pedicle. A further pedicle occurs at the distal end of the ventral mesogastrium and contains the portal vein, hepatic artery and common bile duct. The upper and lower pedicles arose during embryological development after invagination of the upper region of mesentery created an opening into a diverticulum that ultimately became the bursa. The neck or opening of the bursa is demarcated by mesenteric boundaries of the pedicles form the upper and lower boundary (Fig. 3.5).

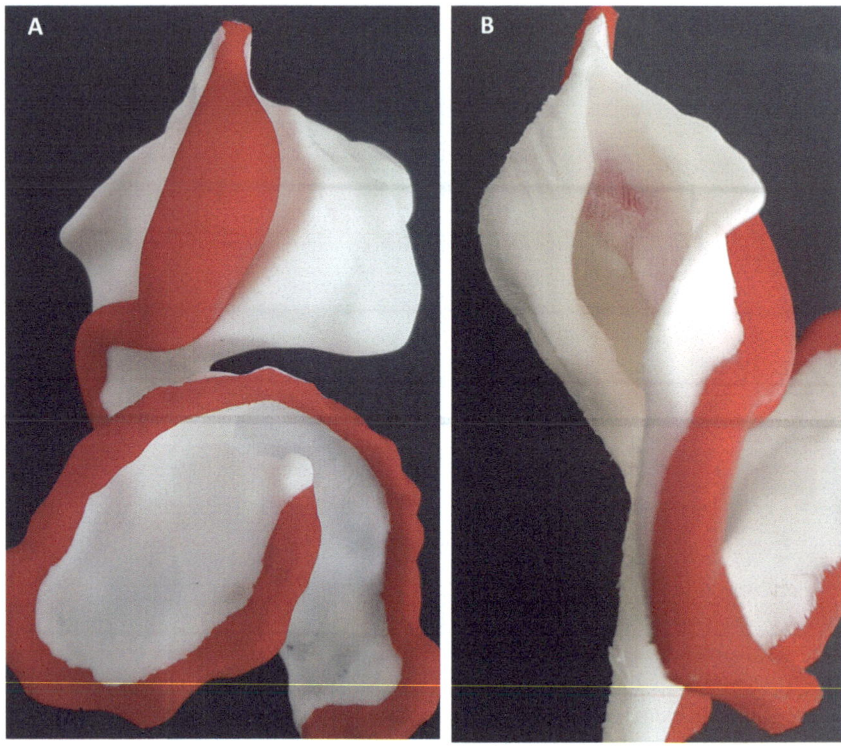

Fig. 3.5 Images of 3D printed intestine (red) and regions of mesentery (white). **a** image demonstrating the upper region of mesentery prior to expansion of the greater omentum. **b** view of the same model from the right side, demonstrating the opening into the bursa

The greater omentum extends distally and currently is best understood if one considers embryological development. Early during embryological development of the upper mesentery, this formed a bubble-like structure that extended to the left and distally. The distal extension of this "bubble" is the forerunner of the greater omentum (Fig. 3.5). The bubble continues to extend distally. The anterior wall was retained in the adult as the greater omentum, while the anatomical floor of the bubble attached to the cephalad (upper) surface of the transverse mesocolon. This attachment is retained as the reflection that occurs between the transverse colon and the greater omentum in the adult (see **MPGS**).

There are two important regions of the greater omentum that need particular focus. The first occurs at the splenic flexure. Here, the greater omentum condenses and densely attached to the upper surface of the mesenteric component of the flexure. This obscures the flexure from view. This is an important anatomical relationship that causes some confusion for surgeons during mobilization of the splenic flexure.

The second region occurs at the head of the pancreas where both it and the mesenteric component of the hepatic flexure/transverse mesocolon are apposed. Here the greater omentum is densely attached to the cephalad aspect of the mesentery which in turn overlaps the second part of the duodenum and pancreas. This anatomical relationship is important to bear in mind during extended right hemicolectomy or total colectomy.

The Lower Region of Mesentery

As with the upper region, the lower region is continuous and extends, in a spiral conformation, around the middle colic mesenteric pedicle. The small intestinal region of mesentery continues as the right mesocolon and the latter continues around the hepatic flexure into the transverse mesocolon. The transverse mesocolon arises as the confluence between the splenic and hepatic mesenteric components of the flexures, and the middle colic vascular pedicle. It narrows in the region of the middle colic pedicle and then fans out considerably at its periphery.

Small Bowel Mesentery and Right Mesocolon

The relationship between the small intestinal and right mesocolic region of mesentery is relevant. Classic texts describe this region, the transverse mesocolon and the mesosigmoid, as attaching directly into the posterior abdominal wall. An anatomical correlate for this is not apparent as the mesentery is continuous throughout. Unfortunately, this erroneous model became indoctrinated in arguably all reference texts until recently, when continuity was observed.

Mesocolon Distal to Transverse Mesocolon

As the left lateral region of the transverse mesocolon correlates with the mesenteric component of the splenic flexure, it follows that the transverse mesocolon in turn continues distally as the left mesocolon (Fig. 3.6). The later continues distally as the mesosigmoid and in turn the mesorectum. The mesosigmoid should be described in a little more detail as it is of central importance in most colorectal surgery. The medial region of the mesosigmoid (i.e. the region near the midline) is attached. Laterally, the lateral border of the attached region detaches to continue as the mobile component of the mesosigmoid. In this regard the mesosigmoid resembles the small intestinal region of mesentery and right mesocolon. The small intestinal region of mesentery corresponds to the lateral and mobile region of the mesosigmoid while the right mesocolon correlates with the attached medial region of the mesosigmoid.

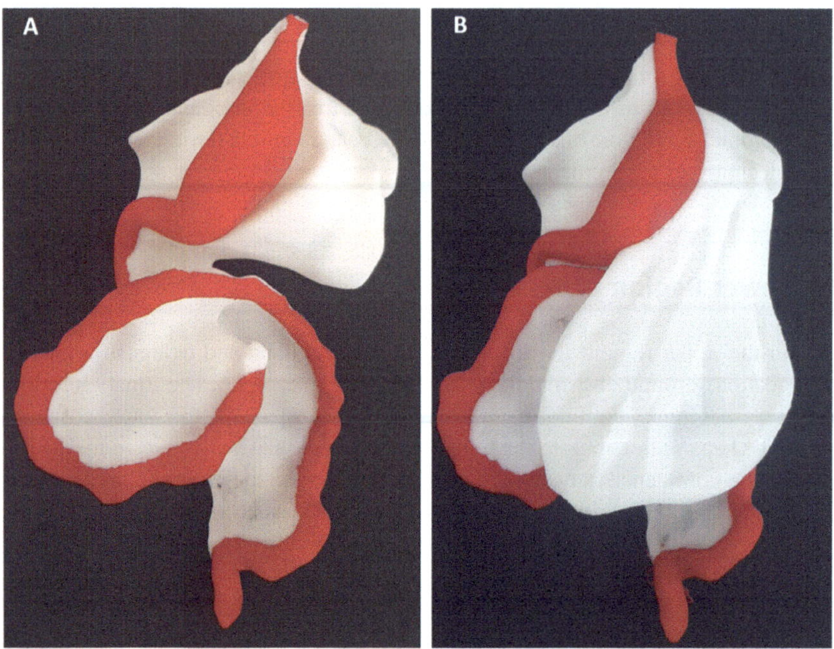

Fig. 3.6 Images of 3D printed intestine (red) and regions of mesentery (white). **a** Image demonstrating the upper region of mesentery prior to expansion of the greater omentum. **b** Image demonstrating the upper region of mesentery after expansion of the greater omentum

Mesosigmoid and Mesorectum

Both mobile and attached regions of the mesosigmoid converge at the rectosigmoid junction to continue distally into his pelvis as the mesorectum. The mesorectum surrounds and is contiguous with the posterior and lateral aspect of the upper and mid-rectum. At the level of the distal rectum, the mesorectum comes around to form a complete collar around the distal rectum. The distal region of this collar tapers sharply and represents the anatomical termination of the mesentery.

Mechanisms of Attachment

As the mesentery collectively maintains all abdominal digestive organs in position, it too must be maintained in position. There are three anatomical mechanisms by which this is achieved. Their disruption forms the technical basis of all visceral and in particular colorectal, surgery.

Central Mechanisms of Attachment

The mechanisms of attachment are best considered by subdividing them according to anatomical location into central, intermediate and peripheral. The central mechanism comprises vascular points of suspension. These are where the coeliac trunk, and superior mesenteric arteries enter mesentery from behind. The inferior mesenteric artery also acts as a mechanism of suspension. The last of the central mechanisms of attachment are the hepatic veins, where these enter the inferior vena cava. Division of these vascular connections is one of the final activities in disconnecting regions of mesentery to allow removal of an organ. It can only be achieved after the other two mechanisms of attachment have been disrupted.

Intermediate Mechanism of Attachment

The intermediate mechanism of attachment involves apposition of Toldt's fascia and the posterior surface of regions of the mesentery. Attached regions include the dorsal mesogastrium, the right and left mesocolon, the medial region of the mesosigmoid (the lateral region is not attached) and the mesorectum. These regions are described above and as mentioned Toldt's fascia occurs between them and posterior abdominal wall. Separation of the mesentery from fascia is an important activity during organ resection. The interface between both can only be identified after the reflection has been divided.

The Peritoneal Reflection

The final mechanism of attachment is the first to be disrupted during surgery. It is the peritoneal reflection. The reflection is the term given to junction of peritoneum between parietal and visceral peritoneum. This is the region where the peritoneum is reflected from the abdominal wall to the organ in question. The reflection is continuous, and once it is divided then the surgeon can identify the junction between the mesentery and underlying fascia.

Abdominal Digestive System Surgery

Given the above, resectional surgery on the abdominal digestive system involves three steps. The first two, division of the peritoneum and separation of the mesentery from underlying fascia allow *detachment* (but not *disconnection*) of the mesentery from the posterior abdominal wall. Division of the central vascular trunks permits *disconnection* of the mesentery and complete removal of that region (and associated organ) from the abdomen.

Suggested Reading

1. Byrnes KG, Walsh D, Dockery P, McDermott K, Coffey JC. Anatomy of the mesentery: current understanding and mechanisms of attachment. Semin Cell Dev Biol. 2018. pii: S1084–9521(18)30205–2. https://doi.org/10.1016/j.semcdb.2018.10.004. [Epub ahead of print].
2. Treves F. Lectures on the anatomy of the intestinal canal and peritoneum in man. Br Med J. 1885;1(1264):580–3.
3. Standring S. Gray's anatomy: the anatomical basis of clinical practice. London, U.K.: Elsevier Health Sciences;2015. pp. 1098–111, 1124–60.
4. Blackburn SC, Stanton MP. Anatomy and physiology of the peritoneum. Semin Pediatr Surg. 2014;23(6):326–30.
5. Sinnatamby CS. Last's anatomy: regional and applied. London, U.K.: Elsevier Health Sciences;2011. pp. 234–8, 247–59.
6. Snell RS. Clinical anatomy by regions. Lippincott Williams & Wilkins;2008. pp. 216, 226, 228, 233, 237, 240.
7. Byrnes KG, Walsh D, Lewton-Brain P, McDermott K, Coffey JC. Anatomy of the mesentery: historical development and recent advances. Semin Cell Dev Biol. 2018. pii: S1084–9521(18)30204–0. https://doi.org/10.1016/j.semcdb.2018.10.003. [Epub ahead of print].
8. Coffey JC, O'Leary DP. The mesentery: structure, function, and role in disease. Lancet Gastroenterol Hepatol. 2016;1(3):238–47.
9. Byrnes KG, McDermott K, Coffey JC. Development of mesenteric tissues. Semin Cell Dev Biol. 2018. pii: S1084–9521(18)30234–9. https://doi.org/10.1016/j.semcdb.2018.10.005. [Epub ahead of print] Review.
10. Coffey JC. Surgical anatomy and anatomic surgery—clinical and scientific mutualism. Surgeon. 2013;11(4):177–82.
11. Coffey JC, et al. Mesenteric-based surgery exploits gastrointestinal, peritoneal, mesenteric and fascial continuity from duodenojejunal flexure to the anorectal junction—a review. Dig Surg. 2015;32(4):291–300.
12. Coffey JC, Lavery I, Sehgal R. Mesenteric principles of gastroinstestinal surgery. CRC Press, FL, 2017. ISBN 9781498711227.
13. Byrnes KG, McDermott KW, Coffey JC. Mesenteric organogenesis. Semin Cell Dev Biol. 2018. pii: S1084–9521(18)30246–5. https://doi.org/10.1016/j.semcdb.2018.10.006. [Epub ahead of print].
14. Culligan K, et al. The mesocolon: a histological and electron microscopic characterization of the mesenteric attachment of the colon prior to and after surgical mobilization. Ann Surg. 2014;260(6):1048–56.
15. Culligan K, et al. Review of nomenclature in colonic surgery—proposal of a standardised nomenclature based on mesocolic anatomy. Surgeon. 2013;11(1):1–5.
16. Coffey JC, et al. Terminology and nomenclature in colonic surgery: universal application of a rule-based approach derived from updates on mesenteric anatomy. Tech Coloproctol. 2014;18(9):789–94.

Part II
Anatomy and Physiology

Vascular Anatomy of the Mesentery

4

Ahmed Khattab, Saad Hashmi, Eli D. Ehrenpreis, and J. Calvin Coffey

Introduction

To best appreciate the role that the vascular system contributes to the overall structure of the mesentery, it is essential to first understand the anatomy of the mesentery itself. The mesentery is a continuous and substantive organ that extends from the esophagogastric to anorectal junction. The mesentery is connected to all abdominal digestive organs. In turn, the mesentery affixes these organs centrally in the within body cavity. This is an important point as conventional models of abdominal anatomy suggest that the vascular system, (with few exceptions), was responsible for the bridge between individual digestive system organs and the musculoskeletal abdominal mainframe.

The mesentery is both directly attached to the body (see Chapter 3 on Mesenteric Anatomy) via Toldt's fascia and the peritoneal reflection, while it is indirectly coupled via the major vascular trunks (celiac, superior mesenteric, and inferior mesenteric) and also via the hepatic veins. The major arterial trunks are the arterial inflow of the mesentery (and all contained organs) while the hepatic veins represent the major venous drainage of the mesentery. These drain into the inferior vena cava.

Given that all abdominal digestive organs first connect to the mesentery, and then to the abdominal wall, their individual vascular, lymphatic and neuronal circuitries are also contained in the mesenteric organ. The celiac trunk and superior mesenteric artery arise from the anterior surface of the aorta and are relatively

A. Khattab · S. Hashmi · E. D. Ehrenpreis
Department of Medicine, Advocate Lutheran General Hospital, 1675 Dempster St, Park Ridge, IL 60068, USA

J. C. Coffey (✉)
School of Medicine and Department of Surgery, University of Limerick Hospital Group, Raheen, Dooradoyle, Limerick, Ireland
e-mail: Calvin.coffey@ul.ie

E. D. Ehrenpreis et al. (eds.), *The Mesenteric Organ in Health and Disease*,
https://doi.org/10.1007/978-3-030-71963-0_4

closely positioned. They almost immediately enter the substance of the mesentery in which they course towards their anatomical destinations. On this pathway, these arteries subdivide into respective branches. These branches utilize the mesenteric frame to travel to and access individual digestive system organs. This concept is important as the mesenteric frame provides the anatomical platform for both the arterial supply and venous drainage of abdominal digestive organs.

The venous drainage of the abdominal digestive system is entirely contained within the mesentery. The inferior mesenteric vein arises in the mesosigmoid, travels proximally in the left mesocolon and reaches the central region of the mesentery. Tributaries of the inferior mesenteric vein include the left colic vein, sigmoid veins, superior rectal vein and rectosigmoid veins. The central or root region of the mesentery is the location where the small intestinal, right and transverse mesocolic regions of mesentery converge. The superior mesenteric vein is positioned in the central region where it receives the inferior mesenteric vein. The right gastro-omental vein, anterior and posterior inferior pancreaticoduodenal veins, jejunal vein, ileal vein, ileocolic vein, right colic vein and middle colic veins all drain into the superior mesenteric vein. Alternatively, the inferior mesenteric vein may enter the splenic vein as it travels from the spleen, within the dorsal mesogastrium, towards the superior mesenteric vein. The superior mesenteric vein and splenic vein merge to form the portal vein, within the mesoduodenum, posterior to the head of the pancreas (Fig. 4.1). The portal vein travels towards the liver, in the mesoduodenum and then enters the portal pedicle (previously called the hepatoduodenal ligament). The portal pedicle is the continuation of the mesentery towards the liver and contains the portal vein, common bile duct and hepatic artery. The portal vein travels in the portal pedicle until it reaches the liver.

As mentioned above, the branches of the celiac trunk, the superior and the inferior mesenteric arteries are all located within the mesentery (Fig. 4.2). There are many anatomical variations in the pattern of branching of each, but all branches utilize the mesenteric frame to access their destination organ in the abdominal digestive system. The following description is a commonly observed pattern. After entering the dorsal mesogastric region of the mesentery, the celiac trunk divides into the common hepatic, left gastric and splenic arteries (Fig. 4.3). At first, all of these divisions are located in the dorsal mesogastrium. The splenic artery remains in the dorsal mesogastrium throughout its course as it passes laterally towards the hilum of the spleen. The dorsal mesogastrium is located posterior to the body of the pancreas. These anatomical observations explain the relationship of the splenic artery to the pancreas. The left gastric artery enters a mesenteric pedicle at the

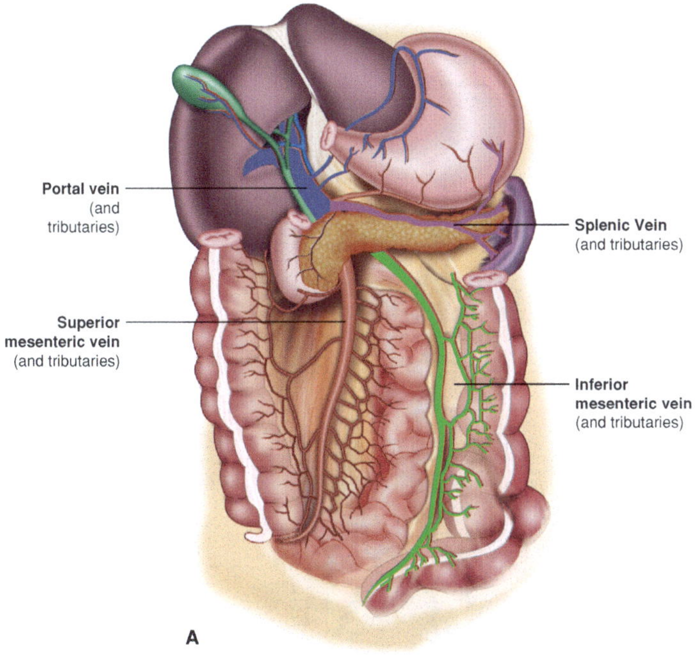

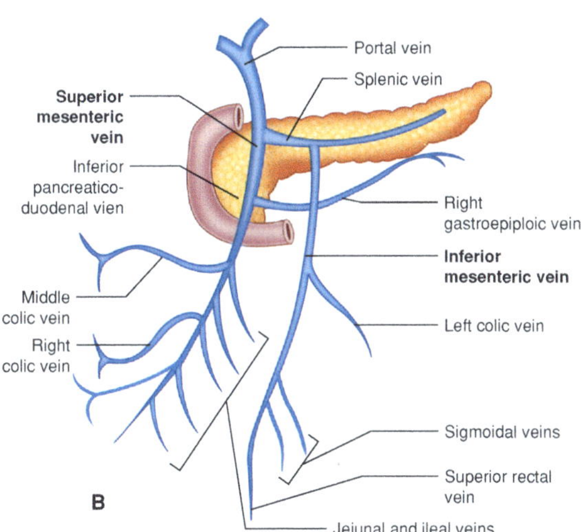

Fig. 4.1 **a** Formation of the portal vein from convergence of superior mesenteric, inferior mesenteric and splenic veins. **b** Tributaries of the superior and inferior mesenteric veins

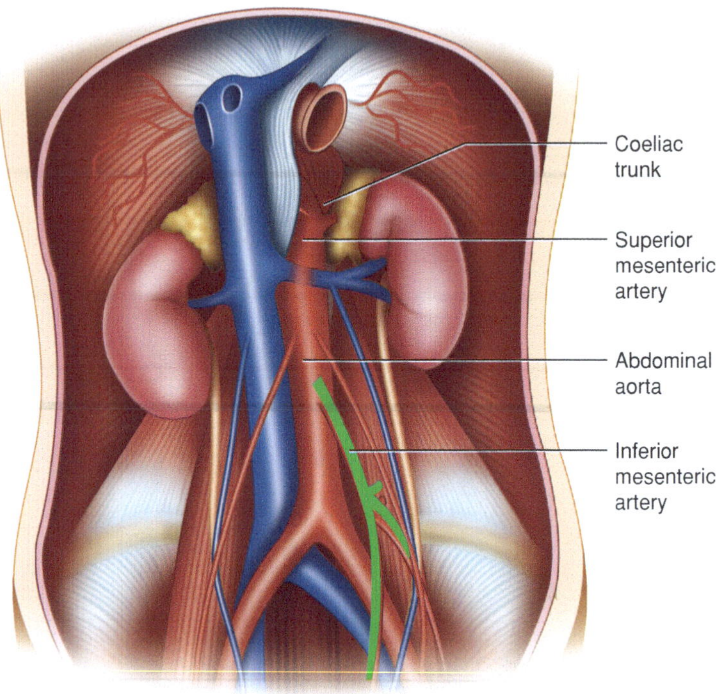

Fig. 4.2 Celiac trunk, superior mesenteric artery and inferior mesenteric arteries branch points from the abdominal aorta

esophagogastric junction. It uses this to access the ventral mesogastrium at the lesser curvature of the stomach. It travels along the lesser curvature, in the ventral mesogastrium, distributing branches to the stomach. The common hepatic artery travels inferolaterally in the dorsal mesogastrium to reach the portal pedicle. Here, it gives off the right gastric artery and then enters the portal pedicle in which it travels until it reaches the liver.

The superior mesenteric artery enters the mesentery soon after it arises from the anterior surface of the aorta. It is incorporated into the substance of the mesentery in which it travels towards the right iliac fossa. It sequentially gives off jejunal and ileal branches and also the ileocolic artery (Fig. 4.4). The jejunal and ileal arteries

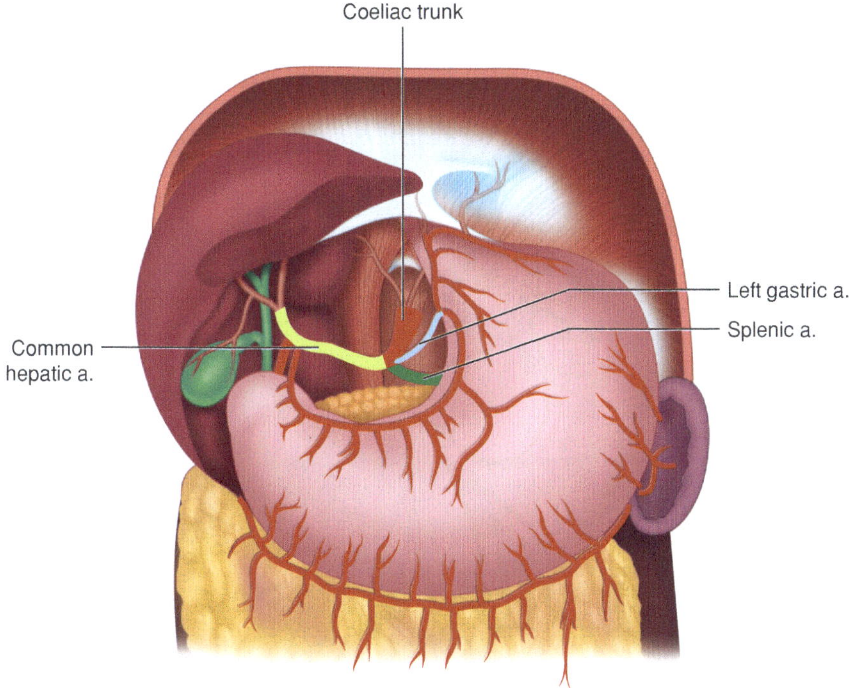

Fig. 4.3 Branches of the celiac trunk

are numerous arteries that form anastomotic arcades which lead to multiple smaller arteries called vasa recta supplying the ileum and jejunum. The jejunum has long and few vasa recta while the ileum has numerous, short vasa recta. The ileocolic artery is virtually an anatomical constant and travels, within the mesentery, to the ileocaecal junction. In the region of the junction, it sometimes gives off the appendicular artery which enters the mesoappendix and passes towards the appendix. The mesoappendix is a posterior appendage of the main body of the mesentery. Another branch of the superior mesenteric artery is the inferior pancreaticoduodenal artery which supplies the head and uncinate process of the pancreas along with the ascending and inferior duodenum.

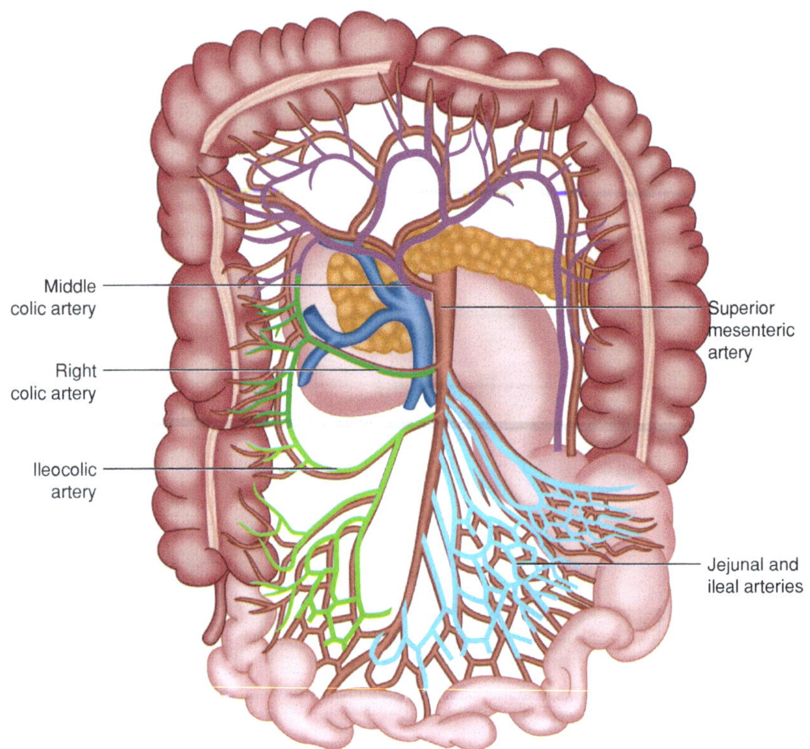

Fig. 4.4 Branches of the superior mesenteric artery. *Note* the inferior pancreaticoduodenal artery is not depicted in this image

The middle colic arterial pedicle is important. It arises from the upper surface of the superior mesenteric artery, in the central region of the mesentery. It subdivides into branches that also travel in the transverse mesocolon to the colon. A right colic artery, when present, arises from either the main trunk of the middle colic artery, or a branch of this. It exploits mesenteric continuity to travel in the mesentery and access the ascending colon.

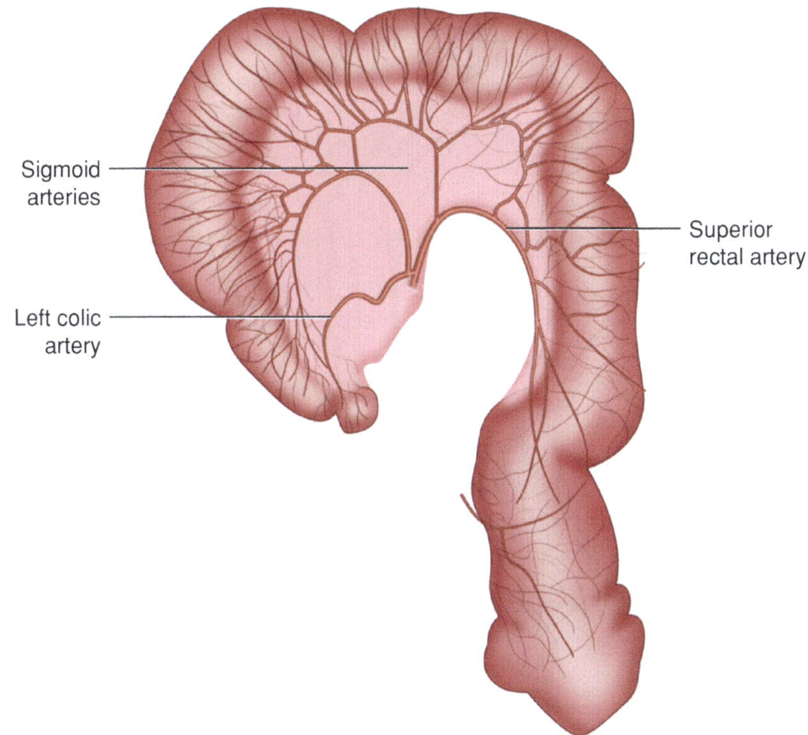

Fig. 4.5 Main branches of the inferior mesenteric artery

The inferior mesenteric artery arises from the anterior surface of the aorta and immediately enters the mesentery at a mesenteric pedicle termed the inferior mesenteric arterial pedicle. Soon after, this artery gives off the left colic branch which enters the left mesocolon and travels in the latter towards the descending colon. Once the left colic artery has been given off the inferior mesenteric artery is called the superior rectal artery. It travels in the mesosigmoid and thereafter in the mesorectum, and branches in each region to supply these components of the intestine (Figs. 4.5 and 4.6).

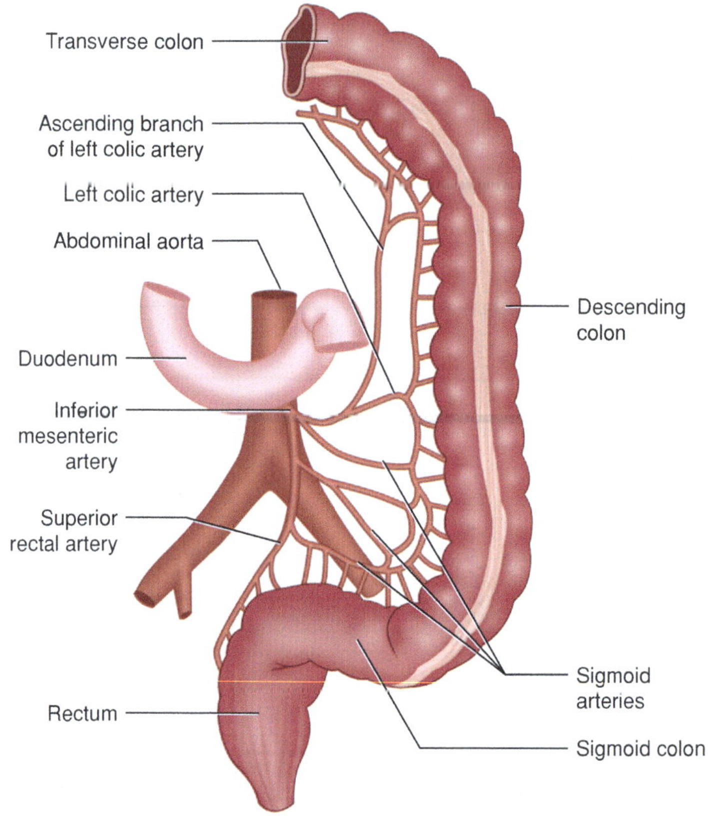

Fig. 4.6 Detailed view of the branches of the inferior mesenteric artery

Suggested Reading

1. https://teachmeanatomy.info/abdomen/vasculature/arteries/superior-mesenteric/.
2. https://teachmeanatomy.info/abdomen/vasculature/arteries/superior-mesenteric/.
3. https://teachmeanatomy.info/abdomen/vasculature/arteries/inferior-mesenteric/.
4. https://www.earthslab.com/anatomy/superior-mesenteric-vein/.
5. https://teachmeanatomy.info/abdomen/vasculature/venous-drainage/.
6. https://s3.amazonaws.com/teachmeseries/tmanatomy/wp-content/uploads/20171222215727/Inferior-Mesenteric-Artery-arising-from-the-Abdominal-Aorta-1024x622.jpg.
7. https://slideplayer.com/slide/4356935/14/images/22/Inferior+mesenteric+artery.jpg.

Introduction to the Physiology of the Mesentery

5

Karl Andersen, Assad Munis, and Eli D. Ehrenpreis

Anatomy

The mesentery is an organ within the abdomen that has multiple purposes, including suspending the intestines and preventing their collapse into the pelvis, acting as a conduit for arteries, veins, nerves, and lymphatics to the intestines, and being a reservoir for adipose tissue and immune cells. The mesentery was traditionally described as a fragmented structure that helped suspend the small intestine, transverse colon, and mesosigmoid, but new research has demonstrated that it is a contiguous structure that extends from the duodenojejunal flexure to the mesorectum. This has facilitated the development of uniform anatomical nomenclature, and provided the foundation to investigate the pathophysiological role that the mesentery plays in different disease processes.

Histologically, the mesentery is derived from the mesoderm germ layer and is composed primarily of two layers of mesothelium: an interposed loose connective tissue layer and adipocytes intermixed in the connective tissue. During embryological development, a network of neurons, blood vessels, and lymphatic channels grow within it and connect the developing gut to the rest of the body. Proper embryological formation of the mesentery ensures correct anatomical placement within the adult body and prevents the development of potentially devastating conditions, such as volvulus and malrotation, that can occur when the gut is not properly anchored to the posterior abdominal wall. These conditions can occur when the mesentery remains suspended at the vascular insertion points of the superior and inferior mesenteric arteries but fail to be attached to the posterior wall, allowing for twisting and occlusion of the arteries. In a normal adult, the root of the mesentery, which is approximately 15 cm in length, is the attachment point to the

K. Andersen · A. Munis · E. D. Ehrenpreis (✉)
Department of Medicine, Advocate Lutheran General Hospital, 1775 Dempster St, 6 South, Park Ridge, IL 60068, USA
e-mail: E2bioconsultants@gmail.com

E. D. Ehrenpreis et al. (eds.), *The Mesenteric Organ in Health and Disease*,
https://doi.org/10.1007/978-3-030-71963-0_5

posterior abdominal wall. It is here where it flattens and becomes the retroperitoneum. It extends inferiorly and obliquely to the right, and from the duodenojejunal junction to the ileocolic junction on the right, crossing over the aorta and inferior vena cava. Thus, the mesentery maintains the right and left colon in their normal anatomic configuration by attaching to the posterior abdominal wall and overlying the retroperitoneum.

With increasing research directed at the function of the mesenteric organ, new understanding of the important role of the mesentery in health and disease. Functioning portions of the mesentery are outlined in this chapter and summarized in Table 5.1.

Vascular Physiology

Coursing through the mesentery are a series of arteries and veins that carry oxygen-rich blood from the heart to the small and large intestine and nutrient-dense blood back to the liver. The primary artery that supplies blood to the mesenteric circulation is the superior mesenteric artery, which brings blood to the small intestine, ascending colon, and the proximal two-thirds of the transverse colon. Regulation of blood flow through the mesentery is dynamic and determined by a complex set of neurologic, hormonal, and metabolic signals that ensure adequate intestinal blood flow supply. During periods of high metabolic demand, such as the

Table 5.1 Functional Components of the Mesentery

Component	Function
Vascular	• Complex system of neurologic, hormonal, and metabolic signals that ensure adequate intestinal blood flow supply • Intrinsic and extrinsic systems are major regulators of blood flow through the mesentery
Immunologic	• Visceral adipocytes are being considered as a major contributor in the pathogenesis of pro-inflammatory states • Antigen presenting cells located within the intestines play a pivotal role in maintaining the balance in the gut microbiome
Neurologic	• Described as the "second brain" due to its complexity and ability to function independently from the rest of the body • Neurogenic vasoconstriction within the mesentery is mediated mainly by the sympathetic postganglionic cells • Mesentery serves as major reservoir for the total effective circulating volume and plays a crucial role in maintaining adequate systemic perfusion
Mesenteric Adipose	• A dynamic endocrine and immunologic component that is involved in multiple disease states including metabolic syndrome, inflammatory bowel disease,and coronary artery disease • Mesenteric adipocytes play big role than previously thought in fat wrapping • Increase inflammation through release of C-reactive protein

post-prandial state, these signaling networks will respond by causing vasodilation and increasing blood flow through the mesentery to the intestines. Conversely, these networks will shunt blood away from the intestines when other organs have greater metabolic demands, such as during periods of exercise, hypovolemia, shock, and hemorrhage.

The major regulators of blood flow through the mesentery and gut are divided into intrinsic and extrinsic systems. The intrinsic system is further divided into the metabolic pathway, which involves adjusting blood flow during periods of digestion and hypoxia, and the myogenic pathway, which involves vasoconstriction and vasodilation via smooth muscle cells to maintain a constant blood flow. The extrinsic system is composed of neurohumoral, autonomic, and cardiovascular signals that augment the blood flow through the mesentery. The neurohumoral control is a mixture of systemic hormones, such as catecholamines, angiotensin II, and vasopressin, and local gastrointestinal hormones, such as gastrin and secretin, all of which bind to their respective receptors and augment blood flow during different physiological conditions. The autonomic control is divided into the sympathetic and parasympathetic branches, with the sympathetic branch having the greater effect on blood flow. Cardiovascular signals, such as cardiac output and effective circulating volume, also play a pivotal role in regulating mesenteric blood flow.

Immunologic Physiology

The digestive system is the largest interface between the outside world and the internal body. The immune system is responsible for patrolling pathogens that attempt to traverse the mucosal barrier. The immune system lining the gut is also responsible for recognizing beneficial micro- and macronutrients and separating them from potentially harmful antigens. The failure to differentiate between harmful and beneficial antigens, or commensal bacteria from pathogenic bacteria, can result in an inappropriate inflammatory state with subsequent development of multiple medical conditions, including Crohn's disease and Celiac disease. Maintenance of a functioning intestinal immune system relies largely upon a network of lymph nodes and immune cells, including fibrocytes, lymphocytes, macrophages, dendritic cells, and neutrophils. The role of visceral adipocytes, which have a large concentration in the mesentery, is currently being investigated as a major contributor to the pathogenesis of pro-inflammatory states.

The process of differentiating harmless from harmful molecules begins with antigen presenting cells (APCs). These cells are located within the lamina propria of the intestines or within lymphoid follicles within the submucosa of the gut (Peyer's patches). The antigen-laden APCs in the lamina propria will travel through lymphatic channels to mesenteric lymph nodes and present directly to naive CD4+ T cells. Similarly, APCs in Peyer's patches will present to naive CD4+ T cells within the submucosa, or will travel through lymphatic channels to mesenteric lymph nodes. The naive CD4+ T cells will then differentiate into type 1 helper T-cells

(Th1) that generate local and systemic inflammation, or into type 3 helper T-cells (Th3) and regulatory T cells (Tregs) that will lead to systemic immunity, IgA secretion, and local tolerance. The development of inflammatory diseases of the gut, such as those seen in inflammatory bowel disease, are not fully understood but likely involve a combination of genetic susceptibility, dysbiosis of the gut, sustained innate immunity, and defective regulation. The relationship of the mesenteric function to gastrointestinal immune regulation is currently being explored and it remains to be seen what relationship exists amongst these entities.

Neuronal Physiology

The enteric nervous system is frequently described as the "second brain" of humans due to its size and ability to function independently from the rest of the body. It connects to the rest of the body through the parasympathetic and sympathetic branches of the autonomic nervous system. The primary parasympathetic innervation for the small intestine, right colon, and the first half of the transverse colon comes from the vagus nerve. The sympathetic innervation to these segments of the gut comes from the greater, lesser, and least splanchnic nerves, which originate from thoracic segments T5-T12. The parasympathetic innervation to the remaining transverse colon, the descending colon, and the sigmoid colon is derived from the lumbar splanchnic nerves, which come from lumbar segments L1-L3. Sympathetic innervation to the second half of the colon comes from sacral splanchnic nerves that originate from segments S2-S4.

Within the mesentery, sympathetic postganglionic cells accomplish neurogenic vasoconstriction of vessels. In animal studies, there is little evidence that the parasympathetic nervous system plays a role in vasodilation. The primary neurotransmitter responsible for vasoconstriction from the sympathetic nerves is noradrenaline, with ATP and neuropeptide Y as co-transmitters. Immunohistochemical staining of human mesenteric and submucosal vessels have demonstrated that a subset of neurons are positive for substance P (SP). Additionally, SP might be the primary vasodilatory neuropeptide responsible for sensory nerve-mediated effector function based on animal studies that revealed SP could cause edema and plasma extravasation.

Innervation of the mesenteric veins and arteries is quite unique compared to the rest of the body. Typically, arteries require more innervation to efficiently allow the body to adjust vascular tone to maintain an adequate blood pressure. In the mesentery, however, the mesenteric veins need greater innervation because the mesentery serves as a major reservoir of the total circulating volume and requires a meticulous control to allow for volume to flow into the effective circulatory system when the blood pressure needs to be adjusted. Innervation of the large mesenteric lymphatic vessels originates from noradrenergic NPY-positive nerve fibers that arise from periarterial plexuses of the mesentery, but the extent to which that the innervation affects lymphatic flow is yet to be determined.

Currently, the mesenteric nervous system is being investigated for its role in acute decompensated heart failure. During a heart failure exacerbation, there is an elevated sympathetic tone that causes the inappropriate movement of fluid out of the mesentery into the effective circulatory system, accelerating the decompensation. Medications that reduce sympathetic tone are now being utilized to hold more blood in the splanchnic circulation to decrease the rate of decompensation.

Role of Mesenteric Adipose

One of the major mesenteric contributions to the pathogenesis of inflammation is through the interaction of the gut with mesenteric adipocytes. Adipose tissue is no longer considered to be a passive repository of stored fuel or a thermoregulatory layer, but rather a dynamic endocrine and immunologic component that is involved in multiple disease states including metabolic syndrome, inflammatory bowel disease, and coronary artery disease. It is thought to contribute to systemic and local disease through the expression of cytokines, i.e. adipokines. It is important to note, however, that not all adipose tissue has the same pro-inflammatory contribution. Subcutaneous adipose has been shown to be the least associated with provoking inflammation. Visceral fat, which is the most abundant in the mesentery, has a greater association with generating inflammation through the release of adipose-specific cytokines such as leptin and resistin. Adipose cells also secrete common cytokines such as tumor necrosis factor alpha (TNF-α), interleukin 6, and interleukin 8.

In regards to Crohn's disease, it has been demonstrated that adipocytes are significant producers of TNF-α, which is strongly associated with its pathogenesis. Mesenteric adipocytes have also been shown to have the ability to dedifferentiate into preadipocytes. Preadipocytes have the capacity to differentiate into macrophages and help sustain a proinflammatory state. Another example of the role that mesenteric adipocytes play in Crohn's disease is the phenomenon known as fat wrapping. This phenomenon is known to occur when greater than 50% of a segment of the large or small intestine is enveloped in mesenteric fat. Interestingly, fat wrapping is commonly used as a landmark by surgeons to identify more diseased sections of the gut. It is felt that these regions of the intestine is where majority of luminal narrowing and stricture formation occur as a consequence of muscularis propria hyperplasia and intestinal fibrosis.

Mesenteric adipose also contributes to systemic inflammation and frequently precedes many common medical conditions including metabolic syndrome, diabetes, and coronary artery disease. A recent study demonstrated that adipocytes from subcutaneous, mesenteric, omental, or periaortic tissue have different cytokine profiles. Mesenteric adipocytes having the greatest inflammatory profile and can contribute to metabolic dysfunction that precedes coronary artery disease. Adipose cells were found to increase inflammation through the release of C reactive protein (CRP), a non-specific marker of inflammation that has been positively correlated with the pathogenesis and severity of Crohn's disease and coronary artery disease.

The liver primarily releases C-reactive protein, but many studies have shown that obesity, especially visceral obesity, is linked to a rise in CRP production as well. A proposed mechanism for greater CRP release is that that visceral adipocytes produce larger amounts of inflammatory adipokines that cause the liver to release greater amounts of CRP. Another hypothesis is that with increasing obesity there is more macrophage and monocyte invasion of the adipose tissue, and that these same macrophages and monocytes help sustain a proinflammatory state. Furthermore, another link has been proposed that connects the role of adipose tissue and systemic inflammation. There appears to be an inverse relationship between adiponectin, an immunomodulatory adipokine, and obesity. The Coronary Artery Risk Development in Young Adults (CARDIA) study demonstrated that increased amounts of central obesity correlated with lower levels of adiponectin. Since adiponectin increases insulin sensitivity, low levels of adiponectin will result in more insulin resistance and the subsequent development of metabolic disorders. Nonetheless, as the role of mesenteric adipose tissue in local and systemic diseases is further scrutinized, it is expected that more effective therapeutic regimens will come to fruition to alleviate the burden of mesenteric adipose- generated inflammatory conditions.

Suggested Reading

1. D'Andrea V, Panarese A, Taurone S, et al. Human lymphatic mesenteric vessels: morphology and possible function of aminergic and NPY-ergic nerve fibers. Lymphatic Res Biol. 2015;13:170–5.
2. De Fontgalland D, Wattchow DA, Brookes SJH. Immunohistochemical characterization of the innervation of human colonic mesenteric and submucosal blood vessels. Neurogastroenterol Motil. 2008;20:1212–26.
3. Birch DJ, Turmaine M, Boulos PB, Burnstock G. Sympathetic innervation of human mesenteric artery and vein. J Vasc Res. 2008;45:323–32.
4. Kreulen DL. Activation of mesenteric arteries and veins by preganglionic and postganglionic nerves. Amer J Physiol. 1986;251:H1267–75.
5. Kreulen DL. Properties of the venous and arterial innervation in the mesentery. J Smooth Musc Res. 2003;39:269–79.
6. Fudim MF, Hernandez AF, Felker M. Role of volume redistribution in the congestion of heart failure. J Amer Heart Assoc. 2017;6:1–9.
7. Coffey JC, O'Leary DP. The mesentery: structure, function, and role in disease. Lancet Gastroenterol Hepatol. 2016;1:238–337.
8. Rivera ED, Coffey JC, Walsh D, Ehrenpreis ED. The mesentery, systemic inflammation, and Crohn's disease. J Inflamm Bowel Dis. 2018;0:1–9.
9. Li Y, Zhu W, Zuo L, et al. The role of the mesentery in crohn's disease: the contributions of nerves, vessels, lymphatics, and fat to the pathogenesis and disease course. J Inflamm Bowel Dis. 2016;22:1483–95.
10. Fink C, Karagiannides I, Bakirtzi K, et al. Adipose tissue and inflammatory bowel disease pathogenesis. J Inflamm Bowel Dis. 2012;18:1550–7.
11. Kranendonk MEG, van Herwaarden JA, Stupkova T, et al. Inflammatory characteristics of distinct abdominal adipose tissue depots relate differentially to metabolic risk factors for cardiovascular disease. Atherosclerosis 2015;419–42.

12. Brooks GC, Blaha MJ, Blumenthal RS. Relation of c-reactive protein to abdominal adiposity. Amer J Cardiol. 2010;106:56–61.
13. Mowat AM. Anatomical basis of tolerance and immunity to intestinal antigens. Nature Rev Immunol. 2003;331–41.
14. Mao R, Kurada S, Gordon IO, Baker ME, et al. The mesenteric fat and intestinal muscle interface: creeping fat influencing stricture formation in crohn's disease. J Inflamm Bowel Dis. 2019;25:421–6.
15. Maloy KJ, Powrie F. Intestinal homeostasis and its breakdown in inflammatory bowel disease. Nature. 2011;474:298–306.
16. Coffey JC, O'Leary DP. Defining the mesentery as an organ and what this means for understanding its role in digestive disorders. Expert Rev Gastroenterol Hepatol. 2017;11:703–5.
17. Coffey JC, Dockery P, Moran BJ, et al. Chapter 2: Mesenteric and peritoneal anatomy. In: Mesenteric Principles of Gastrointestinal Surgery: Basic and Applied Science, 1st ed. Boca Raton: CRC Press, Taylor & Francis Group;2017. p. 11–39.
18. Moore KL, Dalley AF, Agur AMR. Chapter 2: Abdomen. In: Clinically Oriented Anatomy, 6th ed. Philadelphia: Lippincott Williams & Wilkins;2010. p. 226–301.

Emergence of the Human Gut Microbiota as an Influencer in Health and Disease

6

Miranda G. Kiernan, Suzanne S. Dunne, and Colum P. Dunne

The Gut Microbiome

Gut microbiome profiles were determined initially through identification of microbial communities associated with faecal matter, luminal contents and mucosal tissue. Early studies utilised culture methods, which provided limited information. Reports from studies investigating luminal contents and faeces concluded that non-sporing, Gram-positive anaerobes dominate throughout the length of the intestine. *Bacteroides* were found to be ubiquitous, mainly *Bacteroides fragilis*, with bifidobacteria, streptococci and lactobacilli also present in relatively high numbers. Luminal contents, taken from various locations along the small and large intestine, and faecal matter, displayed differing microbial profiles. This variation was thought initially to be influenced by location along the intestine. The duodenum and the jejunum contained low numbers of bacteria, suggesting that colonisation of the proximal small intestine by bacteria may be transient. Of the bacteria present in the proximal small intestine, *Bacteroides* and bifidobacteria were the most abundant while enterobacteria, lactobacilli, and streptococci were detected in lower numbers. In contrast, whilst the large intestine was colonised by low numbers of enterobacteria and staphylococci, it also contained enterococci, clostridia, *Veillonella*, *Streptococcus salivarius* and *Bacteroides melaninogenicus*. The microbial profile of the terminal ileum had more in common with that of the large intestine than the small intestine.

Studies investigating mucosal tissue microbiota promptly followed, providing an arguably more accurate representation of the human gut microbiota. Healthy duodenal mucosa was sterile in general, whilst the jejunum and ileum were sterile in some healthy individuals. In contrast, studies evaluating colonic mucosa consis-

M. G. Kiernan · S. S. Dunne · C. P. Dunne (✉)
School of Medicine and Centre for Interventions in Infection, Inflammation and Immunity, University of Limerick, Castleroy, Limerick V94 T9PX, Ireland
e-mail: Column.dunne@ul.ie

E. D. Ehrenpreis et al. (eds.), *The Mesenteric Organ in Health and Disease*,
https://doi.org/10.1007/978-3-030-71963-0_6

tently demonstrated bacterial colonisation. Gorbach et al. [14] cultured streptococci and lactobacilli from small intestinal mucosa, while others found *E. coli*, *Enterobacter* and *Klebsiella* in ileal mucosa only. The most frequent and abundant bacteria in the large intestinal mucosa were streptococci, enterococci, and *Bacteroides*. However, more recent studies have confirmed differences in the microbiota detected in patients' faecal matter compared with their intestinal mucosa. Most importantly, while studies using culture methods provided us with preliminary information relating to the gut microbiota, it is estimated that up to 80% of intestinal bacteria cannot be cultured.

The advent of culture-independent methods, specifically next generation sequencing, has advanced a far more accurate understanding of the human gut microbiome. Inter-individual variation and geographical factors are now recognised as confounders in the search for a definitive microbial signature of healthy individuals. Others influencers include general hygiene, sanitation, the prevalence of breast-feeding, use of antibiotics and vaccines, and ethnicity. Despite these challenges, consensus is developing regarding what a normal healthy gut microbiota profile may be. As more countries adopt westernised lifestyle and diet, both Firmicutes and Bacteroidetes phyla have begun to dominate the microbiome (approximately 50% and 40%, respectively). The remaining 10% comprise other phyla, including mainly Proteobacteria (includes bacteria from Enterobacteriaceae, Pseudomonadaceae, Burkholderiaceae and Desulfobacteraceae) and Actinobacteria (includes bacterial from Streptomycetaceae, Corynebacterineae and Micrococcaceae).

Bacterial Translocation to Blood and the Mesentery

Bacterial translocation is the movement of viable or non-viable microbes and their metabolites from the intestinal tract to extra-intestinal sites such as the mesentery, bloodstream, mesenteric lymph nodes (MLNs), or other organs such as the liver, spleen and kidneys. Translocation has been described in the context of various types of cancers, sepsis, acute pancreatitis, liver cirrhosis and inflammatory bowel disease.

In IBD specifically, increased intestinal permeability has been described well. For example, increased permeability due to loss of E-cadherin expression allows bacteria or their byproducts translocate to the mesentery, MLNs, Peyer's patches and the bloodstream. MLNs harvested from patients with IBD have been examined for the presence of translocated bacteria. As early as 1994, Sedman et al. found MLNs from patients with Crohn's disease to be free of bacteria but successfully cultured *Citrobacter* spp from nodes of patients with ulcerative colitis. Since then, others have successfully cultured bacteria from MLNs of patients with IBD, including *E. coli*, *C. freundii* and coagulase negative staphylococci.

Despite advances in molecular technology, few studies have characterised the microbial ecology of blood and MLNs using culture independent techniques. Notably, Vrakas et al. (2017) found that patients with IBD had increased concentrations of bacterial DNA in their blood, particularly during active disease, when compared to healthy controls. O'Brien et al. (2014) investigated bacterial translocation to MLNs in patients with IBD. This study focussed primarily on patients with Crohn's disease where *E. coli* was the most common organism found. While using considerably different approaches, both of these studies described a dysbiosis in blood and MLNs from patients with IBD compared to controls. However, it is difficult to evaluate from these studies as to whether translocation is actually increased in IBD patients as their control subjects were, in some cases, undergoing surgery for alternative or additional pathologies. Furthermore, it is estimated that bacterial translocation occurs normally in 5–10% of healthy individuals.

Irrespective, bacterial translocation to blood in patients with Crohn's disease is associated with increased serologic levels of Th1-derived pro-inflammatory cytokines, such as IL-21, TNFα and IFNγ. Translocation to MLNs elicits an adaptive immune response. Bacteria have also been detected in the mesenteric fat of patients with Crohn's disease. Previous studies have demonstrated that adipocytes, when cultured with bacteria or their products, such as lipopolysaccharide (LPS), increase expression of IL-6, TNFα, IL-1β, IL-8, RANTES, CCL11 (also known as Eotaxin-1), MCP-1, adiponectin, resistin, visfatin and leptin at protein and mRNA levels. In patients with Crohn's disease, CRP mRNA and protein levels are increased in mesenteric adipocytes stimulated with *E. coli* or LPS. Given the potential consequences of bacterial translocation to the mesentery, MLNs, or blood in patients with IBD, the mesenteric nodal microbiome appears to be of critical importance.

The Microbiome of Mesenteric Fat

Although numerous studies have demonstrated that bacteria can be isolated from mesenteric fat, no study has determined definitive bacterial profiles of the mesentery fat component in disease states. Zulian et al. (2013) specifically investigated and confirmed the presence of *Enterococcus faecalis* in mesenteric and omental fat in both Crohn's disease and ulcerative colitis. Numbers of *E. faecalis* were significantly increased in fat deposits in Crohn's disease, but not ulcerative colitis. Peyrin-Biroulet et al. (2012) reported that the rate of bacterial translocation to mesenteric fat is similar to that of the MLNs. They demonstrated this is also higher for patients with Crohn's disease than healthy individuals and those with ulcerative colitis.

Mesenteric adipocytes, irrespective of disease state, express a wide range of pattern recognition molecules (PRMs), such as TLR2, TLR4, NOD1 and NOD2, indicating they may respond to bacterial stimulation or challenge. Zulian et al. (2013) also demonstrated that co-culture of adipocytes with *E. faecalis* results in a significant increase in their rate of proliferation. Bacterial or LPS stimulation of adipocytes has also been shown to increase expression of pro-inflammatory and pro-fibrotic cytokines and adipokines. These may contribute to increased production of pro-inflammatory and pro-fibrotic cytokines and adipokines observed in the mesentery in Crohn's disease. The knowledge that bacteria or their components occur in mesenteric fat, and that adipocytes are responsive to these elements, could provide a novel pathogenic mechanism, and therefore novel therapeutic targets, in IBD.

The Microbiome of Mesenteric Lymph Nodes

Since development of molecular anlysis techniques, only two studies have utilised culture-independent methods to investigate bacteria present in MLNs in IBD patients. O'Brien et al. (2014) focused on patients with Crohn's disease. Mirroring the outcomes of earlier studies, *E. coli* was commonly found in Crohn's disease MLNs. The bacterial profile of MLNs resembled that of the associated mucosa, suggesting a general movement of bacteria from the mucosa to the MLNs and not selective bacterial translocation. The overall microbial profile of MLNs from patients with Crohn's disease reflected a dominance of Firmicutes and Bacteroidetes phyla, slightly favouring Bacteroidetes, which was similar to the profile observed in MLNs from non-IBD controls. Although species diversity was decreased in MLNs in Crohn's disease, a unique bacterial profile was not evident. Differences in MLN microbial profiles stemmed from inter-patient variation, rather than location of sampling.

The second study, Kiernan et al. [21], reported collection of MLNs from patients with Crohn's disease or ulcerative colitis undergoing resection (Figs. 6.1 and 6.2). Crohn's disease MLNs presented a distinctly different microbial profile to that observed in ulcerative colitis (Fig. 6.3). The relative abundance of Firmicutes were found to have greater relative abundance in nodes from ulcerative colitis patients, whereas Proteobacteria were more abundant in Crohn's disease. While species diversity was reduced in the MLNs of patients with Crohn's disease, these lymph nodes contained greater numbers of less dominant phyla, especially Fusobacteria. In summary, this study confirmed distinct differences between the Crohn's disease and ulcerative colitis MLN microbiota suggesting potential for diagnosis of Crohn's disease or ulcerative colitis, particularly in cases of indeterminate colitis at time of resection.

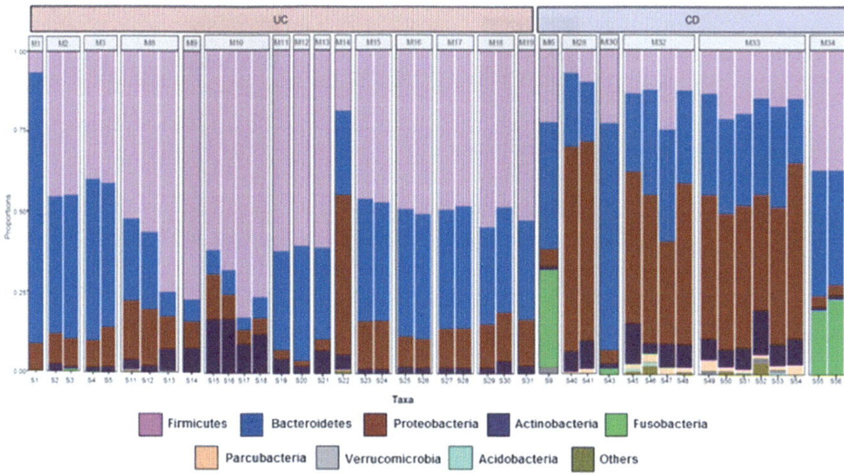

Fig. 6.1 Relative abundance of predominant bacterial phyla in mesenteric lymph nodes [MLNs] from inflammatory bowel disease [IBD] patients. There was a distinct difference in the profile of phyla from MLNs of Crohn's disease [CD] and ulcerative colitis [UC] patients. Adapted from: Kiernan, M.G. et al., Journal of Crohn's and Colitis 2019;13(1): 58–66. Copyright © Oxford University Press

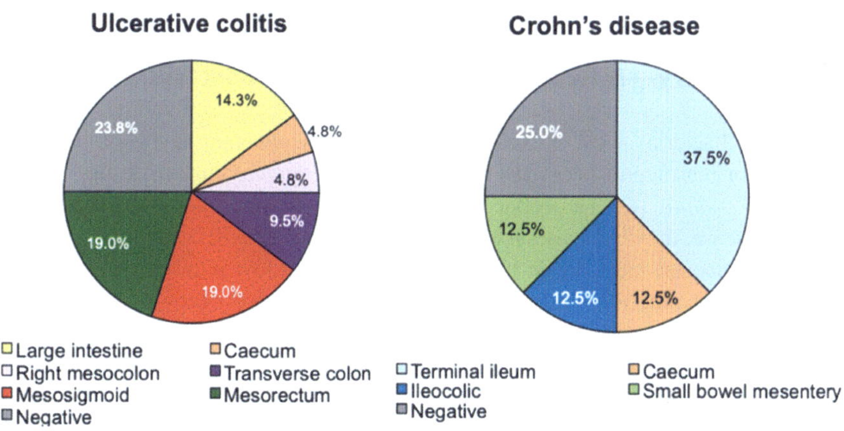

Fig. 6.2 Mesenteric lymph node [MLN] mapping. Proportions of 16S PCR–positive MLNs taken from each location of UC or Crohn's disease-associated mesentery. Adapted from: Kiernan, M.G. et al., Journal of Crohn's and Colitis 2019;13(1): 58–66. Copyright © Oxford University Press

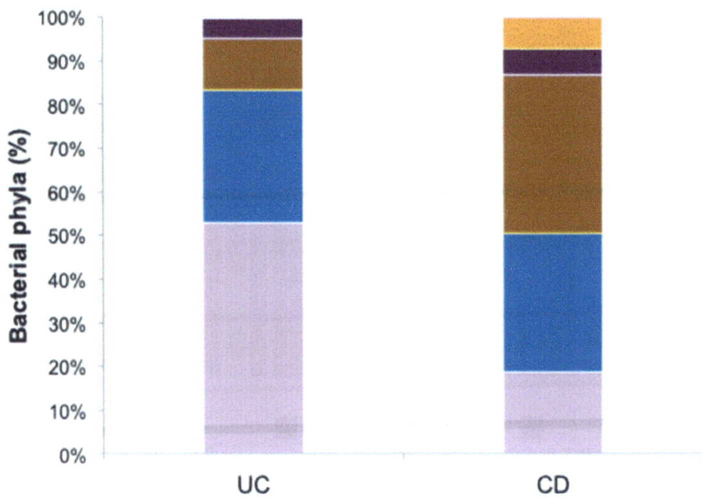

Fig. 6.3 Abundance of phyla in pooled mesenteric lymph nodes [MLNs] of Crohn's disease [CD] and ulcerative colitis [UC] patients. Abundance [%] of major bacterial phyla Firmicutes (52.8%/18.6%), Bacteroides (30.6%/32%), Proteobacteria (11.8%/36.1%), and Actinobacteria (4.6%/6.1%)in pooled MLNs of UC and CD patients, respectively Adapted from: Kiernan, M.G. et al., Journal of Crohn's and Colitis 2019;13(1): 58–66. Copyright © Oxford University Press

The Mesentery—A Reservoir for Pathogenic Bacteria?

Pathogenic bacteria, such as *Salmonella* or *Campylobacter*, have not been yet identified in mesenteric fat in IBD. Although *Escherichia/Shigella*, Campylobacteraceae, *Yersinia*, and Helicobacteriaceae have been identified from MLNs in Crohn's disease, they have also been found in MLNs of non-IBD subjects. *Yersinia*, Helicobacteriaceae and *Listeria* have been detected (in low numbers) in MLNs of patients with Crohn's disease, and warrant further investigation to determine whether they play a role in IBD.

Mesenteric lymphadenitis raises interesting questions. Associated with gastroenteritis (and respiratory tract infections, appendicitis, diverticulitis, pancreatitis and cholecystitis), mesenteric lymphadenitis during childhood or adolescence results in reduced risk of developing ulcerative colitis. MLNs harvested from patients with mesenteric lymphadenitis can contain *E. coli*, *Salmonella*, *Shigella*, *Yersinia enterocolittica*, mycobacteria, *Staphylococcus* and beta-haemolytic *Streptococcus*. This may point to an immunisation effect and perhaps represent an opportunity to identify microbial agents associated with ulcerative colitis onset.

Mesenteric Abscess—A Special Microenvironment for Pathogenic Bacteria in Inflammatory Bowel Disease?

Crohn's disease is commonly complicated by the development of intra-peritoneal abscesses. These can be located between different regions of mesentery (so-called inter-loop abscesses), in the pelvis, or in the flanks. They usually occur following bowel perforation, when a collection of infected fluid forms. Such fluid will most likely contain multiple bacteria, and these are frequently pathogenic. Gram negative bacilli, Gram positive cocci and anaerobic bacteria are typically detected from abscesses in Crohn's disease. *E. coli* and *B. fragilis* are the most common species identified. *Mycobacterium* spp have been detected in a mesenteric abscess in a patient with HIV, indicating at least potential as a reservoir for *Mycobacterium avium spp. paratuberculosis*. It is feasible that further characterisation of microbial ecology associated with mesenteric abscesses could yield important information related to IBD.

Conclusion

It seems reasonable to conclude that given the abundance and diversity of microbes borne in the human intestine that these cells may influence health and disease. Molecular technologies enable better insight into microbial community profiles characteristic of specific disease; and understanding of how such microbes penetrate and influence the mesentery and its component parts are likely to develop in tandem. Although entirely conjecture, it seems rational to suggest that microbial interaction with mesenteric adipocytes and MLNs may prove influential in disease initiation and progression. Elucidating the mechanisms that mediate these processes will involve research with sufficient scale to credibly determine definitive mesenteric microbiome profiles in each discrete disease setting. Accepting the confounding factors associated with geography and diet, among other factors, such scale will require collaboration of international investigators in multicentre studies.

Suggested Reading

1. Ambrose NS, Johnson M, Burdon DW, Keighley MR. Incidence of pathogenic bacteria from mesenteric lymph nodes and ileal serosa during Crohn's disease surgery. Br J Surg. 1984;71 (8):623–5.
2. Bonnardel J, Da Silva C, Henri S, Tamoutounour S, Chasson L, Montanana-Sanchis F, et al. Innate and adaptive immune functions of Peyer's patch monocyte-derived cells. Cell Rep. 2015;11(5):770–84.
3. Chanchlani R. Clinical profile and management of mesenteric lymphadenitis in children - our experience. Int J Orthop Traumatol Surg Sci. 2015;1(1):1–4.
4. Cregan J, Hayward NJ. Bacterial content of healthy small intestine. BMJ. 1953;1 (4824):1356–9.

5. De Filippo C, Cavalieri D, Di Paola M, Ramazzotti M, Poullet JB, Massart S, et al. Impact of diet in shaping gut microbiota revealed by a comparative study in children from Europe and rural Africa. Proc Natl Acad Sci USA. 2010;107(33):14691–6.
6. Diehl GE, Longman RS, Zhang JX, Breart B, Galan C, Cuesta A, et al. Microbiota restricts trafficking of bacteria to mesenteric lymph nodes by CX3CR1hi cells. Nature. 2013;494 (7435):116–20.
7. Dunne C. Adaptation of bacteria to the intestinal niche: probiotics and gut disorder. Inflamm Bowel Dis. 2001;7(2):136–43.
8. Dunne C, Shanhan F. Role of probiotics in the treatment of intestinal infections and inflammation. Curr Opin Gastroenterol. 2002;18(1):40–5.
9. Dunne C, Kelly P, O'Halloran S, Soden D, Bennett M, von Wright A, Vilpponen-Salmela T, Kiely B, O'Mahony L, Collins JK, O'Sullivan GC, Shanahan F. Mechanisms of adherence of a probiotic *Lactobacillus* strain during and post-*in vivo* assessment in ulcerative colitis patients. Microb Ecol Health Dis. 2004;16(2–3): 96–104.
10. Faith JJ, Guruge JL, Charbonneau M, Subramanian S, Seedorf H, Goodman AL, et al. The long-term stability of the human gut microbiota. Science. 2013;341(6141):1237439–39.
11. Finegold SM. Intestinal bacteria—the role they play in normal physiology, pathologic physiology, and infection. Calif Med. 1969;110(6):455–9.
12. Frisch M, Pedersen BV, Andersson RE. Appendicitis, mesenteric lymphadenitis, and subsequent risk of ulcerative colitis: cohort studies in Sweden and Denmark. BMJ. 2009;338: b716.
13. Glick-Bauer M, Yeh M-C. The health advantage of a vegan diet: Exploring the gut microbiota connection. Nutrients. 2014;6(11):4822–38.
14. Gorbach SL, Plaut AG, Nahas L, Weinstein L, Spanknebel G, Levitan R. Studies of intestinal microflora. II. Microorganisms of the small intestine and their relations to oral and fecal flora. Gastroenterology. 1967;53(6):856–67.
15. Helbling R, Conficconi E, Wyttenbach M, Benetti C, Simonetti GD, Bianchetti MG, et al. Acute nonspecific mesenteric lymphadenitis: More than "no need for surgery". Biomed Res Int. 2017;2017:9784565.
16. Hold GL, Smith M, Grange C, Watt ER, El-Omar EM, Mukhopadhya I. Role of the gut microbiota in inflammatory bowel disease pathogenesis: what have we learnt in the past 10 years? World J Gastroenterol. 2014;20(5):1192–210.
17. Human Microbiome Project Consortium. Structure, function and diversity of the healthy human microbiome. Nature. 2012;486(7402):207–14.
18. Ji B, Nielsen J. From next-generation sequencing to systematic modeling of the gut microbiome. Front Genet. 2015;6:219.
19. Kawabe T, Suzuki N, Yamaki S, Sun SL, Asao A, Okuyama Y, et al. Mesenteric lymph nodes contribute to proinflammatory Th17-cell generation during inflammation of the small intestine in mice. Eur J Immunol. 2016;46(5):1119–31.
20. Kelly P, Maguire PB, Bennett M, Fitzgerald DJ, Edwards RJ, Thiede B, Treumann A, Collins JK, O'Sullivan GC, Shanahan F, Dunne C. Correlation using proteomic analysis of probiotic *Lactobacillus salivarius* growth phase with presence of a cell wall-associated adhesin. FEMS Letters. 2005;252(1):153–9.
21. Kiernan MG, Coffey JC, McDermott K, Cotter PD, Cabrera-Rubio R, Kiely PA, Dunne CP. The human mesenteric lymph node microbiome differentiates between Crohn's disease and ulcerative colitis. J Crohn's Colitis. 2019;13(1):58–66. https://doi.org/10.1093/ecco-jcc/jjy136.
22. Lichtman SM. Bacterial translocation in humans. J Pediatr Gastroenterol Nutr. 2001;33(1):1–10.
23. Marteau P, Pochart P, Doré J, Béra-Maillet C, Bernalier A, Corthier G. Comparative study of bacterial groups within the human cecal and fecal microbiota. Appl Environ Microbiol. 2001;67(10):4939–42.

24. Michielan A, D'Incà R. Intestinal permeability in inflammatory bowel disease: Pathogenesis, clinical evaluation, and therapy of leaky gut. Mediat Inflamm. 2015;2015:628157
25. Mohar SM, Saeed S, Ramcharan A, Depaz H. Small bowel obstruction due to mesenteric abscess caused by *Mycobacterium avium* complex in an HIV patient: A case report and literature review. J Surg Case Rep 2017;2017(7):rjx129.
26. O'Brien CL, Pavli P, Gordon DM, Allison GE. Detection of bacterial DNA in lymph nodes of Crohn's disease patients using high throughput sequencing. Gut. 2014;63(10):1596–606.
27. Peyrin-Biroulet L, Chamaillard M, Gonzalez F, Beclin E, Decourcelle C, Antunes L, et al. Mesenteric fat in Crohn's disease: A pathogenetic hallmark or an innocent bystander? Gut. 2007;56(4):577–83.
28. Schirmer M, Franzosa EA, Lloyd-Price J, McIver LJ, Schwager R, Poon TW, et al. Dynamics of metatranscription in the inflammatory bowel disease gut microbiome. Nat Microbiol. 2018;3(3):337–46.
29. Sedman PC, Macfie J, Sagar P, Mitchell CJ, May J, Mancey-Jones B, et al. The prevalence of gut translocation in humans. Gastroenterology. 1994;107(3):643–9.
30. Shanahan F. The colonic microbiota in health and disease. Curr Opin Gastroenterol. 2013;29 (1):49–54.
31. Takahashi Y, Sato S, Kurashima Y, Lai C-Y, Otsu M, Hayashi M, et al. Reciprocal inflammatory signaling between intestinal epithelial cells and adipocytes in the absence of immune cells. EBioMedicine. 2017;23:34–45.
32. Talley NJ, Abreu MT, Achkar JP, Bernstein CN, Dubinsky MC, Hanauer SB et al. An evidence-based systematic review on medical therapies for inflammatory bowel disease', Am J Gastroenterol. 2011;106 Suppl 1:S2–25; quiz S26.
33. Teeples TJ, Tabibian JH. Mesenteric abscess in Crohn's disease. New Engl J Med. 2013;368 (7):663–63.
34. Thadepalli H, Lou SMA, Bach VT, Matsui TK, Mandal AK. Microflora of the human small intestine. Am J Surg. 1979;138(6):845–50.
35. Von Wright A, Vilpponen-Salmela T, Pagès Llopis M, Collins K, Kiely B, Shanahan F, Dunne C. The survival and colonic adhesion of *Bifidobacterium longum infantis* in patients with ulcerative colitis. Int Dairy J. 2002;12:197–200.
36. Vrakas S, Mountzouris KC, Michalopoulos G, Karamanolis G, Papatheodoridis G, Tzathas C, et al. Intestinal bacteria composition and translocation of bacteria in inflammatory bowel disease. PLoS ONE. 2017;12(1):e0170034.
37. Zoetendal EG, von Wright A, Vilpponen-Salmela T, Ben-Amor K, Akkermans ADL, de Vos WM. Mucosa-associated bacteria in the human gastrointestinal tract are uniformly distributed along the colon and differ from the community recovered from feces. Appl Environ Microbiol. 2002;68(7):3401–7.
38. Zulian A, Cancello R, Ruocco C, Gentilini D, Di Blasio AM, Danelli P et al. Differences in visceral fat and fat bacterial colonization between ulcerative colitis and Crohn's disease. An *in vivo* and *in vitro* study. PLoS One 2013;8(10).

Cellular Anatomy of the Mesentery

7

Miranda G. Kiernan and J. Calvin Coffey

Introduction

There has been relatively little focus in the histological composition of the mesentery. The mesentery is now defined as the collection of tissues that maintains all abdominal digestive organs in position and in continuity with other systems. Given this definition, it is important to describe the tissues that make up the mesentery. Any such description should also include a description of the histology of associated structures. These include the peritoneal reflection a Toldt's fascia, both of which maintain the mesentery in position.

Mesenteric Histology

The following description focuses on the detached and mobilized adult mesentery. The first study to characterize the histological appearance of the mesentery at multiple levels was conducted by Culligan et al. In this study, full thickness biopsies were taken at multiple points, through cadaveric mesentery. The overall histological components and organization, is similar at all points in the mesentery from duodenojejunal flexure to anorectal junction. Both surfaces of the mesentery are lined with a sheet of elongated mesothelial cells. The surface of the mesentery (as seen on scanning electron microscopy) is remarkably smooth for the most part but is occasionally interrupted by grooves (Fig. 7.1). The surface is markedly different in disease states such as Crohn's disease.

M. G. Kiernan · J. C. Coffey (✉)
School of Medicine and Department of Surgery, University of Limerick Hospital Group, Raheen, Dooradoyle, Limerick, Ireland
e-mail: Calvin.coffey@ul.ie

E. D. Ehrenpreis et al. (eds.), *The Mesenteric Organ in Health and Disease*,
https://doi.org/10.1007/978-3-030-71963-0_7

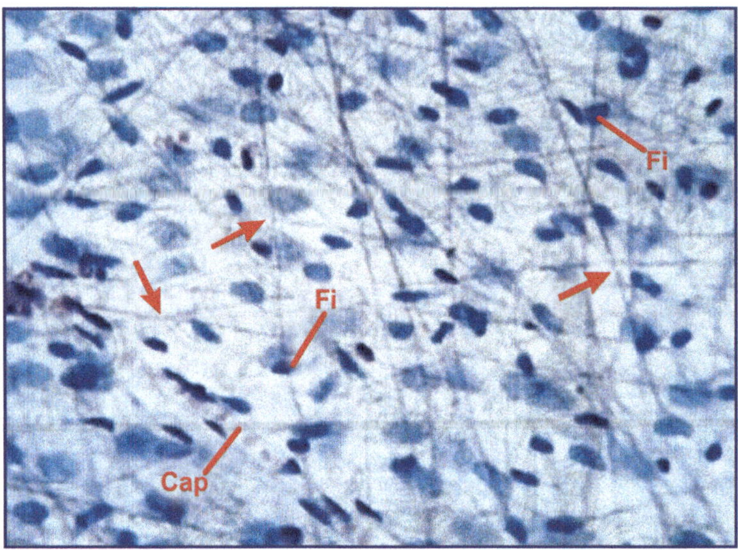

Fig. 7.1 Light microscopy of the mesentery. Elastic fibers (arrows) stain selectively as dark think branching fibers in loose connective tissue. Fibroblast nuclei (Fi) and a capillary (Cap) with erythrocytes are also seen

A connective tissue layer occurs immediately under surface mesothelium of the mesentery. This is variable in thickness and in some regions, it appears absent whilst in others it is well formed. Clusters of mesenchymal cells are intermittently observed within the connective tissue layer. On occasion these clusters are seen in association with connective tissue septations in the mesentery. On other occasions they appear entirely separate, based on examination of serial sections. The nature and function of the cells located in these clusters remains to be characterized.

Lymphatic channels occur in the submesothelial connective tissue. It has been suggested that stomata (openings between surface mesothelial cells) expose these lymphatics to peritoneal fluid and thereby facilitate the absorption of the latter. The channels are in turn continuous with lymphatic channels in mesenteric connective tissue septations. Collectively, these comprise a lymphatic network.

Gene Expression and Protein Synthesis in the Mesentery

These above are qualitative studies and functional studies are required to compare different regions of the mesentery. There are some suggestions, based on gene expression profiles, that regions of mesentery behave in different manners. For example, gene expression profiles from the small bowel region of mesentery differ from those in the mesogastric region of the mesentery. Histology is reflected in functionality and it has been suggested that the mesentery is a site of CRP production. Whilst CRP production was historically thought limited to the liver, emerging data point to mesenteric production and relate this to the control of systemic glucose levels.

Mesenteric Derived Mesothelial Cells

In separate studies, we generated mesenteric cells, cultured these in vitro and characterized the resultant cell types. When this is done they express—after—in culture. These studies were the first to generate and characterize mesenteric derived mesenchymal cells. They proliferate adhere and migrate in a manner that is similar to murine fibroblasts. We also compared these properties with mesenteric mesenchymal cells derived from patients with Crohn's disease. Interestingly, marked differences emerged indicating that each of these properties was increased in mesentery derived mesenchymal cells. Preliminary data also incase that some characteristics may be attenuated through ambition of TGF-beta.

Connective Tissue Continuity

Characterization of the mesentery demonstrated a distinct connective tissue lattice. This is continuous with submesothelial connective tissue. It is also continuous, at the surface of the intestine, with connective tissue septations in the intestinal wall. In addition, mesenteric connective tissue coalesces around intra-mesenteric blood vessels. This means there is continuous network of connective tissue extending from the submucosa of the intestine to contiguous mesentery and thereafter to vessels within it.

Mesenteric Adipocytes

Adipocytes are the main cellular contributor to the mesentery. Importantly, they are organised in a honeycomb-like distribution. It is not possible to distinguish individual adipocytes as their lipid surfaces are closely apposed within the mesentery. The overall appearance of the mesenteric adipocyte fraction, is that of an extensive system of cells. At the surface of intra-mesenteric blood vessels, this system is contiguous with surface connective tissue. This could be of considerable importance as the metabolic activity of the adipose system may be conducted to that of adjacent vessels, and thereafter to blood, thus providing mammals when an excellent mechanism of heat regulation.

Adipocytes within the mesentery are highly active. This is reflected in multiple studies comparing gene expression studies as well as studies examining individual cytokine production.

Fibrocytes

Mesenteric fibrocytes have received little overall attention to date. These are generally described as blood-borne and derived from hematopoietic progenitors. In a study of patients with Crohn's disease, we noted these cells were present in increased numbers in blood vessels in the mesentery, and in the tissue surrounding vessels. We also noted they were present in clusters at the surface of the intestine. This led our group to suggest the mesentery may provide a platform into which fibrocytes may migrate, from blood vessels, and within which they could exploit the connective tissue continuity of the mesentery and intestine, to access the surface of the intestine.

Fibrocytes are not the only cells to utilize the mesentery as a platform for migration. As mentioned in the section on the embryological development of the mesentery, primordial germ cells use the mesentery to migrate to the genital ridge. In addition, neural crest cells migrate across the mesentery from the neural crest. Findings related to fibrocytes are particularly relevant as these cells can differentiate into either fibroblasts or adipocytes. In Crohn's disease, the mesentery advances over the surface of the intestine as fat wrapping. The cellular basis of fat wrapping in unknown but it is tempting to think that fibrocyte exploit and migrate along mesentery, access the surface of the intestine and migrate to give rise to fat wrapping.

The Peritoneal Reflection

The peritoneal reflection must be incised in order to access the plane where the mesentery is attached to Toldt's fascia. Separation of the mesentery and Toldt's fascia is an essential component of mesenteric detachment.

The reflection is the region of peritoneum where parietal peritoneum is continuous with visceral peritoneum. An excellent example of this occurs where the small intestinal region of mesentery continues as the right mesocolon. This anatomical region is not directly apparent when one enters the abdomen, as it is obscured from view by the reflection in this region. Here, peritoneum leaves the posterior abdominal wall and reaches across to seamlessly continue as the visceral peritoneum.

Numerous terms have been applied to the reflection including ligaments (coronary and triangular ligament), folds (bloodless fold of Treves), and membranes (i.e. Jackson's membrane). These are all examples of where the parietal and visceral peritoneum are continuous.

Histological examination of the reflection demonstrates that it comprises two continuous layers of mesothelial cells with variable amounts of connective tissue located between both. Scanning electron microscopy (SEM) demonstrates that the surface facing the peritoneal cavity is smooth in normality, but disrupted and irregular in disease settings (i.e. Crohn's disease). SEM studies also demonstrate the connective tissue is primarily composed of elastin helices. It is feasible the helical shape of these facilitates return to a normal conformation following stretch.

No studies to date have examined the regions where the reflection arises from the parietal peritoneum and continues as the visceral. One of the reasons for this lies in the fact that the peritoneum is continuous, and there are no anatomical landmarks that precisely define specific regions.

Histology of Toldt's Fascia

Wherever an organ comes into close proximity with the abdominal wall it does not fuse directly with the wall. Instead, a connective tissue occurs between the organ and the abdominal wall. This is Toldt's fascia. Although flimsy and variable in composition it is a universal finding that likely contributes to maintaining the organ in position.

Few studies have examined the fascia in detail. The connective tissue composition remains to be confirmed but it would appear from early findings that it is comprised of loose connective tissue. Culligan et al. examined the fascia and found that in 30% of cases it contained lymphatic channels. This is highly relevant as it indicates that it plays a role in fluid absorption.

The relationship between the fascia and mesentery extends beyond simple adhesion. It appears to coalesce with the connective tissue surrounding the major vessels as these enter the mesentery. Given that this connective tissue is also continuous with that of the lattice contained within the mesentery, it means there is a continuity between mesenteric, adventitial and fascial connective tissue. The scientific and clinical relevance of this continuity remains to be determined.

The Mesentery in Disease States

Changes in disease states are described in detail elsewhere but will be briefly discussed here. Several disease states lead to an alteration in the histology of the mesentery. Again, findings here can also be considered as preliminary and requiring further characterization. For example, the surface of the mesentery is altered in peritoneal carcinomatosis in which peritoneal deposits arise in a sporadic distribution. These deposits extend into the mesentery to a limited degree but never appear to penetrate deep into the substance of the mesentery (i.e. they are not locally invasive). It is not known whether peritoneal deposits are distributed in this manner nor is it known why (despite having an apparent capacity of spread) they are not locally invasive.

The surface mesotheluim of the mesentery is substantially altered in mesenteric sclerosis in which there is pronounced fibrosis and thickening. The factors involved are under intensive investigation as this phenomenon frequently leads to failure of peritoneal dialysis in patients with renal failure. Mesenteric panniculitis is characterized by mesenteric infiltration with acute and chronic inflammatory cells. There are a number of subtypes but as with sclerosis, the etiology and pathophysiology remains to be elucidated.

Future Directions

It is essential to characterize the region where the mesentery intersects all abdominal digestive organs. This is particularly important as, given continuity, any signal that passes from one to another must pass through this region. As such to date it is virtually unchartered in terms of what is known about it.

Suggested Reading

1. Culligan K, Walsh S, Dunne C, Walsh M, Ryan S, Quondamatteo F, Dockery P, Coffey JC. The mesocolon: a histological and electron microscopic characterization of the mesenteric attachment of the colon prior to and after surgical mobilization. Ann Surg. 2014;260(6):1048–56. https://doi.org/10.1097/SLA.0000000000000323. PMID: 24441808.

Part III
Diagnostic Procedures

Radiography of the Mesentery

8

Scott Sorensen and Abraham H. Dachman

Plain Films of the Abdomen

Description of Procedure

Plain films of the abdomen are often the first exam done for abdominal symptoms. They are readily available, inexpensive and can be tailored to the clinical question.

Indications

Common indications are abdominal pain, clinical concern for ileus versus mechanical obstruction, constipation to evaluate for stool burden, to search for signs of ischemia or perforation and position and location of radiopaque objects such as tubes and catheters.

Complementary Procedures

Since plain films are often non-specific or non-diagnostic, CT is the most common alternate complimentary procedure.

Contraindications

There are no absolute contraindications to plain radiography. However it is advisable to minimize the use of ionizing radiation in general and a greater level of

S. Sorensen · A. H. Dachman (✉)
The University of Chicago, MC 2026 5841 S. Maryland Ave., Chicago, IL 60637, USA
e-mail: ahdachma@uchicago.edu

E. D. Ehrenpreis et al. (eds.), *The Mesenteric Organ in Health and Disease*,
https://doi.org/10.1007/978-3-030-71963-0_8

caution should be used for children, women of child bearing age, and all young individuals. When doing portable radiography, there is often a small amount of scatter radiation to surrounding patients, e.g., in the setting of a shared room such as an open ICU. Views such as cross table lateral views should be avoided in circumstances where adjacent patients will be exposed.

Preparation of Patient

No patient preparation is needed for plain radiographs. Overlying radiopaque objects should be removed such as clothing and jewelry.

How the Procedure Is Performed

The procedure is normally performed by a technologist. Images can be obtained in various ways depending on the indication. Supine, prone, right and left side down decubitus views, recumbent or erect lateral views, cross-table lateral views are some common examples.

Typical Abnormal Findings

Depending on the disease process, plain films might show ileus, obstruction, intramural air, free intraperitoneal air, extrinsic compression from a mesenteric mass or abnormal location due to internal hernia. Several examples are shown throughout the book.

Complications

None.

Computed Tomography (CT)

Description of Procedure

Computed tomography (CT), has become the mainstay for evaluation of peritoneal and mesenteric disease. X-rays passing through the body to a set of detectors is used to create helical images that are reconstructed in the axial, coronal and sagittal planes. Three-dimensional reconstructions can be used to depict disease and the relationship of mesenteric structures.

Indications

Most suspected mesenteric diseases can be imaged with CT.

Complementary Procedures

Plain films and MRI are the main complementary procedures, although ultrasound can be complimentary to look for fluid collections in the peritoneal cavity. Compared to MRI, CT is less expensive, and less sensitive to patient movement. It can be performed in patient with implanted medical devices that might preclude MRI examinations.

Contraindications

There are no absolute contraindications.

Relative Contraindications

Radiation should be minimized or avoided in children, pregnant women and young patients. Some patients are claustrophobic and have trouble if the face or upper chest is in the gantry. This can be avoided for most CT of the abdomen and pelvis.

Intravenous contrast is relatively contraindicated in patients with reduced renal function, old age or known long standing conditions that may diminish renal function. In some cases, a reduced volume of contrast may be used.

Intravenous contrast is iodine based and patients may have allergic reactions. When necessary, patients can be pre-medicated with steroids and antihistamines. This pre-medication usually commences 12 hours prior the administration of contrast.

Preparation of Patient

Metal objects that might affect the scan should be removed. Most patients are asked to not eat or drink for several hours prior to the scan. When oral contrast is used, it is administered about 1 or more hours prior to scan.

How the Procedure Is Performed

CT can be performed in different phases: pre-intravenous contrast, post intravenous contrast in standard portal venous phase or as a CT angiogram in the arterial phase. Delayed excretory phase can be performed when needed. Oral contrast can be either omitted or used as needed, although most non-emergent CT is done with oral contrast. CT enterography is a specialized study using low density contrast to distend the entire small intestine.

Scans are fast and require only a short breath hold. The patient must lie still to avoid motion artifacts. The contrast injection lasts for 10–30 seconds and might produce a warm, flushed sensation and a metallic taste in the mouth that lasts a minute or two.

Typical Abnormal Findings

Specific diseases are shown through this book.

Complications

Intravenous contrast can result in allergic reactions, usually mild. Severe allergic reactions are uncommon. Extravasated contrast in the arm may occur when there is poor venous access. Transient diminished renal function can occur especially in the elderly and in patient with underlying renal damage.

Additional Comments

Note that much has been written about the theoretical risk of radiation from CT. There is no conclusive evidence that radiation at small amounts delivered by a CT scan causes cancer. Nevertheless, the good practice principles of using radiation doses "as law as is reasonably achievable" aka "ALARA" should be employed. Modern scanners now have many different methods to reduce patient radiation dose compared to those just a few years earlier.

Magnetic Resonance Imaging (MRI)

Description of Procedure

MRI uses magnetic fields and electric filed gradients to generate images of the body. Unlike CT it does not employ ionizing radiation. It has the advantage of providing good soft tissue contrast. It is therefore commonly used to evaluate organs of the abdomen and in the context of this book, inflammatory bowel disease and vasculitis. Often, intravenous contrast agents are used.

Contraindications

There are many contraindications including certain types of metal in the body that might move when exposed to strong magnetic fields.

Relative Contraindications

Claustrophobic patients may not be able to tolerate an MRI without sedation. Some so-called 'open' MRI scanners can be used at the expense of image quality. Patients with impaired renal function may be unable to undergo contrast enhanced scans. Generally pregnant women in the first 3–4 months of pregnancy should avoid MRI.

Breast feeding women who receive intravenous contrast should pump breast milk before the study and wait about 24 hours before breast feeding.

Preparation of Patient

Some items need to be removed before going into the MRI room, e.g., cards with magnetic strips, electronic devices, hearing aids, watches, metal jewelry, hairpins, etc. Some cardiac pacemakers, aneurysm clips and neurostimulators are MRI compatible and some are not.

How the Procedure Is Performed

The patient lies on a table with a narrow bore gantry and undergoes several scans that can take 15–45 minutes (sometimes up to an hour) depending on the nature of the scan. Common pulse sequences are T1 weighted, T2 weighted, proton density weighted, gradient echo, inversion recovery diffusion and perfusion weighted and MR angiography (MRA). MRI fluoroscopy sequences can be used in some cases.

Typical Abnormal Findings

Several examples of mesenteric disease depicted with MRI are shown in this book.

Complications

Renal function can be impaired by intravenous contrast agents such as Gadolinium. Inadvertent use of MRI in patients who have internal metallic objects can also have consequences.

Conventional Angiography

Description of Procedure

Catheter angiography is performed suing X-ray guidance on fluoroscopy and catheterization of a vessel to permit injection of iodinated contrast. In the case of abdominal disease, a femoral route is commonly used to access the aorta or inferior vena cava depending on whether an arterial or venous injection is needed. Rapid sequential images are obtained during the injection of contrast.

Indications

Angiography can be done for diagnostic or therapeutic purposes. In the case of mesenteric disease the vascularity and organ of origin of a mass can be determined. Vasculitis, aneurysms, dissection and arteriovenous malformations involving single or multiple vessels can be diagnosed. Bleeding can be treated with coil embolization.

Complementary Procedures

CT and MR angiography often supplant the need for conventional angiography.

Contraindications

Severe bleeding diathesis.

Preparation of Patient

Patients should not eat for about 12 hours prior to the exam. They should remove metallic jewelry, etc. Medications that affect bleeding are usually withheld. The same concerns for reactions to iodinated contrast for CT apply. Likewise, breast-feeding should be withheld for about 24 hours after the test. Sedation is often used and thus patients should be accompanies and not drive themselves home.

How the Procedure Is Performed

A sterile room is used to perform angiography. The radiologists wear gowns, masks, hats and gloves similar to an operating room. A catheter is inserted via a small skin incision into a blood vessel. The machine consists of a table for the patient and one or two X-ray tubes, which can be positioned around the patient. Video monitors display the images produces.

A technologist and radiologist work together to insert an IV line, sterilely prepare the skin at the site of catheter placement and place the catheter. The radiologist will manipulate the catheter into the desired position before the contrast material is injected. After the procedure the catheter is removed and pressure is applied to the site for 10–20 minutes. The procedure may take 1–2 hours.

Complications

As with surgery, complications can be mild or sever and include pain, bleeding and infection. There may be allergic reactions to the iodinated contrast and effects on renal function. Blood clots can form on catheters and dislodge.

PET Scan

Description of Procedure

Positron emission tomography (PET) uses radioactive agents (radiotracers), a detection camera and computer to study the tissue function of organs and tumors.

Indications

In the case of mesenteric disease, there are few specific disease processes that can benefit from PET scanning, in particular searching for carcinomatosis or metastatic disease elsewhere for primary tumors known to be PET avid.

Complementary Procedures

Conventional imaging can often stage neoplastic disease without the need for PET.

Relative Contraindications

Some claustrophobic patients have anxiety.

Preparation of Patient

Do not eat or drink for several hours prior to the scan. Some medications and vitamins can affect scans Diabetics may have special instructions. The agent may be excreted in breast milk. Metal objects should be removed.

How the Procedure Is Performed

A PET scanner is a large machine with a doughnut shaped home in the middle. Some PET scans are combined with CT in one unit called "PET/CT". The PET images are processed and fused with the CT images.

The desired radioactive agent in injected into a vein. It takes about an hour for the agent to be absorbed by the body organs and by tumor. Scanning for the CT if fast, but the PET scanning takes 20–30 minutes. Patients must remain still for the scan. Sometimes delayed scans about to a few hours might be needed.

Typical Abnormal Findings

See examples in the book for carcinomatosis.

Complications

Allergic reactions are rare. There is a low level of radiation within the patient until the agent is excreted. Women should not be pregnant or breastfeeding.

Additional Comments

The exam is expensive and time consuming but gives unique information in the proper clinical context.

Ultrasound

Description of Procedure

Ultrasound of the abdomen and pelvis in a non-invasive test that uses a transducer to generate sound waves and view the internal structures.

Indications

Ultrasound is commonly available and relatively inexpensive as compared to CT and MRI. It is often used for the evaluation of focal or diffuse abdominal pain, swelling or infection. It is helpful in looking for ascites and at the solid organs and their blood flow. Some other examples of abdominal conditions evaluated by US are gallstones, kidney stones, abdominal aortic aneurysm, organomegaly and abnormal liver function. US can be used for guidance of percutaneous biopsies.

Complementary Procedures

The relative contraindications below to not apply to CT and most do not apply to MRI. These two exams are the principle complimentary tests.

Contraindications

There are no absolute contraindications.

Relative Contraindications

The sound waves must penetrate the body. Thus patients with very high body mass index (BMI) may have non-diagnostic exams. Also gas interferes with transmission of sound waves, thus patient with ileus or bowel obstruction may have gas filled loops obscuring abdominal anatomy.

Preparation of Patient

Patients should be fasting to help avoid bowel gas that might interfere with the exam. When the gallbladder is also evaluated, the patients should avoid any activity that might cause gallbladder contraction, e.g., smoking.

How the Procedure Is Performed

Ultrasound units come in many sizes and levels of sophistication. Most units are portable and consist of a console, computer, video display and assorted transducers. Patients lie recumbent and may be asked to turn on their sides. A warm water-based gel is use to interface the skin with the surface of the transducer. The transducer is moved to produce images. Breath holding is usually needed. Most exams take 30 minutes or less. Doppler US can used to evaluate blood flow with organs or large blood vessels.

Typical Abnormal Findings

As regards peritoneal disease, US can help determine if ascites is present and if it is of sufficient volume to tap. US can characterize masses as solid or cystic and sometimes determine the organ of origin.

Complications

None.

Additional Comments

US intravascular contrast agents are approved for use in the USA but are not commonly used.

Suggested Reading

1. Musson RE, Bickle I, Vijay RK. Gas patterns on plain abdominal radiographs: a pictorial review. Postgrad Med J. 2011;87(1026):274–87.
2. Rydberg J, Buckwalter KA, Caldemeyer KS, Phillips MD, Conces DJ, Aisen AM, Persohn SA, Kopecky KK. Multisection CT: scanning techniques and clinical applications. RadioGraphics. 2000;20(6):1787–806.
3. Mattoon JS, Berry CR, Nyland TG. Abdominal ultrasound scanning techniques. In: Ultrasound, 3rd ed. St. Louis. 2014. https://books.google.com/books?hl=en&lr=&id=ot7TBQAAQBAJ&oi=fnd&pg=P94&ots=WbJEPa-KqR&sig=hRO1ASQ3BVc32flF_-4m5ZjBkx0#v=onepage&q&f=false.
4. Arraiza M, Metser U, Vajpeyi R, Khalili K, Hanbidge A, Kennedy E, Ghai S. Primary cystic peritoneal masses and mimickers: spectrum of diseases with pathologic correlation. Abdom Imaging. 2015;40(4):875–906.
5. Dillman JR, Smith EA, Morani AC, Trout AT. Imaging of the pediatric peritoneum, mesentery and omentum. Pediatr Radiol. 2017;47(8):987–1000.
6. Copin P, Zins M, Nuzzo A, Purcell Y, Beranger-Gibert S, Maggiori L, Corcos O, Vilgrain V, Ronot M. Acute mesenteric ischemia: a critical role for the radiologist. Diagn Interv Imaging. 2018;99(3):123–34.
7. Taffel MT, Khati NJ, Hai N, Yaghmai V, Nikolaidis P. De-misty-fying the mesentery: an algorithmic approach to neoplastic and non-neoplastic mesenteric abnormalities. Abdom Imaging. 2014;39(4):892–907.
8. Dubuisson V, Voïglio EJ, Grenier N, Le Bras Y, Thoma M, Launay-Savary MV. Imaging of non-traumatic abdominal emergencies in adults. J Visc Surg. 2015;152(6 Suppl):S57–64.

9 Mesenteric Biopsy

Fons F. van den Berg and John C. Alverdy

Indications

Indications for a diagnostic mesenteric biopsy include the presence of a mesenteric lesion suspected for advanced metastatic disease in confirmed cancer or a de novo lesion of the mesentery without a history of cancer. In patients with significant symptoms and unclear imaging, performance of a mesenteric biopsy is considered acceptable. Mesenteric biopsies are performed are image-guided using ultrasound (US) or Computed Tomography (CT). Tissue is collected using percutaneous core biopsies and/or fine-needle aspiration (FNA). Reported technical and diagnostic success rates are high. Core biopsies yield higher diagnostic accuracy and are often needed for sub classification of the lesion. As an alternative, a surgical approach under general anesthesia can be performed. Although open surgical biopsy has historically been considered the gold standard method for this procedure, a diagnostic laparoscopy and mesenteric biopsy can be safely performed and is generally preferred, especially in young, healthy patients. Reasons for these approaches when obtaining a mesenteric biopsy include the need for visual inspection of the peritoneal cavity due to suspected peritoneal metastasis, or the necessity for urgent an pathological diagnosis and chemotherapy due to malignant disease (i.e. high-grade lymphoma). An open mesenteric biopsy is also preferable when mesenteric lesions are in close proximity to vital structures, such as major vessels, hollow organs of the digestive tract, spleen, liver and pancreas.

F. F. van den Berg
Department of Surgery, Amsterdam UMC, University of Amsterdam, Gastroenterology, Endocrinology, Metabolism, Amsterdam, North Holland, The Netherlands

J. C. Alverdy (✉)
Department of Surgery, University of Chicago, 5841 S. Maryland, Chicago, IL 60637, USA
e-mail: jalverdy@surgery.bsd.uchicago.edu

E. D. Ehrenpreis et al. (eds.), *The Mesenteric Organ in Health and Disease*,
https://doi.org/10.1007/978-3-030-71963-0_9

Contraindications

Contraindications for mesenteric biopsy are related to the risk of perforation or excessive bleeding. Absolute contraindications include a platelet count less than 50×10^9/L and an International Normalized ratio (INR) above 1.6. Relative contraindications are the presence of an uncorrected coagulopathy, absence of a safe needle path or a patient that is uncooperative or has an uncontrolled movement disorder.

Description of the Procedure

Preparation of Patient

Routine screening for coagulation status is advised. Patients are kept nil per oral (NPO) after midnight of the procedure and undergo general anesthesia in cases that an emergency laparotomy may occur. A single dose of antibiotics is given in when a solid organ, such as the colon, is traversed. Bowel preparation with purgative agents and oral antibiotics is generally not required.

Image-Guided Percutaneous Mesenteric Biopsy

It is advised to use a coaxial technique with a 17-gauge introducer and 18-gauge cutting needle for core biopsy, or 22-gauge needle for FNA specimens. The needle is advanced into the target lesion under US or CT guidance. Immediately after the biopsy, the patient should be monitored for signs of post-biopsy bleeding. An example of a CT-guided mesenteric biopsy is shown in Fig. 9.1.

Diagnostic Laparoscopic Mesenteric Biopsy

The patient is positioned in Trendelenburg position and, depending on the size and location of the mesenteric lesion, one 12-mm and two or more 5-mm trocars are placed under general anesthesia. Core biopsies of the lesion are taken under visual inspection, while suspect lymph nodes are removed in their entirety.

Potential findings from mesenteric biopsy are shown in Table 9.1.

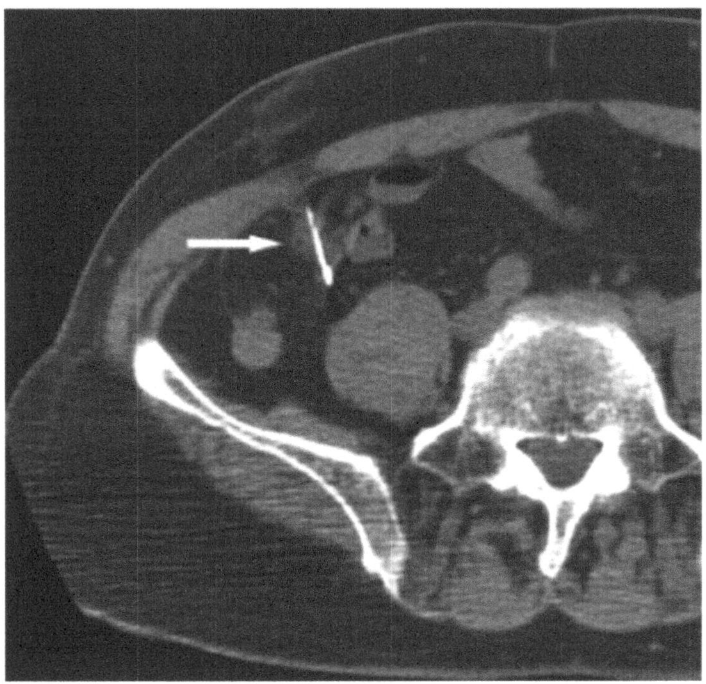

Fig. 9.1 A 67-year-old male with a history of cecum adenocarcinoma underwent a partial colectomy. A new 16 mm omental nodule was seen on follow up CT scan. CT performed during the procedure showing a 16 mm target omental nodule (arrow) with a coaxial needle through it. Biopay pathology showed metastatic adenocarcinoma. Reused with permission from: Vadvala, H. V., et al. (2017). "Image-Guided Percutaneous Omental and Mesenteric Biopsy: Assessment of Technical Success Rate and Diagnostic Yield." J Vasc Interv Radiol 28(11): 1569–1576. Copyright © Elsevier

Table 9.1 Neoplastic and non-neoplastic findings on mesenteric lymph node

Neoplastic	Non-neoplastic
Metachronous peritoneal carcinomatosis	Mesenteric panniculitis/Sclerosing mesenteritis
Lymphoma (follicular, Hodgkin's, B-cell)	Mesenteric lymphadenitis
Desmoid tumor	Anastomosing hemangioma
Methotrexate-associated lymphoproliferative disorder	Peritoneal tuberculosis
	Inflammatory granulation tissue/Suture granuloma

Complications

The complication rate for image-guided mesenteric biopsy is generally low. Reported complications include localized pain, visceral perforation, intra-abdominal hemorrhage and abscess formation. There is small possibility of needle tract seeding in the case of locally aggressive tumor biology.

Suggested Reading

1. Vadvala HV, Furtado VF, Kambadakone A, Frenk NE, Mueller PR, Arellano RS. Image-guided percutaneous omental and mesenteric biopsy: assessment of technical success rate and diagnostic yield. J Vasc Interv Radiol. 2017;28(11):1569–76.

Diagnosing Mesenteric Diseases: Laparoscopy

10

John Anagnostakos and Joshua H. Wolf

Indications

There are a variety of conditions that may lead to laparoscopic exploration of the abdominal mesentery. These will be discussed below.

Traumatic Injury

Traditionally, in the setting of blunt abdominal trauma the standard of care for treating suspected intraabdominal injury has been through exploratory laparotomy. However, as surgeon comfort with laparoscopy has increased over the years, more authorities have been advocating for a preliminary diagnostic laparoscopy. Laparoscopy has been shown to be especially effective in diagnosing and treating blunt, hollow viscous injuries and mesenteric injuries. A mesenteric injury grading system was developed by Bekker et al. on a scale of 1–5 to describe the injury during laparoscopy. A high score indicates a need for further intervention. High grade mesenteric injuries or multiple injuries often require conversion to laparotomy to survey the abdomen to ensure there are no other injuries and to treat the mesenteric injury. However, some injuries can be managed entirely laparoscopically with intracorporal suturing and laparoscopic bowel resection utilizing endoscopic staplers. In the hands of a skilled laparoscopic surgeon, laparoscopy can be an effective way of treating acute traumatic abdominal injuries while avoiding the morbidity associated with laparotomy (Fig. 10.1).

J. Anagnostakos
Department of Surgery, Sinai Hospital of Baltimore, Baltimore, MD, USA

J. H. Wolf (✉)
Department of Surgery, LifeBridge Health, 2435 Belvedere Ave, Baltimore, MD 21215, USA
e-mail: yehoshuawolf@gmail.com

E. D. Ehrenpreis et al. (eds.), *The Mesenteric Organ in Health and Disease*,
https://doi.org/10.1007/978-3-030-71963-0_10

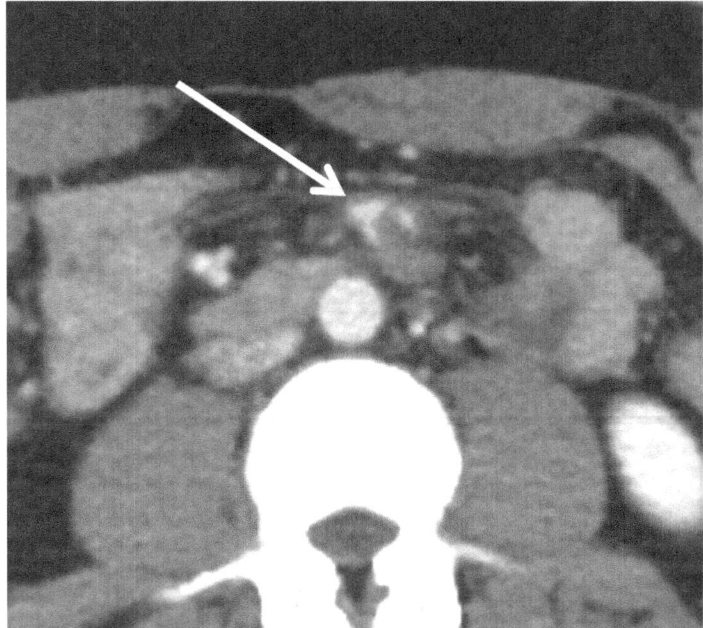

Fig. 10.1 Trauma extravasation

Inflammatory Bowel Disease

The laparoscopic technique has been well described in diagnosis and treatment of Crohn's disease. Crohn's disease can affect not only the bowel but also the surrounding mesentery. One of the more classic features of Crohn's disease is "creeping fat," which is extension of the mesenteric adipose tissue along the external circumference of the bowel wall. Some have argued that more complete mesenteric resections can lead to reduced rates of recurrence in Crohn's disease, but this remains to be demonstrated more definitively in randomized prospective data.

Neoplastic Disease

Laparoscopy can also play a critical role in the treatment of abdominal malignancy. Primary cancers of the abdomen can often metastasize locally to create peritoneal carcinomatosis in which the cancer can deposit into multiple intraabdominal structures including the mesentery. This problem may be a relative contraindication for resection of primary tumors such as in pancreatic adenocarcinoma. Performing a diagnostic laparoscopy to survey the abdominal cavity, including the mesentery, for malignant deposits before performing a major elective tumor resection can prevent the patient from undergoing an extensive procedure that would offer them no benefit. Other indications for laparoscopy in the setting of malignancy include determining the extent of resection for cytoreductive surgery, restaging after neoadjuvant/adjuvant chemotherapy, and obtaining biopsies to determine primary tumor origin (Fig. 10.2).

Neoplasms arising from the mesentery itself are not common. There are a variety of cellular subtypes that can give rise to primary mesenteric neoplasms, including peritoneum, adipose tissue, endothelium, neural tissue and inflammatory cells. Exploratory laparoscopy can aid in diagnosis, excision, enucleation or biopsy as indicated by tumor type.

Pediatric Applications

Although laparoscopy was first widely accepted for adult surgery, it also has many applications for pediatric surgery. The technique for trocar placement is like that in adults, however the size of the ports is usually smaller. Most commonly 4 mm or 5 mm trocars and instruments are used. Less abdominal insufflation pressure is usually required as well, usually 8–9 mm Hg for pediatrics compared to about 15 mm Hg for adults. Laparoscopy in infants and children has several applications including the treatment and diagnosis of mesenteric diseases. For example, congenital mesenteric malrotation, a disease process that involves the failure of the intestines to successfully rotate clockwise during gestation around the axis of the

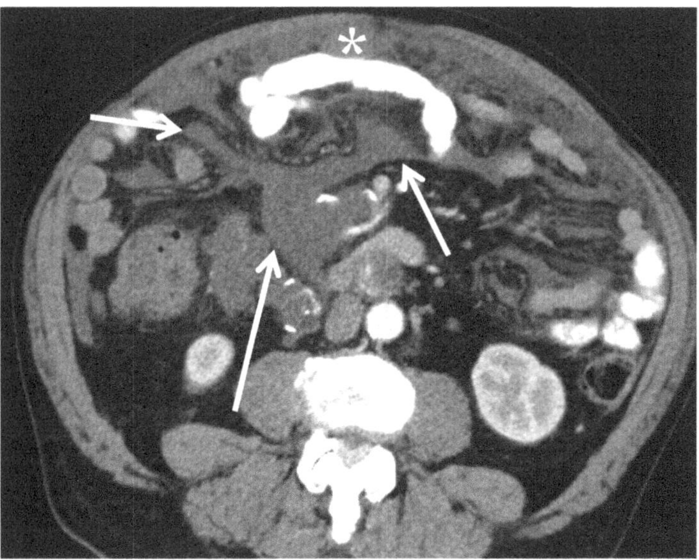

Fig. 10.2 Peritoneal carcinomatosis

superior mesenteric artery. This condition usually results in a narrowed mesenteric root, which predisposes the gut to a clockwise volvulus. The splenic flexure and pylorus often remain fixed, however. The treatment of choice has classically been open Ladd's procedure during which four steps are performed including inspection of the entire bowel, de-torsion of volvulus counterclockwise, lysis of Ladd bands and appendectomy. Nonviable bowel can be resected as well. Some authors have described a laparoscopic technique for treating this disease. The procedure involves, introducing trocars into the abdomen, running the bowel and reducing the volvulus laparoscopically, as well as dividing Ladd's bands with laparoscopic shears or energy devices.

Ischemia

Patients who have suspected mesenteric ischemia are often critically ill and not suitable for minimally invasive approaches. In select patients, however, who are hemodynamically stable and in whom the clinical suspicion for bowel necrosis is low, exploratory laparoscopy can be an effective diagnostic approach. Laparoscopy can be used for a planned "second look" operation in patients after laparotomy for acute mesenteric ischemia or non-occlusive mesenteric ischemia. Several groups have described this as a possible bedside procedure in the intensive-care unit.

Internal Hernia

Defects in the mesentery, either congenital or iatrogenic, can lead to internal herniation of small bowel. Laparoscopy can be performed to diagnose and reduce the hernia and repair the defect. This is appropriate in patients who have no signs of clinical deterioration to suggest bowel necrosis. If an internal hernia arises from a minimally invasive operation, laparoscopy can often be effective in diagnosis and repair without requiring additional abdominal incisions.

Contraindications

Laparoscopy is not indicated in patients with hemodynamic compromise or who at baseline have severe cardiopulmonary dysfunction, uncorrected coagulopathy, or abdominal wall infection or evisceration. A relative contraindication is a history of multiple major open surgeries, or of a recent abdominal operation, due to the likely prohibitive burden of intraabdominal adhesions.

Complementary Procedures

Intra-operative Perfusion Assessment with Fluorescence

For several of the indications mentioned above, in particular mesenteric ischemia and internal hernia, the use of intra-operative perfusion assessments can help assess bowel viability laparoscopically. This can be performed as an adjunctive procedure, by administering an intravenous fluorescent agent, such as indocyanine green, and using a specialized laparoscope that can detect the emission wavelength from the selected fluorophore.

Preparation of the Patient

Patients should be positioned supine in the operating room and placed under general anesthesia with endotracheal intubation. The abdomen should be prepped and a draped in a standard fashion.

Technical Details

Abdominal Entry

One of the most important steps in laparoscopy is safely entering the abdomen without damaging intraabdominal structures. Two commonly used techniques are a Hasson cut-down and utilization of a Veress needle. The authors' preference is Hasson cut-down technique, which involves creating an incision around the umbilicus and then dissecting down to the level of the rectus abdominus muscle under direct visualization. Then an incision is created in the fascia, and subsequently, the peritoneum. A 12 mm trocar is inserted and the gas is connected to it causing the abdomen to insufflate. It is important to start on low flow and then proceed to high flow once the pressure readings reflect that one is in the peritoneal cavity and not the preperitoneal space. The other trocars are then inserted under direct visualization.

The second method involves utilizing a Veress needle, which is a spring-loaded needle usually with a diameter of 2–3 mm. First, a small skin puncture is created. The safest area for insertion is in the left upper quadrant, 3 cm below the costal margin in the midclavicular line (Palmer's point), to avoid inadvertent bowel puncture. The needle is then introduced at a 90-degree angle to the contour of the abdominal wall. The technique of insertion involves listening for three audible clicks which represents penetration through firs the anterior rectus sheath then the posterior sheath and the peritoneum. Confirmation of entrance into the peritoneal cavity is confirmed by instilling saline through the top of the needle and watching for a positive “drop test”. This is when the plastic ball which is located at the top of the needle drops in the plastic housing representing free flow of the saline into the potential space of the peritoneal cavity. The gas is then connected to the needle and once insufflation is confirmed the other trocars are placed. Many surgeons elect to place the first trocar using the “Opti-View” technique, which introduces the scope and trocar through each layer of the abdominal wall under direct visualization until it is confirmed that the trocar is in the abdominal cavity.

Exploration of the Peritoneal Cavity and Mesentery

Port placement will depend somewhat upon the suspected location of the target pathology. In general, the distal small bowel mesentery will be easily accessible by triangulating from two working ports in the left abdomen. The proximal small bowel mesentery can be explored with two ports in the right abdomen. The bowel and mesentery should be handled with atraumatic bowel graspers. The full length of the mesentery can be inspected by sequentially “running the bowel” with hand-over-hand technique. If pathology is encountered, it may require biopsy or treatment depending upon the finding.

Abdominal Closure

Upon completion of the procedure, it is important to approximate the fascia to try to prevent incisional hernias. If the trocar is less than 10 mm, the fascia does not usually have to approximated since the size of the defect is small. However, if it is 10 mm and over, the fascia must be closed. Several techniques are available to accomplish this. One of the most commonly used techniques is to place 0 vicryl sutures on a UR-6 needle through the fascia at the beginning of the case in a figure-of-eight fashion and to tie it down at the completion of the case. Another technique is to use a Carter-Thomason device. This is a device with a grasper in it which is deployed to grab the suture and has a needle at the end which can penetrate the fascia. Each of these methods requires some extra time however is a crucial step in preventing post-operative complications.

Suggested Reading

1. Parajuli P, Kumar S, Gupta A, Bansal VK, Sagar S, Mishra B, Singhal M, Kumar A, Gamangatti S, Gupta B, Sawhney C. Role of laparoscopy in patients with abdominal trauma at level-I trauma center. Surg Laparosc Endosc Percutaneous Tech. 2018;28(1):20–5.
2. Li Y, Xiang Y, Wu N, Wu L, Yu Z, Zhang M, Wang M, Jiang J, Li Y. A comparison of laparoscopy and laparotomy for the management of abdominal trauma: a systematic review and meta-analysis. World J Surg. 2015;39(12):2862–71.

3. Lin HF, Chen YD, Lin KL, Wu MC, Wu CY, Chen SC. Laparoscopy decreases the laparotomy rate for hemodynamically stable patients with blunt hollow viscus and mesenteric injuries. Am J Surg. 2015;210(2):326–33.
4. Garofalo A, Valle M. Laparoscopy in the management of peritoneal carcinomatosis. Cancer J. 2009;15(3):190–5.
5. Ooms N, Matthyssens LE, Draaisma JM, de Blaauw I, Wijnen MH. Laparoscopic treatment of intestinal malrotation in children. Eur J Pediatr Surg. 2016;26(4):376.

Immunologic Function of the Mesentery

11

David H. Kruchko and Eli D. Ehrenpreis

Introduction

The gastrointestinal (GI) tract has two primary roles that are critical for the preservation of internal stability of living organisms. Aside from functions related to digestion and absorption of nutrients, fluids and electrolytes, the GI tract also plays a critical role in the homeostasis of the immune system. Immune homeostasis is characterized by the identification and immune response to potentially harmful microbes while simultaneously avoiding an immune response to non-harmful microbes. In immunocompromised patients, failed immune responses to harmful pathogens can result in potentially life-threatening infections. Conversely, inappropriate immune response to benign, commensural organisms and food proteins in the GI tract can lead to the development of autoimmune conditions including Crohn's disease and ulcerative colitis. The delicate balance between immune tolerance and inflammatory responses secondary to inappropriate immune reactions has been a subject of much basic science research and clinical investigation. This research has generally been focused on the intraluminal component of the GI tract and has largely assumed that breeches in the mucosal barrier provide an entry site for pathogens. Mucosal-associated lymphoid tissue (MALT) consists of lymphatic cells and epithelial lining of the intestines and other mucosa throughout the body. As shown in Fig. 11.1, the architecture of MALT provides an intimate connection between the mucosal barriers and immunologic cells. It also contains a variety of cell types with specialized roles in the recognition and reaction to antigens and other foreign invaders.

D. H. Kruchko
Department of Internal Medicine, Advocate Lutheran General Hospital, Chicago, IL, USA

E. D. Ehrenpreis (✉)
Department of Medicine, Advocate Lutheran General Hospital, 1775 Dempster St, 6 South, Park Ridge, IL 60068, USA
e-mail: e2bioconsultants@gmail.com

E. D. Ehrenpreis et al. (eds.), *The Mesenteric Organ in Health and Disease*,
https://doi.org/10.1007/978-3-030-71963-0_11

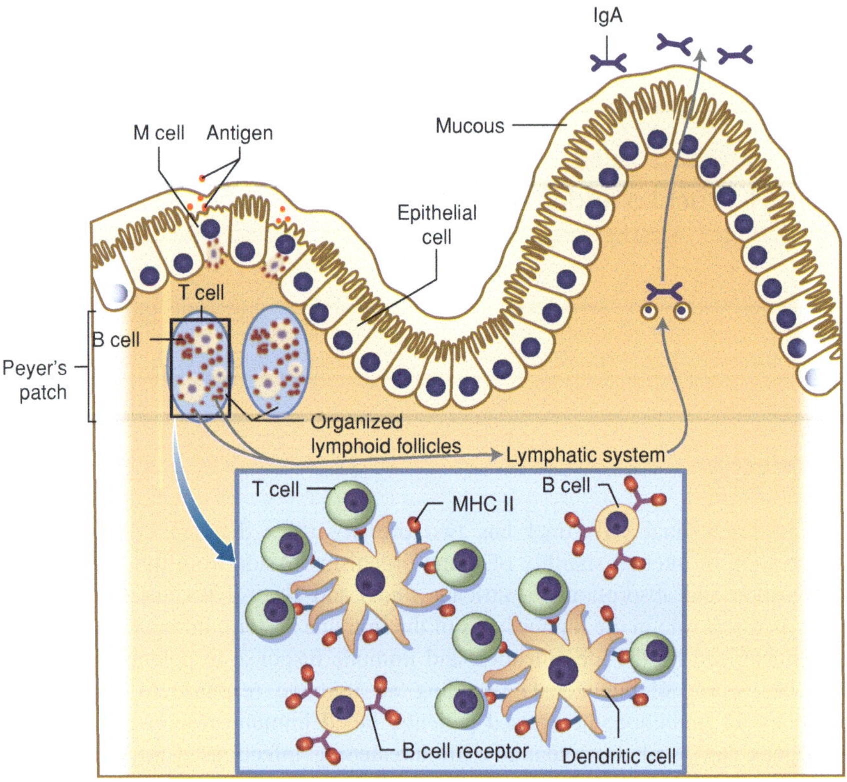

Fig. 11.1 Mucosal-Associated Lymphoid Tissue (MALT)—Functional anatomy and interplay of immune cells. In this figure, the important first step that M cells play in antigen update is first demonstrated. The inner surface of the M cell encapsulates the antigen so that antigen-presenting cells (such as dendritic cells) can present the antigens onto the cell surface using MHC II molecules. Antigen presenting cells then migrate to nearby tissue collections (Peyer's patches.) Within the Peyer's patch, there is T cell, B cell, and antigen presenting cell aggregation forming organized lymphoid follicles. This allows for T and B cell activation which then allows for antigen presenting cells to be loaded with the antigen and migrate throughout the lymphatic system. Subsequent formation of antibodies such as IgA allows for secretion into the intestinal lumen

More recently, attention has focused on the mesenteric components of the immune system and its relationship to gastrointestinal function. The mesentery can be defined as an individual organ system by its contiguous nature and the integrated components of the intramesenteric lymphatic, vascular, and neurological systems. This view also defines the mesentery as an organ of the immunologic system.

The mesoderm is one of three germ layers that comprises embryonic development. The mesodermal germ layer surface is lined with mesothelial cells on both a ventral and a dorsal side of the mesentery. The ventral aspect of the mesentery is effectively lost during maturation at the midgut, but the dorsal aspect persists and makes up the majority of the adult mesentery. The architecture of primitive intestinal loops is created by the embryonic mesentery and the mesothelial cells that are created by this mesodermal origin begin to function as primitive immune cells. Most notably, inflammatory cells such granulocytes are formed as a first line of defense for innate immunity responses. The unique ability of the mesentery to simultaneously be both a fixed and freely moving organ provides ideal positioning throughout the gastrointestinal tract for environmental sampling. When an invading foreign substance is sampled by the mesentery, it is well positioned to mediate local responses by utilizing granulocytes and other inflammatory cells. It is also capable of mediating systemic responses by coordination with the nearby organs that will effectively utilize local lymph nodes to prompt an adaptive immune response. The adaptive immune response is a component of the immune system that is composed of more specialized cells that can travel systemically in order to respond to and eliminate invading pathogens. Local lymph nodes are known as mesenteric lymph nodes and are scattered throughout the mesentery, thereby playing a critical role in sampling the commensal bacterial organisms within the GI tract as well as those from the environment. If the organisms are recognized as invaders, the mesenteric lymph nodes then regulate migration of B cells, T cells, natural killer (NK) cells, and dendritic cells to the adjacent mucosa.

Innate and Adaptive Responses of the Immune System

The mesentery, like many of the body's organs, can prompt both an innate, or a rapid but non-specific, immune response, as well as an adaptive, or a slower but more directed immune response (Fig. 11.2). Innate immunity is an important first defense and primarily relies on epithelial and mucosal linings that develop during embryogenesis. The physical barriers that these linings provide against pathogens and invaders rely on macrophage activity. Macrophages play a vital role in identifying bacterial or viral components and secrete cytokines (small molecules that recruit inflammatory and immune cells locally) in response to foreign invaders. Macrophages do so by using Toll-like receptors (TLRs) to identify the bacterial lipopolysaccharide (LPS), which are specifically seen on gram-negative bacteria found in the GI tract. TLRs are also helpful in identifying double-stranded RNA (dsRNA) that are specific viral features. Once TLRs identify these specific characteristic bacterial and/or viral components, they secrete cytokines to request assistance from other cells but also activate macrophage phagocytosis, or the process of ingesting foreign molecules.

Other cells of the innate immune system include mast cells, neutrophils (or polymorphonuclear neutrophils, PMNS), macrophages, eosinophils, basophils, natural killer cells (NK cells; can be classified as either innate or adaptive), gut-associated lymphoid tissue (GALT), and dendritic cells, as outlined in Table 11.1.

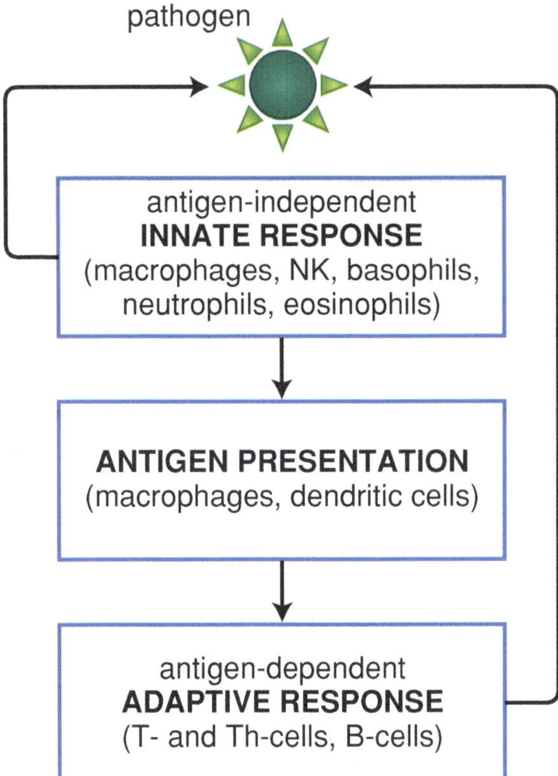

Fig. 11.2 Fundamental immune response outline. Activation of the immune system begins with the antigen-independent innate immune system after exposure to a pathogen. This response can be described as nonspecific attempts at detecting and destroying the pathogen. If this is unsuccessful, antigen presentation occurs and serves to activate the adaptive response. As it relates to the mesentery, this step is critical and relies on the dendritic cells. The adaptive immune system can be described as an antigen-dependent response as it responds to and destroys pathogenic cells that have specific antigens. It is important to note, however, that these are not disparate roles of the immune system, but rather successive and sequential components that are interdependent on one another

The cellular differentiation of both innate immune cells and adaptive immune cells is demonstrated in Fig. 11.3. Mast cells provide an inflammatory response by releasing granules and histamine to defend against pathogens and can be found in mucous membranes connective tissues. PMNs are phagocytic, as macrophages are, and act as the first line of defense to mount immune responses by releasing granules and other toxic substances to bacteria and fungi. As mentioned at the beginning of this section, macrophages have a more specific role than PMNs and can activate the adaptive immune response via cytokine secretion as well as phagocytosing cells infected with bacterial/viral components. Eosinophils specifically directly target

Table 11.1 Immune cell types and their functions and locations

Innate immune cells	Function	Location
Mast cells	Inflammatory response via histamine release	Mucous membranes of connective tissues
Neutrophils	First line of defense via phagocytic mechanisms	Found in all tissues, shortest lifespan, recycled frequently
Macrophages	Recognize bacterial/viral components and secrete cytokines and phagocytose infected cells	Peripheral blood and all tissues
Eosinophils	Release toxic free radicals targeted at parasites and some bacteria	Found in all tissues
Basophils	Release histamine like mast cells but target parasites like eosinophils	Found in all tissues
Natural killer cells (innate and/or adaptive)	Destroy the entire host containing cell instead of pathogen only	Peripheral blood and all tissues
Gut-associated lymphoid tissue	Mucosal barrier that acts as the first line of defense	Throughout the intestinal mucosal lining
Dendritic cells (DC)	Antigen-presenting cells, responsible for immune response initiation and activation	Found both in the external environment, e.g. epidermis, and internal mucosal lining

parasites and secrete toxic proteins as well as free radicals to kill parasites as well as some bacteria. Basophils similarly attack multi-cellular parasites but release histamine in a manner that is similar to mast cell histamine release. NK cells do not attack pathogens directly but rather destroyed the host cell that contains the pathogen.

Aside from functioning as a mucosal barrier, first line of defense, and integration with mesenteric lymph nodes to sample and initiate responses to foreign invaders, GALT also contains clinically important lymphatic tissue arranged anatomically as Peyer's patches. Peyer's patches are aggregated lymphoid follicles that are primarily present in the wall of the ileum. They function to monitor intestinal bacteria to allow for the growth of commensal organisms but also to prevent the proliferation of pathogenic organisms. Dendritic cells are a form of antigen presenting cell that are responsible for direct contact with the external environment via the inner mucosal lining of the GI tract. Furthermore, dendritic cells effectively act as a bridge from locations where the innate immune system is less effective, but the adaptive immune system has yet to be activated. Dendritic cells have an important in the function within the mesentery. Lastly, the complement system is complex and multifactorial, but essentially works as another bridge between the innate and adaptive immune system and allows for removal of antigens by multiple mechanisms.

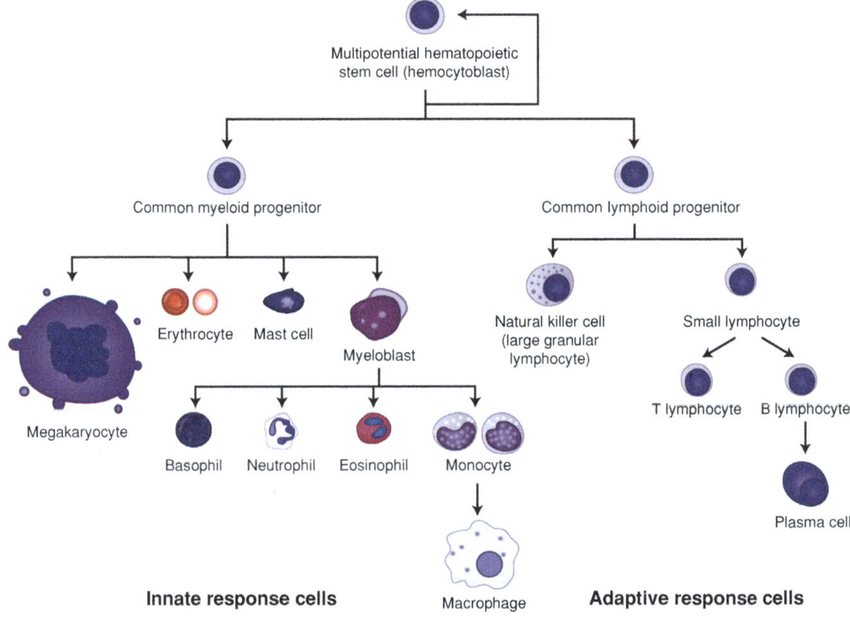

Fig. 11.3 Cell differentiation in the innate and adaptive immune response. Cell lineage of the innate immune response cells starts with a common myeloid progenitor and then further develops as shown above. There are many commonalities between cells in the innate immune system and all perform nonspecific functions. Cell lineage of the adaptive immune response shows a common lymphoid progenitor cell. The lymphoid components of these cells allow for differentiation into either B or T cells for a more tailored immune response. Natural killer (NK) cells are listed on the adaptive immune response side, but really they can be identified as both types of immune responses any many texts will associate them with the innate response cells

The innate immune system can function independently for certain pathogens, but frequently the activation of the adaptive immune system is also required to provide a complete immunologic response. The adaptive immune system relies on lymphocytes that are activated by cytokines. Lymphocytes are made up of B and T cells. B cells secrete antibodies or (immunoglobulins), molecules made up of protein that bind to specific antigens that are present on pathogens. Antigens are can either be identified in an extracellular location on a specific pathogen or they can be found unbound and free-floating within the host circulation. The binding of antibodies and antigens prompts attack mechanisms to destroy invading pathogens. After this antibody-antigen binding, B-cell subtypes go on to become memory cells to prevent re-infection and/or shorten the duration of symptoms during re-infection.

T cells are either identified as cytotoxic T cells or helper T cells, both of which rely on T-cell receptors to bind to pathogens by sensing specific protein sequences. Helper T cells interact with B cells by activating them, attracting macrophages, as well as secreting cytokines. Cytotoxic T cells effectively kill pathogenic cells by drilling into cells via ligands and other surface molecules that allow introduction of chemicals to trigger apoptosis.

Appropriate responses to pathogens and tolerance to commensal organisms constitutes immune homeostasis and is vitally important for effective immune function. Immune homeostasis, both innate and adaptive, relies on four primary components of the immune system. Responses to pathogens is ultimately modeled as a negative feedback within the components of the immune system (Fig. 11.4). By contrast, tumor responses are characterized by positive feedback system that are beyond the scope of this chapter.

The model is made up the interplay of four units; sensory, regulator, effector, and rehabilitator molecules. Sensory molecules consist of cells such as macrophages and lymphocytes that recognize a pathogenic cell. Regulators represent antigen presenting cells that investigate the sensory signal and work with the sensor cells to

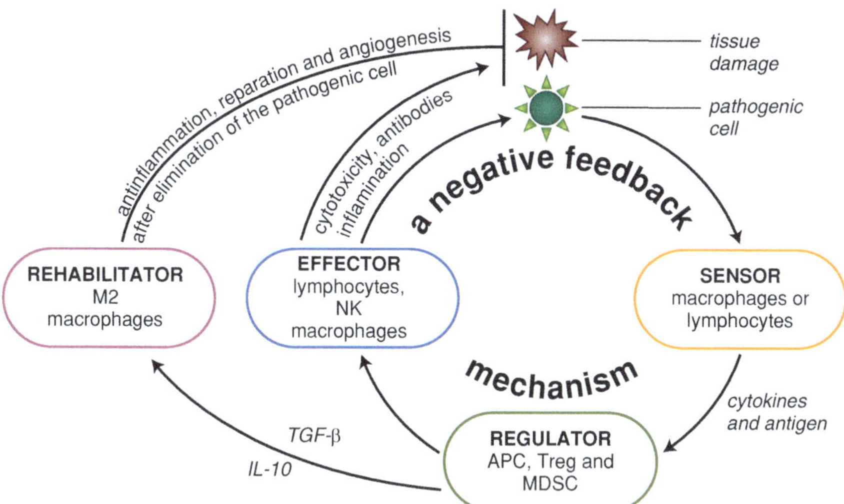

Fig. 11.4 Negative feedback mechanism for immune homeostasis. Lymphocytes and macrophages initiate the immune response by performing the sensor function, i.e. initiating a warning signal via cytokines and antigen presenting cells (APCs). The regulator consists of these APCs as well as T regulatory cells (Tregs), and myeloid-derived suppressor cells (MDSCs). APCs determine the type of cytokine that was released and program T helper (Th) cells to the effector cells, thereby dictating the action that these effectors take. The effector cells then destroy pathogens and/or host cells and surrounding tissue containing the pathogens as well. The final step, and one which has recently been recognized, is the action of recovering damaged tissue by repair, or rehabilitation as labeled in this diagram. Rehabilitator cells consist of M2 macrophages by mechanisms discussed on the previous page

appropriately program effector cells. Effector cells consist of programmed macrophages and lymphocytes that function in the elimination of pathogens. Rehabilitator molecules, representing specific types of macrophages known as M2 macrophages, restrict inflammation and provide functions for repair of tissue damage and restoration of homeostasis. Tissue repair is a relatively newly identified role of M2 macrophages but considered as critical in restoration of morphological and functional structure after macrophage phagocytosis and destruction of local tissue. Once tissue damage has been repaired, there are negative feedback signals to the sensor cells stop the cycle, thus preventing ongoing tissue damage. If this response is turned off, or positive instead of negative feedback to the sensor cells occurs, the risk of immune overactivity in the setting of either autoimmunity or carcinogenesis is increased. Specific macrophages, labeled M2 macrophages, are important components in this transformation to pathogenic positive feedback mechanisms. The mechanism by which this transformation takes place is hypothesized to be a response to inappropriate cytokine secretion leading to an influx of immune cells responding to what appears to be a foreign attack. The model suggests that similar pathologic positive feedback mechanisms occur in tumor development and autoimmunity.

Mesenteric-Specific Immune Cells

The pivotal and most extensively studied immune cells of the mesentery are dendritic cells, mesenteric lymph nodes, and T cells – specifically CD4+ T cells. Dendritic cells are some of the most active antigen presenting cells within the mesentery. There are three primary signals by which dendritic cells utilize antigen presenting mechanisms and prepare naive T cells in mesenteric lymph nodes and intestines. The first signal is T-cell receptor identification on Major Histocompatibility Complex (MHC) molecules. MHC provides a means for identification of large proteins within the immune system to identify foreign proteins. This is a vital role of the mesenteric immune system since it is repeatedly exposed to food proteins and other ingested molecules. The second signal is co-stimulation, which relies on the expression of several co-stimulatory molecules that are collectively referred to as cluster of differentiation (CD) molecules. CD28 is a specific molecule found on both CD4+ T cells as well as CD8+ T cells. There is also an important peripheral membrane protein known as B7 that is found on the surface of activated dendritic and other antigen presenting cells. B7 plays a critical role for either co-stimulation or co-inhibition of the MHC-TCR signal and allows for dendritic cells to begin the process of priming naïve T cells as depicted in Fig. 11.5. The third signal is naive T-cell differentiation by dendritic cells. This determines the differentiation of T-cells into specific effector cells such as CD4+ Th cells or CD8+ cells, which are labeled as cytotoxic T lymphocytes (CTLs) in Fig. 11.5. CTLs facilitate killing of infected cells (labeled tumor cells in Fig. 11.5) by antigen-specific

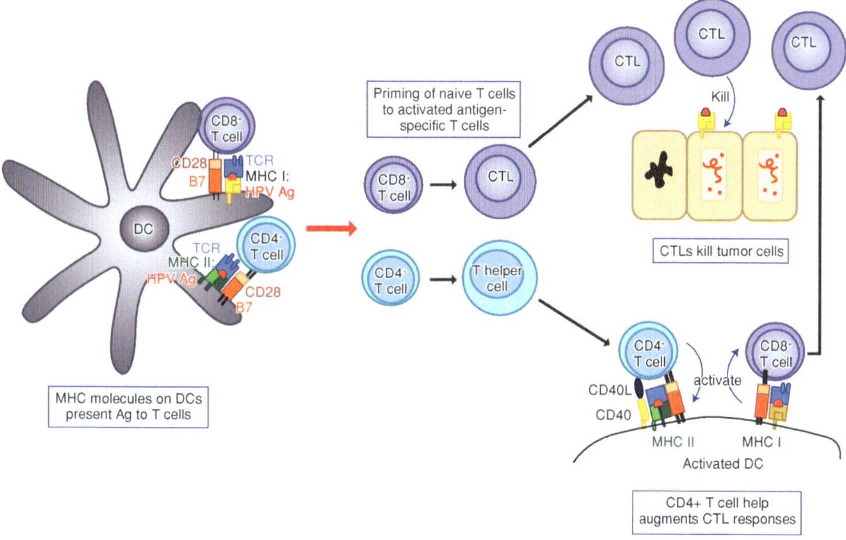

Fig. 11.5 The role of dendritic cells (DCs) and T cells in cellular immunity activation. As described in the text, this is a depiction of cellular immunity generation after uptake of the Human papillomavirus antigen (HPV Ag) by the DC. Key cells of differentiation (CDs) and other surface proteins (e.g. B7 and CD28) are labeled above and are pivotal in T cell recognition and HPV Ag presentation. Next, both CD8+ and CD4+ naïve T cells are primed by DCs in order to become effector cells to participate in killing of infected cells

mechanisms, while Th cells can differentiate into naïve CTLs and assist with antigen-specific killing of infected cells.

Lastly, CD80 is a specific molecule found on the surface of DCs and has been found to play a critical role in immune response. CD80 expression determines whether the immune system will attack or ignore commensal and/or foreign organisms. Once CD80 molecules are expressed, priming of naïve T cells begins which will ultimately promote anergy, or the lack of an immune response by ignoring a pathogen, as seen in Fig. 11.6. Conversely, overexpression and inappropriate expression of CD80 cells can lead to inapt immune responses to non-pathogenic, commensal organisms which effectively leads to autoimmunity. Crohn's disease is an example of the interplay between mesenteric lymph nodes and DCs producing CD molecules that ultimately lead to dysregulation of Th cells. Specifically, Th1 and Th17 dysregulation as a result of any abnormality in the three aforementioned signals, is a key feature of Crohn's disease.

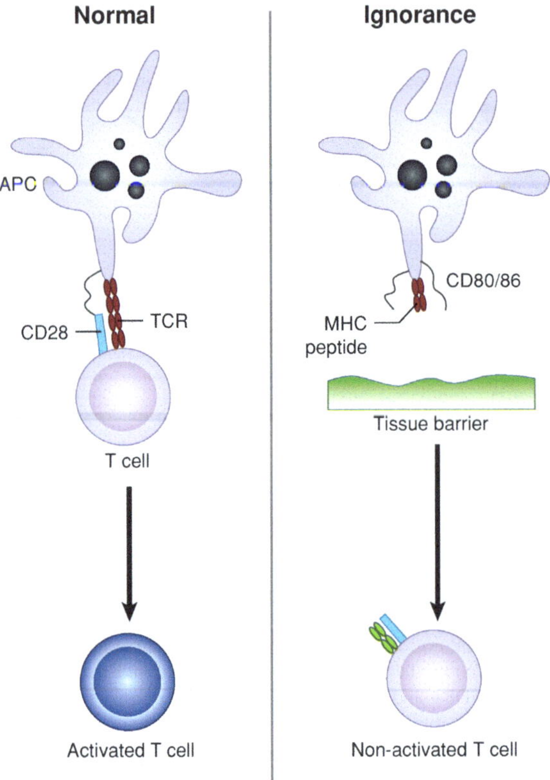

Fig. 11.6 T cell peripheral tolerance mechanisms. The column labeled "Normal" relies on the now well described interaction between MHC molecules, co-stimulatory molecules, and peptides that ultimately lead to T cell activation. The next column, labeled "Ignorance" represents anergy, described previously as the lack of an immune response by way of ignoring the molecule. This the prominent CD molecule at play in anergy is CD80 and this has been well documented in the literature. In order to avoid overactivity from the immune system, CD80 needs to be appropriately expressed to prevent the inappropriate T cell activation

Mesenteric Immunity and Crohn's Disease (Also See Chap. 15 Crohn's Disease)

Recent studies have focused on the interplay between mesenteric lymph nodes and the pathogenesis of Crohn's disease. The recruitment of T helper cells, also known as CD4+ T cells, is largely dependent on proper mesenteric lymph node response. In normal immunology, mesenteric lymph node recruit CD4+ T cells and thereby enhance the intercellular communication by production of cytokines (e.g. interferon-gamma, or γ-interferon) and glycoproteins (e.g. interleukin-17, IL-17). In the setting of Crohn's disease, there is also an increased production of γ-interferon.

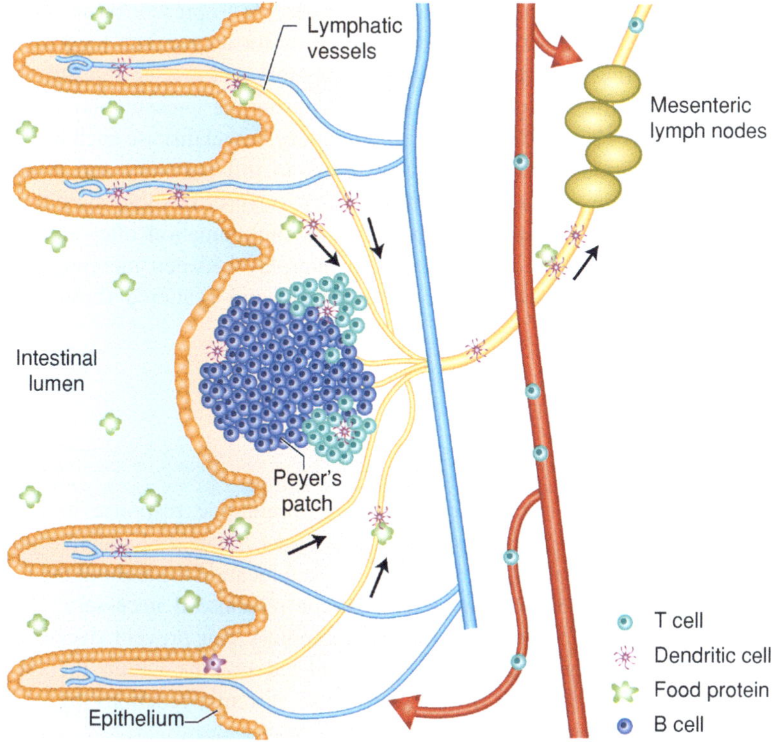

Fig. 11.7 Food tolerance and generation of immune response involving mesenteric lymph nodes after exposure to benign food proteins. This is a great depiction of initial, nonspecific immune responses after ingestion of benign food proteins. Dendritic cells can be seen bringing the food proteins to the Peyer's patch via lymphatic channels in order for sampling and determination if it warrants immune response. There are two separate clusters of T cell proliferation illustrated in this figure to show that although this is a benign particle, the immune system will frequently err on the side of caution and mount a mild initial response. Despite these small clusters of T cell proliferation, this is an accurate depiction of systemic tolerance and the lack of any real immune response. The food-containing DCs then travel to the mesenteric lymph nodes, which mediate systemic tolerance via further sampling and are then circulated freely via efferent channels of the lymphatic system

Increased γ-interferon has been implicated in mesenteric fibrosis, fat wrapping, and transmural inflammation. Another potential relationship between the mesenteric system and the pathogenesis of Crohn's disease is the mesenteric production of C-reactive protein (CRP). CRP production by mesenteric adipocytes may be triggered by local inflammation and bacterial translocation to mesenteric fat. Subsequent mesenteric fat hyperplasia may contribute to the inflammatory response in Crohn's disease.

Lastly, CD4+ T cells determine the MHC molecule to present on its surface in order to promote hyposensitivity to food and allow for appropriate food tolerance. This effectively prevents overwhelming inflammatory states after ingesting benign food particles as illustrated in Fig. 11.7. Inappropriate presentations of MHC molecules by CD4+ T cells, as seen in inflammatory bowel disease such as Crohn's disease, leads to food hypersensitivity resulting in inflammatory and potentially ongoing autoimmune states ranging from mild to severe. In these circumstances, mesenteric lymph nodes are vulnerable to acute inflammation after exposure to benign particles. This is due to the intimate interplay between mesenteric lymph nodes and the GI tract and can be seen grossly and scattered throughout the mesentery.

Summary

The mesentery plays a vital role in the host's immune response to invading pathogens. It is the anatomic site of cells of both the innate and adaptive immune system. Its sprawling anatomy provides the optimal location throughout the GI tract to sample foreign objects and mount an immune response if necessary. With the assistance of mesenteric lymph nodes, the mesentery can prevent overwhelming infections by recruiting systemic cells to localized regions within the GI tract. Mesenteric lymph nodes can also prevent such systemic reactions by correctly identifying benign organisms.

Suggested Reading

1. Wu HJ, Wu E. The role of gut microbiota in immune homeostasis and autoimmunity. Gut Microbes. 2012;3(1):4–14.
2. Coffey JC, O'Leary DP. The mesentery: structure, function, and role in disease. Lancet Gastroenterol Hepatol. 2016;1(3):238–47.
3. Thomason RT, Bader DM, Winters NI. Comprehensive timeline of mesodermal development in the quail small intestine. Dev Dyn Off Publ Am Assoc Anat. 2012;241(11):1678–94.
4. Hikspoors JPJM, Kruepunga N, Mommen GMC, Peeters JPWU, Hulsman CJM, Eleonore Kohler S, Lamers WH. The development of the dorsal mesentery in human embryos and fetuses. Semin Cell Dev Biol. 2018.
5. Isaza-Restrepo A, Martin-Saavedra JS, Velez-Leal JL, Vargas-Barato F, Riveros-Dueñas R. The peritoneum: beyond the tissue—a review. Front Physiol. 2018;9:738.
6. Macpherson AJ, Smith K. Mesenteric lymph nodes at the center of immune anatomy. J Exp Med. 2006;203(3):497–500.
7. Janeway CA Jr, Travers P, Walport M, et al. Immunobiology: the immune system in health and disease. 5th ed. New York: Garland Science; 2001.
8. Malyshev IY, Manukhina EB, Malyshev YI. Physiological organization of immune response based on the homeostatic mechanism of matrix reprogramming: implication in tumor and biotechnology. Med Hypotheses. 2014;82(6):754–65.

9. Sakuraba A, Sato T, Kamada N, Kitazume M, Sugita A, Hibi T. Th1/Th17 immune response is induced by mesenteric lymph node dendritic cells in Crohn's disease. Gastroenterology. 2009;137(5):1736–45.
10. Peyrin-Biroulet L, et al. Mesenteric fat as a source of C reactive protein and as a target for bacterial translocation in Crohn's Disease. Gut. 2012;61(1):78–85.

Neurophysiologic Function of the Mesentery

12

Amir Patel, Jeffrey Prochot, and Eli D. Ehrenpreis

Anatomy

A basic understanding of the neuroanatomy of the abdomen is required to fully appreciate the physiologic functions of the mesenteric nervous system. autonomic nervous system supplies the entire gastrointestinal tract and runs through the two peritoneal layers of the mesentery. A representation of the sympathetic and parasympathetic nervous systems is shown in Fig. 12.1.

The Sympathetic Nervous System

Sympathetic trunks track a path from the base of the skull to the coccyx in a parallel fashion along both sides of the spinal columns. Preganglionic fibers exit the anterior (ventral) branch of the spinal cord and run through the white rami to reach the sympathetic trunk. Extending off the sympathetic trunk are the preganglionic thoracic splanchnic nerves, which consist of the greater, lesser, and least splanchnic nerves. These will continue until they reach the network of nerves called the abdominal prevertebral plexus.

A. Patel
Department of Gastroenterology and Hepatology, Medical College of Wisconsin/Froedtert Hospital, Milwaukee, WI, USA
e-mail: ampatel@mcw.edu

J. Prochot
Department of Internal Medicine, Advocate Lutheran General Hospital, 1775 Dempster St, 6 South, Park Ridge, IL 60068, USA

E. D. Ehrenpreis (✉)
Department of Medicine, Advocate Lutheran General Hospital, 1775 Dempster St, 6 South, Park Ridge, IL 60068, USA
e-mail: e2bioconsultants@gmail.com

E. D. Ehrenpreis et al. (eds.), *The Mesenteric Organ in Health and Disease*,
https://doi.org/10.1007/978-3-030-71963-0_12

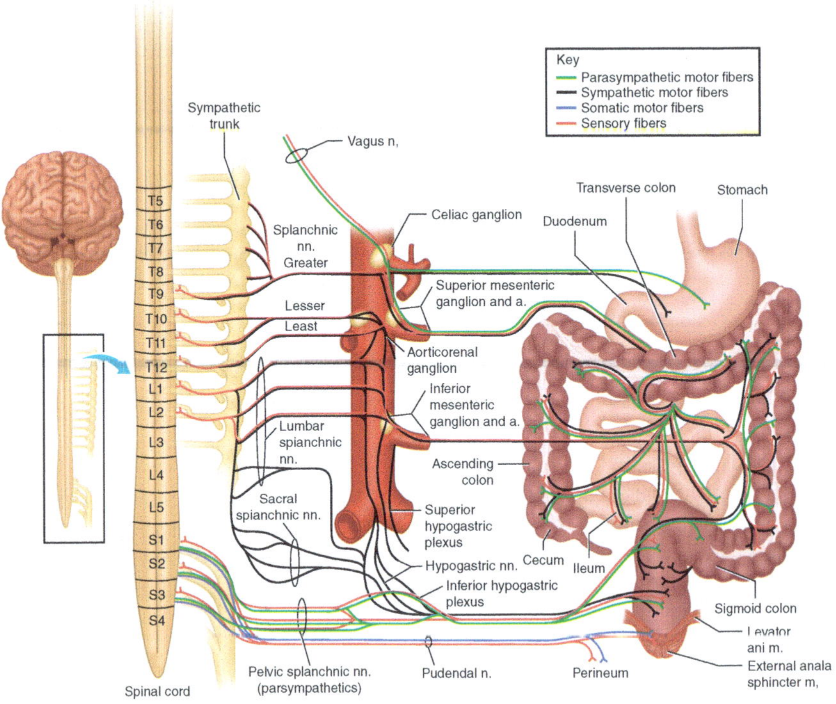

Fig. 12.1 The components of the sympathetic and parasympathetic nervous systems

The abdominal prevertebral plexus has three major divisions. The celiac plexus contains the celiac ganglion, superior mesenteric ganglion and aoritco-renal ganglion. The aortic plexus contains the inferior mesenteric ganglion and root of the inferior mesenteric artery. The superior hypogastric plexus comprises the third division.

Preganglionic nerve fibers synapse in their corresponding ganglion and are distributed through the mesentery to their destination along the intestine and abdominal compartments. The celiac ganglion distributes sympathetic nerve impulses to stomach, liver, pancreas, adrenals, and upper half of duodenum. The superior mesenteric ganglion supplies the lower half of the duodenum as well as jejunum, ileum, ascending and transverse colon. The inferior mesenteric ganglion distributes to descending colon and upper portion of rectum.

The Parasympathetic Nervous System

Parasympathetic innervation to the abdominal organs is supplied by the vagus nerve and pelvic splanchnic nerves. However, innervation through the mesentery consists of solely the vagus nerve. The vagus nerve enters the abdomen through the diaphragm and sends preganglionic parasympathetic branches to the prevertebral plexus. From this point, they are distributed to their respective sites in the midgut including the myenteric plexus within the smooth muscle of gut wall and Meissner's plexus to the glands of the mucosae in the intestine.

Neurologic Regulation of Mesenteric Blood Flow

Arterial and Venous Innervation

An important aspect of the neurologic function of the mesentery is regulation of blood flow within the mesentery itself. The splanchnic circulation receives about sixty percent of cardiac output and can hold up to one third of the total blood volume. Thus, the splanchnic circulation is a large reservoir for blood. With the aid of neurologic regulation, large amounts of blood can be allocated during a stress response when high cardiac output is required, or for digestion after a meal.

The splanchnic circulation is innervated by both the sympathetic division of the autonomic nervous system and by spinal sensory nerves. Postganglionic nerve fibers from the celiac, superior, and inferior mesenteric ganglion synapse on the vasculature. The sympathetic and sensory innervation of the mesenteric circulation consists of prevertebral sympathetic ganglion and dorsal root ganglion neurons, respectively. The axons of the neurons travel to the mesenteric arteries and veins in the paravascular nerves, which divide in the adventitia of the blood vessels to form the perivascular nerve plexus. Ultimately these axons distribute into terminal axons, loose their Schwann cell sheath, and form neuroeffector junctions with vascular smooth muscle cells.

Sympathetic stimulation of the vasculature is accomplished by the release of norepinephrine-mediated alpha-adrenergic vasoconstriction. Sympathetic nerves are primarily vasoconstrictor in their action while the sensory nerves are vasodilatory. Sympathetic nerve stimulation increases peripheral resistance and mobilizes up two thirds of the reserve blood volume in the veins. Local activation of sensory neurons has vasodilatory effects.

Parasympathetic stimulation of vasculature is less well understood. While postganglionic nerves from the vagus and pelvic splanchnic nerves synapse on the walls of the intestine themselves, it is not clear that they produce a direct parasympathetic stimulation of the vasculature. It is thought that the stimulation of intestinal wall and secretions in the myenteric and Meissner's plexus indirectly causes an increase in blood flow by release of acetylcholine in the endothelial layer. This activates muscarinic receptors to release nitric oxide causing smooth muscle relaxation and vasodilation.

Arterial and Venous Components of Blood Pressure Regulation

Innervation of the mesenteric vasculature is essential for vasoconstriction and dilation. These functions are required for regulation of blood pressure as well as potentially liberating large quantities of blood for use elsewhere i.e. fight or flight instincts. The splanchnic circulation is innervated by both the sympathetic division of the autonomic nervous system and by spinal sensory nerves.

The differences in regulation of vascular resistance and capacitance cannot be accounted for solely based on the sympathetic innervation. Primary sensory neurons also provide substantial innervation to the vasculature as well as to the sympathetic neurons that innervate the blood vessels. Furthermore, the vascular endothelium provides a modulatory influence on neurogenic contractions of artery and vein. The interaction of sensory and sympathetic innervation of arteries and veins with each other and with other modulatory influences such as endothelium are an important component of understand blood pressure regulation.

Neuronal Control of Mesenteric Arteries and Veins

When juxtaposing mesenteric artery and vein neuroeffector transmission, two major differences are apparent. Relatively rapid excitatory junction potentials can be recorded from arteries, but not from veins. This excitatory junction potential is mediated by ATP released from sympathetic nerves acting on P2X receptors on vascular smooth muscle cells. In addition, repetitive nerve stimulation produces a slow arterial and venous depolarization. However, the effect of this depolarization on the resulting contraction of the vasculature is proportionally greater for a given frequency of nerve stimulation in the veins than in arteries. In the mesenteric arteries, repetitive nerve stimulation also produces a long-lasting inhibitory junction potential that mediates vasodilation. This response is mediated by spinal sensory nerves. This inhibitory junction potential and its associated vasodilatation can be evoked in the mesenteric arterial supply of the colon and results in colonic distension (Fig. 12.2).

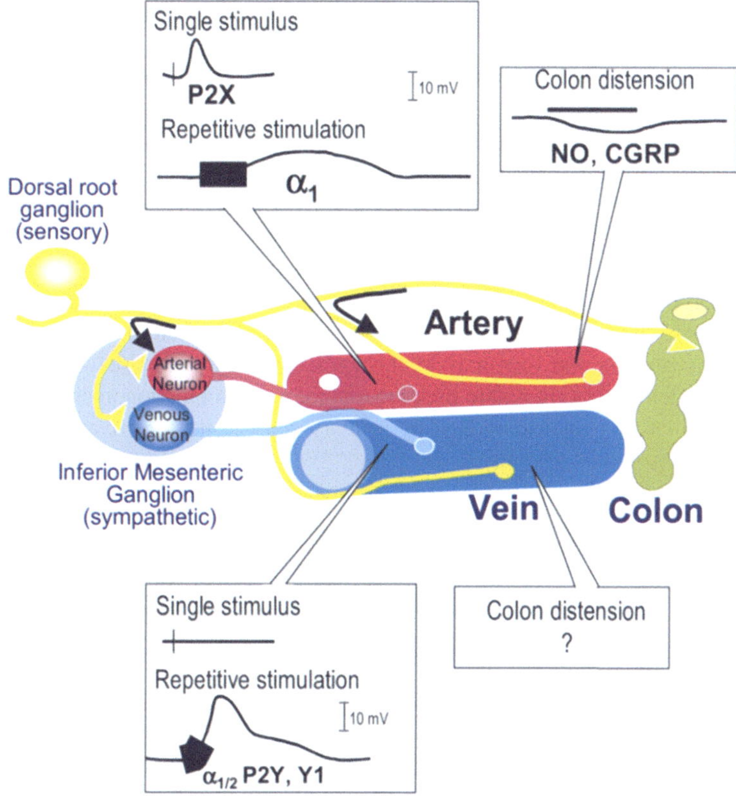

Fig. 12.2 The above schematic demonstrates how mesenteric arteries and veins are provided sensory innervation. The artery and vein react differently, as indicated by the stimulus curves, to single and repetitive stimuli. Cellular responses to nerve stimulation and distension of the colon are indicated inside the boxes. Reused from Kreulen, DL. Properties of the Venous and Arterial Innervation in the Mesentery. Journal of Smooth Muscle Research 2003;39(6), 269–279. Copyright © Japanese Society of Smooth Muscle Research

Anatomic Organization of Sympathetic Innervation of the Large Intestine

The mesenteric arteries that supply the large intestine are innervated by sympathetic neurons in the inferior mesenteric ganglion. Using retrograde tracers placed in the lumen of mesenteric arteries and veins that supply the large intestine, it has been shown that neurons in the inferior mesenteric ganglion that innervate arteries are different from those that innervate veins. Furthermore, the arterial neurons are localized in the core of the inferior mesenteric ganglion whereas the venous neurons are localized near the boarders.

Activation of Vascular Innervation by Peripheral Reflexes

Peripheral reflexes are mediated solely outside of the central nervous system and involve pathways that travel through various ganglia. Distension of the intestine activates primary afferent neurons that innervate the mesenteric blood vessels as well as sympathetic ganglia. Thus, intestinal distension can evoke an inhibitory junction potential in mesenteric blood vessels. This inhibitory junction potential is an important promoter of vasodilation. Sensory nerves that are activated by intestinal distension include primary sensory nerves whose cell bodies are within the dorsal root ganglia and intestinal afferents with the cell bodies in the myenteric plexus of the intestine. Both types of afferents depolarize neurons in prevertebral sympathetic ganglia and thus may evoke vasoconstrictor responses via this pathway. Nerve depolarization within the prevertebral ganglia function to counteract the direct vasodilatory response evoked by primary sensory nerve activation.

Innervation of the Lymphatic Vessels

While there is significant data available for innervation of arteries and veins of the mesentery, information regarding neurologic interactions with lymphatic vessels is much more limited. It is generally believed that the walls of mesenteric lymphatic vessels have decreased innervation compared to blood vessels. Most nerve fibers are confined to the adventitia of lymphatic vessels while nerve fibers are typically in the tunica media of arteries and veins. In human anatomy, cholinergic nerve fibers, though few in quantity, have been described and are localized in the wall of large mesenteric lymphatic vessels. Although the exact mechanism responsible for lymph flow is unclear, two predominating hypotheses have been described. The first is that lymphatic flow is a passive phenomenon without the role of actively contracting lymphatic vessels. Propulsion of lymphatic fluid is mediated by extrinsic tissues compressing lymphatic vessels. This is supported by the relatively small quantity of cholinergic nerve fibers found in lymphatic vessels. The second hypothesis is that active contraction of the smooth muscle cells surrounding the wall of the lymphatic vessel that is controlled by sympathetic stimulation and cholinergic action on mesenteric arteries. Release hormonal peptides such as substance P are hypothesized to cause lymphatic vessel wall contraction. This phenomenon occurs following an increase in intraluminal pressure or chemical stimuli.

The Mesenteric Nervous System and Inflammation

New research is emerging that highlights the function of the mesenteric nervous system in the control of inflammation. The vagus nerve has an important role in maintaining homeostasis between pro and anti-inflammatory mechanisms in the mesentery and the bowel. During times of stress and inflammation, vagal afferent neurons are activated in response to cytokine and endotoxin release. Inflammation enhances neuroplasticity and signaling sensitivity via upregulation of dendritic receptors, modulating sodium ion channel density and function, and inducing reactive gliosis of satellite glial cells. The resulting central effects of these processes include fever, nausea, vomiting and weight loss. Peripheral effects occur after sympathetic neuronal-mediated release of catecholamines and glucocorticoids. The vagus nerve also plays an anti-inflammatory role. It has been established that the release of acetylcholine from vagal efferent nerves acts to inhibit pro-inflammatory cytokines such as tumor necrosis factor (TNF). There is also evidence of communication between the autonomic nervous system and T cells that results in the induction of a cholinergic response. Activated sympathetic neurons in the mesenteric ganglia promote the release of noradrenalin (norepinephrine) and substance P (SP). Enteric neurons containing SP are located in myenteric and submucosal plexuses throughout the gastrointestinal tract. The effects of SP begin with ligation of G-protein-coupled NK-1R. NK-1R induces proinflammatory signaling characterized by the production of a considerable quantity of proinflammatory molecules. Direct neuromodulation of immune response is also demonstrated by acetylcholine. The acetylcholine receptor alpha 7 subunit (α7nAChR) is the component of acetylcholine that is responsible for inhibiting macrophage production of TNF-alpha production. Transmembrane proteins (TLRs) are present in both gastrointestinal tissue and nervous tissue. TLRs play a role in stimulating the production of proinflammatory cytokines. Gelatinase, a protein secreted by E. faecalis, is associated with increased colonic epithelial permeability. Proteinase-activated receptors (PARs) are expressed throughout the gastrointestinal tract. PAR2 has been is involved in neuronally-induced colonic inflammation. These factors and cholinergic activity combine to regulate proinflammatory cytokines that are released from macrophages and other immunocytes (see Fig. 12.3).

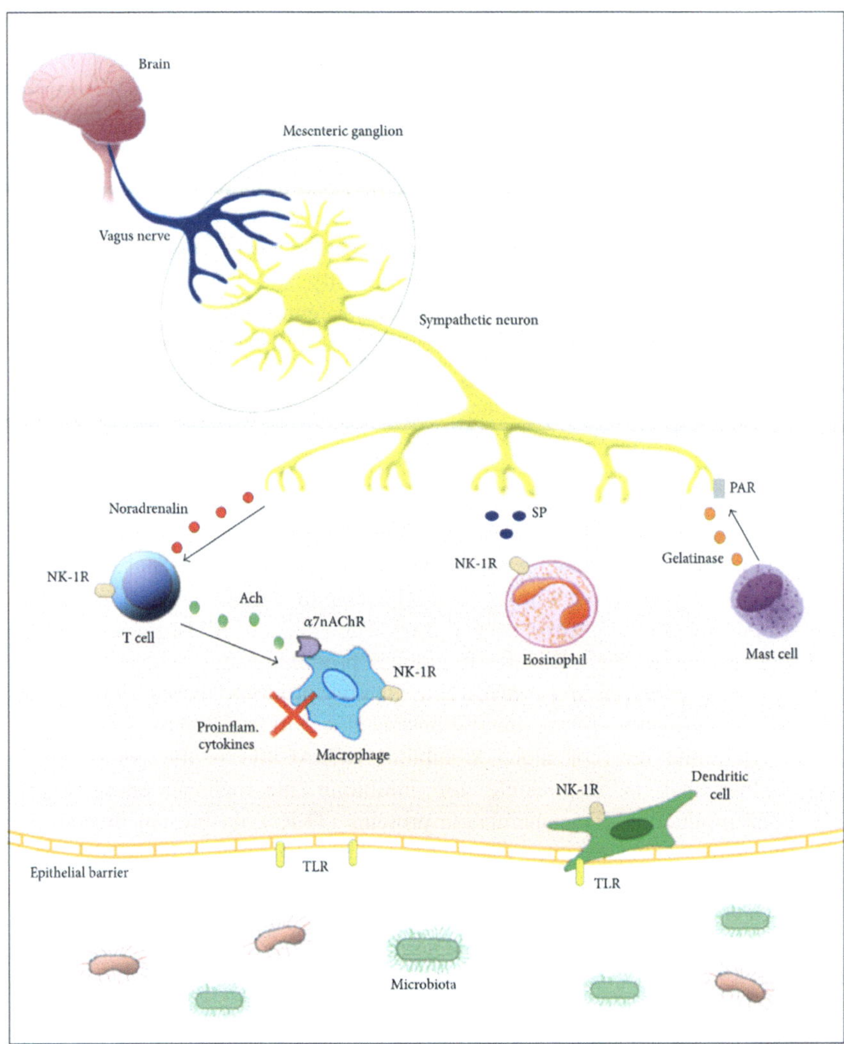

Fig. 12.3 Interaction between the enteric neuron system (ENS) and mucosal immune cells. Upon vagus nerve stimulation, sympathetic innervation produces neurotransmitters that can modulate immune cells and the inflammatory response. The intestinal microbiota also participates in the inflammatory response, resulting in fine-tuning of the interaction between the ENS and mucosal immune cells. Ach = acetylcholine, SP = substance P, α7nAChR = acetylcholine receptor alpha 7 subunit, TLR=Toll-like receptors. Reused from Bernardazzi C, BPêgo B, de Souza HSP. Neuroimmunomodulation in the Gut: Focus on Inflammatory Bowel Disease. Mediators of Inflammation 2016; Article ID 1363818 https://doi.org/10.1155/2016/1363818. Copyright © 2016 Claudio Bernardazzi et al. This is an open access article distributed under the Creative Commons Attribution License, which permits unrestricted use, distribution, and reproduction in any medium, provided the original work is properly cited

Suggested Reading

1. Standring S, Gray H. Grays anatomy: the anatomical basis of clinical practice. Philadelphia: Elsevier; 2016.
2. Li Y, Zhu W, Zuo L, Shen B. The role of the mesentery in Crohn's disease: the contributions of nerves, vessels, lymphatics, and fat to the pathogenesis and disease course. Inflamm Bowel Dis. 2016;22(6):1483–95. https://doi.org/10.1097/MIB.0000000000000791.
3. Sahin A, Artas H, Eroglu Y, Tunc N, Oguz G, Demirel U, Bahcecioglu IH, et al. A neglected issue in ulcerative colitis: mesenteric lymph nodes. J Clin Med. 2018;7(6):142. https://doi.org/10.3390/jcm7060142.
4. Kiernan MG, Calvin Coffey J, McDermott K, Cotter PD, Cabrera-Rubio R, Kiely PA, Dunne CP. The human mesenteric lymph node microbiome differentiates between Crohn's disease and ulcerative colitis. J Crohn's Colitis. 2019;13(1):58–66. https://doi.org/10.1093/ecco-jcc/jjy136.
5. Mao R, Kurada S, Gordon IO, Baker ME, Gandhi N, McDonald C, Calvin Coffey J, Rieder F. The mesenteric fat and intestinal muscle interface: creeping fat influencing stricture formation in Crohn's disease. Inflamm Bowel Dis. 2019;25(3):421–426. https://doi.org/10.1093/ibd/izy331.
6. Drake RL, Vogl W, Mitchell AWM, Gray H. Gray's anatomy for students. Philadelphia: Elsevier/Churchill Livingstone; 2005.
7. Rivera ED, Calvin Coffey J, Walsh D, Ehrenpreis ED. The mesentery, systemic inflammation, and Crohn's disease. Inflamm Bowel Dis. 2019;25(2):226–234. https://doi.org/10.1093/ibd/izy201.
8. Harper D, Chandler B. Splanchnic circulation. BJA Educ. 2016;16(2):66–71. https://doi.org/10.1093/bjaceaccp/mkv017.
9. Kreulen DL. Properties of the venous and arterial innervation in the mesentery. J Smooth Muscle Res. 2003;39(6):269–79. https://doi.org/10.1540/jsmr.39.269.
10. Browning KN, Verheijden S, Boeckxstaens GE. The vagus nerve in appetite regulation, mood and intestinal inflammation. Gastroenterology. 2017;152(4):730–44.
11. Bernardazzi C, Pêgo B, de Souza HSP. Neuroimmunomodulation in the gut: focus on inflammatory bowel disease. Med Inflamm. 2016;2016(3):1–14. Article ID 1363818. https://doi.org/10.1155/2016/1363818.
12. Pavlov VA, Wang H, Czura CJ, Friedman SG, Tracey KJ. The cholinergic anti-inflammatory pathway: a missing link in neuroimmunomodulation. Mol Med. 2003;9(5–8):125–34.
13. Nguyen C, Coelho A-M, Grady E, et al. Colitis induced by proteinase-activated receptor-2 agonists is mediated by a neurogenic mechanism. Can J Physiol Pharmacol. 2003;81(9):920–7.

Physiology of the Mesenteric Circulation

13

Saad Hashmi, Ahmed Khattab, and Eli D. Ehrenpreis

Introduction

Although the terms "mesenteric circulation" and "splanchnic circulation" are frequently used interchangeably, they are not synonymous. The "mesenteric circulation" stringently describes the blood supply to the small intestine, ascending colon and the proximal two-thirds of the transverse colon. The "splanchnic circulation" is a broader term that refers to the vasculature of the gastrointestinal tract and its associated organs (liver, pancreas and spleen).

The superior mesenteric artery is the major supplying vessel to the mesenteric circulation. It receives approximately 12% of the cardiac output. Under normal conditions, the flow rate of the superior mesenteric artery is 700 mL min. However, this flow rate can vary from 500 to 1300–1400 mL min (Figs. 13.1 and 13.2).

Several factors interplay to achieve such a wide range of blood flow. Such mechanisms are broadly categorized as either intrinsic or extrinsic and are described in more detail below. To better understand those mechanisms, it is necessary to review the structure and function of the mesenteric microcirculation as well as the physiologic properties of fluid dynamics.

S. Hashmi
Department of Internal Medicine, Advocate Lutheran General Hospital, Skokie, IL, USA

A. Khattab
Department of Internal Medicine, Division of Gastroenterology, Advocate Lutheran General Hospital, 1775 Dempster St, 6 South, Park Ridge, IL 60068, USA

E. D. Ehrenpreis (✉)
Department of Medicine, Advocate Lutheran General Hospital, 1775 Dempster St, 6 South, Park Ridge, IL 60068, USA
e-mail: e2bioconsultants@gmail.com

E. D. Ehrenpreis et al. (eds.), *The Mesenteric Organ in Health and Disease*,
https://doi.org/10.1007/978-3-030-71963-0_13

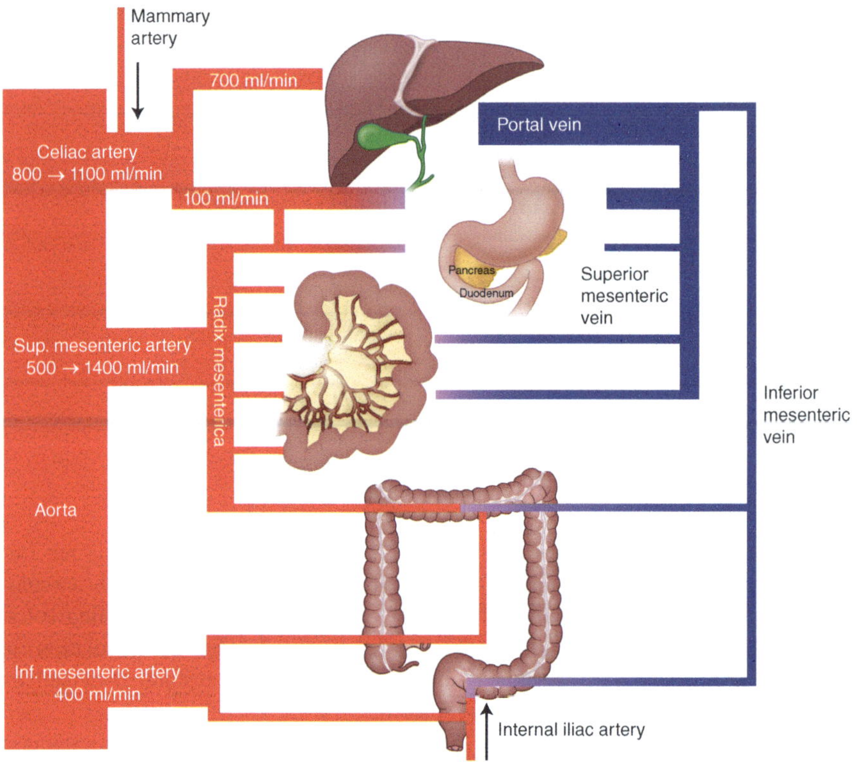

Fig. 13.1 Schematic representation of blood supply of celiac, superior mesenteric and inferior mesenteric arteries with corresponding average flow rates

Review of Mesenteric Microcirculation and Fluid Dynamics

Mesenteric Microcirculation

There are three components to the mesenteric microcirculation: microscopic arteries and arterioles (resistance vessels), capillaries (exchange vessels) and microscopic veins (capacitance vessels) (Figs. 13.3 and 13.4). The mesenteric mucosal circulation has an in-series relationship with the circulation of the submucosal layer of the intestine and an in-parallel relationship with the circulation of the muscular layer. This allows for fluctuations in blood flow to the mesentery only, without affecting total intestinal blood flow. The microscopic arteries and arterioles are also known as resistance vessels because they have the capability to constrict or dilate as a response to hormonal or neuronal influences. Changes in vascular diameter produce increases and decreases in local blood pressure, which in turn influence local blood flow. Sympathetic neuronal output in the form of catecholamines and

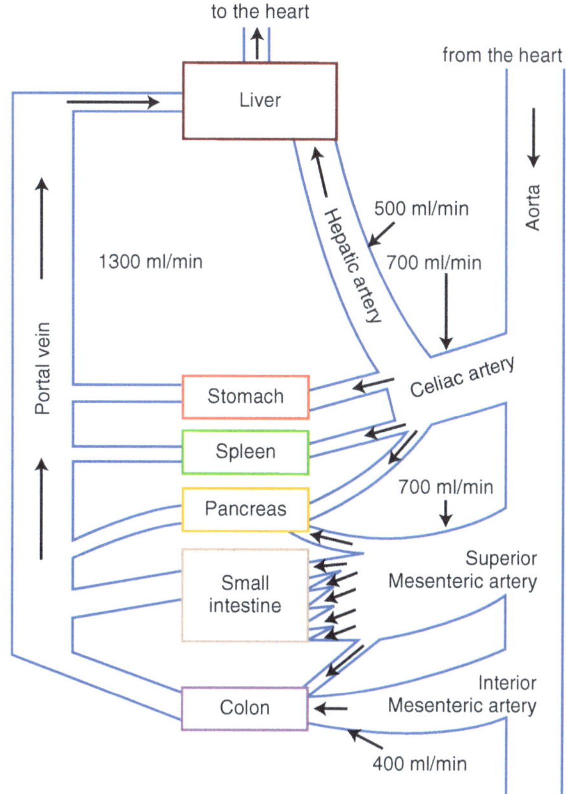

Fig. 13.2 Another schematic representation of blood supply of celiac, superior mesenteric and inferior mesenteric arteries with corresponding average flow rates

hormones such as vasopressin have the largest influence on the mesenteric circulation by producing vasoconstriction of these resistance vessels. It is important to note that under pathologic conditions, such as long-standing hypertension, resistance vessels can undergo hypertrophic remodeling resulting in a decrease in the diameter of the vessels leading to decreased blood flow.

The capillaries are the site where oxygen, nutrients and fluids move into cells from the arterial circulation and metabolic byproducts, CO_2 and heat are transported out of cells into the venous circulation. Thus, these capillaries are known as exchange vessels. Blood flow to the exchange vessels is regulated by precapillary sphincters. When the precapillary sphincters contract, blood flow to the exchange vessels is shut down, resulting in a reduction of the exchange process. Following relaxation of the precapillary sphincters, blood flow to the capillaries resumes and there is an increase in exchange of arterial and venous systems. Blood drained from the capillaries ultimately returns via the venous system. The venous system is characterized by thin-walled vessels having smooth muscle layers that respond to external stimuli. Venous contraction accentuates blood return to the heart. Nearly 80% of the total blood flow of the mesenteric microcirculation is contained within

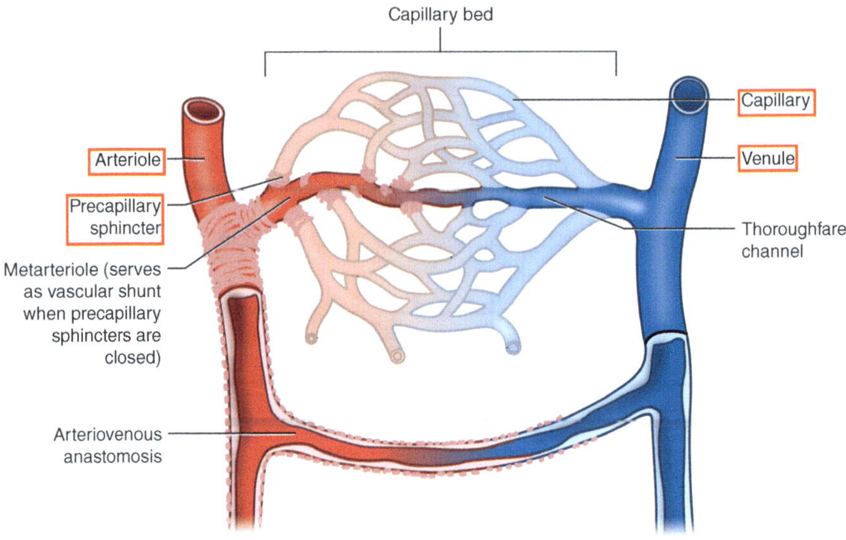

Fig. 13.3 The mesenteric microcirculation is comprised of three major components: microscopic arteries and arterioles, capillaries and microscopic veins. Blood flow to the capillaries is regulated by the precapillary sphincters

the microscopic veins. Because storage of blood occurs within the venous system, vessels within this system are known as capacitance vessels.

Fluid Dynamics

Poiseuille's equation states that flow within a vessel (in this case blood flow) is equal to the change in pressure divided by resistance and is written as such:

$$\text{Blood flow} = \Delta P/\text{Resistance}$$

Resistance is affected by three main variables: Fluid viscosity, vessel length and vessel radius. The relationship of these variables is demonstrated by the following equation:

$$\text{Resistance} = 8\eta L/\pi r^4$$

In this equation, 8 and π are constants, η = fluid (blood) viscosity, L = length of the vessel, and r = radius of the vessel (Fig. 13.5).

Practically speaking, since viscosity and blood vessel length do not readily change, the main factor determining resistance to blood flow is the diameter of the vessel. Therefore, despite the variety of factors interplaying to alter blood flow, the

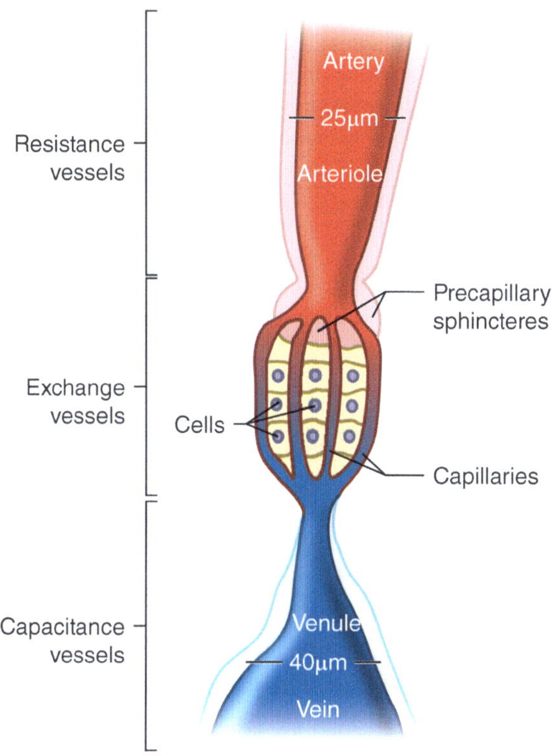

Fig. 13.4 The mesenteric microcirculation is comprised of three major components: microscopic arteries and arterioles (resistance vessels), capillaries (exchange vessels) and microscopic veins (capacitance vessels). Blood flow to the capillaries is regulated by the precapillary sphincters

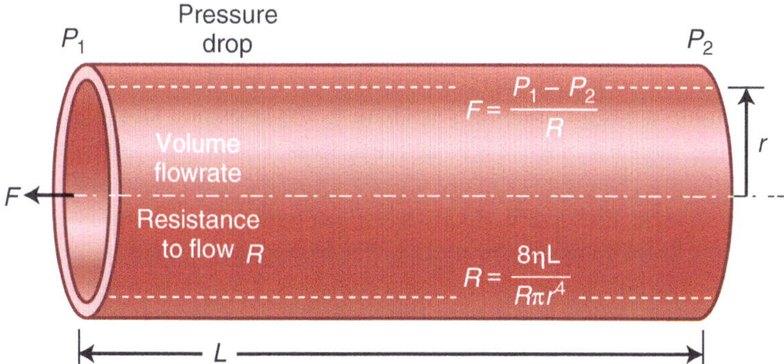

Fig. 13.5 Derivation of Poiseuille's equation with Blood flow (F) equal to the change in pressure (P1–P2) divided by resistance (R). In addition, resistance is equal to $8\eta L/\pi r^4$ where 8 and π are constants, η represents fluid (blood) viscosity, L the length of the vessel, and r the radius of the vessel

most distinctive effects occur as a result of increasing or decreasing blood vessel diameter, i.e. vasodilation or vasoconstriction, respectively. As the equation above demonstrates, vasoconstriction and vasodilation have a profound effect on vascular resistance. This becomes a clinically important factor in the setting of low mesenteric blood pressure from systemic hypotension in which blood flow is maintained at an adequate range by vasoconstriction.

The Intrinsic System

The intrinsic system autoregulates the mesenteric blood flow via the metabolic and myogenic pathways.

Metabolic Pathway

This mainly describes the direct vasodilator effect on mesenteric blood vessels caused by tissue hypoxia and accumulation of metabolic byproducts.

Oxygen Demand and Supply

The major variable regulating blood flow in the metabolic pathway is tissue oxygenation. As in most vascular beds in the body, tissue hypoxia causes an increase in mesenteric blood flow via direct vasodilation. When tissue oxygenation is restored, blood vessel diameter returns to the resting state. Augmented oxygen demand in the mesentery occur when there is increased secretion, absorption, or motility of the gastrointestinal tract. Conversely, factors such as temperature reduction causes a decrease in tissue metabolism and oxygen demand, thus leading to vasoconstriction.

Oxygen supply in the mesentery may be pathologically low as a result of decreased cardiac output or mesenteric ischemia.

Metabolic Byproducts

The accumulation of metabolic byproducts—especially CO_2 and adenosine—leads to vasodilation within the mesentery. During digestion, the transport of nutrients such as glucose amino acids and free fatty acids from the intestinal lumen to the enterocytes lining the intestinal mucosa is an active process requiring oxygen. As a result of energy metabolism required for many forms of nutrient absorption and other small intestinal processes, lactic acid, CO_2, hydrogen ions, prostaglandins, histamine, adenosine and bradykinin, are produced. These products cause vasodilation via relaxation of precapillary sphincters and a decrease in vascular resistance. This in turn leads to an increase in blood flow allowing for improved delivery of oxygen to the tissues. The production of these metabolites also results in local hyperemia. This form of hyperthermia should not to be confused with reactive hyperemia, a transient increase in blood temperature occurring after a period of occlusion of blood flow.

Intraluminal Hyperosmolarity

The intraluminal fluid containing dietary-derived nutrients is hyperosmolar and has a high sodium content. Sodium absorption leads to a high intracellular sodium concentration. This in turn activates Na^+/Ca^{2+} exchanger, increasing intracellular Ca^{2+} concentration which in turn stimulates nitric oxide-mediated relaxation of vascular smooth muscle. Mesenteric vasodilation is mediated via activation of nitric oxide synthase (Figs. 13.6 and 13.7).

Myogenic Pathway

a. The myogenic pathway involves the phenomenon known as the Bayliss effect in which vasoconstriction is produced by stretching of vascular smooth muscle cells (Fig. 13.8). More specifically, an elevated intravascular pressure causes a reactive stretching of vascular smooth muscle cells. This leads to depolarization of mechanosensitive sodium (Na^+) channels and subsequent activation of voltage-gated calcium (Ca^{2+}) channels. This allows the influx of calcium ions triggering vascular smooth muscle contraction and vasoconstriction.

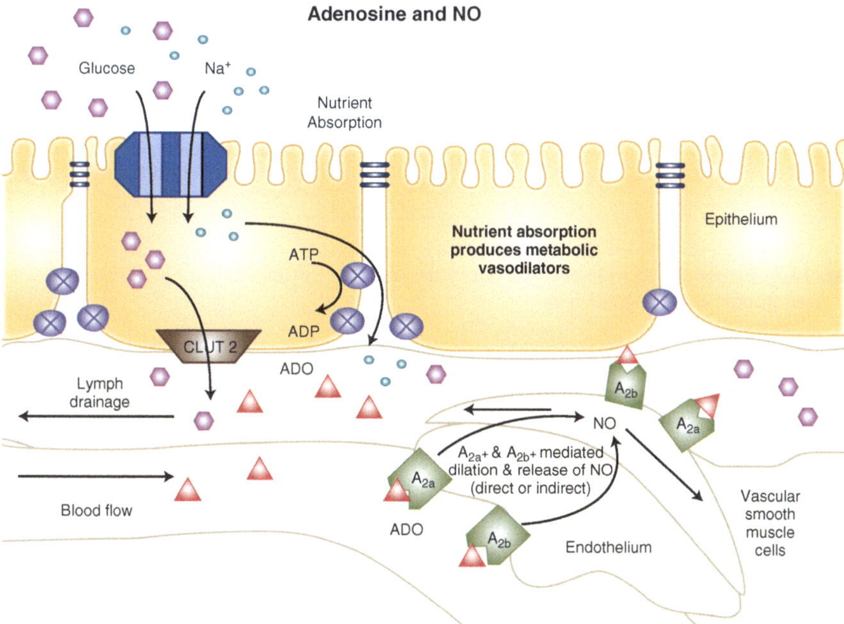

Fig. 13.6 A schematic representation of dietary factors in mesenteric vasodilation. Dietary sodium and glucose absorption increase ATP utilization stimulating production of vasodilators such as adenosine and nitric oxide. This is a proposed mechanism for post-prandial mesenteric vasodilation

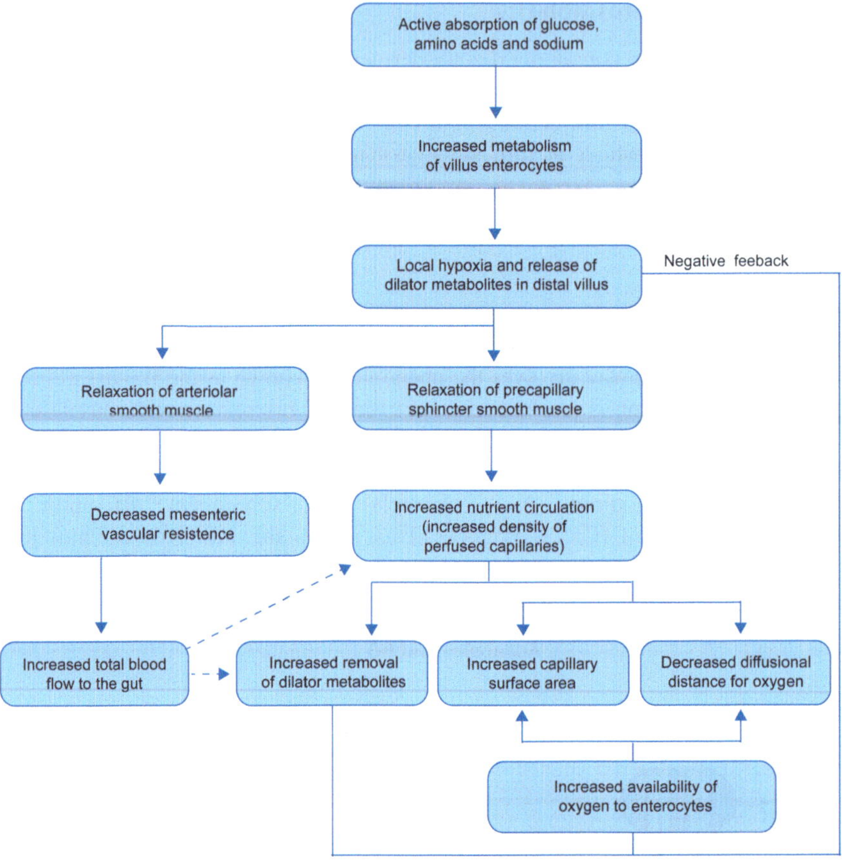

Fig. 13.7 Feedback mechanisms involved in the regulation of mesenteric blood flow based on nutrient absorption and metabolism

In contrast, a low intravascular pressure will lead to a reactive vasodilation, restoring blood flow. This effect is based on the Law of Laplace which proposes that smooth muscle tension is directly proportional to the intravascular pressure and vessel radius. Hence, an increase in blood pressure leads to a decrease in vessel radius (vasoconstriction) in order to maintain the same amount of tension.

$$\text{Laplace's equation: } T = p \times r$$

T = smooth muscle tension, p = intravascular pressure and r = vessel radius.

In the setting of blood pressure fluctuations, the function of the myogenic pathway is to maintain constant blood flow to the mesentery via vasoconstriction or vasodilation.

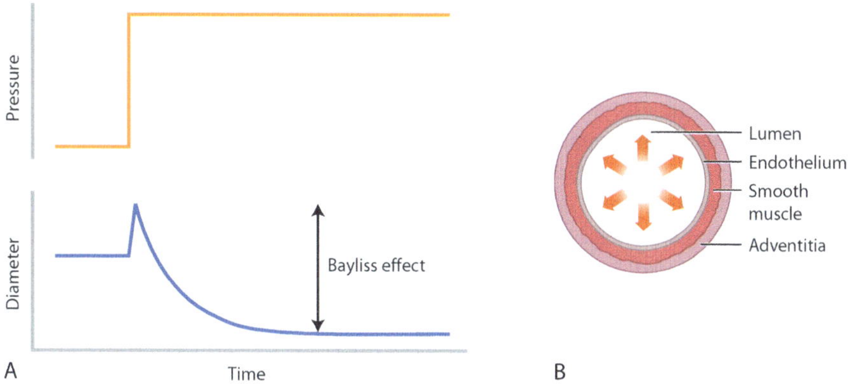

Fig. 13.8 **a** Graphic representation of the Bayliss effect in which an increase in pressure in blood vessels leads to a decrease in diameter of the vessel (vasoconstriction) **b** The Bayliss effect is mediated by smooth muscle cells within blood vessels

Extrinsic System

The extrinsic system functions as an additional regulator of the mesenteric circulation. It comprises central cardiovascular control, autonomic autoregulation, and neurohumoral control.

Central Cardiovascular Control

The components of cardiac cardiovascular control include cardiac output, systemic arterial blood pressure, venous return, and circulating blood volume. Any significant change in one or more of these parameters will cause a resultant change in mesenteric blood flow. For example, in cases with severe blood loss diminished circulating blood volume and subsequent decrease in venous return will occur. These factors produce a reduction of mesenteric blood flow.

Autonomic Neuro-Regulation

Sympathetic Nervous System (SNS) Stimulation

Postganglionic vasomotor fibers originate from the celiac, superior mesenteric, and inferior mesenteric ganglia. These are distributed along the branches of the corresponding mesenteric arteries. These terminate on vascular smooth muscle of specific portions of the microcirculation (such as the microscopic arteries, arterioles, venules, and microscopic veins). Physiological and pathological stresses, such as exercise and hemorrhage result in potent stimulation of the sympathetic neurons.

Stimulation of postganglionic sympathetic fibers cause the release of adenosine triphosphate (ATP) and noradrenaline. The resulting transient vasoconstriction leading to expulsion of blood from capacitance vessels and reduction of mesenteric blood flow. Strangely, within minutes of prolonged stimulation, the resistance vessels dilate leading to return to near-normal levels of flow. This is followed by a third phase consisting of reactive hyperemia causing a transient increase in blood flow. This triphasic pattern describes a phenomenon called "autoregulatory escape." It is believed to be caused by intrinsic metabolic mechanisms (see above) producing an overriding of the sympathetic neurogenic constriction. The overall effect is reduction in blood flow to the mesentery while maintaining adequate perfusion to organs such as the brain, heart and lungs.

Parasympathetic Nervous System (PNS) Stimulation

The PNS has a less prominent role in controlling mesenteric blood flow. Activation of the parasympathetic system occurs via innervation from the vagal and pelvic nerves resulting in increases in intestinal motility and secretion. Thus, fibers from the PNS do not have direct vasomotor effects. However, activation of muscarinic receptors via acetylcholine triggers release of nitric oxide (NO). Stimulation of NO receptors results in vascular smooth muscle relaxation which in turn produces an increase in mesenteric blood flow. Activation of visceral muscle and secretory units may also have an indirect effect of increasing mesenteric blood flow.

Neurohumoral Control

There are two main groups of vasoactive substances contributing to neurohumoral control.

The first group includes classical vasoconstrictor agents of the body. They can be endogenously released or exogenously administered as pharmacologic agents.

Catecholamines

These are released in response to stresses, such as strenuous exercise. They can also be administered exogenously in cases of shock. Higher levels of catecholamines cause an increase in vascular resistance in the same manner as sympathetic activation.

Angiotensin II

The renin-angiotensin system is activated in response to reduction in the extracellular fluid volume. This leads to increased production of angiotensin II, which in turn causes a direct vasoconstriction of the mesenteric vasculature.

Vasopressin

Loss of blood volume and hyperosmolarity result in the release of vasopressin (anti-diuretic hormone) from the posterior pituitary gland. This leads to mesenteric vasoconstriction and reduction of portal venous pressure.

The second group includes circulating paracrine and autocrine mediators. Histamine, bradykinin and prostaglandins (A, D, E and I) cause direct vasodilation as mentioned above in the metabolic control section. Prostaglandin $F_{2\alpha}$, however, causes direct vasoconstriction. In contrast to the direct effect of the aforementioned vasoactive substances, the gastrointestinal peptides cholecystokinin, secretin, and gastrin mediate an indirect vasodilatory effect through prompting gastrointestinal motility. Those peptides are endogenously and locally released during digestion (Table 13.1).

The algorithm in Fig. 13.9 summarizes intrinsic and extrinsic control mechanisms of mesenteric blood flow.

Table 13.1 Endogenous vasoactive mediators of the mesenteric circulation

Constrictors	Dilators
I. Vasoactive metabolites	
$\downarrow PCO_2$	$\uparrow PCO_2$
$\uparrow PO_2$	$\downarrow PO_2$
$\uparrow pH$	$\downarrow pH$
	Lactic acid
	Adenosine
	Nitric oxide
II. Autonomic neuro-regulation	
Sympathetic stimulation	Parasympathetic stimulation
III. Neurohumoral control	
Catecholamines (adrenaline and noradrenaline)	
Angiotensin II	
Vasopressin	
IV. Circulating paracrine and autocrine mediators	
Constrictor prostaglandin ($F_{2\alpha}$)	Dilator prostaglandins (A, D, E and I)
	Histamine
	Bradykinin
	Gastrointestinal peptides (cholecystokinin, secretin and gastrin)[a]

[a]Gastrointestinal peptides indirectly mediate mesenteric vasodilation by promoting gastrointestinal motility. Reused with permission from Matheson PJ, Wilson MA, Garrison RN. Regulation of intestinal blood flow. *J Surg Res*. 2000;93(1):182–196. Copyright © Elsevier 2000

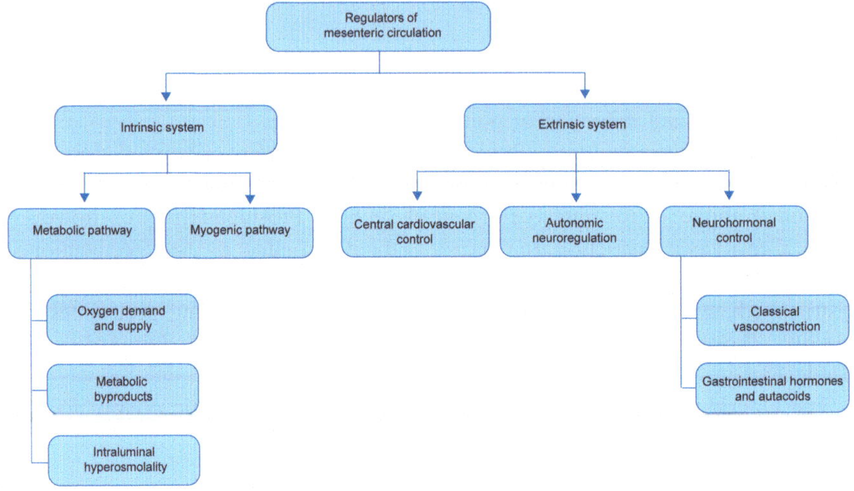

Fig. 13.9 Summarizes the main classes of intrinsic and extrinsic control mechanisms of mesenteric blood flow

Suggested Reading

1. Harper D, Chandler B. Splanchnic circulation. BJA Educ. 2016;16(2):66–71. https://doi.org/10.1093/bjaceaccp/mkv017.
2. Parks DA, Jacobson ED. Physiology of the splanchnic circulation. Arch Intern Med. 1985;145(7):1278–81. https://doi.org/10.1001/archinte.1985.00360070158027.
3. https://pdfs.semanticscholar.org/2e72/4e438d00c4c7b278301706c4fe1489f63c4b.pdf.
4. Rosenblum J, Boyle CM, Schwartz L The mesenteric circulation. Anatomy and physiology. Surg Clin N Am. 1997;77:289–306. https://doi.org/10.1016/S0039-6109(05)70549-1.
5. Kvietys PR. The gastrointestinal circulation. San Rafael (CA): Morgan & Claypool Life Sciences; 2010. https://www.ncbi.nlm.nih.gov/books/NBK53092/.
6. https://coherence.com/physiology_of_the_mesenteric_circulation.pdf.
7. https://www.ncbi.nlm.nih.gov/pmc/articles/PMC2595408/pdf/yjbm00138-0072.pdf.
8. https://www.gastrojournal.org/article/S0016-5085(76)80374-5/pdf.
9. https://www.medschool.lsuhsc.edu/physiology/docs/regulation_splanchnicBF.pdf.
10. https://onlinelibrary.wiley.com/doi/pdf/10.1002/cphy.cp060139.
11. https://annals.org/aim/article-abstract/680534/physiology-splanchnic-circulation.
12. Johansson B. Myogenic tone and reactivity: definitions based on muscle physiology. J Hypertens Suppl. 1989;7:S5–S8. https://doi.org/10.1097/00004872-198901000-00002. discussion S9. [PubMed][Cross Ref]. https://www.ncbi.nlm.nih.gov/pubmed/2681596.
13. https://www.clinsci.org/content/96/4/313.full-text.pdf.
14. Sun D, Messina EJ, Kaley G, Koller A. Characteristics and origin of myogenic response in isolated mesenteric arterioles. Am J Physiol. 1992;263:H1486–H1491. [PubMed] https://www.ncbi.nlm.nih.gov/pubmed/1443200.
15. https://pdfs.semanticscholar.org/876d/0bc5f96110ccc531ac1d7d74e2bdd6cf9514.pdf.
16. https://www.researchgate.net/profile/Lewis_Schwartz/publication/14071203_The_mesenteric_circulation_Anatomy_and_physiology/links/547df3da0cf241dc999260ae/The-mesenteric-circulation-Anatomy-and-physiology.pdf.

17. https://citeseerx.ist.psu.edu/viewdoc/download?doi=10.1.1.545.8615&rep=rep1&type=pdf.
18. https://www.gastrojournal.org/article/0016-5085(80)90692-7/fulltext (https://www.gastrojournal.org/article/0016-5085(80)90692-7/pdf).
19. https://www.sciencedirect.com/topics/medicine-and-dentistry/resistance-vessel.
20. https://opentextbc.ca/anatomyandphysiology/chapter/20-2-blood-flow-blood-pressure-and-resistance/.
21. Berne RM, Levy MN, Koeppen BM, Stanton BA. Physiology. 5th ed. St. Louis, MO: Mosby; 2004.
22. Young KA, Wise JA, DeSaix P, Kruse DH, Poe B, Johnson E, Johnson JE, Korol O, Gordon Betts J, Womble M. Anatomy and physiology, 1st ed.
23. https://gut.bmj.com/content/60/5/722.short.
24. https://hyperphysics.phy-astr.gsu.edu/hbase/ppois.html *HyperPhysics* from the Department of Physics and Astronomy at Georgia State University.
25. https://www.researchgate.net/publication/23764566_TRPCs_GPCRs_and_the_Bayliss_effect.

Part IV
Role of Mesentery in Systemic Medical Diseases

The Role of the Mesentery in Metabolic Syndrome and Diabetes Mellitus

14

Natasha Shah, David H. Kruchko, and Eli D. Ehrenpreis

Introduction

Metabolic syndrome is a term that originated in the 1920s and has since been used to describe an individual with multiple medical conditions including central obesity, hyperglycemia, hypertension, and hypertriglyceridemia. These conditions can ultimately increase an individual's risk for Type II diabetes mellitus, cardiovascular disease and overall mortality. Aging, stress, diet, a sedentary lifestyle, genetic predisposition and systemic inflammation have been identified as causative factors for metabolic syndrome, as outlined in Fig. 14.1. Systemic inflammation is the result of the release of C-reactive protein (CRP), interleukin-6 (IL-6), and tumor necrosis factor alpha (TNF-α).

A key finding of metabolic syndrome is central obesity as defined by an increase in waist size due to augmented deposition of visceral adipose tissue. Visceral adipose tissue consists of fat present in the mesentery and omentum. Visceral fat is distinguished from subcutaneous fat by its location and association with pathophysiologic states. Increased visceral fat has been associated with a greater risk of developing Type II diabetes mellitus and cardiovascular disease and is a strong predictor of overall mortality in men. In the setting of obesity, adipose tissue becomes dysfunctional and is characterized by adipose hypertrophy and macrophage infiltration. These, in turn can lead to insulin signaling impairment, and insulin resistance both of which increase the likelihood of developing Type 2 DM (Fig. 14.2).

N. Shah
Department of Gastroenterology, Advocate Lutheran General Hospital,
1775 Dempster St, 6 South, Park Ridge, IL 60068, USA
e-mail: shah.natasha49@gmail.com

D. H. Kruchko
Department of Internal Medicine, Advocate Lutheran General Hospital,
1775 Dempster St, 6 South, Park Ridge, IL 60068, USA
e-mail: david.kruchko@gmail.com

E. D. Ehrenpreis (✉)
Department of Medicine, Advocate Lutheran General Hospital,
1775 Dempster St, 6 South, Park Ridge, IL 60068, USA
e-mail: e2bioconsultants@gmail.com

E. D. Ehrenpreis et al. (eds.), *The Mesenteric Organ in Health and Disease*,
https://doi.org/10.1007/978-3-030-71963-0_14

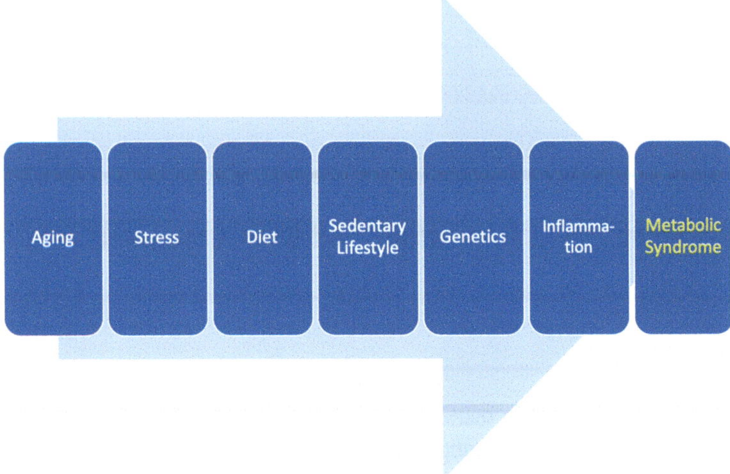

Fig. 14.1 Flow chart for causative factors that all contribute to metabolic syndrome. Notably, although these risk factors often occur simultaneously, there is no particular order and one can occur without the other

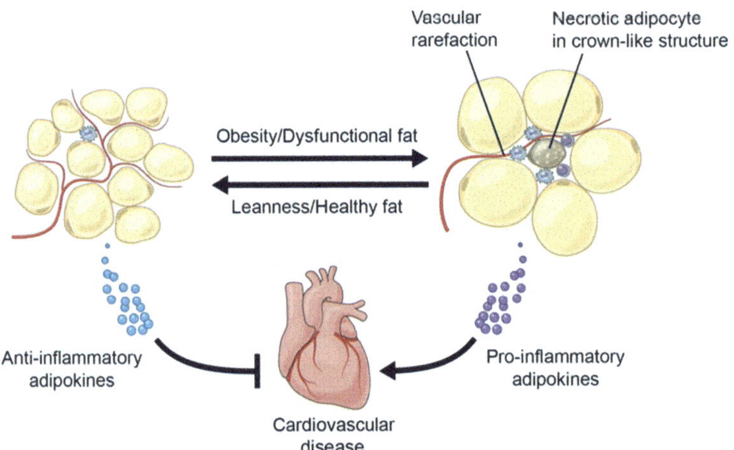

Fig. 14.2 Obesity and resultant dysfunctional adipocytes produce pro-inflammatory markers such as macrophages. In addition, vasculature injury with loss of arterioles and capillaries occurs in a process known as vascular rarefaction. This local inflammation leads to cardiovascular disease with a root cause that is traceable to central obesity and increased visceral fat. The left side of Fig. 2 shows healthy adipocytes producing anti-inflammatory adipokines preventing the progression of cardiovascular disease. Reused with permission from AHA Journals; Circulation Research; https://www.ahajournals.org/doi/full/10.1161/circresaha.115.306885. Copyright © Wolters Kluwer

The exact role of the mesentery in these disorders remains unclear. However, several investigative studies have found associations between the mesentery adipose tissue and metabolic syndrome. Yang et al., conducted a study where three adipose depots (subcutaneous, omental and mesenteric) were obtained from obese patients undergoing gastric bypass surgery. Using real-time polymerase chain reaction (PCR), lipolysis was measured to determine adipocyte function. The findings suggested that obesity-related genes in the mesenteric adipose tissue of patients with diabetes were upregulated and could play a pivotal role in insulin resistance in Type 2 diabetes and those with metabolic syndrome.

Pathogenesis

The mesentery is comprised of tightly packed adipocytes within a framework of tissue. The mesentery is the single greatest contributor to visceral adiposity, and thus is highly suspected to be a key regulator in the pathogenesis of these conditions. The mesentery is no longer considered to be made up of "innocent bystander cells" or a "passive energy depot" but studies have shown it to potentially be a key organ that regulates inflammatory processes that can produce or worsen diabetes and obesity.

Subcutaneous deposition of fat occurs when there is excess caloric intake with limited energy expenditure. In this scenario, storage of triglycerides and free fatty acids occurs. Fat accumulates outside of subcutaneous tissue in the viscera when it is no longer able to store or generate new adipocytes. Inadequate storage results from dysfunctional adipocytes being too large and inflamed to be deposited within the proper visceral storage sites. The overflow of dysfunctional adipocytes into the subcutaneous tissues is the initial step in central adiposity and resultant mesenteric fat deposition (Fig. 14.3).

Visceral adipose tissue is made of a large number of adipocytes as well as nonfat cells that make up the connective tissue matrix. Vascular and neural structures are also present in visceral adipose tissue. Visceral adipocytes are larger compared to adipocytes in subcutaneous adipose tissue, and thus can be more dysfunctional. Abnormalities in visceral adipocytes enhances the development of insulin resistance. Inability to store free-fatty acids (FFAs) and triglycerides (TGs) results in their deposition in non-adipose tissue. The tendency for visceral adipocytes to have an increased susceptibility to cellular dysfunction can be explained by the fact that visceral adipose tissue contains a high proportion of inflammatory and immune cells. Prolonged reactive inflammatory and immune response in the setting of increased adipose tissue will eventually lead to fibrosis and scaring, thereby making these visceral adipocytes more dysfunctional. Furthermore, the venous system of visceral fat blood drains directly into the liver. This is hypothesized to allow for increased free fatty acids to accumulate in the liver leading to hyperinsulinemia, hypertriglyceridemia, and glucose intolerance (Fig. 14.4).

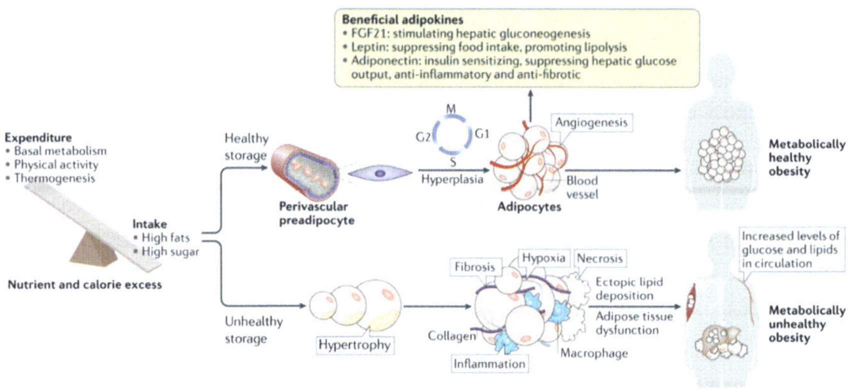

Fig. 14.3 Components of visceral adipose tissue. The balance of nutrient and calorie intake versus expenditure impacts whether or not adipocytes can perform at a normal physiologic level or become hypertrophic and thereby dysfunctional. Following the pathway of healthy storage, the preadipocytes affect the perivasculature and undergo mitosis to develop adipocytes with health blood vessels. Furthermore, there are physiologic signals that help build the components of adipocytes such as the connective tissue matrix and vasculature. These signals are listed as the beneficial adipokines at the top of the figure. Following the unhealthy storage pathway, the hypertrophy and resultant inflammatory state leads to fibrosis, inadequate oxygen delivery due to damaged vasculature, and harmful signals leading to lipid deposition in the subcutaneous tissues and surrounding viscera. The end result is an increase in the levels of glucose and lipids in circulation and an unhealthy state of obesity due to dystrophic adipocytes. Reused with permission from Nature Reviews, Molecular Cell Biology; https://www.nature.com/articles/s41580-018-0093-z?WT.feed_name=subjects_stem-cells. Copyright © Springer Nature

Hypoxemia may also play a role in the alteration of adipocytes in obesity. Hypoxemia leads to cellular hypertrophy and macrophage infiltration. In addition, hypoxemia triggers the release of several cytokines and acute phase proteins. This results in the combination of an inflammatory state, increased free fatty acid deposition, and increased insulin resistance. Visceral adipose tissue secretes a large number of cytokines including anti-TNF, IL-1, 6, 8, angiotensin II and plasminogen activator inhibitor-1 in comparison to subcutaneous adipose tissue (Fig. 14.5). Also see Chap. 11 Immunologic Function of the Mesentery.

Treatment

Studies have demonstrated that losing weight is associated with reductions in CRP level and produces a greater reduction of visceral adipose tissue rather than subcutaneous adipose tissue; (Fig. 14.6). Decreasing caloric intake and weight reduction can prevent adipose tissue dysfunction and may thus prevent subsequent its inflammatory consequences. Since visceral adipose tissue in the mesentery is associated with inflammatory and immune cells, sensitivity to lipolysis, and insulin

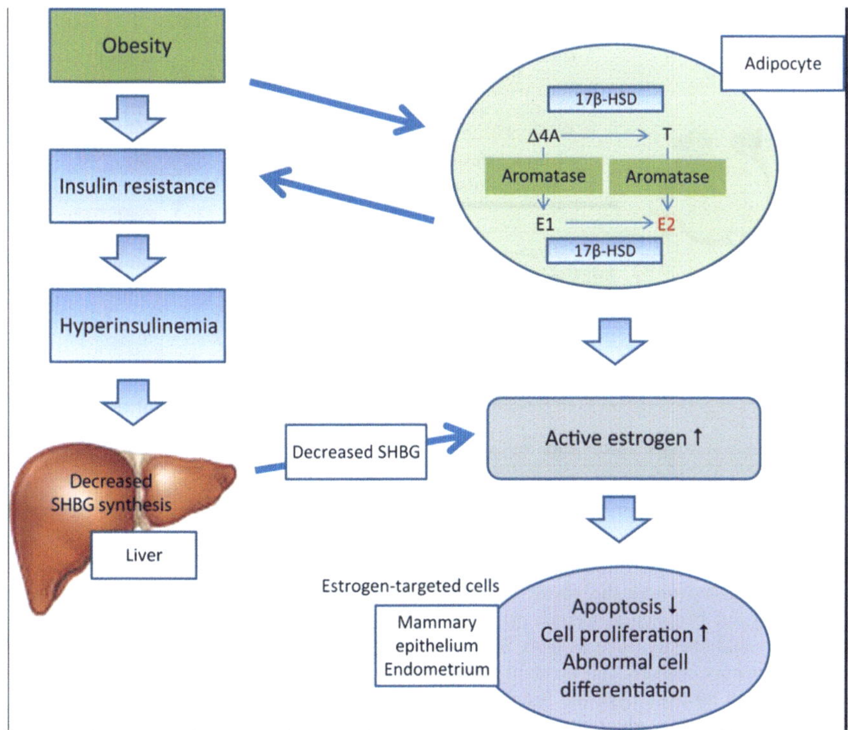

Fig. 14.4 The interplay of hormonal changes on obesity and resulting hyperinsulinemia resulting in adipocyte proliferation. Obesity results in an increase in aromatase with increasing estrogen levels while simultaneously leading to hyperinsulinemia. High insulin levels in the blood lead to decreased synthesis of sex-hormone binding globulin (SHBG); which also increased estrogen production. Increased levels of active estrogen increase cell proliferation and leads to unintentional and harmful increases in the number of adipocytes in local tissue. Reused with permission from ResearchGate; https://www.researchgate.net/figure/Hypothetical-mechanism-of-oncogenesis-as-mediated-by-active-estrogen-in-hyperinsulinemia_fig6_251567860. Copyright © John Wiley and Sons

resistance a strong associated between the mesentery, obesity and diabetes is suggested. Mesenteric adipose tissue-directed therapy holds promise in treating these disease processes.

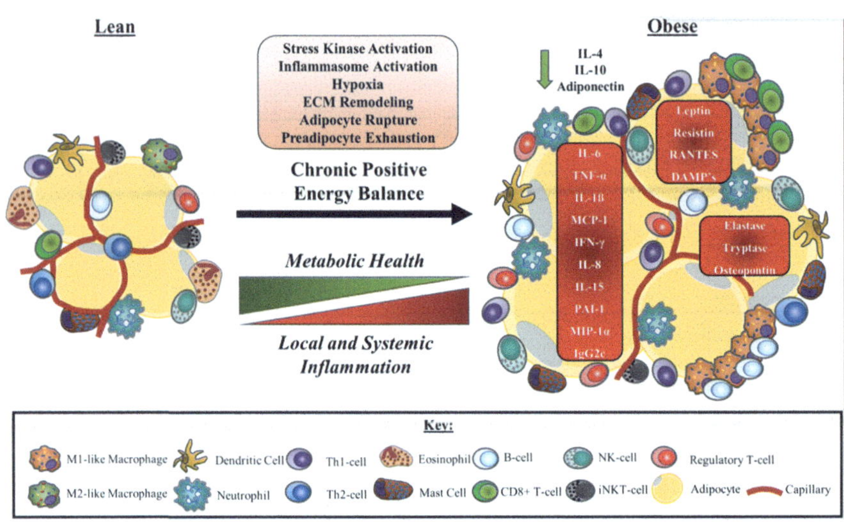

Fig. 14.5 Immunological changes within adipose tissue due to obesity. Local inflammatory mediators as listed under the obese depiction of adipose cells play a large roll in adipocyte hypertrophy and dysfunction. Certain cytokines have been identified to play a more pathogenic roll in adiposity and obesity, such as IL-6, IL-8, and TNF- α. The end result of all of these cell lines and inflammatory markers can be summarized by local tissue hypoxia due to inadequate tissue blood flow and resultant hypoxia. Adipocytes can then rupture and release lipid components into the blood stream and nearby tissue resulting in the release of damage-associated molecular pattern (DAMPs). This results in the activation of stress kinase and remodeling of the extracellular membrane. This ongoing inflammatory process in obese individuals results in fibrosis and scarring in the local tissue. Adapted from Frontiers in Immunology; https://www.frontiersin.org/articles/10.3389/fimmu.2018.00169/full. ***Abbreviations***: CD, cluster of differentiation marker; DAMP, damage-associated molecular pattern; ECM, extracellular matrix; IFN-γ, interferon gamma; Ig, immunoglobulin; IL, interleukin; iNKT-cell, invariant natural killer T cell; MCP-1, monocyte chemotactic protein 1 (CCL-2); MIP-1α, macrophage inflammatory protein 1 alpha (CCL-3); PAI-1, plasminogen activator inhibitor 1; RANTES, regulated on activation, normal T-cell expressed and secreted (CCL-5); Th cell, helper T cell; TNF-α, tumour necrosis factor-α

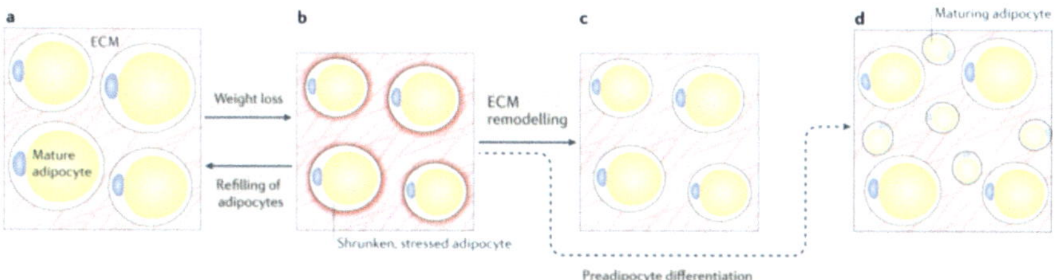

Fig. 14.6 Benefits of weight loss on adipocytes at a cellular level. This illustration demonstrations the reduction in adipocyte diameter after weight loss that can also lead to remodeling of the xtracellular matrix. This remodeling helps to resolve the stress that was initially experienced by the adipocytes due to alterations in size and functionality. Weight loss also encourages the growth of new adipocytes and provides an environment for optimal maturation. Reused with permission from Nature; https://www.nature.com/articles/s41574-018-0148-4. Copyright © Springer Nature

Suggested Reading

1. Kaur J. A comprehensive review on metabolic syndrome. Cardiol Res Pract. 2014;2014:943162. https://doi.org/10.1155/2014/943162.
2. Coffey JC, O'Leary DP. The mesentery: structure, function, and role in disease. Lancet Gastroenterol Hepatol. 2016;1(3):238–47.
3. Li Y, Zhu W, Zuo L, Shen B. The role of the mesentery in Crohn's disease: the contributions of nerves, vessels, lymphatics, and fat to the pathogenesis and disease course. Inflamm Bowel Dis. 2016;22(6):1483–95.
4. Desreumaux P, Ernst O, Geboes K, Gambiez L, Berrebi D, Müller-Alouf H, Hafraoui S, Emilie D, Ectors N, Peuchmaur M, Cortot A, Capron M, Auwerx J, Colombel JF. Inflammatory alterations in mesenteric adipose tissue in Crohn's disease. Gastroenterology. 1999;117(1):73–81.
5. Cusi K. The role of adipose tissue and lipotoxicity in the pathogenesis of type 2 diabetes. Curr Diab Rep. 2010;10(4):306–15.
6. Ibrahim MM. Subcutaneous and visceral adipose tissue: structural and functional differences. Obes Rev. 2010;11(1):11–8.
7. Liu KH, Chan YL, Chan JC, Chan WB, Kong WL. Mesenteric fat thickness as an independent determinant of fatty liver. Int J Obes (Lond). 2006;30(5):787–93.
8. Doucet E, St-Pierre S, Alméras N, Imbeault P, Mauriège P, Pascot A, Després JP, Tremblay A. Reduction of visceral adipose tissue during weight loss. Eur J Clin Nutr. 2002;56(4):297–304.

Crohn's Disease and the Mesentery

15

Eli D. Ehrenpreis

Introduction

The classical view of Crohn's disease focuses on the gastrointestinal tract rather than the mesentery. In this interpretation, the initiation and amplification of Crohn's disease occurs at the lumen of the bowel and the process extends into deeper layers of the bowel wall. The relationship of the mesentery to this process is confined to its proximity to the site of inflammation, and its function as a structural conduit for blood and lymphatic vessels and the enteric nervous system. This view eliminates the possibility that the mesentery functions as a cofactor in the pathogenesis of the disease. By contrast, recent evidence suggests that the mesentery likely plays a critical role in the integration of inflammation, healing, and autoimmunity and is a probable factor in the pathogenesis of inflammatory bowel disease.

Animal models of intestinal inflammation have demonstrated that the interaction of mesenteric adipocytes, fibroblasts and lymphocytes can result in intestinal inflammation resembling Crohn's disease. Translation of these data is likely to influence disease management in the future. Resection of inflamed portions of the mesentery is already being investigated in the surgical management of Crohn's disease.

Mesenteric Immunity and Crohn's Disease

Recent studies have focused on the interplay between mesenteric lymph nodes and the pathogenesis of Crohn's disease. The recruitment of T helper cells, also known as CD4 + T cells, is largely dependent on proper mesenteric lymph nodes responses. In normal immunology, mesenteric lymph nodes recruit CD4 + T cells and enhance

E. D. Ehrenpreis (✉)
Department of Medicine, Advocate Lutheran General Hospital, 1775 Dempster St, 6 South, Park Ridge, IL 60068, USA
e-mail: E2bioconsultants@gmail.com

E. D. Ehrenpreis et al. (eds.), *The Mesenteric Organ in Health and Disease*,
https://doi.org/10.1007/978-3-030-71963-0_15

intercellular communication by production of cytokines (e.g. interferon-gamma, or γ-interferon) and glycoproteins (e.g. interleukin-17, IL-17). In Crohn's disease, upregulation of γ-interferon production occurs. Increased γ-interferon may be responsible for mesenteric fibrosis, fat wrapping, and transmural inflammation. Another potential relationship between the mesenteric system and the pathogenesis of Crohn's disease is the mesenteric production of C-reactive protein (CRP). CRP production by mesenteric adipocytes may be triggered by local inflammation and translocation of bacteria from the intestinal lumen to mesenteric fat. Subsequent mesenteric fat hyperplasia may contribute to inflammatory responses in Crohn's disease.

Lastly, appropriate food tolerance requires MHC molecules that are present on the surface of CD4 + T cells. This system decreases the likelihood of inflammatory states developing as a response to ingested food particles (See Fig. 15.1).

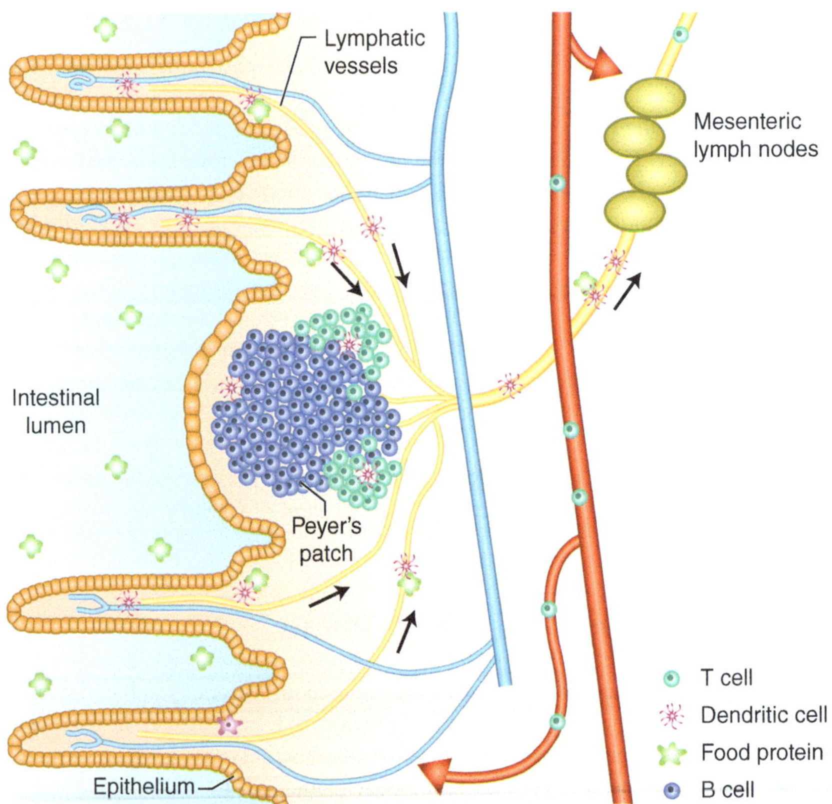

Fig. 15.1 Effect of food particle exposure on mesenteric lymph nodes. Food proteins arrive to a Peyer's patch via lymphatic channels. Here they are sampled for possible immune responses. T cell proliferative response to benign food particles may involve a mild inflammatory response. Food-containing dendritic cells (DCs) mediate systemic tolerance via further sampling in mesenteric lymph nodes and circulate in efferent of the lymphatic channels

Inappropriate presentations of MHC molecules by CD4 + T cells, can lead to food hypersensitivity, inflammation and may have the potential to produce autoimmune states such as inflammatory bowel disease. According to this model, inflammation of mesenteric lymph nodes may occur after exposure to food-borne peptides. The proximity of the lymph nodes that are scattered throughout the mesentery and the adjacent gastrointestinal tract suggests that mesenteric lymph nodes are a likely source for food-induced intestinal inflammation.

Crohn's Disease and Fat Wrapping

Fat wrapping (also known as creeping fat) refers to inflamed and thickened mesentery that is located in adjacent to parts of the small and large intestine that are affected by active Crohn's disease (Fig. 15.2).

The amount of fat wrapping correlates with the severity of local intestinal inflammation. Intestinal edema, adipocyte hyperplasia and infiltration of the mesenteric tissue with inflammatory cells are characteristic findings. Collagen accumulation can produce local mesenteric fibrosis. Extension of this process to the affected portion of the intestine is associated with development of intestinal strictures. Fat wrapping of the mesentery is seen near separate areas (or skip lesions) of intestinal inflammation while normal mesentery is seen at the intervening portions of normal intestine (Fig. 15.3).

Similar changes occur on the microscopic level with blurring of the interface between inflamed portions of the intestine and surrounding mesentery. Recent evidence suggests that abnormalities in adipose tissue, adipocyte hyperplasia and

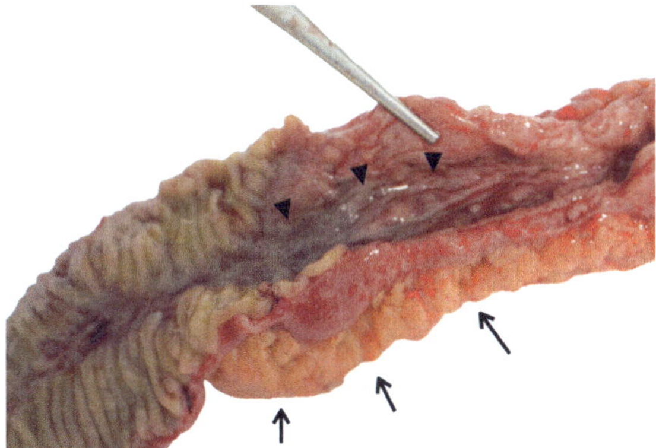

Fig. 15.2 Demonstration of fat wrapping in a surgically resected portion of the small intestine from a patient with Crohn's disease. A segment of active Crohn's disease with luminal narrowing is seen (arrowheads). Adjacent to the area of intestinal involvement, inflamed and thickened mesentery is seen (arrows). Photographs courtesy of Dr. J Calvin Coffey

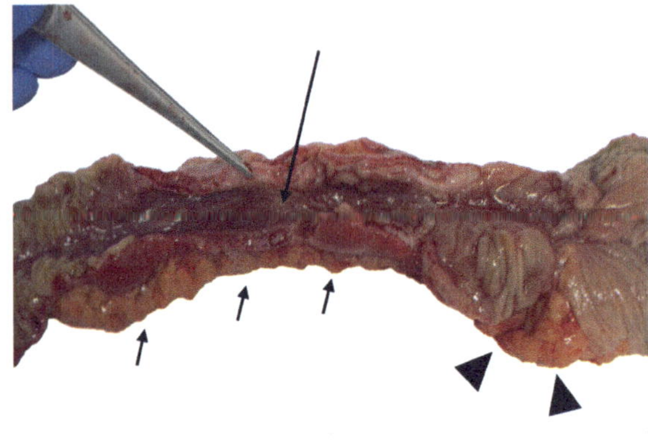

Fig. 15.3 Fat wrapping association with skip areas in a surgically resected portion of the small intestine from a patient with Crohn's disease. Fat wrapping is seen adjacent to an area of inflamed and narrowed small intestine (arrows) and also is seen in adjacent to a second affected area or skip lesion(arrowheads) with a small area of normal small intestine in between the two affected areas. Photographs courtesy of Dr. J Calvin Coffey

fibroblast differentiation may produce an "outside in" phenomenon that are associated with both inflammation and fibrosis of the intestine. B cells and innate lymphocytes infiltrate mesenteric lymphatics in areas of fat wrapping. Bacterial translocation and lymphaticslymphatics infiltration from the inflamed bowel wall may initiate these events.

Imaging Studies of the Mesentery in Crohn's Disease

Imaging studies demonstrate the relationship between the mesenteric and intestinal activity in patients with Crohn's disease. CT imaging in patients with Crohn's disease may demonstrate mesenteric edema with hypervascularity and stretching of the mesenteric vessels, known as the Comb sign (Fig. 15.4).

Other CT findings include increased fat density, fibrofatty proliferation, and mesenteric lymphadenopathy (Figs. 15.5, 15.6, and 15.7). Engorgement of mesenteric vasculature and mesenteric edema are also seen (Fig. 15.8). Stenosis and sacculation in the mesentery may also be seen on small intestinal radiography (Fig. 15.9).

Radiographic images can demonstrate synchronicity between intestinal organs and the mesentery, predicting disease activity and anatomical changes such as fibrosis. For example, Sakurai et al. recently demonstrated that the findings of mesenteric hypervascularity and enlarged mesenteric lymph nodes correlated with endoscopically visualized mucosal ulceration. Gale et al. compared CT

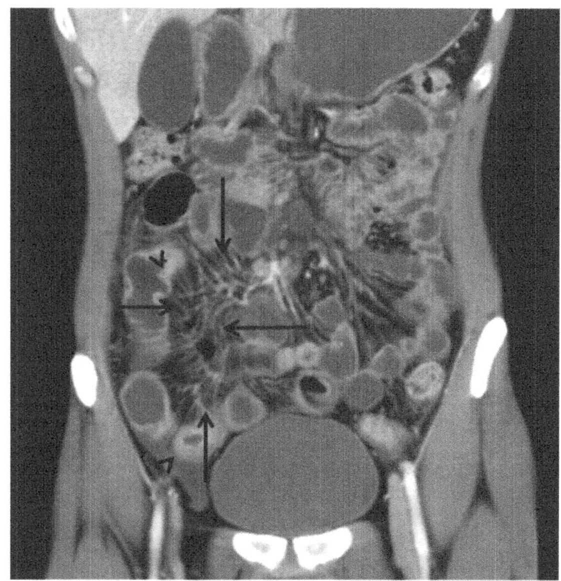

Fig. 15.4 Abdominal CT (sagittal view) of a patient with active Crohn's disease. Stretching of the mesenteric vessels, also known as the Comb sign is seen (arrows). Diffuse ileal involvement and wall thickening is present (arrowhead)

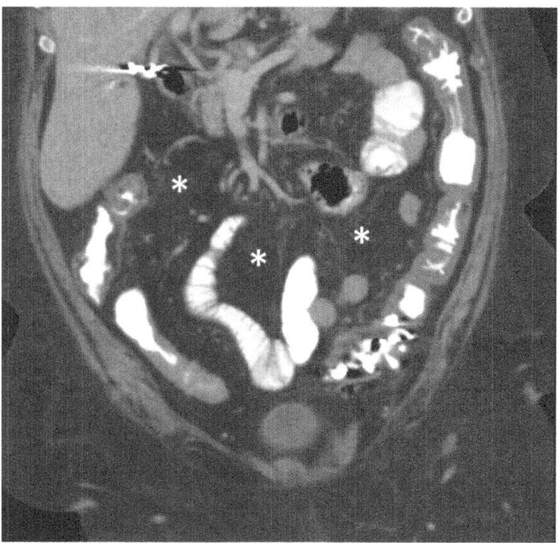

Fig. 15.5 Abdominal CT in a patient with Crohn's disease demonstrating fibofatty proliferation (asterisk)

enterography and MR enterography in 84 children and adolescents with active Crohn's disease. In the pediatric population, mesenteric inflammatory changes including mesenteric hypervascularity, edema, fibrofatty proliferation and lymphadenopathy correlate closely with active mucosal Crohn's disease.

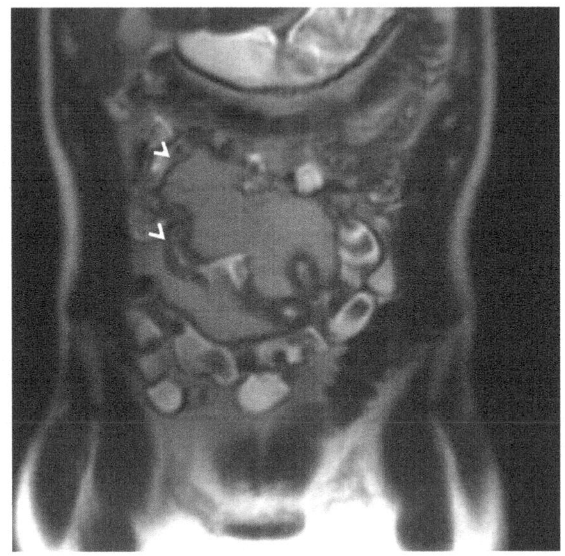

Fig. 15.6 Coronal MRI of the abdomen in a patient with Crohn's disease demonstrating fatty proliferation and increased fat density on a T2 imaging (arrowheads)

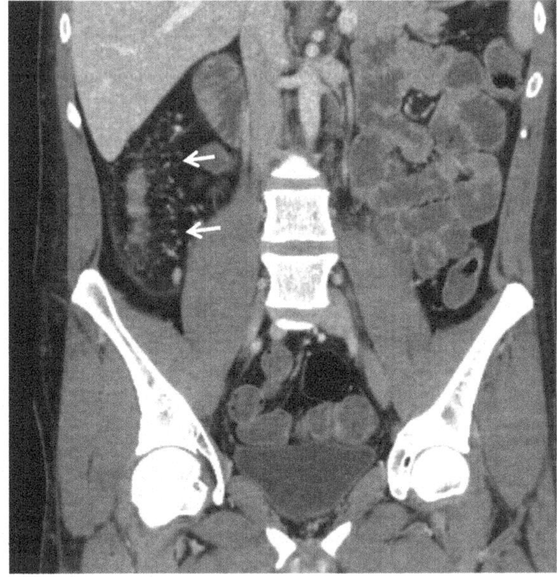

Fig. 15.7 Abdominal CT (coronal view) of a patient with active Crohn's disease in the right lower quadrant of the abdomen. Multiple small lymph nodes (arrows) and stretched, hyperemic mesenteric vessels are demonstrated

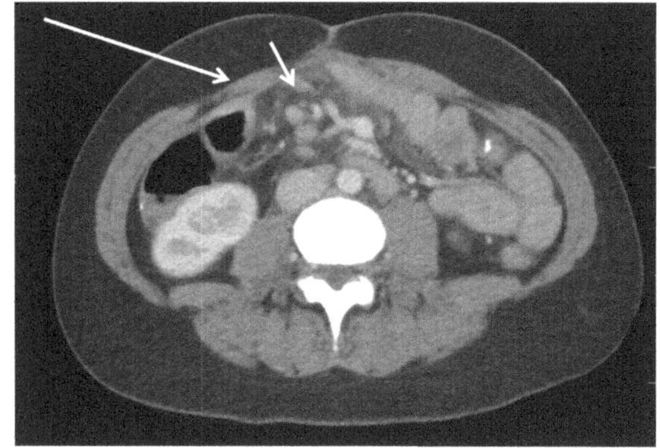

Fig. 15.8 Axial view near the level of the umbilicus showing mesenteric edema (large arrow) and engorged mesenteric vasculature (white arrow)

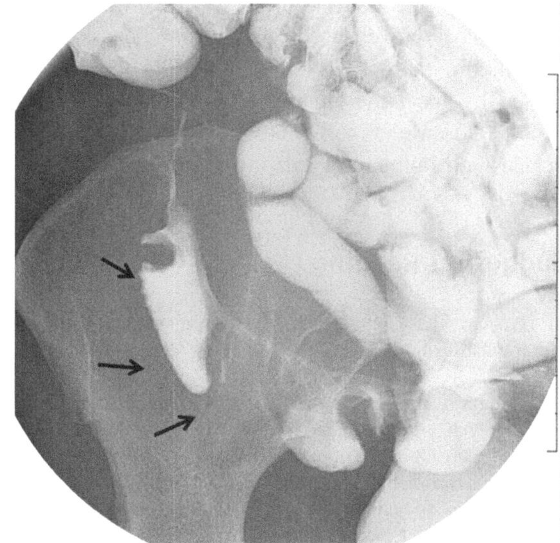

Fig. 15.9 Spot film of the distal ileum on small intestinal radiography. The is a long segment stricture of the distal ileum. A segment with asymmetric sparing along the non-mesenteric boarder results in an area of relative outpouching (black arrows). This area of outpouching is referred to as pseudosacculation. Images are courtesy of Dr. Abraham Dachman

Mesenteric Disorders that Are Associated with Crohn's Disease

Arterial and venous thrombi within the mesenteric vasculature and mesenteric venous congestion are associated with Crohn's disease. These may be complicated by bowel ischemia and perforation. Small intestinal lymphoma is more common in patients with Crohn's disease and may present as a mesenteric mass with some thickening of surrounding tissues.

Inflammatory Disorders of the Mesentery and Inflammatory Bowel Disease

The most common primary inflammatory disorder of the mesentery is mesenteric panniculitis (also known as sclerosing mesenteritis). (See Chap. 20 Mesenteric Panniculitis). Mesenteric panniculitis shares common features with inflammatory bowel disease, especially Crohn's disease. Typically, patients with mesenteric panniculitis have a waxing and waning clinical course, abdominal signs and symptoms, fever and weight loss, and response to anti-inflammatory agents. Some cases of mesenteric panniculitis have ileocolic involvement. Both conditions demonstrate elevated CRP and erythrocyte sedimentation rates (ESR). As in Crohn's disease, inflammatory masses in mesenteric panniculitis can produce obstructive signs and symptoms from extrinsic compression of the small intestine.

Longstanding inflammation in the mesentery may produce mesenteric fibrosis and scarring, similar to fibrotic bowel seen in Crohn's disease. A family history of inflammatory bowel disease is seen in some patients with mesenteric panniculitis. In addition, mesenteric panniculitis-like findings on abdominal imaging studies caused by local mesenteric inflammation have been described in patients with Crohn's disease. Finally, both Crohn's disease and mesenteric panniculitis are treated with immunomodulating agents.

Suggested Reading

1. Lakhan SE, Kirchgessner A. Neuroinflammation in inflammatory bowel disease. J Neuroinflammation. 2010;7:37.
2. Burke JP, Velupillai Y, O'connell PR, Coffey JC. National trends in intestinal resection for Crohn's disease in the post-biologic era. Int J Color Dis 2013;28:1401–1406.
3. Coffey JC, O'Leary DP, Kiernan MG, Faul P. The mesentery in Crohn's disease: friend or foe? Curr Opin Gastroenterol. 2016;32:267–73.
4. Peyrin-Biroulet L, Chamaillard M, Gonzalez F, et al. Mesenteric fat in Crohn's disease: a pathogenetic hallmark or an innocent bystander? Gut. 2007;56:577–83.
5. Paeschke A, Erben U, Kredel LI, Kühl AA, Siegmund B. Role of visceral fat in colonic inflammation: from Crohn's disease to diverticulitis. Curr Opin Gastroenterol. 2017;33(1):53–8.
6. Peyrin-Biroulet L, Gonzalez F, Dubuquoy L, Rousseaux C, Dubuquoy C, Decourcelle C, Saudemont A, Tachon M, Béclin E, Odou MF, Neut C, Colombel JF, Desreumaux P. Mesenteric fat as a source of C reactive protein and as a target for bacterial translocation in Crohn's disease. Gut. 2012;61(1):78–85.
7. Batra A, et al. Mesenteric fat—control site for bacterial translocation in colitis? Mucosal Immunol. 2012;5(5):580–91.
8. Li Y, Zhu W, Zuo L, Shen B. The role of the mesentery in Crohn's disease: the contributions of nerves, vessels, lymphatics, and fat to the pathogenesis and disease course. Inflamm Bowel Dis. 2016;22:1483–95.
9. Randolph GJ, Bala S, Rahier JF, Johnson MW, Wang PL, Nalbantoglu I, Dubuquoy L, Chau A, Pariente B, Kartheuser A, Zinselmeyer BH, Colombel JF. Lymphoid Aggregates Remodel Lymphatic Collecting Vessels that Serve Mesenteric Lymph Nodes in Crohn Disease. Am J Pathol. 2016;186(12):3066–73.

10. von der Weid PY, Rehal S, Ferraz JG. Role of the lymphatic system in the pathogenesis of Crohn's disease. Curr Opin Gastroenterol. 2011;27(4):335–41.
11. Park MJ, Lim JS. Computed Tomography Enterography for Evaluation of Inflammatory Bowel Disease. Clin Endosc. 2013;46(4):327–66.
12. Sakurai T, Katsuno T, Saito K, Yoshihama S, Nakagawa T, Koseki H, Taida T, Ishigami H, Okimoto KI, Maruoka D, Matsumura T, Arai M, Yokosuka O. Mesenteric findings of CT enterography are well correlated with the endoscopic severity of Crohn's disease. Eur J Radiol. 2017;89:242–8.
13. Gale HI, Sharatz SM, Taphey M, Bradley WF, Nimkin K, Gee MS. Comparison of CT enterography and MR enterography imaging features of active Crohn disease in children and adolescents. Pediatr Radiol. 2017;47(10):1321–8.
14. Zezos P, Kouklakis G, Saibil F. Inflammatory bowel disease and thromboembolism. World J Gastroenterol. 2014;20(38):13863–78.
15. https://rarediseases.org/rare-diseases/mesenteric-panniculitis/. Accessed 3 Jan 18
16. Ehrenpreis ED, Roginsky G, Gore RM. Clinical significance of mesenteric panniculitis-like abnormalities on abdominal computerized tomography in patients with malignant neoplasms. World J Gastroenterol. 2016;22(48):10601–8.
17. Miranda-Bautista J, Fernández-Simón A, Pérez-Sánchez I, Menchén L. Weber-Christian disease with ileocolonic involvement successfully treated with infliximab. World J Gastroenterol. 2015;21(17):5417–20.

The Role of the Mesentery in Pancreatic Diseases

16

Charles Broy, Chloe Lee, and Eli D. Ehrenpreis

Embryology of the Pancreas and Pancreatic Mesentery

Between the gestational ages of 30–47 days, the endodermal lining of the foregut forms two outgrowths caudal to the developing liver. These are the ventral pancreatic bud and the dorsal pancreatic bud. Within each pancreatic bud, the endoderm develops into branched tubules that are attached to secretory acini, thus forming the exocrine portion of the pancreas. The endocrine pancreas, including of the islets of Langerhans, arise from stem cells at the duct branch points. These develop into discrete islands of vascularized endocrine tissue within the parenchyma of the exocrine glandular tissue. During the seventh week of gestation, the ventral and dorsal buds merge together to form a single organ. Within the newly formed pancreas, the uncinate process of the head of the pancreas is derived from the ventral pancreatic bud and the remaining portion of the head, body and tail of the pancreas is derived from the dorsal pancreatic bud. It is during the process of fusion of the ventral and dorsal pancreatic buds that heterotopic pancreas tissue (pancreatic tissue in extra-pancreatic sites) is believed to occur. Exocrine function of the pancreas begins after birth, while endocrine function, specifically the release of hormones, can be measured after the 10th to 15th week of gestation.

C. Broy
Department of Gastroenterology, Edward Hines Jr VA, Hines, IL 60141, USA
e-mail: Charles.Broy@va.gov

C. Lee
Department of Internal Medicine, Advocate Lutheran General Hospital, Park Ridge, IL 60068, USA

E. D. Ehrenpreis (✉)
Department of Medicine, Advocate Lutheran General Hospital, 1775 Dempster St, 6 South, Park Ridge, IL 60068, USA
e-mail: e2bioconsultants@gmail.com

E. D. Ehrenpreis et al. (eds.), *The Mesenteric Organ in Health and Disease*,
https://doi.org/10.1007/978-3-030-71963-0_16

The embryology of the mesentery is closely intertwined with its attendant organs. The dorsal mesentery of the upper abdomen develops synchronically with the rapid and asymmetric development of the stomach. As the stomach and duodenum expand laterally, the lesser sac develops. The pancreas grows into this area and is the part of the fusion of the mesogastrium with the dorsal peritoneum. The dorsal mesentery later develops into the gastrosplenic and splenorenal ligaments.

Aside from its function as a conduit for blood vessels, nerves and lymphatics, the mesentery has been shown to influence the normal function of adjacent organs and can also be involved with the development of diseases within these organs. It is probable that the mesentery has a role in normal pancreatic function as well as pancreatic disease.

Pancreatic Diseases

The pancreas is a retroperitoneal organ with anatomic contiguity with other peritoneal organs in the abdomen including the mesentery. It is known that pancreatic cancer as well as pancreatitis can spread via the portion of the mesentery known as the transverse mesocolon. The root of the mesentery extends from the horizontal portion of the duodenum to the iliac fossa, creating an avenue the spread of pancreatic cancer in the subperitoneum. In addition, fat necrosis and inflammation may traverse from the mesentery from its mesocolic to small bowel mesenteric sites due to their direct connection.

Acute Pancreatitis

While the initiation of acute pancreas involves the release of pancreatic digestive enzymes, complications of pancreatic inflammation may occur from primary responses within the mesentery. Multiple animal studies suggest that mesenteric lymph has toxic factors that may be the prime driver of multiple organ dysfunction. One study of hemorrhagic shock in male rats compared trauma from laparotomy to no trauma with or without mesenteric lymph duct ligation. CD11b and CD18 expression was increased in rats with trauma but prevented by ligation of the mesenteric lymph duct. Since these integrins increase myeloid phagocytic cells, these data provide evidence that neutrophil activation and sequestration are carried by the mesenteric lymph. MicroRNAs (miRNAs) regulate gene expression associated with acute pancreatitis and are being studied as a clinical biomarker for the disease. Blenkiron, et al performed a study in rats having three groups of control, mild and moderate taurocholate-induced acute pancreatitis. Seven different miRNAs, comprising miR-375, -217, -148a, -216a, -122, -214, and -138, were increased in the mesenteric lymph of rats with acute pancreatitis. The degree of increase of these miRNAs also correlated with the severity of disease.

Translocated bacteria may also function as a cofactor in these processes.

Vascular complications of pancreatitis including hemorrhage can affect local organs including the mesentery. Computerized tomography (CT) of the abdomen in patients with acute pancreatitis often demonstrate concomitant inflammatory changes occurring in the mesentery (Fig. 16.1).

In one study of MRI findings in the transverse mesocolon, 40 patients without pancreatic disorders were compared with 210 patients with acute pancreatitis. One hundred and thirty patients (61.9%) with acute pancreatitis demonstrated involvement of the transverse mesocolon portion of the mesentery. The grading of transverse mesocolon effect was strongly correlated with the MRI severity index and the APACHE-II score (r = 0.759 and 0.384 respectively).

Mesenteric Panniculitis Involving the Pancreas and Pancreatic Panniculitis

Mesenteric panniculitis is a rare condition in which a mass-like lesion of the mesentery or less organized "misty mesentery" occurs. Mesenteric panniculitis is characterized by chronic inflammation and fat necrosis. It commonly involves the

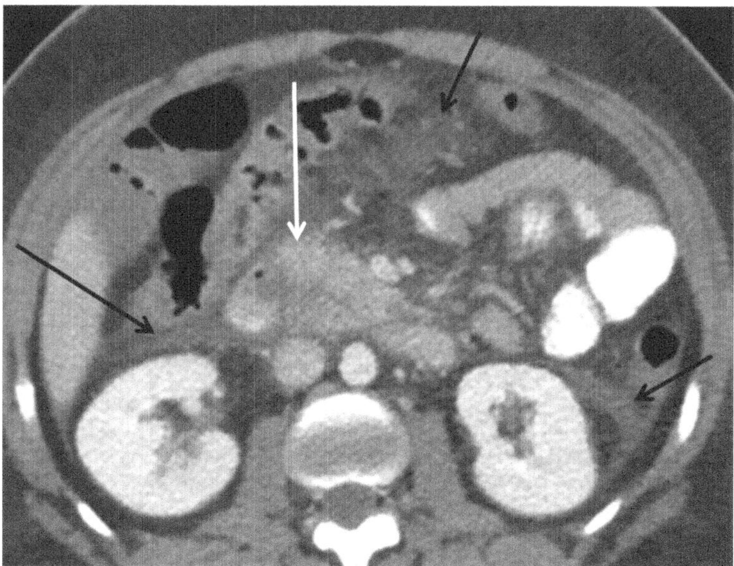

Fig. 16.1 Necrotizing pancreatitis with mesenteric involvement. Decreased enhancement seen on contrast-enhanced CT images in the head and neck of the pancreas indicate areas of necrosis (white arrows). Mesenteric fat stranding and free fluid in the upper abdomen are common sequelae (black arrows). Image is courtesy of Dr. Abraham Dachman and Dr. Justin Ramirez

portion of the mesentery that is proximal to the small intestine. One retrospective review of patients with mesenteric panniculitis that was performed at the Mayo Clinic suggested an association between elevated IgG4 (with possible autoimmune pancreatitis) and mesenteric panniculitis, but this finding has not been repeated in other studies. Mesenteric panniculitis has also been described in the peripancreatic region. (Also see Chap. 20. Mesenteric Panniculitis and Chap. 17 The Mesentery and IgG4-Related Diseases).

Panniculitis can be a manifestation of a number of local and systemic diseases, with erythema nodosum being the most common form of panniculitis. Panniculitis has been divided into different etiologic subtypes that guide management: inflammatory, infectious, enzymatic destruction, traumatic, malignant and deposition (Table 16.1). Systemic forms of panniculitis may at times also involve the mesentery.

Symptoms of panniculitis depend on the area impacted, but the most common presentations of these conditions are the development of painful, tender nodules in the lower extremities. Pancreatic panniculitis (also known as enzymatic panniculitis) is an extremely rare disease and is most commonly associated with acute or chronic pancreatitis. Pancreatic panniculitis may occur in 2–3% of patients with a variety of pancreatic disorders. At present, little is known about the pathophysiology of pancreatic panniculitis. One study that included 15 patients with necrotizing pancreatitis showed increased lipase in fat that was adjacent to the pancreas. Furthermore, free fatty acid levels in visceral fat impact the severity of acute pancreatitis. A direct effect of circulating pancreatic enzymes may be an important factor in the development of pancreatic panniculitis. However, other factors are needed for pancreatic panniculitis to occur, since acute pancreatitis in general is characterized by elevated levels of lipase and amylase, and most patients do not develop pancreatic panniculitis. Additionally, some causes of pancreatic panniculitis are not associated with elevated lipase such as pancreatic cancer. Other potential factors leading to the development of pancreatic panniculitis include trypsin activation and deposition of immune complexes.

Histopathology may reveal subcutaneous lobular fat necrosis with anuclear adipocytes called ghost cells without vasculitis. Fat saponification and dystrophic calcification may be present. Although these histologic findings can be pathognomonic of pancreatic panniculitis, often histology evaluation results in an equivocal diagnosis.

Metastatic fat necrosis is a known entity that can involve the heart, joints, lung, bone marrow and the mesentery. One case series describes six patients with mesenteric panniculitis and pancreatic disease. The clinical and radiographic findings in these patients mimicked primary pancreatic cancer. In this series, all patients reported abdominal pain and three reported weight loss. The pancreatic histopathology was typical for panniculitis with fat necrosis, lymphocytes, plasma cells and rare eosinophils. Four of 5 of these patients had at least a partial response to therapy which consisted of corticosteroids, tamoxifen, azathioprine, resection or a combination of these.

Table 16.1 Causes of Panniculitis

Enzymatic
Pancreatic
Alpha-1 antitrypsin deficiency
Inflammatory
Erythema nodosum
Erythema nodosum migrans
Erythema induratum
Steroid use
Systemic lupus erythematosus
Dermatomyositis
Connective tissue disease
Rheumatoid Nodules
Sarcoidosis
Vasculitis
Lipodermatosclerosis
Superficial Migratory Thrombophlebitis
Deposition
Gout
Calciphylaxis
Hyperoxaluria
Malignancy
Lymphoma
Leukemia
Cytophagic Histiocytic (occurs with benign disease or subcutaneous panniculitis-like T-cell lymphoma)
Infection
Bacterial
Mycobacterial
Fungal
Parasitic
Arthropod
Trauma
Blunt Trauma
Iatrogenic
Post Radiation Exposure
Cold Exposure

Reused with permission from Requena L, Sanchez Y. Panniculitis. Part II. Mostly lobular panniculitis. *J Am Acad Dermatol.* 2001;45(3):325-364. Copyright © Elsevier

A series of two cases of patients presenting with pancreatic and a mesenteric mass appearing to be pancreatic cancer was also published. These patients were diagnosed with mesenteric panniculitis and both had resolution of their symptoms with corticosteroid therapy.

A series of the imaging findings in 17 patients with mesenteric panniculitis includes one patient with pancreatitis and pancreatic panniculitis.

Heterotopic Pancreas

Heterotopic pancreas (HP) is defined as the presence of pancreatic tissue without vascular or ductal connection to the orthotopic pancreas. It is a congenital anomaly most commonly located in the stomach, duodenum and jejunum. HP may also be present in the mesentery and mesenteric involvement was found in 3.8% in one study of 184 consecutive histology-proven HP cases in China. It has been hypothesized that HP occurs when deposits of pancreatic tissue are "dropped" into the developing gastrointestinal system during fusion of the ventral and dorsal buds. A study of the endoderm of chicken embryos, compared different concentrations of cyclopamine, an inhibitor of the sonic hedgehog signaling pathway during development, that causes birth defects to controls. Stomach and mesenchyme were then isolated at stage 15 and cell cultures were studied. 1.0 microMolar cyclopamine exposure produced signs of pancreatic tissue growth whereas control and exposure to 0.1 microMolar cyclopamine showed no growth of pancreatic tissue. This study suggests that the sonic hedgehog gene has a role in the development of HP.

HP is usually an incidental finding during surgery or autopsy and the majority of patients with HP are asymptomatic. However, since HP is composed of pancreatic tissue, pancreatic diseases can affect areas of HP as well. A variety of complications of HP have been described (See Table 16.2).

On imaging studies including computerized tomography (CT) and magnetic resonance imaging (MRI), HP appears as small, homogenous, well enhancing lesions with an appearance that is similar to pancreatic tissue. However, imaging studies may not be able to differentiate mesenteric HP from metastatic tumors or lymphoma. In one series of 17 patient with surgically confirmed HP, only one was correctly diagnosed as HP before surgery. In one case report, an MRCP scan was used to make the diagnosis of HP.

Histological examination of areas of HP reveals similar elements of the orthotopic pancreas, including pancreatic ducts, islet cells and acini. No treatment is recommended for asymptomatic disease while surgical excision is performed for HP-associated malignancy or other complications of HP.

There are a few case reports describing HP in the mesentery. Kok et al. reported a 58-year-old woman who presented with clinical suspicion of acute pancreatitis. Imaging studies revealed a large lesion found to be HP in the small bowel

Table 16.2 Complications of Heterotopic Pancreas

Complications
Cancer
Pancreatitis
Pseudocyst Formation
Abnormal Hormone Secretion
Bowel Obstruction
Intussusception
Common Bile Duct Obstruction
Gastrointestinal Bleeding

mesentery. Endoscopic ultrasound-guided fine needle aspiration of the lesion showed benign cells and inflammation characteristic of acute pancreatitis. Conservative treatment for one week helped to relieve the patient's symptoms and laboratory abnormalities. Exploratory laparoscopy was performed and excision of the lesion revealed HP in the mesentery of the proximal jejunum. Histology confirmed the presence of pancreatic acini, ducts and islet cells.

Suggested Reading

Embryology

1. Hikspoors JPJM, Kruepunga N, et al. The development of the dorsal mesentery in human embryos and fetuses. Semin Cell Dev Biol. 2018. https://www.sciencedirect.com/science/article/pii/S1084952117305438.

Pancreatitis and Other Diseases

2. Blenkiron C, Askelund K, et al. MicroRNAs in mesenteric lymph and plasma during acute pancreatitis. Ann Surg. 2014;260(2):341–7.
3. Chi XX, Zhang XM, et al. The normal transverse mesocolon and involvement of the mesocolon in acute pancreatitis: an MRI study. PlosOne. 2014. https://doi.org/10.1371/journal.pone.0093687.
4. Fanous MYZ, Phillips AJ, et al. Mesenteric lymph: the bridge to future management of critical illness. JOP. 2007;8(4):374–99.
5. Caruso JM, Feketeova E, et al. Factors in intestinal lymph after shock increase neutrophil adhesion molecule expression and pulmonary leukosequestration. J Trauma. 2003; 55:727–33. Van MinnenLP, Besselink MG, et al. Colonic involvement in acute pancreatitis: a retrospective study of 16 patients. Dig Surg 2004; 21: 33–38; discussion 39–40.
6. Mittal A, Phillips ARJ, et al. The proteome of mesenteric lymph during acute pancreatitis and implications for treatment. JOP. 2009;10(2):130–42.

Panniculitis

7. Laureano A, Mestre T, Ricardo L, et al. Pancreatic panniculitis - a cutaneous manifestation of acute pancreatitis. J Dermatol Case Rep. 2014;8(1):35–7.
8. Acharya C, Navina S, et al. Role of pancreatic fat in the outcomes of pancreatitis. Pancreatology. 2014;14(5):403–8.
9. Akram S, Pardi DS, Schaffner JA, Smyrk TC. Sclerosing mesenteritis: clinical features, treatment, and outcome in ninety-two patients. Clin Gastroenterol Hepatol. 2007;5:589.
10. Requena L, Sanchez Y. Panniculitis: part II mostly lobular panniculitis. J Am Acad Dermatol. 2001;45(3):325–364.
11. García-Romero D, Vanaclocha F. Pancreatic panniculitis. Dermatol Clin. 2008;26:465–70.

12. Camilleri MJ, Daniel Su WP. Disorders of subcutaneous tissue: panniculitis. In Freedberg IM, Eisen AZ, Wolff K (eds) Fitzpatrick's dermatology in general medicine, 6th edn, pp 1047–1067. New York: Mc Graw Hill; 2003.
13. Price-Forbes AN, Filer A, et al. Progression of imaging in pancreatitis panniculitis polyarthritis (PPP) syndrome. Scand J Rheumatol. 2006;35:72–4.
14. Scudiere JR, Shi C, et al. Sclerosing mesenteritis involving the pancreas as mimicker of pancreatic cancer. Am J Surg Pathol. 2010;34(4):447–53.
15. Phillips RH, Carr RA, Preston R, et al. Sclerosing mesenteritis involving the pancreas: two cases of a rare cause of abdominal mass mimicking malignancy. Eur J Gastroenterol Hepatol. 1999;11:1323–9.

Heterotopic Pancreas

16. Tang XB, Liao MY, Wang WL, Bai YZ. Mesenteric heterotopic pancreas in a pediatric patient: A case report and review of literature. World J Clin Cases. 2018;6(14):847–53.
17. Wong J, Robinson C, et al. Recurrent ectopic pancreatitis of the jejunum and mesentery over a 30-year period. Hepatobiliary Pancreat Dis Int. 2011;10(2):218–20.
18. Canbaz H, Colak T, et al. An unusual cause of acute abdomen: mesenteric heterotopic pancreatitis causing confusion in clinical diagnosis. Turk J Gastroenterol. 2009;20(2):142–5.
19. Hsia CY, Wu CW, et al. Heterotopic pancreas: a difficult diagnosis. J Clin Gastroenterol. 1999;28:144–7.
20. Silva AC, Charles JC, et al. MR Cholangiopancreatography in the detection of symptomatic ectopic pancreatitis in the small-bowel mesentery. Am J Roentgenol. 2006;187:W195–7.
21. Kim SK, Melton DA. Pancreas development is promoted by cyclopamine, a hedgehog signaling inhibitor. Proc Natl Acad Sci. 1998;95:13036–41.
22. de Kok BM, de Korte FI, et al. Acute clinical manifestation of mesenteric heterotopic pancreatitis: a pre-and postoperative confirmed case. Case Reports in Gastrointestinal Medicine; 2018.

IgG4-Related Diseases and the Mesentery

17

Ahmed Khattab, David H. Kruchko, and Eli D. Ehrenpreis

Introduction

IgG4 is one of four subcomponents of IgG, or immunoglobulin G. IgG is the most common immunoglobulin in the circulation, representing almost 75% of circulating immunoglobulins in human blood. IgG is produced in plasma cells. The role of IgG is to bind to foreign invaders such as bacteria, fungi, and viruses to prevent these organisms from producing disease states. Of the four subgroups of IgG, IgG4 represents the lowest concentration in human serum. Tissue level of IgG4 is directly related to the presence of acute inflammation. There is evidence to suggest that IgG4 has a regulatory rather than a direct inflammatory effect, and with this model, IgG4 has the physiologic function of protecting other molecules of the IgG subclasses. IgG4 produces a pathologic state when deviation from its physiologic action occurs. In this setting, rearrangement of the antibody structure of IgG4 occurs across a labile disulfide bond (Fig. 17.1). The figure demonstrates downward-facing red arrows that indicate the disulfide bonds. This is referred to as the hinge region of the molecule. Occasionally, these labile disulfide bonds enable inappropriate transformation of the IgG4 molecule, resulting in pathogenic IgG4 autoantibodies that inappropriately attack self-antigens.

A. Khattab
Department of Internal Medicine, Division of Gastroenterology, Advocate Lutheran General Hospital, Park Ridge, IL 60068, USA

D. H. Kruchko
Department of Internal Medicine, Advocate Lutheran General Hospital, Chicago, IL 60068, USA

E. D. Ehrenpreis (✉)
Department of Medicine, Advocate Lutheran General Hospital, 1775 Dempster St, 6 South, Park Ridge, IL 60068, USA
e-mail: e2bioconsultants@gmail.com

E. D. Ehrenpreis et al. (eds.), *The Mesenteric Organ in Health and Disease*,
https://doi.org/10.1007/978-3-030-71963-0_17

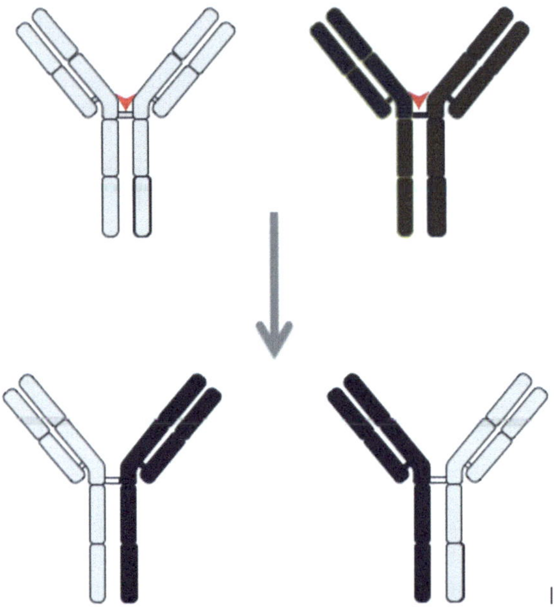

Fig. 17.1 Characteristic feature of Fab (Antigen-binding Fragment) arm exchange across the disulfide bond of IgG4. The hinge region, (downward-facing red arrowhead), contains the disulfide bond. These disulfide bonds are labile and thereby permit inappropriate splitting and exchange of the Fab structure found in IgG4. This is what allows IgG4 molecules to transform from regulatory to pathogenic antibodies. Reused from Kawa S. Immunoglobulin G4-related Disease: An Overview. JMA J. 2019;2(1):11–27. Attribution 4.0 International (CC BY 4.0)

These disease states are referred to as IgG4-Related Disease, or IgG4-RD. IgG4-RD are characterized by excessive inflammatory response leading to fibrosis. For this reason, IgG4-RD are referred to as a fibroinflammatory conditions. Most organs are potential targets of IgG4-RD, including the mesentery, and gastrointestinal (GI) tract organs, including the biliary tree and the pancreas.

Pathogenesis

Pathogenetic development of IgG4-RD can ultimately be summarized as lymphoplasmacytic infiltration of IgG4+ plasma cells. The term lymphoplasmacytic indicates a concentration of lymphoid cells and plasma cells. Interestingly, IgG4-containing plasma cells may be confined to affected tissues and frequently are not circulating in the blood; therefore, IgG4 levels may not be elevated when measured in the peripheral blood. When local inflammation is present, lymphoid cells are generally recruited along with the plasma cells. Specifically, regulatory T (Treg) cells are most commonly associated with IgG4-RD as well as other autoimmune diseases. Treg cells are responsible for secreting specific cytokines in normal immune physiology. In IgG4-RD, dysregulation of Treg cells occurs with associated inappropriate cytokines secretion, (see Fig. 11.4 in Chap. 11 Immunologic Function of the Mesentery). The most notable cytokines secreted by Treg cells are IL-4, IL-5, IL-10, IL-13, and TGF-beta. These cytokines play an important role in the production and recruitment of eosinophils. This ultimately leads to eosinophilic deposition in target tissues and organs. Peripheral eosinophilia and elevated serum IgE levels also occur. IL-10 was found to have the most intimate relationship with IgG4 production and this ultimately leads to local fibrosis via TGF-beta expression. Local fibrosis can best be explained by the fact that TGF-beta is primarily responsible for cell growth, proliferation, and apoptosis (cell death). Therefore, if any of these three roles of TGF-beta are dysregulated, inappropriate cell growth and proliferation and inadequate cell death follow.

Storiform fibrosis is a key histologic feature of IgG4-RD and is defined as "swirling fibrotic tissue encasing cellular infiltrate in collagen fibrils." When seen on tissue biopsy, storiform fibrosis is highly suggestive, although not pathognomonic, of IgG4-RD. Storiform fibrosis is commonly seen in Type I Autoimmune Pancreatitis, an IgG4-RD affecting the pancreas.

Autoimmune pancreatitis (AIP) is the most well studied IgG4-RD of the GI tract. AIP can present as an acute pancreatitis, but more often, it manifests as a more insidious onset. It is estimated that AIP accounts for approximately 5% of all cases of chronic pancreatitis. IgG4-related sclerosing cholangitis, has attracted recent investigation, as have studies of the pathogenesis of retroperitoneal fibrosis and other mesenteric manifestations of IgG4-RD. Human leukocyte antigen (HLA) genes have been implicated in the development of IgG4-RD. Specifically, HLA subtypes DQB1_0401 and DRB1_0405 were found to be highly expressed in patients with AIP as compared to healthy patients. Furthermore, AIP patients also have higher levels of DQB1_0401 and DRB1_0405 expression compared to other patients with chronic pancreatitis.

Non-HLA genes that are associated with AIP consist of single nucleotide polymorphisms (SNPs). SNPs have been increasingly found to have clinical significance as it relates to the development, response to treatment, and prognosis of a variety of diseases. The SNPs associated with AIP are CTLA-4_49A, Cytotoxic T-Lymphocyte Antigen-4, and TNF-alpha promoter_863, a Tumor Necrosis Factor-alpha promoter gene. Studies performed in Chinese and Japanese patients have found these SNPS to be more highly expressed in AIP patients than in patients without known disease. In terms of clinical significance, CTLA-4_49A was a marker for a higher likelihood of relapse after seemingly adequate initial treatment of AIP.

Clinical Manifestations

IgG4-RD has been identified in virtually all organ systems. In most cases (60–90%) it affects more than one system simultaneously by causing subacute painless enlargement of the affected organs. As expected, clinical manifestations vary depending on the organ system involved and commonly include complications secondary to compression of nearby structures. Thus, IgG4-RD should be considered in patients presenting with unexplained enlargement of one or more organs. At the time of diagnosis, the patients are usually well and have no systemic symptoms, especially fever. However, significant weight loss in the months preceding diagnosis is not uncommon. Due to an increased awareness of the IgG4-RD, these more commonly being diagnosed as an incidental finding seen on imaging or histopathology in otherwise asymptomatic patients.

IgG4-RD include type 1 autoimmune pancreatitis (AIP)*, sclerosing cholangitis*, sclerosing mesenteritis*, Mikulicz disease, sclerosing sialadenitis, Inflammatory orbital pseudotumor, chronic sclerosing dacryoadenitis, idiopathic retroperitoneal fibrosis*, Chronic sclerosing aortitis and periaortitis*, and Riedel's thyroiditis.

Because diseases with an (*) are closely related to the mesentery, they will be discussed in more detail.

Type 1 Autoimmune Pancreatitis (AIP)

Type 1 AIP was the first identified form of IgG4-RD and is considered to be the prototypical form. It accounts for approximately 5% of patients with chronic pancreatitis. Almost all patients are older than 45 at the time of diagnosis. Type 1 AIP often presents with a pancreatic mass or painless jaundice and can therefore be mistaken for pancreatic cancer. It has been estimated that 3–9% of patients who undergo pancreatic resection are discovered to have had Type 1 AIP. Patients with Type 1 AIP develop diabetes mellitus due to destruction of endocrine portions of the pancreas. It is commonly associated with other IgG4-related conditions. Radiologic features of AIP include a "sausage-shaped" pancreas with a surrounding halo of edema around the organ. It is very difficult to distinguish AIP from pancreatic adenocarcinoma on imaging. Some authors have suggested that diffusion-weighted magnetic resonance (MR) imaging findings maybe more useful than CT in making such a distinction.

IgG4-Related Sclerosing Cholangitis (SC)

IgG4-related SC is the second most common manifestation of IgG4-RD. Although IgG4-related SC affects 60–80% of patients with type 1 AIP, it can occasionally occur as an isolated manifestation. Because it leads to focal or diffuse bile duct wall thickening IgG4-related SC tends to present as obstructive jaundice. The most commonly affected segment is the intra-pancreatic portion of the common bile duct. IgG4-related SC and primary sclerosing cholangitis have similar magnetic resonance cholangiopancreatographic (MRCP) and endoscopic retrograde cholangiopancreatographic (ERCP) findings. Primary sclerosing cholangitis is usually a disease of young and middle age males while IgG4-related SC typically affects middle aged and elderly me. IgG4-related SC exhibits a good response to glucocorticoid therapy.

Sclerosing mesenteritis (SM), mesenteric lipodystrophy and mesenteric panniculitis are terms that have been used interchangeably in the literature (Table 17.1). There have been some case reports relating SM to IgG4-RD. However, to this date, there is not enough evidence to determine this association (see Chap. 20 Mesenteric Panniculitis).

SM likely represents a specific form of mesenteric panniculitis that is predominated by fibrosis. As in other IgG4-RD, SM may present with weight loss fever and malaise. In these cases, an abdominal mass can be palpated on examination, and occurs in one-third to one-half of patients. Abdominal tenderness, distension, signs of peritoneal irritation and ascites are less common. Some patients present with bowel obstruction, obstructive uropathy or renal failure, chylous ascites, and vascular occlusion. Laboratory findings include elevated erythrocyte sedimentation rate and C-reactive protein. The natural history of SM associated with IgG4-RD has not been fully established in the literature.

Table 17.1 Summary of terminology used to describe sclerosing mesenteritis among different disciplines

Discipline	Mesenteric lipodystrophy	Mesenteric panniculitis	Retractile mesenteritis	Sclerosing mesenteritis
Pathology	Predominant fat necrosis	Predominant chronic inflammation	Predominant fibrosis	Umbrella term for any benign fibroinflammatory disease of the mesentery with any degree of fat necrosis, chronic inflammation and fibrosis
Radiology	Not frequently used	Meets at least 3 of 5 criteria (i) Fatty mesenteric mass; (ii) Hyperattenuated mesenteric fat; (iii) Mesenteric small (<10 mm) lymph nodes; (iv) Fat halo sign; (v) Pseudocapsule	Both terms used interchangeably and are not clearly defined. Often describe more aggressive disease with signs of obstruction or encasement of bowel or vessels	
Clinical	All 4 terms have been used interchangeably throughout the literature			

From Danford CJ, Lin SC, Wolf JL. Sclerosing Mesenteritis. Am J Gastroenterol. 2019 Jun;114 (6):867–873. Copyright © Wolters Kluwer.

Retroperitoneal Fibrosis, Chronic Sclerosing Aortitis and Periaortitis

Retroperitoneal fibrosis (RF), chronic sclerosing aortitis and periaortitis are other forms of IgG4D. It has been suggested that previously described "idiopathic" RF is actually an IgG4-mediated disease. RF typically affects the infrarenal portion of the abdominal aorta with the simultaneous involvement of the iliac arteries. Progressive fibro-inflammatory changes in the distal aorta lead to obstructive uropathy. On imaging, IgG4-related RF is seen as a soft-tissue mass encasing the abdominal aorta and its branches. Hydronephrosis and hydroureter are not uncommon. As with AIP, RP is responsive to glucocorticoid therapy. IgG4-related periaortitis, may lead to aneurysmal formation and possible rupture.

Diagnosis of IgG4-RD

The diagnostic process of IgG4-RD begins by clinical recognition of the symptoms and physical exam findings suggestive of the disease. Alternatively, these conditions are suggested by incidental findings found on imaging studies. Biopsy findings are the cornerstone of diagnosing IgG4-RD. These findings include lymphoplasmacytic tissue infiltration of IgG4-positive plasma cells and lymphocytes in association with storiform fibrosis. These findings are often accompanied by obliterative phlebitis and tissue eosinophilia. IgG4 lymphoplasmacytic infiltrates are not specific to IgG4-RD. They can be found in malignancies and autoimmune disorders. It is important to rule out those disorders first as they are much more common than IgG4-RD. After the diagnosis of IgG4-RD has been established, the extent of the disease should be determined. Imaging of the chest, abdomen and pelvis via CT or MRI is highly effective in achieving that purpose. Characteristic findings include diffuse and focal organ infiltration and encasement by inflammatory and fibrotic tissue.

Treatment

The location, severity and duration of patient symptomology and clinical findings determine the treatment regimen for IgG4-RD. Gastrointestinal disease (e.g. mesentery, biliary tree, and pancreatic) often presents acutely and requires urgent treatment. Treatment principles focus on reducing inflammation and host immune response. The initial treatment focuses on glucocorticoid therapy, often administered at high doses followed by a slow taper. However, some patients may not be able to tolerate corticosteroids or may have ongoing symptoms and disease activity despite corticosteroids. In these patients, other immunosuppressive agents are administered and these treatments may be used in combination with glucocorticoids. Rituximab monotherapy is has shown to eliminate relapses within a six-month follow up period in over 75% of patients with a variety of IgG4-RD. Other immunosuppressive agents that are commonly used (sometimes in combination with glucocorticoids) include mycophenolate, azathioprine, 6-mercaptopurine methotrexate, and cyclophosphamide. If a patient enters disease remission, maintenance therapy is considered. The decision for use of maintenance therapy focuses on prior history of relapsing disease and whether the patient's initial presentation resulted in severe organ damage. In Japan, AIP maintenance therapy consists of low-dose glucocorticoids. If relapses occur, other immunosuppressive agents are added to the treatment regimen.

Suggested Reading

1. Crescioli S, Correa I, Karagiannis P, Davies AM, Sutton BJ, Nestle FO, Karagiannis SN. IgG4 Characteristics and Functions in Cancer Immunity. Current Allergy Asthma Rep. 2016;16(1):7.
2. Van der Neut KM, Schuurman J, Losen M, et al. Anti-inflammatory activity of human IgG4 antibodies by dynamic Fab arm exchange. Science. 2007;317(5844):1554–7.
3. Stone JH, Zen Y, Deshpande V. IgG4-related disease. N Engl J Med. 2012;366(6):539–51.
4. Janeway CA Jr, Travers P, Walport M, et al. Immunobiology: The Immune System in Health and Disease. 5th ed. New York: Garland Science; 2001.
5. Hamano H, Kawa S, Horiuchi A, Unno H, Furuya N, Akamatsu T, Fukushima M, Nikaido T, Nakayama K, Usuda N, Kiyosawa K. High serum IgG4 concentrations in patients with sclerosing pancreatitis. N Engl J Med. 2001;344(10):732–738.
6. Nishimori I, Tamakoshi A, Otsuki M, Research committee on intractable diseases of the pancreas, Ministry of Health, Labour, and Welfare of Japan. Prevalence of autoimmune pancreatitis in Japan from a nationwide survey in 2002. J Gastroenterol. 2007;42(18):6.
7. Zen Y, Nakanuma Y. Pathogenesis of IgG4-related disease. Curr Opin Rheumatol. 2011;23 (12):114–118.
8. Stone JH. IgG4-Related Disease: Harrison's Principles of Internal Medicine, 19e. 2017 November; Part 14: Section 391e.
9. Khosroshahi A, et al. International Consensus Guidance Statement on the Management and Treatment of IgG4-Related Disease. Arthritis Rheum. 2015;67:1688–99.
10. Carruthers MN, et al. Rituximab for IgG4-related disease: a prospective, open-label trial. Ann Rheum Dis. 2015;74:1171–7.
11. Kubo K, Yamamoto K. IgG4-related disease. Int J Rheum Dis. 2016;19:747–62. https://doi.org/10.1111/1756-185X.12586.
12. Yadlapati S, Verheyen E, Efthimiou P. IgG4-related disease: a complex under-diagnosed clinical entity. Rheumatol Int. 2018;38(2):169–77. https://doi.org/10.1007/s00296-017-3765-7.
13. Anxo Martínez-de-Alegría mailto:anxomartinezdealegria@gmail.com, Sandra Baleato-González, Roberto García-Figueiras, Anaberta Bermúdez-Naveira, Ihab Abdulkader-Nallib, José A. Díaz-Peromingo, Carmen Villalba-Martín: IgG4-related Disease from Head to Toe. RadioGraphics 2015;35:2007–2025.
14. Danford CJ, Lin SC, Wolf JL. Sclerosing mesenteritis. Am J Gastroenterol. 2019;114(6):867–73.

Role of the Mesentery in Systemic Inflammation Response Syndrome (SIRS) and Multiple Organ Dysfunction Syndrome (MODS)

18

Cindy G. Pulido and Eli D. Ehrenpreis

Introduction

The Systemic Inflammatory Response Syndrome (SIRS) is a nonspecific defensive reaction that is stimulated by the presence of exogenous or endogenous insults. The SIRS response is proposed to occur for the purpose of localizing and eliminating malicious sources of injury, which in turn is designed to prevent further harm to the organism. SIRS is initiated by a variety of pathologic states including ischemia, trauma, infection, or a combination of these. The SIRS involves an intricate and more complex reaction than expected to occur from acute phase responses and is characterized by complex disturbances of homeostasis with many potential consequences. An overactive SIRS response results in organ system compromise, producing SIRS shock and SIRS related Multiple Organ Dysfunction Syndrome (MODS).

The extent of the SIRS response is determined by the balance between pro-inflammatory and anti-inflammatory processes. The Compensatory Anti-inflammatory Response Syndrome (CARS) is an immunologic phenomenon that halts the inflammatory processes in order to restore homeostasis. The stimuli that induce CARS are the same as those that influence SIRS, and both occur as simultaneous processes.

Reversal of inflammation by the CARS response occurs by several means. These include apoptotic reduction of lymphocytes and a decrease in the number of human leukocyte antigen (HLA) receptors on monocytes. The stimulation of anti-inflammatory cytokines production, (such as IL-10 and IL-4) that suppresses the expression of TNF-α, IL-1, IL-6, and IL-8 are also involved in restraining the development of a magnified protective response.

C. G. Pulido · E. D. Ehrenpreis (✉)
Department of Medicine, Advocate Lutheran General Hospital, 1775 Dempster St, 6 South, Park Ridge, IL 60068, USA
e-mail: e2bioconsultants@gmail.com

E. D. Ehrenpreis et al. (eds.), *The Mesenteric Organ in Health and Disease*,
https://doi.org/10.1007/978-3-030-71963-0_18

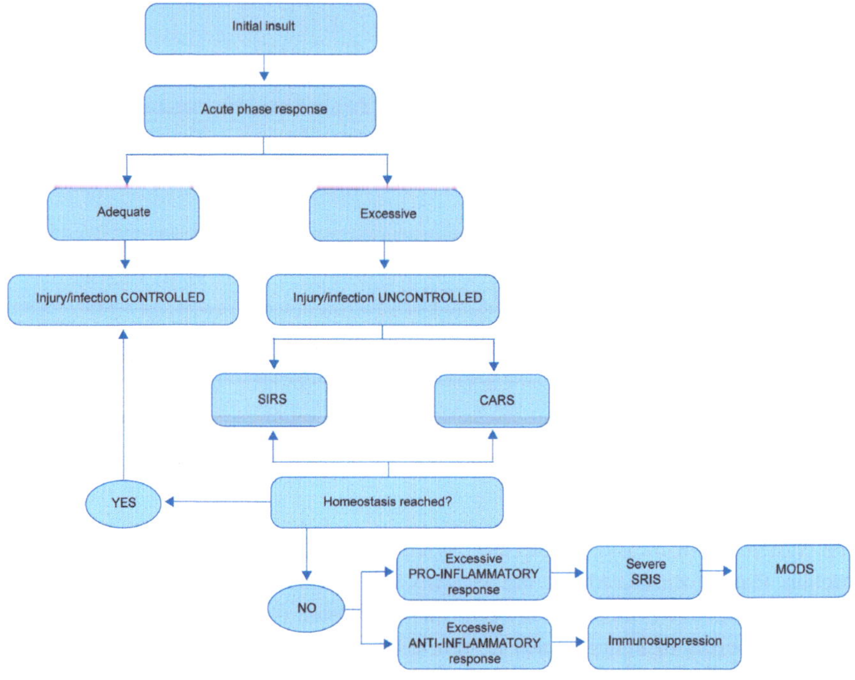

Fig. 18.1 SRIS and SRIS-related multiple organ dysfunction syndrome (MODS)

Pro-inflammatory reactions rise to eliminate bodily insults. However, sometimes, the mechanisms leading to the development of the protective response often cause harm to the host. CARS response and the anti-inflammatory reactions attempt to restrain the damage while not delaying the elimination of the offensive source. The consequences of SIRS and CARS can be damaging to the host due to lack of effect restriction and exaggerated or poorly timed events (Fig. 18.1).

Pathophysiology

The inflammatory sequence involved in the development of SRIS is a multifactorial process that integrates humoral and cellular responses. These include the participation of the complement cascade, and de novo generation and unrestricted proliferation of pro-inflammatory and anti-inflammatory cytokines. SIRS, independent of its etiology, has similar pathophysiologic manifestations. Bone et al. have described a three-stage process leading to the development of SIRS (Table 18.1).

Table 18.1 Summary of the Relationship Between Inflammatory Interactions and the Development of the Systemic Inflammatory Response Syndrome (SIRS)

Stages of SIRS development	
STAGE 1	**In response to injury/infection, the acute phase reaction leads to**: • Recruitment of inflammatory cells (macrophages, platelets) • Acute phase mediator participation (TNFα, IL-1, IL-6, IL-8, CRP) • Control of inflammatory response by endogenous anti-inflammatory products (IL-10, PGE2, antibodies, cytokines receptor antagonists)
STAGE 2	**Local cytokines are released into the circulation leading to enlarged recruitment of cellular protective species**: • Stimulation of coagulation cascade • Stimulation of complement cascade (C3a, C5a) • Stimulation of nitric oxide production • Vasodilation and increased vascular permeability
STAGE 3	**Failure to control inflammation due to inability of homeostasis restoration and persistent seeding of the inflammatory stimuli into the systemic circulation leads to destruction rather than protection**: • Loss of capillary integrity (prostaglandins, leukotrienes) • Maldistribution of microvascular blood flow • Organ injury and organ dysfunction

Reference: Bone RC. Toward a theory regarding the pathogenesis of the systemic inflammatory response syndrome: what we do and do not know about cytokine regulation. Crit Care Med 1996; 24(1):163–72

MODS Pathophysiology

MODS is an unintended consequence of SIRS. The most common organs affected by MODS are the central nervous system, heart, lungs, liver and kidneys. Processes responsible for the development of SIRS-related MODS are shown in Fig. 18.2.

Mesenteric Lymph, SIRS and MODS

The mesenteric lymph is proposed to be an essential component in the development of inflammatory conditions such as SIRS and MODS. Lymph flow from the intestine occurs via a series of lymphatic vessels and nodes. Lymph coming from the hindgut organs drains into the inferior mesenteric lymph nodes. Lymph from the midgut organs and the inferior mesenteric nodes drains into the superior mesenteric lymph nodes, and then empties into the pre-aortic lymph nodes.

The celiac and mesenteric lymph nodes (superior and inferior) occur around the origins of their corresponding arteries. These structures meet each other to form the intestinal lymph trunk, which enters the cisterna chyli, and in turn, drains into the thoracic duct. The thoracic duct transports lymph and chyle from the abdomen into the left brachiocephalic vein.

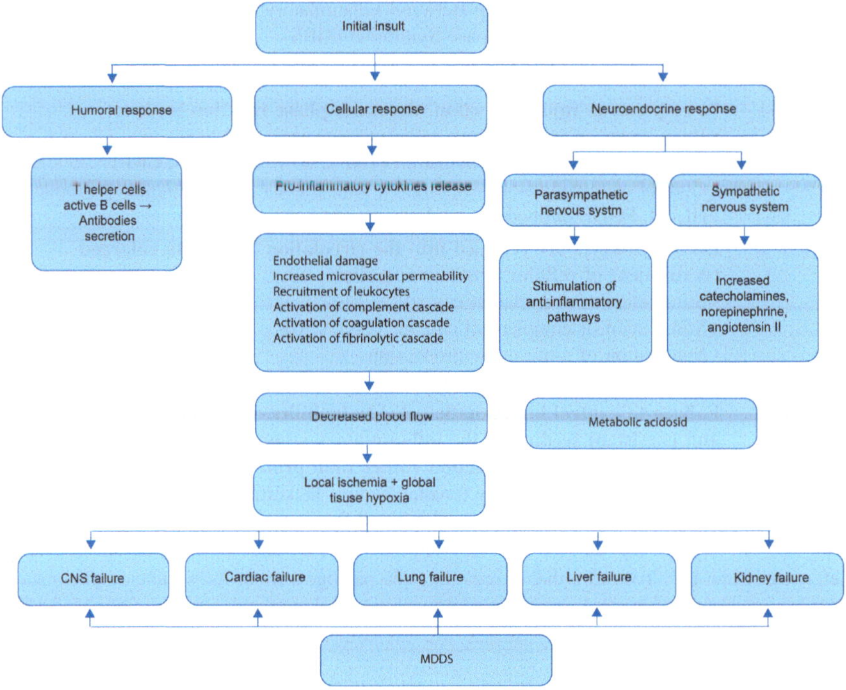

Fig. 18.2 The Mechanisms Associated with the Development of Systemic Inflammatory Response Syndrome (SRS)-Related Multiple Organ Dysfunction Syndrome (MODS)

Mesenteric lymph derives exclusively from the intestines with no contribution from other organs and can reach the systemic circulation through the aforementioned lymphatic pathways. Mesenteric lymph also bypasses the portal circulation and the liver. Intestinal lymph provides the opportunity for SRIS and MODS to develop from failure of the intestinal barrier, translocation of intestinal bacteria, portal bacteremia and the presence of endotoxemia (see Fig. 18.3).

Interventions Directed at Mesenteric Lymph

In vitro studies of exposure to shock-derived mesenteric lymph, have demonstrated that cellular abnormalities (e.g. neutrophil activation, structural changes in the red blood cell and endothelial dysfunction) can be associated with a systemic inflammatory response. Moreover, in vivo studies have shown that exposure to shock-derived mesenteric lymph can lead to acute lung injury and cardiac dysfunction.

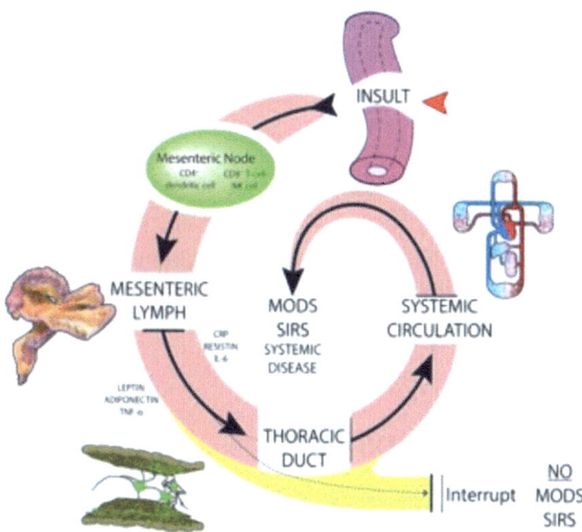

Fig. 18.3 Mesenteric lymph and the Development of SIRS and MODS. In SIRS and MODS, pro-inflammatory factors and toxic intestinal factors reach the systemic circulation via mesenteric lymph while bypassing the portal circulation and liver. Once in the system circulation, they influence other splanchnic organs and induce a systemic response. This description of the process implicates the mesenteric lymph, rather than portal vein blood, s the exit route of gut-derived nonbacterial inflammatory markers. Reused with permission from Rivera ED, Coffey JC, Walsh D, Ehrenpreis ED. The mesentery, Systemic Inflammation, and Crohn's Disease. Inflamm Bowel Dis 2019;25(2)226–234. Copyright © 2019 Oxford University Press

Targeting the mesenteric lymph has been the focus of several laboratory investigations of treatments for SIRS and MODS. Methods to prevent mesenteric lymph from entering the systemic circulation have included mesenteric duct ligation and vagus nerve stimulation.

Mesenteric Duct Ligation and the Inflammatory Response

Mesenteric duct ligation may have a preventative effect on the development of SRIS and subsequent distant organ failure. A study on rats subjected to hemorrhagic shock was performed to examine the effects of mesenteric lymph duct ligation on the renal tissue. Decreased disturbance of the microcirculation and the oxygen consumption in the kidneys was seen following mesenteric ductal ligation. This, in turn, resulted in an improvement in the derangement of energy metabolism, one of the most important mechanisms for injury of the cell during shock. Mesenteric duct ligation reduced hemorrhage-induced renal ischemia and limited the reduction of renal ATP that occurred in rats without mesenteric duct ligation.

Despite favorable experimental results, application of this technique in the clinical setting faces a number of limitations. Among these include anatomic challenges in the performance of mesenteric ductal ligation in humans and lymphovenous communications between mesenteric lymph vessels and the thoracic duct that would allow mesenteric lymph to bypass the ligation.

Vagal Nerve Stimulation (VNS)

Animal studies have also demonstrated that augmentation of the cholinergic anti-inflammatory effects achieved by vagal nerve stimulation results in the limitation of the inflammatory response following injury.

Adipocytes, SIRS and MODS

Multiple lines of evidence have now demonstrated that adipose tissue is a metabolically active structure. Adipose tissue is the source of several hormones and cytokines that are an integral part of numerous physiological processes, as well as several disease states. Adipocytes are cellular elements now considered to contain dynamic endocrine functions in their own right, with the ability to secrete several biologically active proteins. These proteins, known as adipokines, exert their functions through endocrine, autocrine and/or paracrine mechanisms. Adipokines are also produced by the stromal vascular fraction within the adipose tissue. This is comprised of preadipocytes, endothelial cells, fibroblasts, leukocytes and macrophages.

Adipocytes also produce mediators of immunologic function such as adiponectin, leptin, angiotensinogen and retinol-binding protein. Some adipokines, including monocyte chemoattractant protein-1 and apelin, are produced by both fractions, the adipocytes and the stromal elements (preadipocytes, endothelial cells, fibroblasts, leukocytes and macrophages) (see Table 18.2).

Adiponectin, SIRS and MODS

The adipocyte-specific protein adiponectin circulates in high concentrations in human plasma. It has anti-inflammatory, antioxidant, antiatherosclerotic and vasoprotective effects and might be involved in the pathogenesis of the systemic inflammatory response during sepsis. Adiponectin is the prototype of an anti-inflammatory adipokine. Adiponectin is produced almost exclusively by adipocytes. It has antioxidant, anti-atherosclerotic, vasoprotective and insulin-sensitizing effects. On the other hand, it also has anti-inflammatory effects which might also be involved in the pathogenesis of SIRS and SIRS related Multiple Organ Dysfunction Syndrome (MODS).

Table 18.2 Function of Adipokines

Adipokines	Functions
Adiponectin	• Decreases oxidative stress in endothelial cells, therefore exerts an anti-atherogenic effect • Promotes insulin sensitivity by decreasing gluconeogenesis in the liver and pancreatic β-cells apoptosis • Inhibits lipolysis leading to increased triglyceride levels • Exerts anti-inflammatory activity by inhibiting growth of myelomonocytic progenitors, functioning of mature macrophages and NK lymphocytes secretion of inflammatory cytokines (IL-6 and CRP), stimulating secretion of anti-inflammatory cytokines (IL-10)
Leptin	• Decreases appetite • Increases energy stores • Decreases thyroid hormones and female reproductive hormones • Improves insulin sensitivity • Promotes hematopoiesis and lymphopoiesis by stimulating proliferation of stem cells in bone marrow • Exerts pro-inflammatory activity by stimulating secretion of pro-inflammatory cytokines (IL-12, IL-18, IFN-γ, TNF-α, IL-1, IL-6) and production of free radicals and reactive oxygen species by PMN leukocytes, having anti-apoptotic effects on mature T cells; inhibiting anti-inflammatory cytokines secretion (IL-10, IL-4)
Angiotensinogen	• Promotes systemic vasoconstriction • Promotes the release of anti-diuretic hormone from the pituitary gland • Stimulates aldosterone production, resulting in excretion of potassium and retention of sodium and water • Promotes proinflammatory action and immune cell infiltration
Retinol—Binding Protein (RBP)	• Promotes pro-inflammatory action by inducing the secretion of TNF-α, IL-6, MCP-1, IFN-γ, IL-1β, IL-2, IL-12, IL-10, and IL-8 in macrophages, as well as ICAM-1, VCAM-1, E-selectin, IL-6, and MCP-1 in endothelial cells • Stimulates cellular oxidative stress
Monocyte Chemoattractant Protein—1 (MCP-1)	• Promotes recruitment of monocytes/macrophages and lymphocytes into foci of active inflammation
Apelin (APLN)	• Antithrombotic effect by inhibiting thrombin and collagen mediated platelet activation • Vasodilator properties that leads to lowering blood pressure • Insulin resistance • Inhibits the release of vasopressin, which has a diuretic effect • Inhibits the production and release of reactive oxygen species in adipocytes and therefore decreases cellular oxidative stress

Adiponectin's anti-inflammatory effects are principally mediated by the negative feedback of TNF-α, Interleukin-6 (IL-6) and C-reactive protein (CRP), all well-known pro-inflammatory mediators. In the presence of endotoxemia, down-regulation of adiponectin receptor mRNA results in the inhibition of several processes affecting endothelial activation, leukocyte infiltration and chemotaxis. Moreover, several studies have also reported a negative correlation between adiponectin and Sequential Organ Failure Assessment (SOFA) score. Reduced levels of adiponectin in critically ill patients upon admission to ICU, and higher levels of adiponectin during recovery have aslo been described.

In a prospective study performed in the intensive care unit (ICU) at Children Clinical University Hospital, Riga, Latvia, Tretjakovs et al. compared children having SIRS, with and without sepsis, with healthy controls. Their data demonstrated that serum adiponectin levels are decreased in patients that developed SIRS compared to those that did not. Adiponectin levels were also significantly reduced in patients suffering from life-threatening conditions and at higher risk of mortality. The authors suggested that these data provided evidence that supports the use of plasma adiponectin levels as a clinical biomarker in patients with SIRS.

In addition, when marked increases in pro-inflammatory substances occur, levels of adiponectin are diminished. Some authors have suggested that the development of therapies that stabilize or increase adiponectin levels may be useful in severe inflammatory conditions such as SIRS and MODS.

Leptin, SIRS and MODS

Leptin is a cytokine secreted mainly by adipocytes. Measurement levels of leptin have been shown to correlate with total amounts of body fat. Circulating leptin levels serve as a measure for energy reserves and are postulated to direct the central nervous system to adjust appetite and energy expenditure accordingly. Leptin is also involved in neuroendocrine, metabolic, and immune functions.

Leptin acts as a pro-inflammatory mediator. Its secretion is stimulated by lipopolysaccharides and acute secretion of pro-inflammatory mediators such as TNF-α and IL-1. Leptin has a structural similarity to various cytokines. Moreover, leptin specific receptors belong to the class I cytokine receptors on immune cells such as monocytes, lymphocytes and neutrophils. This location of leptin-specific receptors supports the idea that leptin exerts immune-modulating activity. Circulating leptin levels may have a role as a clinical marker for the diagnosis and prognosis of SIRS. In a prospective study performed in the intensive care unit (ICU) at St. Spiridon University Emergency Hospital, Iaæi, Romania, Grigoras et al. compared patients with postoperative SIRS after major elective abdominal surgery with healthy controls. Their data demonstrated that serum leptin levels were increased in patients developing SIRS. Leptin and other markers measured (IL-6 and CRP) had similar dynamics with different times that peak levels were present. Peak IL-6 levels occurred at 6 h, peak leptin levels at 12 h and peak CRP at 24 h after surgery. The authors suggested that serum leptin may serve a useful marker for the diagnosis and prognosis of critical illness in surgical patients.

CRP, Resistin and SIRS/MODS

CRP and resistin are plasma proteins whose circulating concentrations rise in response to inflammatory processes. Besides being involved in the development of inflammatory reactions, CRP is also involved in regulating glycemic and lipid metabolism, while resistin is associated with the establishment of insulin resistance. In the past, the liver was considered to be a unique source of CRP. Recent evidence suggests that adipocytes are an extrahepatic source of CRPproduction in humans. Resistin secretion takes place in several inflammatory cells including monocytes, macrophages and is also produced by adipocytes and preadipocytes. Due to their capability to yield pro-inflammatory products and therefore magnify acute phase responses, all of the aforementioned cells are considered fundamental in the development of SRIS and are potential cofactors in the development of MODS.

After the original initial insult, the development of SIRS involves the release of TNF-α and IL-1. These cytokines promote the cleavage of the NF-κB inhibitor. After NF-κB inhibitor removal, NF-κB can initiate the production of mRNA that will induce the production of other pro-inflammatory cytokines. Macrophages and T cells release IL-6. IL-6, in turn, stimulates adipocyte and hepatocyte secretion of acute phase reactants, such as C-reactive protein (CRP) and resistin.

Overall, CRP and resistin are implicated in the generation of pro-inflammatory cytokines. Furthermore, the balance established by inflammatory and the counter-regulatory mechanisms (CARS) helps to determine a patient's outcome after an insult. The likelihood of developing SIRS and MODS is also dependent upon balancing of these systems. Thus, these two mediators (CRP and resistin) are an integral component of SIRS and MODS.

Fibrocytes, SIRS and MODS

Fibrocytes also have a role in the development of systemic inflammation and autoimmune disorders. Human fibrocytes are hematopoietic-derived cells produced in the bone marrow that migrate to the other tissues. Fibrocytes exist as a unique population of circulating precursors of fibroblasts.

Fibrocytes may appear following a variety of injuries. They are first included in the acute inflammatory response and later in the remodeling process. Fibrocytes contribute to the establishment of the inflammatory reaction through the secretion of pro-inflammatory cytokines. These allow defensive mechanisms to mount an appropriate response to constraint the injury process. Moreover, fibrocytes contribute to the fibrotic response by promoting extracellular matrix production and connective tissue remodeling.

Fibrocytes contribution to the inflammatory process is facilitated by TNF-α and IL-1β (as well as endotoxin in some infectious cases). They secrete a variety of pro-inflammatory cytokines (including IL-6, IL-8, IL-10, and TNF-α). These are involved in the induction of leukocyte entry, survival, and retention during inflammation as well as the recruitment of other cells that promote the expansion of the inflammatory reaction. IL-6 production is associated with the translocation of the NF-κB transcription factor, that was previously described to be related to SIRS development. Fibrocytes also secrete products such as Connective Tissue Growth Factor (CTGF) and Transforming Growth Factor Beta 1 (TGF-β1). These promote tissue remodeling by stimulating extracellular matrix assembly by local fibroblasts. Pilling et al. have postulated that the positive feedback loop involving signals from fibroblasts back to fibrocytes may lead to the persistence of inflammatory and a maladaptive repair process. This mechanism may account for the supporting participation of fibrocytes in the development of SIRS and perhaps with the progression of SIRS to MODS.

Summary

Although inflammation is an essential host response, the onset and progression of infection, trauma, and ischemia can lead to both pro-inflammatory and anti-inflammatory reactions. A subsequent chain of events can result in the establishment of SIRS and, in some cases, to MODS. Mesenteric components, including the mesenteric lymph, adipocytes, fibrocytes and cytokines play a key role in regulating pro-inflammatory and anti-inflammatory reactions that occur during the development of SIRS and MODS. In the future, interventions directed at some of these mesenteric components may represent therapeutic options for SIRS and MODS.

Suggested Reading

1. Rivera ED, Coffey JC, Walsh D, Ehrenpreis ED. The mesentery, systemic inflammation, and Crohn's Disease. Inflamm Bowel Dis. 2019;25(2):226–34.
2. Balk RA. Systemic inflammatory response syndrome (SIRS): Where did it come from and is it still relevant today? Virulence. 2014;5(1):20–6.
3. Ward NS, Casserly B, Ayala A. The compensatory anti-inflammatory response syndrome (CARS) in critically ill patients. Clin Chest Med. 2008; 29(4): 617–viii.
4. García de Alba C, Buendia-Roldán I, Becerril C, Ramírez R, González Y, Checa M, Navarro C, Ruiz V, Pardo A, Selman M (Instituto Nacional de Enfermedades Respiratorias Ismael Cosío Villegas, Mexico City, Mexico). Fibrocytes contribute to inflammation and fibrosis in chronic hypersensitivity pneumonitis through paracrine effects. Am J Respir Crit Care Med. 2015; 19(4):427–436. https://doi.org/10.1164/rccm.201407-1334OC.
5. Pilling D, Vakil V, Cox N, Gomer RH. TNF-α–stimulated fibroblasts secrete lumican to promote fibrocyte differentiation. Proc Natl Acad Sci USA. 2015; 112(38): 11929–11934. https://doi.org/10.1073/pnas.1507387112.
6. Galligan C, Fish E. The role of circulating fibrocytes in inflammation and autoimmunity. J Leukoc Biol 2013; 93: 45–50. https://doi.org/10.1189/jlb.0712365.

7. Balmelli C, Alves MP, Steiner E, Zingg D, Peduto N, Ruggli N, Gerber H, McCullough K, Summerfield A (Institute of Virology and Immunoprophylaxis, Sensemattstrasse 293, CH-3147 Mittelhäusern, Switzerland). Responsiveness of fibrocytes to toll-like receptor danger signals. Immunobiology. 2007; 212(9–10): 693-699. https://doi.org/10.1016/j.imbio.2007.09.009.
8. Du Clos TW (The VA Medical Center and the University of New Mexico School of Medicine, Department of Medicine, Albuquerque, USA. tduclos@unm.edu). Function of C-reactive protein. Ann Med. 2000; 32(4):274–278.
9. Bone RC (Medical College of Ohio, Toledo, USA). Toward a theory regarding the pathogenesis of the systemic inflammatory response syndrome: what we do and do not know about cytokine regulation. Crit Care Med. 1996; 24(1):163–72. https://insights.ovid.com/pubmed?pmid=8565523.
10. Deshmane SL, Kremlev S, Amini S, Sawaya BE. Monocyte chemoattractant protein-1 (MCP-1): an overview. J Interferon Cytokine Res. 2009; 29(6): 313–326. https://doi.org/10.1089/jir.2008.0027.
11. Livingston M. The pathophysiology of multiple organ dysfunction syndrome. https://doi.org/10.13140/rg.2.2.19242.26569. Accessed 1 July 2019.
12. Bone RC. Immunologic dissonance: a continuing evolution in our understanding of the systemic inflammatory response syndrome (SIRS) and the multiple organ dysfunction syndrome (MODS). Ann Intern Med. 1996; 125(8):680–687. https://doi.org/10.7326/0003-4819-125-8-199610150-00009.
13. Fanous MY, Phillips AJ, Windsor JA. Mesenteric lymph: the bridge to future management of critical illness. JOP. 2007; 8(4):374–399. http://pancreas.imedpub.com/mesenteric-lymph-the-bridge-to-future-management-of-critical-illness.pdf.
14. Tretjakovs P, Rautiainena L, Krievina G, Jurka A, Grope I, Gardovska D. Changes in serum high-molecular-weight adiponectin levels in critically ill children with systemic inflammatory response syndrome. Open Med J. 2016; 3:166–170. https://doi.org/10.2174/1874220301603010166.
15. Grigoras I, Branisteanu DD, Ungureanu D, Rusu D, Ristescu I. Early dynamics of leptin plasma level in surgical critically ill patients: a prospective comparative study. Chirurgia (Bucur). 2014; 109 (1):66–72. http://revistachirurgia.ro/pdfs/2014-1-66.pdf.
16. Zhang LM, Jiang LJ, Zhao ZG, Niu CY. Mesenteric lymph duct ligation after hemorrhagic shock enhances the ATP level and ATPase activity in rat kidneys. Ren Fail. 2014; 36(4):593–597. https://doi.org/10.3109/0886022x.2014.882183.
17. Kaplan LJ, Pinsky MR. Medscape. New York (NY): WebMD; 2018. https://emedicine.medscape.com/article/168943-overview.
18. Ruan H, Dong LQ. Adiponectin signaling and function in insulin target tissues. J Mol Cell Biol. 2016; 8(2):101–109. https://doi.org/10.1093/jmcb/mjw014.
19. Paz-Filho G, Mastronardi C, Bertoldi Franco C, Boyang Wang K,Wong ML, Licinio J. Leptin: molecular mechanisms, systemic pro-inflammatory effects, and clinical implications. Arq Bras Endocrinol Metabol. 2012; 56(9). http://dx.doi.org/10.1590/S0004-27302012000900001.
20. Rudemiller NP, Crowley S. Interaction between the immune and renin angiotensin system in hypertension. Hypertension. 2016;68(2):289–296. https://doi.org/10.1161/hypertensionaha.116.06591.
21. Zabetian-Targhi F, Mahmoudi MJ, RezaeiN, Mahmoudi M. Retinol binding protein 4 in relation to diet, inflammation, immunity, and Cardiovascular Diseases. Adv Nutr. 2015; 6 (6):748–762. https://doi.org/10.3945/an.115.008292.
22. Livingston M. The pathophysiology of multiple organ dysfunction syndrome [dissertation on the Internet]. Leicester (England): The University of Birmingham; 2009; 88p. https://doi.org/10.13140/rg.2.2.19242.26569.

Part V
Medical Disorders of the Mesentery

19 Mesenteric Hemorrhage

Ryan T. Hoff and Eli D. Ehrenpreis

Definition

Mesenteric hemorrhage is defined as bleeding occurring from any of the blood vessels supplying the mesentery, including those also supplying the stomach, small intestine, pancreas, and spleen.

Epidemiology and Risk Factors

Mesenteric hemorrhage may be related to trauma, infection, malignancy, or vascular malformations, including rupture of splanchnic aneurysms. Mesenteric hemorrhage may be mild and limited to contusions or small hematomas, or severe with hemodynamic instability and mesenteric devascularization.

Injuries to mesenteric vessels may occur during surgical procedures, including during direct trocar entry prior to laparoscopy. CT guided biopsy of intra-abdominal lesions, including pancreatic tumors and cysts, can lead to mesenteric hemorrhage. Rupture of a mesenteric hematoma may occur following transcatheter aortic valve replacement (TAVR), which can be life threatening if not identified early. Additional rare causes of mesenteric hemorrhage include labor, electroconvulsive shock therapy, and endoscopic ultrasound guided fine needle aspiration of pancreatic lesions. Mesenteric hemorrhage may occur immediately as an intraoperative complication, or in a delayed fashion postoperatively.

R. T. Hoff · E. D. Ehrenpreis (✉)
Department of Medicine, Advocate Lutheran General Hospital, 1775 Dempster St, 6 South, Park Ridge, IL 60068, USA
e-mail: e2bioconsultants@gmail.com

E. D. Ehrenpreis et al. (eds.), *The Mesenteric Organ in Health and Disease*,
https://doi.org/10.1007/978-3-030-71963-0_19

Rupture of aneurysms and pseudoaneurysms may cause mesenteric hemorrhage. Aneurysms of the superior mesenteric artery (SMA) are usually related to infection, whereas aneurysms of the branching arteries of the SMA are usually due to congenital or acquired medial defects. Compared with other visceral artery aneurysms, aneurysms of the SMA have a high rate of rupture. Additional causes of splanchnic aneurysms include infection, atherosclerosis, trauma, pancreatic or biliary tract disease. Rupture of these aneurysms may occur spontaneously or secondary to trauma.

In the absence of trauma or recent surgery, mesenteric hemorrhage may occur spontaneously secondary to an incarcerated inguinal hernia, anticoagulant therapy, or connective tissue disorders such as Ehlers-Danlos syndrome. Rarely, a mesenteric hematoma may form spontaneously as a complication of a Crohn's disease exacerbation. Acute myelogenous leukemia may present with massive mesenteric hemorrhage. Amyloidosis may involve the mesentery, leading to hemorrhage and hematoma formation.

Mesenteric hemorrhage may occur spontaneously, leading to abdominal apoplexy (also known as spontaneous hemoperitoneum), though this is very rare. Historically, abdominal apoplexy has been associated with a high rate of mortality, reported up to 40 percent. However, modern management and surgery have improved outcomes (Table 19.1).

Table 19.1 Causes of Mesenteric Hemorrhage

Iatrogenic	Anticoagulation therapy
	CT guided biopsy
	Electroconvulsive shock therapy
	EUS guided biopsy
	Surgery
	Transcatheter aortic valve replacement
Systemic disorders	Acute myelogenous leukemia
	Amyloidosis
	Ehlers-Danlos syndrome
	Crohn's disease exacerbation
Local factors	Abdominal apoplexy
	Incarcerated inguinal hernia
	Labor
	Rupture of aneurysm or pseudoaneurysm
	Trauma

Pathophysiology

Mesenteric hemorrhage occurs by several mechanisms, including direct injury to blood vessels, inflammation leading to aneurysm and subsequent weakening of vessel walls, and factors that inhibit normal coagulation of blood. Thrombocytopenia and coagulopathy (therapeutic or pathologic) may contribute. Mutations of the type III procollagen gene may increase the risk in spontaneous mesenteric hemorrhage.

Symptoms

Ruptured aneurysms present with abdominal pain, nausea, vomiting, and acute anemia. Rarely, mesenteric hematomas may rupture into the small intestine, leading to hematochezia. Large mesenteric hematomas may occlude segments of bowel, leading to bowel obstruction. Patients with underlying coagulopathy or collagen vascular disease may have bruising, or symptoms of overt bleeding.

Physical Findings

Findings on physical examination may include evidence of acute blood loss, including hypotension, tachycardia and pallor. Shock may develop if blood loss is severe. In the setting of trauma, physical exam findings may include extra-abdominal injuries, such as laceration, ecchymosis, etc. In the event of bowel injury and peritonitis associated with abdominal trauma, peritoneal findings, including rebound tenderness and guarding may be present. Laboratory studies may demonstrate an acute decrease in serum hemoglobin concentration or elevated lactic acid.

Imaging and Diagnosis

Computed tomography angiography (CTA) of the abdomen provides helpful information regarding the presence and anatomic extent of mesenteric hemorrhage and serves as the imaging test of choice. CTA uses high-resolution datasets with rapid acquisition times. These are performed at various time intervals after administration of intravascular contrast material. This results in images obtained as intravascular contrast is distributed throughout the circulation in the form of multiphasic studies. These in turn can be used to create three dimensional images that provide great details of the vasculature. CTA is useful for identifying active bleeding, which is imaged as extravasation of contrast material, while establishing

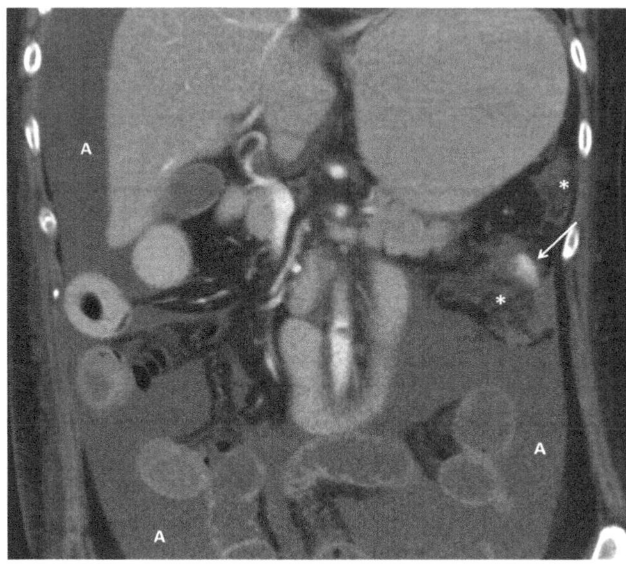

Fig. 19.1 Coronal image from a CT angiogram shows a left upper quadrant hematoma (asterisks) with a focus of contrast "blush" (arrow) indicative of active extravasation. There is also diffuse ascites (A) throughout the abdomen

the source of hemorrhage (See Figs. 19.1 and 19.2). Areas of acute mesenteric hemorrhage are often heterogeneous and hyperattenuated on CT imaging and repeat CTA scans can provide information on continued bleeding, thus guiding therapy. If administered, contrast extravasation into the hematoma indicates ongoing bleeding. The differential diagnosis of a possible mesenteric hemorrhage includes mesenteric lymphangioma, abscess, and aneurysm.

Active bleeding into a hematoma may also be established with Technetium 99 m labeled red blood cell scintigraphy. Acute hemorrhage often exhibits a CT density of about 35 to 45 hounsfield units. If clots develop, then the density may increase to 70 to 90 hounsfield units. As the hematoma matures, this hyperattenuation gradually decreases, due to breakdown and removal of proteins from red blood cells. After 2 to 4 weeks following the initial hemorrhage, mesenteric hematomas may be difficult to distinguish from other similar appearing fluid collections, including abscesses.

Angiography is an endovascular procedure that provides visualization of the mesenteric blood vessels and offers an opportunity for therapeutic intervention. Digital subtraction angiography (DSA) involves the real time digital processing of contrast enhanced fluoroscopic x-ray images, with removal of unwanted signals and isolation of desired images. DSA serves as the reference standard for evaluating the mesenteric vasculature and has demonstrated accuracy in post-pancreatectomy

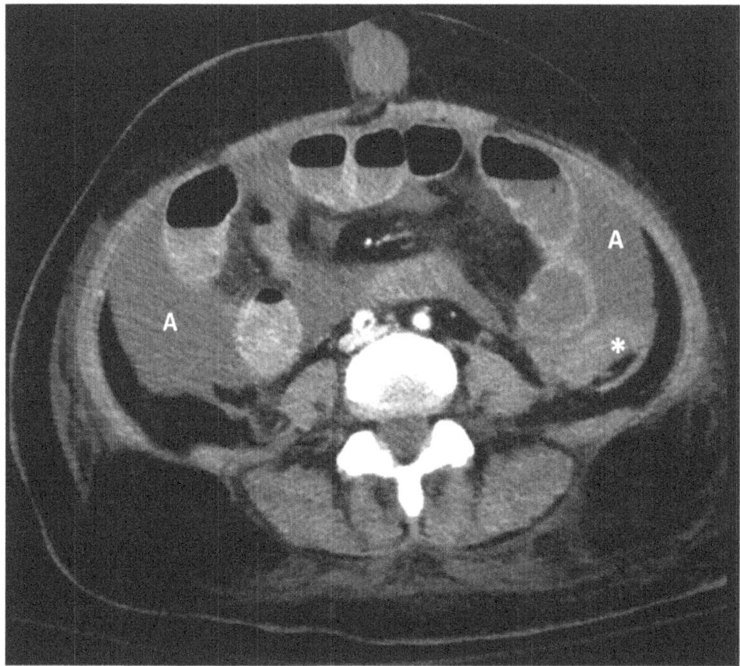

Fig. 19.2 Axial CT image again demonstrates the diffuse ascites (A) with layering hyperattenuating debris (asterisk) in the left paracolic gutter consistent with blood products. Images are courtesy of Dr. Abraham Dachman

bleeding. Active bleeding is demonstrated by the extravasation of contrast medium (Fig. 19.3).

CTA scan is sensitive for identifying mesenteric injuries, including hemorrhage. CTA is widely used due to its availability, high accuracy, and lack of procedure related complications. Oral contrast can assist in evaluating for concurrent underlying injury to the bowel wall, which is particularly helpful in the setting of trauma. However, CT imaging with only intravenous (IV) contrast (without oral contrast), is 76% sensitive for all gastrointestinal tract in injuries in trauma patients, and up to 91% sensitive for major injuries, including mesenteric hemorrhage. Oral contrast may not be necessary in all patients with blunt abdominal trauma but may be useful in select patients or as follow up to evaluate findings on initial CT imaging.

Blunt abdominal trauma can lead to mesenteric hemorrhage, which can be life threatening. Active mesenteric hemorrhage can be diagnosed with multiphasic CTA imaging, which shows extravasation of contrast material into the mesenteric during the arterial or portal phase. Alternatively, contrast extravasation into the mesentery that occurs during the equilibrium phase is usually well contained and amenable to conservative treatment. Bleeding that results in the formation of a hematoma and tamponade effect may not demonstrate extravasation of contrast and usually does not require surgical intervention, unless abdominal compartment syndrome or

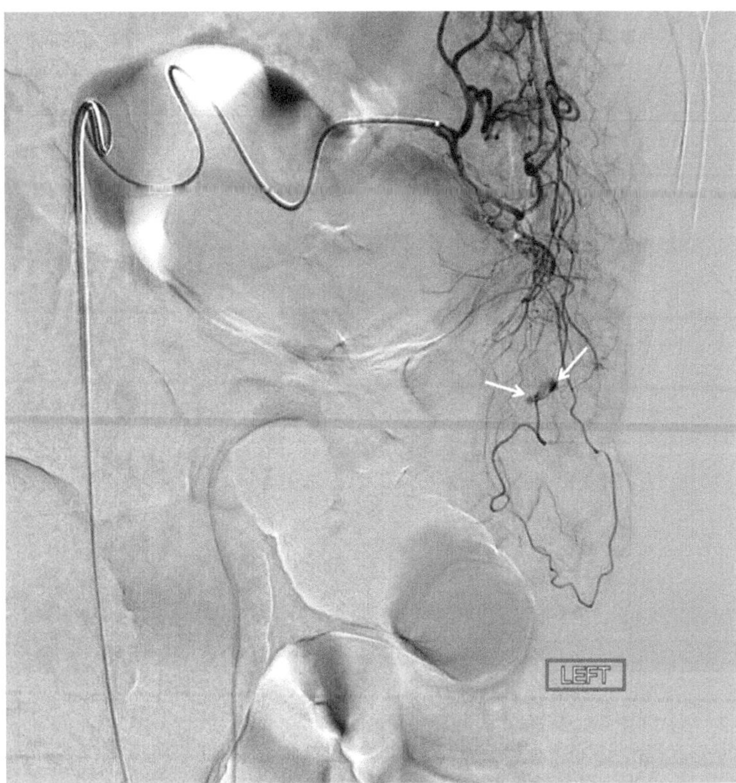

Fig. 19.3 Digital subtraction angiography following superselective catheterization of a distal splenic artery branch in the same patient shows irregular contrast blush (arrows) consistent with active hemorrhage. Images courtesy of Dr. Abraham Dachman and Dr. Justin Ramirez

bowel devascularization develops. Patients with significant extravasation of contrast material observed on CTA imaging often present with shock or hemodynamic instability. However, some patients with mesenteric hemorrhage may appear clinically stable despite active bleeding, and CT evidence of ongoing mesenteric hemorrhage can influence the decision to intervene surgically.

In about one third of cases of spontaneous mesenteric hemorrhage, no active bleeding or source of bleeding is found. In patients with blunt abdominal trauma and hemoperitoneum without a clear source of bleeding, mesenteric hemorrhage should be considered.

Management

For small mesenteric hemorrhages, conservative therapy with monitoring is appropriate. Often, small bleeds resolve spontaneously, with hemostasis achieved via intra-abdominal tamponade. Minor bleeding, particularly bleeding from small vessels and capillaries, may be controlled with vasopressin, which can limit blood loss via vasoconstriction.

For moderate or severe bleeding involving large vessels, transcatheter embolization is effective. This approach is particularly suited for hemodynamically stable patients who have evidence of ongoing bleeding. Transcatheter arterial embolization is an effective treatment for mesenteric hemorrhage and provides a less invasive alternative to surgery for achieving hemostasis. If transcatheter embolization fails to achieve hemostasis, the use of a hydrogel-coated self-expandable coil may be useful. A transcatheter approach is appropriate in cases with isolated hemorrhage, in the absence of additional injuries to the mesentery and bowel. Where clinical suspicion for bowel perforation is high, surgery is preferable. Since not all centers have the capability of transcatheter intervention, transfer to a tertiary care center may be necessary. Intestinal infarction may occur after embolization; however, this complication is rare. Severe bleeding with hemodynamic instability that does not respond to resuscitation efforts requires surgery.

Hemorrhage related to abdominal trauma is often associated with additional injuries, which require surgical intervention. The indicated surgical procedures depends on operative findings, but may include ligation of injured vessels, resection of injured bowel, end-to-end anastomosis, etc. Blunt trauma that primarily leads to active significant blood loss without other major visceral injuries, transcather embolization may be considered. While most cases of vascular injury and mesenteric hemorrhage from laparoscopy require laparotomy with open repair, hemodynamically stable patients may be treated laparoscopically.

Mesenteric hematomas that occur following EUS-FNA should be treated with broad spectrum antibiotics, as the nutrient rich collection of blood easily becomes infected by bacteria seeding through the needle tract.

Mesenteric hemorrhage secondary to Ehlers Danlos syndrome should be managed with the minimal surgical intervention possible to achieve hemostasis. Operative management in these patients is complicated by friability of vessels and surrounding tissue, and poor wound healing (Table 19.2).

Table 19.2 Management of mesenteric hemorrhage

Mild bleeding	Supportive care
	Serial imaging
	Antibiotics (if related to imaging guided biopsy)
Moderate to severe bleeding	Transcatheter embolization
	Surgical repair, ligation of vessels

Suggested Reading

1. Ertugrul I, et al. Comparison of direct trocar entry and veress needle entry in laparoscopic bariatric surgery: randomized controlled trial. J Laparoendosc Adv Surg Tech A. 2015;25 (11):875–9 Epub 2015 Sep 23.
2. Jacobson MT, et al. Laparoscopic control of a leaking inferior mesenteric vessel secondary to trocar injury. JSLS. 2002;6(4):389–91.
3. Hammers LW, et al. Computed tomographic (CT) guided percutaneous fine-needle aspiration biopsy: the Yale experience. Yale J Biol Med. 1986;59(4):425–34.
4. Abbasi D, et al. Diagnosis and management of rare case of mesenteric hematoma rupture after transcatheter aortic valve replacement (TAVR): a case report and review of the literature. Case Rep Vasc Med. 2018;2018:6273538. (eCollection 2018).
5. Barber MC. Intra-abdominal haemorrhage associated with labor. Br Med J. 1909;2 (2534):203–4.
6. Wallace JA. An unusual complication due to electroconvulsive therapy: Mesenteric hemorrhage. J Tenn Med Assoc. 1971;64(7):599.
7. Siddiqui A, et al. Acute mesenteric hemorrhage associated with EUS-guided fine needle aspiration. J Clin Gastroenterol. 2007;41(7):722–3.
8. Sharma G, et al. Ruptured and unruptured mycotic superior mesenteric artery aneurysms. Ann Vasc Surg. 2014;28(8):1931.e5–8 Epub 2014 Jul 10.
9. Hirano K, et al. Spontaneous mesenteric hematoma of the sigmoid colon associated with rivaroxaban: a case report. Int J Surg Case Rep. 2018;44:33–7 Epub 2018 Feb 10.
10. Ashrafian H, et al. Spontaneous mesenteric hematoma complicating an exacerbation of Crohn's disease: report of a case. BMC Surg. 2014;3(14):35.
11. Suzuki H, et al. Death resulting from a mesenteric hemorrhage due to acute myeloid leukemia: an autopsy case. Leg Med (Tokyo). 2014;16(6):373–5 Epub 2014 Jul 14.
12. Kim MS, et al. Amyloidosis of the mesentery and small intestine presenting as a mesenteric haematoma. Br J Radiol. 2008;81(961):e1–3.
13. Reilly EF, et al. Spontaneous colonic mesenteric hemorrhage: report of an unusual case of abdominal apoplexy. Dis Colon Rectum. 2005;48(7):1484–6.
14. Moskowitz R, Rundback J. Middle colic artery branch aneurysm presenting as spontaneous hemoperitoneum. Ann Vasc Surg. 2014;28(7):1797.e15–7 Epub 2014 May 21.
15. Shikata D, et al. Report of a case with a spontaneous mesenteric hematoma that ruptured into the small intestine. Int J Surg Case Rep. 2016;24:124–7 Epub 2016 May 24.
16. Artigas JM, Martí M, Soto JA, Esteban H, Pinilla I, Guillén E. Multidetector CT Angiography for acute gastrointestinal bleeding: technique and findings. Radiographics. 2013;33(5):1453–70.
17. van Dijk LJ, et al. Vascular imaging of the mesenteric vasculature. Best Pract Res Clin Gastroenterol. 2017;31(1):3–14. Epub 2017 Jan 5.
18. Fang Y, et al. Diagnosis and treatment efficacy of digital subtraction angiography and transcatheter arterial embolization in post-pancreatectomy hemorrhage: A single center retrospective cohort study. Int J Surg. 2018;51:223–8 Epub 2018 Feb 8.
19. Orzel JA, et al. Evaluation of traumatic mesenteric hemorrhage in a hemophiliac with Tc-99 m labeled red blood cell scintigraphy. J Trauma. 1986;26(11):1056–7.
20. Holmes JF, et al. Performance of helical computed tomography without oral contrast for the detection of gastrointestinal injuries. Ann Emerg Med. 2004;43(1):120–8.
21. Stuhlfaut JW, et al. Blunt abdominal trauma: performance of CT without oral contrast material. Radiology. 2004;233(3):689–94 Epub 2004 Oct 29.
22. Wu CH, et al. Contrast-enhanced multiphasic computed tomography for identifying life-threatening mesenteric hemorrhage and transmural bowel injuries. J Trauma. 2011;71 (3):543–8.
23. Sher R, et al. Computed tomography detection of active mesenteric hemorrhage following blunt abdominal trauma. J Trauma. 1996;40(3):469–71.

24. Parker SG, Thompson JN. Spontaneous mesenteric haematoma; diagnosis and management. BMJ Case Rep. 2012;2012. pii: bcr2012006624.
25. Gabata T, et al. Transcatheter embolization of traumatic mesenteric hemorrhage. J Vasc Interv Radiol. 1994;5(6):891–4.
26. Borazan E, Yılmaz L, Aytekin A, Gökaslan G, Kervancıoğlu S. A rare case of non-traumatic acute intraabdominal hemorrhage: ruptured superior mesenteric artery aneurysm. Acta Chir Belg. 2018;118(1):64–7 Epub 2017 Jul 6.
27. Hosaka A, et al. Spontaneous mesenteric hemorrhage associated with Ehlers-Danlos syndrome. J Gastrointest Surg. 2006;10(4):583–5.

Mesenteric Panniculitis

20

Eli D. Ehrenpreis

Definition and Nomenclature

Mesenteric panniculitis is an idiopathic, chronic inflammatory disease of the mesentery. Characteristic for mesenteric panniculitis is the occurrence of degeneration and necrosis of mesenteric fat with associated chronic inflammation. In some cases, fibrosis and scarring of fatty tissue within the mesentery occurs as a primary or secondary manifestation of chronic inflammation. The presence of fibrosis resulted in the name as "retractile mesenteritis" when mesenteric panniculitis was first described in the medical literature in 1924. Other names for the disease have included sclerosing mesenteritis, mesenteric lipodystrophy, lobular panniculitis, isolated (or mesenteric) Pfeifer-Weber-Christian Disease, mesenteric fibrosis, mesenteric sclerosis, liposclerotic mesenteritis, mesenteric lipogranuloma, xanthogranulomatous mesenteritis, and inflammatory pseudotumor. These names denote the predominant features of the disease process in the mesentery, but do not likely represent different conditions. For this reason, mesenteric panniculitis is the preferred term for the disease.

Epidemiology

The epidemiology of mesenteric panniculitis has not been fully defined. Our group reported that findings consistent with mesenteric panniculitis occurred in 359 (0.24%) from a total of 147,794 abdominal computed tomography (CT) examinations performed over a five year period in a large community based medical system. Of these, 100 patients (28%) had known malignancy or were later diagnosed with

E. D. Ehrenpreis (✉)
Department of Medicine, Advocate Lutheran General Hospital, 1775 Dempster St, 6 South, Park Ridge, IL 60068, USA
e-mail: e2bioconsultants@gmail.com

E. D. Ehrenpreis et al. (eds.), *The Mesenteric Organ in Health and Disease*,
https://doi.org/10.1007/978-3-030-71963-0_20

cancer. Male predominance of 2:1 has been reported in some studies. Mesenteric panniculitis most often appears during the sixth and seventh decade of life. A few cases of mesenteric panniculitis have been reported in children and adolescents.

Patients at Risk

Our experience suggests that patients with mesenteric panniculitis will frequently have a personal or family history of other autoimmune conditions. The most common of these conditions appear to be multiple sclerosis, psoriatic arthritis, and Crohn's disease.

Pathophysiology

Evidence is highly indicative that mesenteric panniculitis is an autoimmune disorder. Factors supporting this hypothesis include biopsy evidence of chronic inflammation, systemic symptoms that are characteristic of other autoimmune diseases and family histories of other autoimmune diseases. In addition, serum inflammatory markers such as the erythrocyte sedimentation rate (ESR) and C-reactive protein (CRP) are often elevated in patients with mesenteric panniculitis. Medications that alter immune responses appear to improve the symptoms of mesenteric panniculitis as well. Because mesenteric panniculitis is associated with abdominal surgery, trauma and some infectious illnesses, triggering factors are a major occurrence in the disease. Mesenteric panniculitis may follow acute inflammation, infection, or mesenteric ischemia. It is likely that these conditions represent an exposure that produces a secondary autoimmune reaction.

Mesenteric abnormalities consistent with mesenteric panniculitis are associated with neoplastic diseases. In our study, about 28% of patients with an abnormal CT of the mesentery suggestive of mesenteric panniculitis had either a known history of cancer or were newly diagnosed with cancer. The most common cancers with mesenteric panniculitis like abnormalities on CT scan are lymphomas. Solid tumors associated with mesenteric panniculitis include colon, renal and prostate cancers (Table 20.1). Several cases of carcinoid tumor presenting as a mesenteric mass resembling mesenteric panniculitis have been reported.

Although biopsies of abnormal regions of the mesentery may reveal the presence of invasive tumor cells, often, these biopsies show characteristic histologic features of mesenteric panniculitis without the presence of neoplasia (see Histology section below). Because of this, our group and others have hypothesized that mesenteric thickening and inflammation in cancer patients often represents a paraneoplastic syndrome. This is supported by studies that show that the mesenteric process rarely demonstrates uptake of fluorine-18 (F-18) fluorodeoxyglucose (FDG) on positron

Table 20.1 Cancer types in patients with computed tomography (CT) findings consistent with mesenteric panniculitis

Total patients	N = 100
Lymphomas	N = 36
Follicular	17
Diffuse large B cell	9
Small lymphocytic	3
Mantle cell	2
T cell	2
Other lymphomas (Burkitt, Hodgkin, Marginal zone B cell)	3
Other Neoplasms	N = 64
Prostate	7
Renal cell	6
Lung	6
Chronic lymphocytic leukemia	5
Bladder	5
Endometrial	5
Breast	4
Colon	3
Rectal	3
Other (including carcinoid tumor)	20

Adapted from: Ehrenpreis ED, Roginsky G, Gore RM. Clinical significance of mesenteric panniculitis-like abnormalities on abdominal computerized tomography in patients with malignant neoplasms. World J Gastroenterol. 2016;22(48):10,601–10,608

emitting tomographic (PET) scans. Mesenteric panniculitis-like lesions are generally stable in patients with cancer and may resolve following successful antineoplastic therapy.

It has also been suggested that mesenteric panniculitis belongs to a larger spectrum of diseases in which inflammation and fibrosis affect multiple organ systems of the body. Fibrosclerotic disorders that have been reported to occur with mesenteric panniculitis include retroperitoneal fibrosis, Sjögren's syndrome and sclerosing pancreatitis. Elevated IgG4 has been implicated in some fibrosing forms of mesenteric panniculitis. (See Chap. 17 IgG4 and Mesenteric Diseases).

Symptoms and Signs

The clinical presentation of mesenteric panniculitis is highly variable. Symptoms of mesenteric panniculitis fall into several categories. In some patients, abdominal pain is vague or in a location that cannot be explained purely by the presence of the mesenteric mass. The pain is generally located in the middle portion of the abdomen but can be present in other areas of the abdomen or pelvis as well.

Alternatively, abdominal pain may be due to the mass-like effect of mesenteric inflammation, and impingement or involvement of adjacent structures including the small intestine. Another group of symptoms occur in the presence of chronic inflammation and may include weight loss, fever, anorexia and fatigue. In other cases, affected individuals may develop complications such as small bowel obstruction or an acute abdomen. Small bowel obstruction may occur as a result of fibrosing disease or from a mass-like effect. A small subset of patients develops a rapidly progressive, fibrosing form of the disease with multiple areas of obstructive small intestine. This group of patients have a very poor prognosis.

A large percentage of patients with mesenteric panniculitis have the unexplained symptom of nausea. Other common symptoms include early satiety, and altered bowel habits (either constipation or diarrhea).

Diagnosis

Mesenteric panniculitis is usually diagnosed when a "misty mesentery" or a mesenteric mass is seen on a CT scan of the abdomen that is performed for the evaluation of abdominal pain. Alternatively, the CT scan is performed as a general workup in patients with known malignancy. The diagnosis of mesenteric panniculitis can be suspected and is included in the differential diagnosis in patients with abdominal pain and systemic symptoms such as fever, night sweats and weight loss.

Physical Findings

A thorough examination to rule out peripheral lymphadenopathy or other signs of lymphoma or solid neoplasms is necessary in all patients. Breast, gynecologic, rectal and prostate examinations and appropriate screenings are required. The abdominal examination is often nonspecific with ill-defined abdominal tenderness present. In some patients a tender mass may be palpated in the middle portion of the abdomen. Physical findings associated with intestinal obstruction may be present in a small number of patients. Our group reported chylous ascites with associated abdominal distension in a patient with mesenteric panniculitis, and this finding has been described in several additional cases.

Laboratory Testing

Some patients will demonstrate elevated laboratory markers of inflammation such as erythrocyte sedimentation rate (ESR) and C-reactive protein (CRP). Our group has demonstrated that decreases in these markers occur in patients with symptomatic improvement after treatment with immune modulating medications.

Anemia of chronic disease may occur in some patients. Tests that are indictive of underlying malignancies may include elevated prostate specific antigen (PSA), elevated serum tumor markers, iron deficiency anemia, etc.

Imaging

The initial diagnosis of mesenteric panniculitis is usually made by computerized tomographic (CT) scanning of the abdomen. Abdominal CT is also commonly used for patient follow up, although magnetic resonance imaging (MRI) offers another means for follow up imaging without repeated radiation exposure to patients. Characteristic findings on these imaging procedures in mesenteric thickening and calcification. Mild cases are often referred to as showing a "misty mesentery" (Fig. 20.1a and b).

A mesenteric mass originating from the mesenteric root that is loosely or more definitively organized is often seen (Figs. 20.2 and 20.3). The mass is characterized by fat enhancement and multiple soft tissue nodules. Enlargement and calcification of mesenteric and pelvic lymph nodes are often seen. Because mesenteric panniculitis is not an invasive disorder, blood vessels within the mesentery appear to be spared from the inflammatory mass. This is referred to as perivascular sparing or the "halo sign" or "fat-halo sign" and is highly characteristic of mesenteric panniculitis as opposed to neoplastic disease of the mesentery (Figs. 20.2 and 20.4).

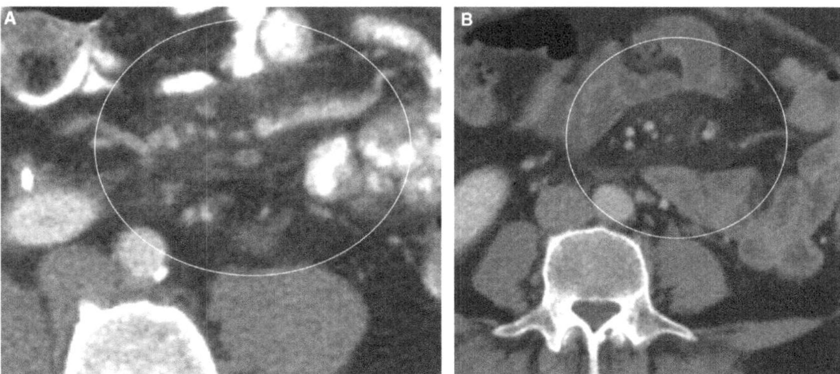

Fig. 20.1 **A** and **B** Hazy increased attenuation in the mesenteric fat (ovals) surrounding vessels in the mesenteric root gives the appearance of a "misty mesentery". Displacement of loops of the small intestine are seen due to a mass effect in B. Although characteristic of mesenteric panniculitis, these are nonspecific findings seen in a number of disease processes including inflammation, edema, hemorrhage or tumor infiltration

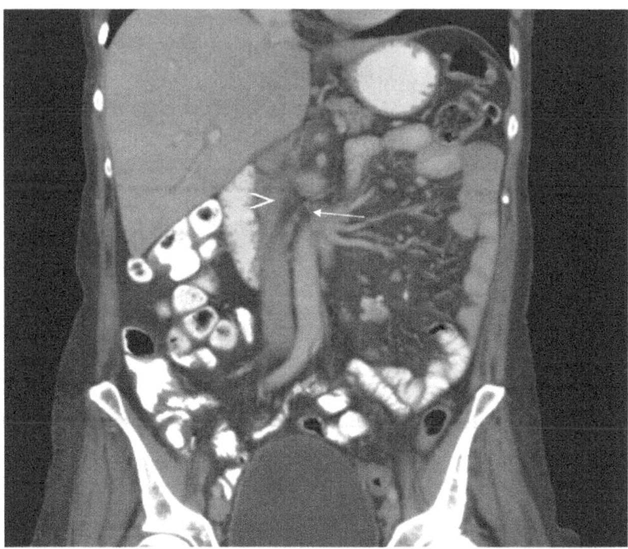

Fig. 20.2 Abdominal CT (coronal view) showing hazy increased attenuation throughout the mesentery with sparing around the mesenteric vessels. There are scattered prominent mesenteric lymph nodes and a developing soft tissue mass (Arrow). Sparing of vessels by the mass (halo sign) are also seen (Arrowhead)

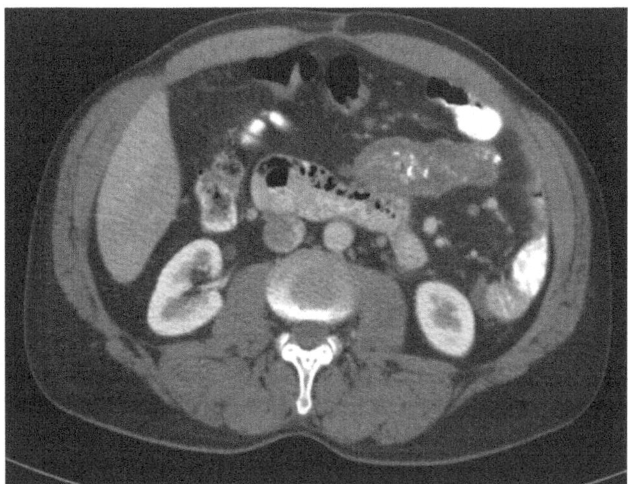

Fig. 20.3 More advanced mesenteric panniculitis in a different patient shows a confluent mesenteric mass with coarse calcifications

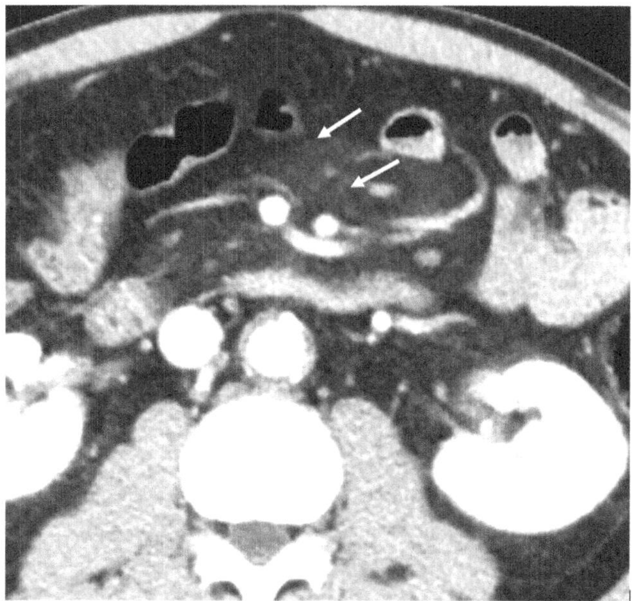

Fig. 20.4 Axial image from a patient with mesenteric panniculitis also showing perivascular sparing, also known as the "fat-halo" (Arrows). Radiographic images are courtesy of Dr. Abraham Dachman and Dr. Justin Ramirez

Biopsies and Histologic Findings

Surgical biopsy and microscopic study of affected tissue is required to completely rule out other conditions and to confirm a diagnosis of mesenteric panniculitis. Our group recommends that biopsies are obtained using a laparoscopic approach. Laparoscopy allows for full visualization of the lesion with adequate sampling of multiple affected areas, including mesenteric lymph nodes. CT directed biopsies have also been described in the literature. Laparoscopy or laparotomy reveals mesenteric thickening, or a uniform appearing fatty mass. Hemorrhagic areas of the mass may be present. Lymphadenopathy is often seen.

Histologic findings include chronic inflammation, fat degeneration and lipid necrosis, a nonspecific and predominantly lymphocytic inflammatory infiltrate, and fibrosis.

Treatments

Medical

The goals of treatment for mesenteric panniculitis are reduction of mesenteric inflammation and the control of symptoms of the disease. Recommended treatments are generally based on case reports or small case series. Individuals with no symptoms are not treated, but are regularly monitored with abdominal imaging.

For symptomatic patients, anti-inflammatory agents, especially corticosteroids, are the usual initial treatment of choice. Other drugs that have been used to treat mesenteric panniculitis include colchicine, azathioprine, cyclophosphamide, infliximab and pentoxifylline. Our group has performed the only prospective clinical trials of treatments for mesenteric panniculitis. These trials demonstrated efficacy of both thalidomide and low dose naltrexone (LDN) as treatments for mesenteric panniculitis. Thalidomide, a tumor necrosis factor alpha (TNF-α) inhibitor improves symptoms and reduces blood levels of ESR and CRP in patients with mesenteric panniculitis. LDN has immune modulating effects, mediated by increased blood levels of enkephalins and endorphins. We have also found efficacy of enteric coated budesonide in treating the disease (unpublished).

Tamoxifen and other hormonal therapies have been proposed to treat patients with mesenteric fibrosis due to their anti-fibrotic effects. Unfortunately, serious side effects may occur with these medications including the development of thromboembolic phenomena and secondary malignancies, and are thus recommended to be avoided. Because of the rarity of mesenteric panniculitis, it is unlikely that controlled clinical studies of medical therapies for this condition will occur in the future.

Surgery

Other than for the purpose of obtaining a biopsy for diagnostic purposes, surgery should be avoided in patients with mesenteric panniculitis. There is no role for surgical debulking or attempted removal of the mesenteric mass for the purpose of controlling or curing the disease. Small intestinal obstruction secondary to mesenteric panniculitis is one potential indication for surgery.

Sugggested Reading

1. Ginsburg PM, Ehrenpreis ED. Mesenteric Panniculitis In: NORD Guide to Rare Disorders. Philadelphia, PA: Lippincott Williams & Wilkins; 2003. p. 350.
2. Cross AJ, McCormick JJ, Griffin N, Dixon L, Dobbs B, Frizelle FA. Malignancy and mesenteric panniculitis. Colorectal Dis. 2016;18(4):372–7.
3. Roginsky G, Mazulis A, Ecanow JS, Ehrenpreis ED. Mesenteric panniculitis associated with vibrio cholerae Infection. ACG Case Rep J. 2015;3(1):39–41.

4. Kerdsirichairat T, Mesa H, Abraham J, et al. Sclerosing mesenteritis and IgG4-related mesenteritis: case series and a systematic review of natural history and response to treatments. Immunogastroenterology. 2013;2:119–28.
5. Roginsky G, Alexoff A, Ehrenpreis ED. Initial findings of an open-label trial of low-dose naltrexone for symptomatic mesenteric panniculitis. J Clin Gastroenterol. 2015;49(9):794–5.
6. Smith ZL, Sifuentes H, Deepak P, Ecanow DB, Ehrenpreis ED. Relationship between mesenteric abnormalities on computed tomography and malignancy: clinical findings and outcomes of 359 patients. J Clin Gastroenterol. 2013;47(5):409–14.
7. Canyigit M, Koksal A, Akgoz A, Kara T, Sarisahin M, Akhan O. Multidetector-row computed tomography findings of sclerosing mesenteritis with associated diseases and its prevalence. Jpn J Radiol. 2011;29(7):495–502.
8. Vlachos K, Archontovasilis F, Falidas E, et al. Sclerosing mesenteritis: diverse clinical presentations and dissimilar treatment options. A case series and review of the literature. Int Arch Med. 2010;4:17. https://www.ncbi.nlm.nih.gov/pubmed/21635777
9. Viswanathan V, Murray KJ. Idiopathic sclerosing mesenteritis in paediatrics: report of a successfully treated case and a review of literature. Pediatr Rheumatol Online J. 2010;8:5. https://www.ncbi.nlm.nih.gov/pubmed/20205836
10. Kasporitakis AN, Rizos CD, Delikoukos, etal. Retractile mesenteritis presenting with malabsorption syndrome. Successful treatment with oral pentoxifylline. J Gastrointestin Live Dis. 2008;17:91–94. https://www.ncbi.nlm.nih.gov/pubmed/18392253
11. Akram S, Pardi DS, Schaffner JA, Smyrk TC. Sclerosing mesenteritis: clinical features, treatment, and outcome in ninety-two patients. Clin Gastroenterol Hepatol. 2007;5:589–596. https://download.journals.elsevierhealth.com/pdfs/journals/1542-3565/PIIS1542356507002248.pdf
12. Ginsburg PM, Ehrenpreis ED. A pilot study of thalidomide for patients with symptomatic mesenteric panniculitis. Aliment Pharmacol Ther. 2002;16:2115–22.
13. Ferrari TC, Couto CM, Vilaça TS, Xavier MA, Faria LC. An unusual presentation of mesenteric panniculitis. Clinics (Sao Paulo). 2008;63(6):843–4.
14. National Organization for Rare Disorders. Mesenteric Panniculitis. https://rarediseases.org/rare-diseases/mesenteric-panniculitis/. Accessed 26 Feb 2019.

PPP Syndrome: Pancreatitis, Panniculitis, Polyarthritis

21

Eli D. Ehrenpreis

Definition

PPP Syndrome is a rare occurrence of extra-pancreatic complications of acute pancreatitis., Panniculitis usually involves the fatty layers of the skin of the lower extremities, but can be involve fatty tissue in any location in the body, including the mesentery. Symmetric polyarthritis generally of the lower extremities is a characteristic occurrence.

Epidemiology

Pancreatic panniculitis is purported to affect about 2–3% of patients with pancreatic disease. About 40 cases of PPP Syndrome have been reported in the medical literature. Only a few cases with mesenteric involvement have been described in the medical literature.

Patients at Risk

This syndrome is generally described in male patients with a history of alcohol abuse between the ages of 40 and 70 years old. It occurs in the setting of acute or chronic pancreatitis. It has also been described in association with cancer of the pancreas.

E. D. Ehrenpreis (✉)
Department of Medicine, Advocate Lutheran General Hospital, 1775 Dempster St, 6 South, Park Ridge, IL 60068, USA
e-mail: e2bioconsultants@gmail.com

E. D. Ehrenpreis et al. (eds.), *The Mesenteric Organ in Health and Disease*,
https://doi.org/10.1007/978-3-030-71963-0_21

Pathophysiology

Although the exact cause of this condition is unknown, it appears to be due to entry of pancreatic enzymes into the systemic circulation. This in turn leads to an inflammatory destruction of fat-containing organs including subcutaneous tissues, bone marrow and the mesentery. Destruction of fatty tissue, primarily by pancreatic lipase results in the development of fat saponification and necrosis. It is presumed that joint inflammation occurs due to circulating inflammatory cytokines, although some have proposed that deposition of antigen–antibody complexes contribute to inflammatory lesions of the disease. Arthritis may also develop as a result of joint infiltration by free fatty acids. Fistulation between a pancreatic pseudocyst and a portion of the mesenteric venous system may be an important factor in the development of the condition.

Signs and Symptoms

The clinical presentation of this condition is variable and dependent on the severity of pancreatic disease, and the extent of involvement of other affected areas. Typically, patients will have abdominal pain related to acute and chronic pancreatitis. The severity of abdominal symptoms may worsen with the subsequent development of mesenteric fat necrosis and the development of an inflammatory mass of the mesentery. Tender, deep erythematous nodular lesions of the skin are present. These may open and extrude an oily brown or creamy material. An inflammatory polyarthritis and interosseous fat necrosis are additional characteristics of this condition, resulting in joint and bone pain. The consequences of severe pancreatic disease is associated with a 25% mortality rate. Untreated disease can result in bone marrow necrosis, scarring of the skin, septic arthritis, chronic debilitating arthritis, and worsening chronic pancreatitis.

Diagnosis

The diagnosis of PPP syndrome is made on the basis of the triad of pancreatitis, panniculitis and polyarthritis occurring simultaneously in an affected patient. Typical physical findings, elevated serum lipase level and characteristic radiographic and histologic findings confirm the diagnosis.

Physical Findings

The presentation of this condition can be variable in severity. In general, patients will have painful nodular lesions of the skin joint pain with effusions and physical findings consistent with acute or chronic pancreatitis. A symmetrical inflammatory polyarthritis may occur. The abdominal examination may reveal a tender palpable mesenteric mass.

Laboratory Testing

Characteristic laboratory findings associated with acute and chronic pancreatitis occur in this condition. However, cases described in the literature demonstrate markedly elevated levels of serum lipase and other pancreatic enzymes.

Imaging

CT scan of the abdomen may reveal acute pancreatitis, pancreatic pseudocysts, chronic pancreatitis and/or pancreatic cancer. When mesenteric involvement occurs, the presence of a mass in the mesenteric fat will be detected. Joint disruption and effusions as well as lytic lesions of the bone may be detected with radiography. Bone marrow abnormalities including necrosis are detected with MRI.

Histologic Findings

Fat necrosis is seen in skin lesions and in biopsies of the mesenteric mass when present. Loss of nucleation and partial destruction of the cell membranes of adipocytes produces the characteristic "ghost cells" seen on biopsy. Analysis of fluid removed from affected joints reveals the presence of inflammation and or sepsis.

Treatments

Medical

Treatment of pancreatic disease appears to result in improvement of the extra-pancreatic complications associated with this condition. Chronic arthritis and chronic pancreatitis may ensue. Nonsteroidal anti-inflammatory drugs and corticosteroids have been used to reduce associated inflammation and may be effective in controlling skin and joint manifestations of this condition. Use of octreotide and plasmapheresis have been described in the literature.

Surgical

One case report described the stenting of a disrupted pancreatic duct and drainage of a necrotic pancreatic cyst using endoscopic ultrasound and endoscopic retrograde cholangiopancreatography. Partial pancreatoduodenectomy and resection of a mesenteric fistula has also been described. Surgical drainage of pancreatic pseudo-cysts may be beneficial. Total pancreatectomy has been described as curative treatment.

Additional Comments

PPP syndrome is one of several complications of acute pancreatitis that may affect the mesentery. Acute pancreatitis may also be associated with thrombosis in the mesenteric veins or arteries.

Suggested Reading

1. Dong E, Attam R, Wu BU. Board review vignette: PPP syndrome: pancreatitis, panniculitis, polyarthritis. Am J Gastroenterol. 2017;112:1215–6.
2. Narvaez J, Bianchi MM, Santo P, et al. Pancreatitis, panniculitis, and polyarthritis. Semin Arthritis Rheum. 2010;39:417–23.
3. Dieker W, Derer J, Henzler T, et al. Pancreatitis, panniculitis and polyarthritis (PPP-) syndrome caused by post-pancreatitis pseudocyst with mesenteric fistula: diagnosis and successful surgical treatment: Case report and review of literature. Int J Surg Case Rep. 2017;31:170–175.
4. Laureano A, Mestre T, Ricardo L, et al. Pancreatic panniculitis-a cutaneous manifestation of acute pancreatitis. J Dermatol Case Rep. 2014;8:35–7.
5. Ferri V, Ielpo B, Duran H, et al. Pancreatic disease, panniculitis, polyarthrtitis syndrome successfully treated with total pancreatectomy: Case report and literature review. Int J Surg Case Rep. 2016;28:223–6.
6. Kang DJ, Lee SJ, Choo HJ, et al. Pancreatitis, panniculitis, and polyarthritis (PPP) syndrome: MRI features of intraosseous fat necrosis involving the feet and knees. Skeletal Radiol. 2017;46:279.

Mesenteric Adenitis

22

Eli D. Ehrenpreis

Definition

Mesenteric adenitis (or mesenteric lymphadenitis) is a benign inflammatory condition involving the lymphatic system located in the right lower quadrant of the abdomen. It is an important entity in the differential diagnosis of patients presenting with right lower quadrant pain. Mesenteric adenitis is defined by the presence of a group of three or more lymph nodes in the right lower quadrant mesentery that are 5 mm or greater when other conditions are ruled out. Mild terminal ileum inflammation may be present. Primary mesenteric adenitis is acute in onset and is likely an infectious condition of origin. Secondary mesenteric adenitis occurs as a complication of a variety of infectious and inflammatory disorders and can occur in patients infected with human immunodeficiency virus (HIV).

Epidemiology

The complaint of abdominal pain makes up about seven percent of all emergency room visits in the United States. Next to acute appendicitis, mesenteric adenitis is the second most common cause of right lower quadrant of the abdomen pain. In the past, mesenteric adenitis was diagnosed in patients undergoing surgery for suspected appendicitis. With the addition of modern imaging techniques, mesenteric adenitis is diagnosed in 2–16% of patients initially suspected of having acute appendicitis.

E. D. Ehrenpreis (✉)
Department of Medicine, Advocate Lutheran General Hospital, 1775 Dempster St, 6 South, Park Ridge, IL 60068, USA
e-mail: e2bioconsultants@gmail.com

E. D. Ehrenpreis et al. (eds.), *The Mesenteric Organ in Health and Disease*,
https://doi.org/10.1007/978-3-030-71963-0_22

Patients At Risk

Primary mesenteric adenitis is most commonly seen in adolescent and young adult patients with an average age of 25 years (range 5–44) years. Secondary mesenteric adenitis is seen in patients belonging to high risk groups. These include patients with coliac disease, diverticulitis, cholecystitis, Crohn's disease, ulcerative colitis, mesenteric panniculitis, pancreatitis, systemic lupus erythematosus and other connective tissue. Patients infected with HIV may develop secondary mesenteric adenitis from mycobacterium avium complex infection. In addition, HIV infected patients are at risk for the immune reconstitution inflammatory syndrome (IRIS). This occurs 3–12 months after initiation of treatment with highly active antiretroviral therapy (HAART). Due to an overall improvement of the immune system in HIV-infected patients treated with HAART therapy, an increase in circulating CD4 T cells results in the development of an inflammatory response to ongoing infections with viruses (such as cytomegalovirus or herpes simplex), mycobacteria, or fungal infections (such as Cryptococcus neoformans or Pneumocystis jerovecii). IRIS may manifest as abdominal pain and abdominal lymphadenopathy.

Pathophysiology

It is hypothesized that mesenteric adenitis occurs when infectious organisms gain entry to intestinal lymph nodes after invasion of intestinal lymphatics. A subsequent inflammatory process is triggered within the lymph nodes, and the intensity of this response varies according to the virulence of the invasive organism. Suppurative lymph node inflammation and/or associated systemic involvement may occur. Cultures of mesenteric lymph nodes and blood have demonstrated a number of infectious causes of the condition (Table 22.1).

Symptoms

The classic presentation of primary mesenteric adenitis is the acute onset of fever and right lower quadrant abdominal pain. Although the majority of patients in the appropriate age group with these symptoms will have acute appendicitis, the additional presence of diarrhea is more consistent with the diagnosis of mesenteric adenitis. Signs and symptoms of an upper respiratory tract infection, other systemic infectious illnesses and peripheral lymph node enlargement may also occur.

Table 22.1 Organisms associated with mesenteric adenitis

Bacteria
Yersinia species (most common causative organisms)
Beta-hemolytic streptococcus
Staphylococcus species
Escherichia coli
Streptococcus viridans
Mycobacterium tuberculosis
Non-*Salmonella* typhoid
Bartonella henselae (Cat Scratch Disease)
Parasites
Giardia lamblia
Viruses
Coxsackieviruses (A and B)
Rubeola virus
Adenovirus (serotypes 1, 2, 3, 5, and 7)
Epstein-Barr virus (EBV)
Acute human immunodeficiency virus (HIV)

Alan S. Putrus and Michael H. Piper. "Mesenteric Lymphadenitis". https://emedicine.medscape.com/article/181162.

Physical Findings

Fever and abdominal tenderness (particularly in the right lower quadrant) are physical signs that occur in patients with either mesenteric adenitis or acute appendicitis. Rebound tenderness and may be present in both mesenteric adenitis and acute appendicitis. Rectal tenderness on the right or middle portion of the digital rectal exam may occur, but does not help to differentiate between the two conditions. Peripheral lymphadenopathy may be present in mesenteric adenitis. The presence of other physical findings may identify causes of secondary mesenteric adenitis.

Laboratory Testing

Leukocytosis is present in both mesenteric adenitis and acute appendicitis and cannot be reliably used to differentiate the two conditions. A white blood cell count (WBCs) of greater than 10,000/μL in present in more than of 50% of with mesenteric adenitis.

Imaging

The initial purpose of performing imaging studies in the appropriate clinical setting is to confirm or rule out the diagnosis of acute appendicitis and other causes of acute abdominal pain. Abdominal ultrasound (US) and computerized scan of the abdomen (CT) are complementary tests for identification of acute appendicitis. The range of sensitivities of abdominal US for diagnosing acute appendicitis is between 71 and 94% and a specificity ranges 81 and 98%. Ultrasound criteria for the diagnosis of appendicitis include an appendiceal diameter of > 6 mm, thickening of the appendiceal wall and non-compressibility of the appendix. Appendiceal inflammation can also be confirmed using Doppler imaging. Due to high sensitivity and specificity, US is the initial imaging procedure of choice to rule out acute appendicitis.

Abdominal CT may identify the presence of multiple lymph nodes in the right lower quadrant that are characteristic of mesenteric adenitis (Figs. 22.1, 22.2, and 22.3). CT is more effective in diagnosing acute appendicitis and has a range of sensitivity between 76–100% and a specificity range from 83–100%. CT criteria for appendicitis include appendiceal enlargement (the range for normal diameter is generally agreed to be between 6–9 mm), thickening of the appendiceal wall and inflammation of the mesenteric fat that is adjacent to the appendix. It should be noted that right sided abdominal lymph node enlargement may be present in patients with acute appendicitis. CT imaging criteria for mesenteric adenitis includes the presence of three or more lymph nodes with a maximal diameter of 5 mm or more in the right lower quadrant of the abdomen. Mild thickening of the wall of the terminal ileum (≤ 5 mm) may also be present.

CT is also beneficial to identify etiologies for the occurrence of secondary mesenteric adenitis. Imaging with abdominal CT is highly effective in identifying other causes of acute abdominal pain in patients presenting with symptoms that are initially suggestive of acute appendicitis such as cholecystitis, diverticulitis, ovarian torsion and a perforated viscous. CT can also be used to identify other causes of right lower quadrant pain that are in the differential diagnosis for mesenteric adenitis including epiploic appendagitis, inflammatory bowel disease, Meckel's diverticulum, neutropenic colitis, right-sided diverticulitis, and omental infarction.

Treatment

Medical

Stabilization of the patient with adequate hydration is a mainstay of therapy for mesenteric adenitis. Pain control may be required on a short-term basis. Use of empiric antibiotics, with coverage for *Yersinia enterocolitica* is considered. Ongoing observation with serial abdominal examinations by the medical and surgical team may be needed if acute appendicitis or other causes of abdominal pain

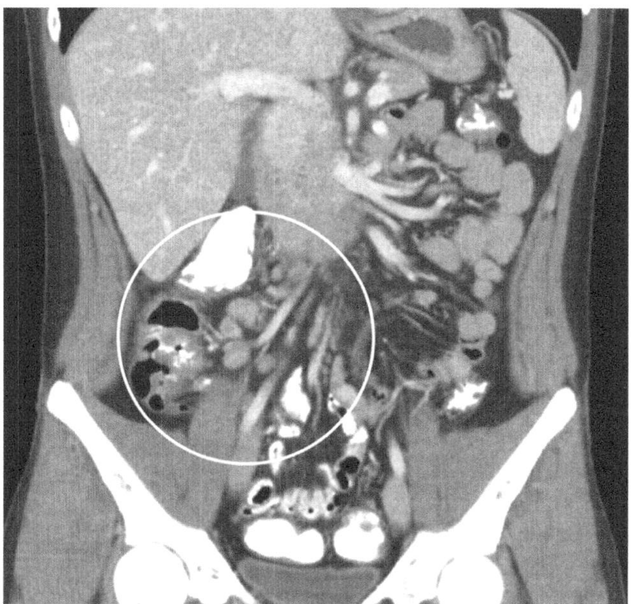

Fig. 22.1 Coronal CT image of a patient with mesenteric adenitis demonstrating enlarged mesenteric lymph nodes. These are most prominent in the ileocolic distribution Associated ileocecal thickening is also visualized (circle)

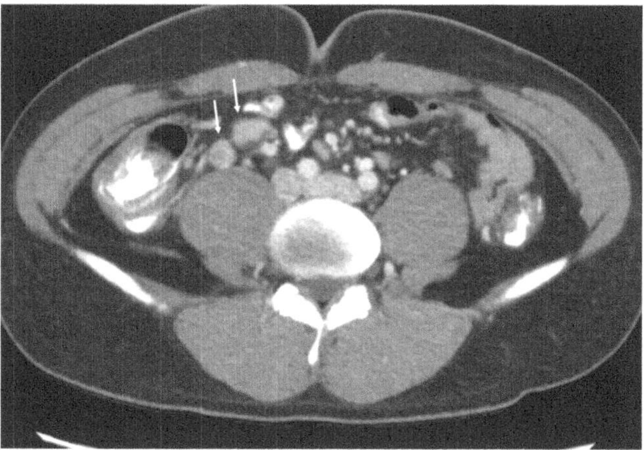

Fig. 22.2 Axial image from the same patient showing enlarged lymph nodes (arrows) anterior to the right psoas muscle with thickening of the cecal wall

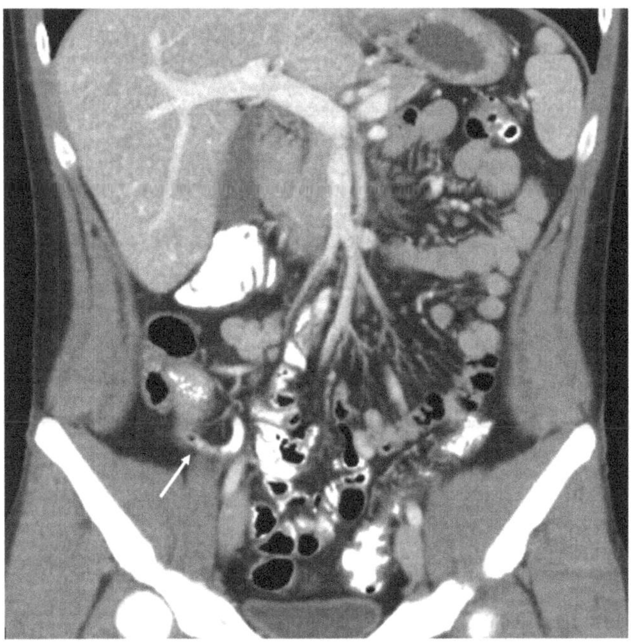

Fig. 22.3 This additional coronal image in the same patient with mesenteric adenitis demonstrates a normal caliber appendix that is partially filled with gas and oral contrast (arrow). Images courtesy of Dr. Abraham Dachman and Dr. Justin Ramirez

have not been completely ruled out. Primary mesenteric adenitis is a self-limited condition with a 1 to 7-day recovery period.

Surgical

Mesenteric adenitis does not require surgical management. When the appendix is visualized and normal appearing using appropriate imaging, the diagnosis of acute appendicitis is essentially ruled out. Under this circumstance, the patient may be managed as a case of mesenteric adenitis. Expedited appendectomy is performed when acute appendicitis is the most certain diagnosis results in better patient outcomes.

Suggested Reading

1. Helbling R, Conficconi E, Wyttenbach M, et al. Acute nonspecific mesenteric lymphadenitis: more than no need for surgery . Biomed Res Int. 2017;2017:9784565. https://doi.org/10.1155/2017/9784565.
2. Patel NB, Wenzke DR. Evaluating the patient with right lower quadrant pain. Radiol Clin North Am. 2015;53(6):1159–70.
3. Natrajan K, Medisetty M, Gawali R, et al. Strongyloidosis hyperinfection syndrome in an hiv-infected patient: a rare manifestation of immune reconstitution inflammatory syndrome. Case Rep Infect Diseases 2018, Article ID 6870768, 4 p, https://doi.org/https://doi.org/10.1155/2018/6870768.
4. Takada T, Nishiwaki H, Yamamoto Y, et al. The role of digital rectal examination for diagnosis of acute appendicitis: a systematic review and meta-analysis. PLoS One. 2015;10(9):e0136996. Published 2015 Sep 2. doi:https://doi.org/10.1371/journal.pone.0136996
5. Putrus AS, Piper MH. Mesenteric lymphadenitis. Medscape. https://emedicine.medscape.com/article/181162-overview#a7
6. Macari M, Hines J, Balthazar E, Megibow A. Mesenteric adenitis. Am J Roentgenol. 2002;178(4):853–8.
7. Mostbeck G, Adam EJ, Nielsen MB, et al. How to diagnose acute appendicitis: ultrasound first. Insights Imaging. 2016;7(2):255–63.
8. Balthazar EA, Birnbaum BY, Yee J, MegibowAJ, Roshkow J, Gray C. Acute appendicitis: CT and US correlation in 100 patients. Radiology 1994;190 31–35
9. Gorter RR, Eker HH, Gorter-Stam MA, et al. Diagnosis and management of acute appendicitis. In: EAES consensus development conference 2015. Surg Endosc. 2016;30(11):4668–4690.

Mesenteric Abscess

23

Ryan T. Hoff and Eli D. Ehrenpreis

Definition

A mesenteric abscess is organized infection occurring within the mesenteric organ (peritoneal connective tissue that attaches the stomach, small intestine, pancreas, and spleen to the abdominal wall). While abdominal abscesses may occur in any location within the abdominal cavity, mesenteric abscesses are confined to the mesenteric connective tissue.

Pathophysiology

Several physiologic mechanisms work to limit infections within the peritoneum and mesentery. The integrity of the lining of the gastrointestinal tract serves as an initial anatomic barrier, while secretory immunoglobulin A prevents pathogens from adhering to the intestinal wall. Once translocation occurs, the majority of bacterial organisms are removed via lymphatic drainage. Lymphatic blockage can occur as a consequence of infections such as cellulitis and filariasis, deep vein thrombosis, neoplasm and radiation injury. Damage and obstruction may result in the development of infections and associated inflammation. Neutrophil infiltration enhances local inflammation. This in turn promotes increased blood flow by dilation of local vasculature.

As this suppurative inflammation continues, liquefactive necrosis occurs and then forms an abscess. A mass of necrotic cells develops in the center, while an outer margin of intact cells persists. The abscess may organize further and become walled off as fibrin is deposited at the margins. The formation of an abscess helps to

R. T. Hoff · E. D. Ehrenpreis (✉)
Department of Medicine, Advocate Lutheran General Hospital, 1775 Dempster St, 6 South, Park Ridge, IL 60068, USA
e-mail: e2bioconsultants@gmail.com

E. D. Ehrenpreis et al. (eds.), *The Mesenteric Organ in Health and Disease*,
https://doi.org/10.1007/978-3-030-71963-0_23

contain the extent of the infection, while the mesentery provides a means of anatomically compartmentalizing the potential spaces of the peritoneum, thereby limiting the spread of infection.

Any factors that promote bacterial translocation, including inflammatory bowel disease, surgery, or trauma, increase the risk for developing a mesenteric abscess. In addition, any host factors that impair immune function, including malignancy, immunosuppressive therapies, malignancy, or diabetes mellitus may also increase the risk of abscess formation. Increased age and malnutrition are additional risk factors.

Epidemiology

The overall incidence of mesenteric abscess has not been reported. The vast majority of mesenteric abscess occur secondary to an underlying disorder, injury, or insult that compromises the integrity of the intestinal mucosa. Primary mesenteric abscesses, i.e. abscesses that occur in the absence of any provoking event or underlying condition, are exceptionally rare. Secondary causes of mesenteric abscess include diverticulitis, appendicitis, Crohn's disease, malignancy, and trauma (Table 23.1). Small intestinal diverticula may become complicated by diverticulitis and perforation, leading to mesenteric abscesses. A Meckel's diverticulum may become occluded from enteroliths, parasites, malignancy, or peptic ulcer related fibrosis. Such an obstruction may lead to Meckel's diverticulitis, with perforation and mesenteric abscess formation. Severe peptic ulcer disease may become complicated by perforation, leading to mesenteric abscess formation. A mesenteric abscess may occur as a postoperative complication of abdominal or pelvic surgeries, such as open radical hysterectomy. A mesenteric abscess as the presenting finding of carcinoma has been reported. Conditions causing immunodeficiency, such as HIV/AIDS, increase the risk of opportunistic infections, including mycobacterium avium complex and mycobacterium tuberculosis, which may cause mesenteric abscesses. In immunosuppressed patients, Yersinia enterocolitica infection of the terminal ileum may spread to lymph nodes, leading to abscess formation. Infection with Enterobius vermicularis, also known as pinworm, has been reported to cause mesenteric abscess, with a presentation similar to perforated appendicitis.

Crohn's disease is of particular relevance to mesenteric abscess formation. Inflammatory disease increases the risk of infection by compromising bowel wall integrity, allowing penetration by bacteria to extraintestinal sites. Immunosuppressive agents and surgery also increase the overall infection risk. Postoperative abdominal abscess occurs as a complication of peritoneal exposure to bacteria during the operation, or secondary to anastomotic leakage. Tobacco use increases the risk of both abscesses and fistulas in Crohn's disease. Corticosteroids are also known to increase the of intra-abdominal abscess in Crohn's disease. Azathioprine increases the risk of postoperative intra-abdominal infectious complications in

Table 23.1 Causes of mesenteric abscesses

Iatrogenic	Post-operative complication
Immune suppression	AIDS/HIV related infections
	Solid organ transplant recipients
	Immunomodulating medications
	Chemotherapy
Infectious	Enterobius vermicularis (Pinworm)
	Mycobacterium avium complex
	Mycobacterium tuberculosis
	Yersinia enterocolitica
	Enterobius vermicularis (Pinworm)
Malignancy	Carcinoma
Perforation, Leading to Abscess	Appendicitis
	Ileal diverticulitis
	Jejunal diverticulitis
	Meckel's diverticulum
	Collagen disorders, Ehlers Danlos Syndrome
	Peptic ulcer
	Trauma
Primary	Idiopathic

Crohn's disease. However, azathioprine does not appear to increase the risk of intra-abdominal infection in non-operative Crohn's disease. Vedolizumab, administered before or after surgery, does not appear to increase the risk of the risk of post-operative abscesses in Crohn's disease, although there may be an increased risk of surgical site infections in patients treated with vedolizumab preoperatively. In patients with Crohn's disease with an intra-abdominal abscess, ongoing treatment with a tumor necrosis factor alpha (TNF) antagonist is currently considered to be safe. Ustekinumab does not increase the risk of post-operative infection when compared with anti-TNF agents.

Symptoms

The presentation of mesenteric abscess varies by location and clinical context. Symptoms may include fever, chills, sweats, and abdominal pain. Much of the symptomatology is determined by the location involved. Perigastric abscesses may cause early satiety, nausea and vomiting. Large abscesses of the small intestine may cause a small bowel obstruction, leading to nausea, crampy abdominal pain, and inability to pass flatus or stool. Perirectal abscesses may cause tenesmus.

Physical Findings

Exam findings for mesenteric abscess may include fever, tachycardia, and a palpable abdominal mass. The abdominal mass may be freely mobile, easily shifted from side to side via palpation. If there is underlying Crohn's disease, a perianal or enterocutaneous fistula may be present. Additional systemic manifestations of infection are often present, such as tachycardia and diaphoresis. In the setting of bowel obstruction, abdominal distension may be observed. Mesenteric abscesses may rupture, leading to peritoneal signs, with guarding and rebound tenderness.

Laboratory Testing

Laboratory results are nonspecific and represent the presence of underlying bacterial infection. Findings may include leukocytosis, bandemia, elevated inflammatory markers (erythrocyte sedimentation rate and C reactive protein), and procalcitonin. Blood cultures may be positive for a single organism, or multiple pathogenic bacteria. Culture of drained purulent material, acquired via radiologic guidance or surgery, is the gold standard for microbiologic diagnosis. Infectious organisms may include gram negative rods such as Escherichia coli, anaerobic bacteria such as Bacteroides fragilis and Peptostreptococcus species, or gram positive cocci including staph aureus, streptococcus, and Enterococcus species. Tuberculosis may be isolated in rare cases. When possible, a microbiologic diagnosis should be obtained, which also may provide sensitivities to antimicrobial therapy. In most cases, abscesses are polymicrobial. In the setting of immunosuppression, staining with acid fast bacilli should be considered to detect mycobacterial infection.

Diagnosis and Imaging

Computerized tomography (CT) of the abdomen and pelvis is the imaging test of choice in most cases of suspected mesenteric abscess. Images demonstrate a well circumscribed mass containing fluid and surrounding inflammatory changes (Fig. 23.1), which may contain pockets of air. Additional findings may include thickening of the intestinal wall, intestinal dilation with air fluid levels, and enlarged mesenteric lymph nodes. If intravenous contrast is used, the mass may be ring enhancing.

Ultrasound provides the benefit of avoiding radiation but offers more limited anatomic information. Plain film radiography may demonstrate evidence of intra-abdominal abscess but is significantly less sensitive and specific than CT imaging.

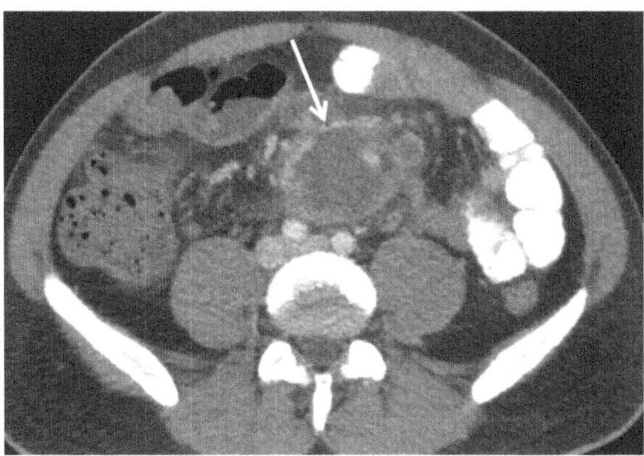

Fig. 23.1 Crohn's abscess. Round, well-circumscribed fluid collection in the lower abdomen with surrounding inflammatory changes compatible with a mesenteric abscess in this patient with a prior diagnosis of Crohn's disease (arrow)

Barium enema may demonstrate the presence of an intra-abdominal mass, with bowel loop separation or displacement. In one series of patients with Crohn's disease, these findings on barium enema (suggesting a mass) indicated a mesenteric abscess in one third of cases. The most common finding in these cases was the presence of fibrofatty changes of the mesentery.

The presence of a mesenteric abscess observed on imaging should prompt investigation regarding the likely etiology of this finding. Potential etiologies include Crohn's disease and complicated jejunoileal diverticulitis. Additional diagnostic considerations include malignancy, foreign body related perforations, traumatic hematomas, and medication induced ulcers. Depending on anatomic location, the initial presentation of mesenteric abscess can be identical to acute appendicitis. Complicated Meckel's diverticular perforation should be considered, particularly if gastrointestinal bleeding is present or if exploratory laparotomy is performed which shows a normal appearing appendix. In women, the presentation can be similar to ectopic pregnancy, pelvic inflammatory disease, tubo-ovarian abscess, and hemorrhagic or ruptured ovarian cysts.

Treatment

In most cases, definitive therapy requires drainage of the abscess. Drainage may be achieved either percutaneously with CT guidance or by a surgical technique. Antibiotics should be administered, with a regimen directed at treating the most likely causative organisms.

If complicated ileal or jejunal diverticulitis is suspected, surgery should be pursued. Surgery may be laparoscopic or open and consists of resection of diseased bowel followed by primary anastomosis. Percutaneous abscess drainage and antibiotics may be considered in select patients with Ehlers Danlos Syndrome, as these patients presents challenges in achieving successful surgical outcomes, including poor wound healing, enterocutaneous fistula formation, and bleeding. There is increasing interest in the use endoscopic techniques for drainage of mesenteric abscesses. Endoscopic ultrasound guided drainage may be effective for postoperative mesenteric abscesses.

Small mesenteric abscesses in Crohn's disease (less than 3 cm) may be treated with antibiotics alone; this often leads to resolution. Abscesses greater than 3 cm in patients with Crohn’s disease may benefit from percutaneous drainage. Percutaneous drainage and antibiotics are effective for the majority of Crohn’s disease related abscesses. In the minority of cases that require surgery, percutaneous drainage in advance may allow for sufficient healing that a primary anastomosis is feasible at the time of operation, thus avoiding a multistep surgery with a diverting ostomy.

Surgery is indicated if medical therapy and percutaneous drainage is unsuccessful or additional abnormalities, such as congenital malrotation or fistula are present. Risk factors that predict the need for surgery in patients with Crohn’s disease include large multifocal and multiloculated abscesses, the presence of colonic disease, and concomitant corticosteroid therapy. Amongst patients requiring ileocecal or ileocolic resections for Crohn's disease, the presence of an abscess is a risk factor for anastomotic-associated complications. Surgery for Crohn's-related abscesses consists of exploration and evacuation of the abscess, irrigation and debridement of the abscess cavity. Resection of diseased bowel is usually necessary; however, excessive resection should be avoided to minimize the risk of short bowel syndrome. Patients with an abscess in the setting of refractory pancolitis or fistula formation may require a complete proctocolectomy with end ileostomy. Additional surgical options for severe disease include segmental colectomy and subtotal colectomy with ileorectal anastomosis; the choice of surgery depends largely on the clinical context and the extent of disease.

Suggested Reading

1. Hokama A, et al. Mesenteric abscess in Crohn's disease. Gastrointest Endosc. 2005;62(2):306; discussion 306.
2. Navarro C. Perforated diverticulum of the terminal ileum. A previously unreported cause of suppurative pylephlebitis and multiple hepatic abscesses. Dig Dis Sci. 1984;29(2):171–7.
3. Sakpal SV, Fried K, Chamberlain RS. Jejunal diverticulitis: a rare case of severe peritonitis. Case Rep Gastroenterol. 2010;4(3):492–7.
4. Seitun S, et al. Perforated Meckel’s diverticulitis on the mesenteric side: MDCT findings. Abdom Imaging. 2012;37(2):288–91.
5. Ko EM. Robotic versus open radical hysterectomy: a comparative study at a single institution. Gynecol Oncol. 2008;111(3):425–30 Epub 2008 Oct 16.

6. Yamagata Y, et al. Poorly differentiated mesenteric carcinoma of unknown primary site detected by abscess formation: case report. World J Surg Oncol. 2014;8(12):4.
7. Luo Y, et al. Primary mesenteric adenocarcinoma covered by abscess of the mesocolon and intestinal obstruction: a case report. Mol Clin Oncol. 2017;6(4):593–6 Epub 2017 Feb 13.
8. Watanabe K, et al. Mesenteric lymph node abscess due to Yersinia enterocolitica: case report and review of the literature. Clin J Gastroenterol. 2014;7(1):41–7 Epub 2014 Jan 9.
9. Mohar SM, et al. Small bowel obstruction due to mesenteric abscess caused by Mycobacterium avium complex in an HIV patient: a case report and literature review. J Surg Case Rep. 2017;2017(7):rjx129. eCollection 2017 Jul.
10. Beeson BB, Woodruff AW. Mesenteric abscess caused by threadworm infection. Trans R Soc Trop Med Hyg. 1971;65(4):433.
11. Agrawal A, et al. Effect of systemic corticosteroid therapy on risk for intra-abdominal or pelvic abscess in non-operated Crohn's disease. Clin Gastroenterol Hepatol. 2005;3 (12):1215–20.
12. Law CCY, et al. Systematic review and meta-analysis: preoperative vedolizumab treatment and postoperative complications in patients with inflammatory bowel disease. J Crohns Colitis. 2018;12(5):538–45.
13. Lightner AL, et al. Postoperative outcomes in vedolizumab-treated patients undergoing abdominal operations for inflammatory bowel disease. J Crohns Colitis. 2017;11(2):185–90 Epub 2016 Aug 19.
14. Ibáñez-Samaniego L, et al. Safety and efficacy of Anti-TNFα treatment in Crohn's disease patients with abdominal abscesses. Hepatogastroenterology. 2015;62(139):647–52.
15. Lightner AL, et al. Postoperative outcomes in Ustekinumab-treated patients undergoing abdominal operations for Crohn's disease. J Crohns Colitis. 2018;12(4):402–7.
16. Cheung HY, et al. Acute abdomen: an unusual case of ruptured tuberculous mesenteric abscess. Surg Infect (Larchmt). 2005;6(2):259–61.
17. Casey MC, et al. Non-operative management of diverticular perforation in a patient with suspected Ehlers-Danlos syndrome. Int J Surg Case Rep. 2014;5(3):135–7 Epub 2014 Jan 8.
18. Strong S, et al. Clinical practice guideline for the surgical management of Crohn's Disease. Dis Colon Rectum. 2015;58(11):1021–36.
19. Raza A, et al. Concomitant laparoscopic ileocolectomy and ladd's procedure for Crohn's ileocolitis with mesenteric abscess and congenital megacolon. Am Surg. 2016;82(9):e284–6.
20. Feagins LA, et al. Current strategies in the management of intra-abdominal abscesses in Crohn's disease. Clin Gastroenterol Hepatol. 2011;9(10):842–50 Epub 2011 May 5.

24 Mesenteric Venous Thrombosis

Nyi Nyi Tun and Eli D. Ehrenpreis

Introduction

Mesenteric venous thrombosis (MVT) is the formation of a blood clot within the mesenteric venous system, including the superior and inferior mesenteric veins or their venous tributaries. The superior mesenteric vein is generally the site of involvement of MVT. Thrombosis of the inferior mesenteric vein thrombosis is a rare occurrence. The presentation of MVT may be acute, subacute or chronic, each of these having different clinical patterns with possible overlap between the various forms. MVT is one of the recognized causes of mesenteric ischemia which refers to inadequate perfusion to meet the metabolic demands of visceral organs, (see Chap. 26 Ischemic Enteropathy).

Epidemiology

MVT may present with symptoms including abdominal pain or may be an asymptomatic incidental finding on abdominal imaging. Despite variation in the literature, MVT is estimated to account for 5–15% of cases of acute mesenteric ischemia. Compared to acute MVT, cases of chronic MVT are reported less frequently, likely because of vague or absent symptoms in a large number of these patients. MVT is the primary diagnosis for 1 in 5,000 to 15,000 hospital admissions and approximately 0.1% of emergency surgical laparotomies for acute abdomen.

N. N. Tun
Department of Internal Medicine, Miami Valley Hospital, Dayton, OH 45409, USA
e-mail: drtun85@gmail.com

E. D. Ehrenpreis (✉)
Department of Medicine, Advocate Lutheran General Hospital, 1775 Dempster St, 6 South, Park Ridge, IL 60068, USA
e-mail: e2bioconsultants@gmail.com

E. D. Ehrenpreis et al. (eds.), *The Mesenteric Organ in Health and Disease*,
https://doi.org/10.1007/978-3-030-71963-0_24

Routine use of abdominal CT scanning has resulted in an increasing number of diagnosed cases of MVT and has been responsible for improved survival in acute MVT. The average age at presentation varies between 45 and 60 years with a slight male predominance. The presenting age varies depending on the underlying cause of MVT.

Anatomy

Mesenteric veins parallel their arterial counterparts and drain into the portal venous system. The superior mesenteric vein (SMV) drains blood from the pancreas, small intestine, and the proximal colon. Initially, blood returns through smaller venous structures called the venae rectae. These coalesce to form ileocolic, middle colic and right colic veins which as previously stated combine to form the superior mesenteric vein. The superior mesenteric vein and the splenic vein join together to form the portal vein (Fig. 24.1). The inferior mesenteric vein (IMV) drains blood from distal colon and rectum. The IMV drains the tributaries of sigmoid and left colic veins, as well as the rectal veins and eventually drains into the splenic vein (Fig. 24.2).

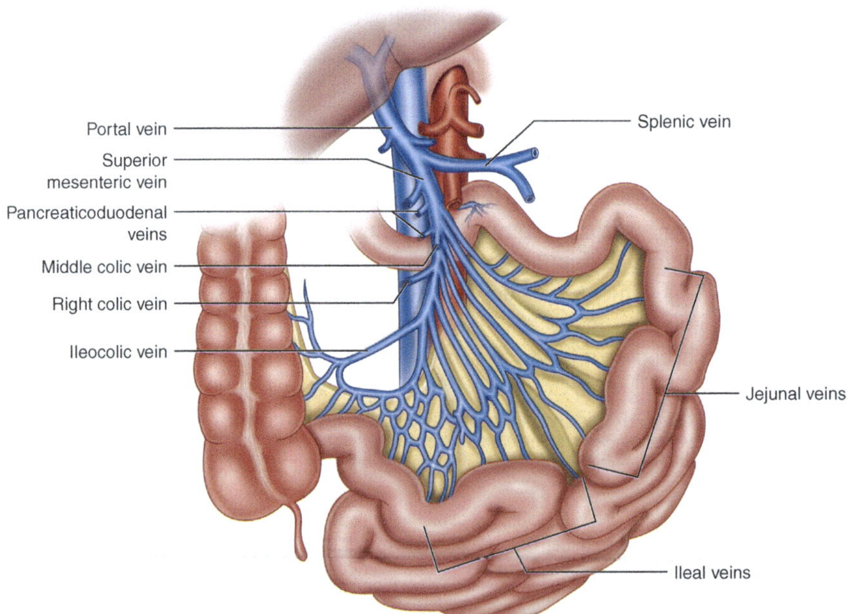

Fig. 24.1 Anatomic view of the superior mesenteric vein

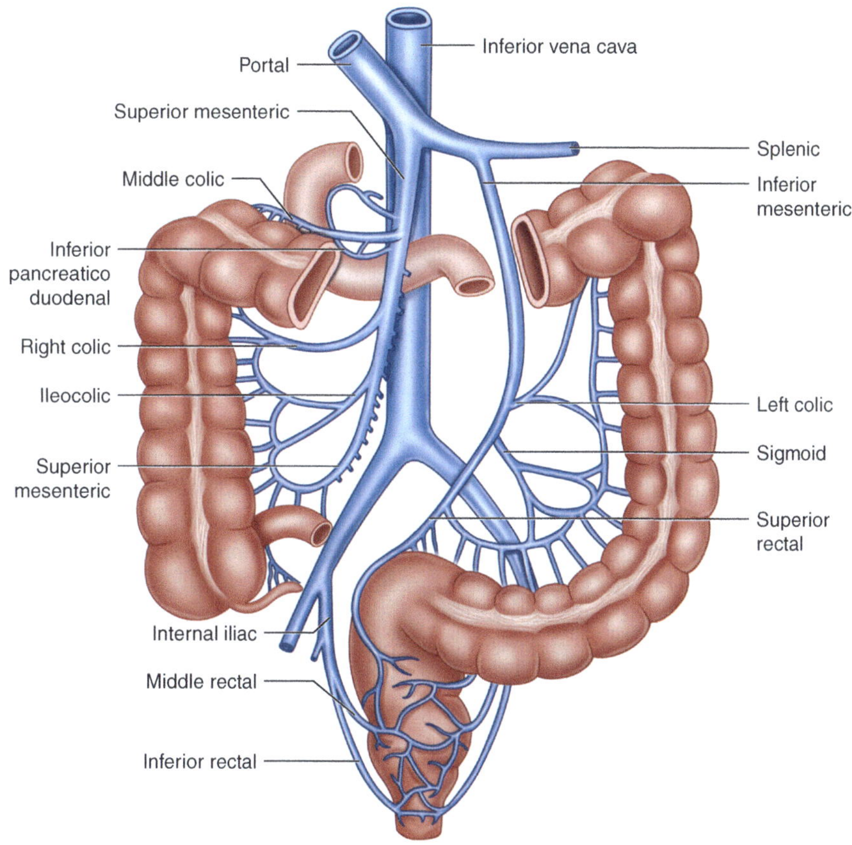

Fig. 24.2 Anatomic view of the inferior mesenteric vein

Pathophysiology

Venous thrombosis results from a combination of vascular endothelial injury, hypercoagulability, and stasis of blood flow. These are classically known as Virchow's triad. As stated previously, MVT almost always involves the superior mesenteric vein distribution whereas the inferior mesenteric vein distribution is rarely involved for reasons which are not clearly understood. Acute MVT may also involve more than one mesenteric vein as well as its branches. The occurrence of a thrombus in a large vein increases venous pressure and vascular resistance within the mesenteric venous bed. A reduction in perfusion pressure follows these processes. Reduced venous flow and translocation of fluid into nearby tissues creates a pathologic state wherein significant bowel wall edema, and hemorrhage of the submucosa occurs. Depending on the degree of venous occlusion and the presence or absence of collateral drainage, MVT can cause progressive changes to intestinal tissue, leading in time to intestinal ischemia. In the most severe cases, intestinal infarction may occur. Milder cases of MVT are characterized by intestinal ischemia without infarction.

The presence of chronic MVT often produces dilated venous collaterals as a consequence of elevated venous pressure. These dilated, pressure-filled veins carry the risk of venous rupture and intestinal bleeding. Patients with chronic MVT may also have concomitant portal vein thrombosis with subsequent portal hypertension.

Patient At Risk for MVT

There are a variety of conditions that predispose to the development of MVT (Table 24.1). Since the initiating event involves the appearance of a clot within the vascular system, any condition that accentuates the risk of coagulation will predispose to MVT. Prothrombotic disorders may be hereditary (such as Factor V Leiden mutation, Protein C deficiency, Protein S deficiency and antithrombin III deficiency) or acquired (as seen in a variety of malignant neoplasms). Hormonal factors including pregnancy and use of oral contraceptive medications are potential risk factors for MVT. Surgery, trauma and inflammatory disorders such as acute pancreatitis, inflammatory bowel disease, and intra-abdominal infections also have assocaitions with the development of MVT. Stasis within the mesenteric venous system can occur in profound cardiac insufficiency, portal hypertension and hyperviscosity (as seen in polycythemia vera and paroxysmal nocturnal hemoglobinuria). A variety of other conditions including nephrotic syndrome and hyperhomocysteinemia may also predispose to MVT. Up to 37% of cases of MVT are classified as idiopathic with no discoverable predisposing factors. Notably, inflammation and other local factors appear to contribute to clot formation in large mesenteric veins, whereas hypercoagulable disorders are more likely to be responsible for clot formation in smaller vessels such as venae rectae and venous arcades.

Clinical Presentation

MVT can be acute, subacute or chronic and its signs and symptoms are often nonspecific, making the initial diagnosis highly challenging. Yet early diagnosis is crucial to increase survival in patients with MVT. The clinical features of MVT depend on the location and acuity of clot formation within the mesenteric vasculature.

Table 24.1 Conditions predisposing to MVT

A. Thrombophlia	B. Inflammation	C. Stasis	D. Idiopathic
Heritable	Intraabdominal	Cirrhosis	
Deficiency of protein C or S or antithrombin II	Pancreatitis	Congenital venous anomaly	
Factor V Leiden mutation Prothrombin gene mutation	Inflammatory bowel disease	Heat failure	
Sickle cell disease	Trauma	Congestive splenomegaly	
Acquired	Surgery (most commonly splenectomy)		
Hematologic conditions:	Infection		
Polycythemia			
Myelofibrosis			
Myeloproliferative disease			
Monoclonal gammopathy			
JAK2 mutation			
Anti-phosphoidal antibodies			
Paroxysmal nocturnal hemoglobinuria			
Disseminated intravascular coagulation			
Heparin-induced thrombocyopenia and thrombosis			

The most common symptom in acute MVT is abdominal pain. The symptom of abdominal pain is reported in 91–100% of MVT cases. Like other causes of acute mesenteric ischemia, acute MVT often presents with severe abdominal pain that is out of proportion to the findings on physical examination of the abdomen. Approximately 50% of patients have nausea and vomiting. Gastrointestinal bleeding occurs in about 15% of patients and is characterized by the presence of hematemesis, hematochezia, or melena. Occult blood in the feces is detectable in up to 50% of the cases. About 75% of patients report having symptoms for more than 48 h before seeking medical attention. Some authors suggested that the arbitrary duration of symptoms of acute MVT is less than 4-weeks. Although findings on abdominal examination can be variable, the presence of peritoneal signs indicate that the disease has progressed to bowel necrosis and infarction. In one study, 6–29% of patients with acute MVT also manifest hemodynamic instability on presentation.

MVT can also present more insidiously with a subacute or chronic pattern. Those patients can be asymptomatic or may present with only intermittent, non-specific abdominal pain, likely as a result of the development of collateral circulation to compensate for diminished flow in the affected large vessel.

Chronic MVT may also present with complications related to portal hypertension such as ascites or variceal bleeding.

Diagnosis

As no clinical features are specific for MVT, it is essential to have a high index of suspicion to arrive at a timely, and potentially life-saving diagnosis. Given a suggestive clinical presentation, thorough history taking can uncover further clinical clues that increase the likelihood that MVT is present. This process, in turn, results in further, appropriate diagnostic testing.

Imaging

Imaging studies play a major role to establish the diagnosis of MVT by revealing the presence of thromboses within the mesenteric venous system.

Contrast-enhanced computerized tomography (CT) is the initial diagnostic modality of choice for MVT. When present, MVT is characterized on contrast-enhanced CT by the presence of a filling defect within a mesenteric vein (Figs. 24.3, 24.4 and 24.5). Other findings on the CT scan include bowel wall thickening, indistinct bowel margins, gas within bowel wall (pneumatosis intestinalis) and the presence of a thickened mesentery (see Figs. 24.3, 24.4, 24.5 and 24.6). The sensitivity and specificity of contrast-enhanced CT scan for the diagnosis of MVT are reported to be 93% and 100% respectively with positive and negative predictive values of between 94 and 100%.

Magnetic resonance venography (MRV) is an excellent diagnostic tool for the diagnosis of acute or chronic MVT with reports of 100 percent sensitivity and specificity in some studies. It can also be used in patients with allergy to iodinated contrast. However, contrast-enhanced CT scan is preferred over MRV because of its widespread availability and shorter testing time unless contrast allergy precludes its use.

In patients with an uncertain diagnosis of MVT with a non-diagnostic CT or MRI where the suspicion remains high, catheter-based mesenteric angiography is an appropriate study. Angiography also offers access for interventional measures such as thrombolysis. Although Doppler ultrasound can be rapidly performed at the bedside and is widely available, it is not as sensitive as CT and MRI and not able to detect blood clots in smaller mesenteric vessels.

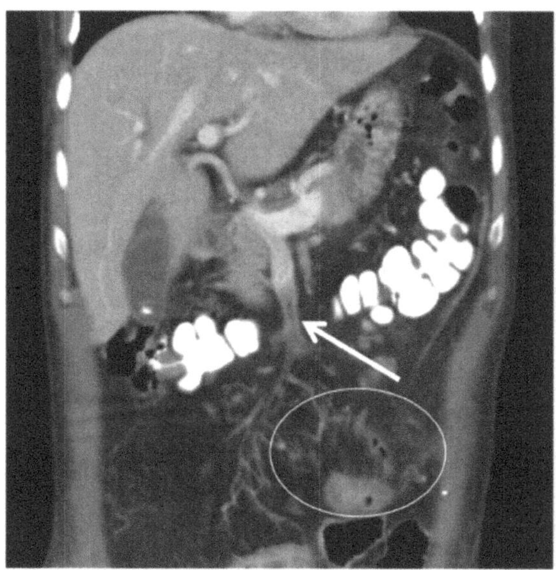

Fig. 24.3 Coronal CT image in a patient with history of ulcerative colitis demonstrates a filling defect within the superior mesenteric vein (arrow) extending distally to the portosplenic confluence and proximally into the jejunal and ileocolic tributaries. Perivascular hazy fat-stranding (arrowheads), one appearance of "misty mesentery", represents mesenteric edema from venous congestion. With permission from Dr. Abraham Dachman and Dr. Justin Ramirez

Laboratory Findings

Although laboratory tests are neither sensitive nor specific for MVT, patients often have abnormal laboratory findings such as leukocytosis, elevated lactic acid level and metabolic acidosis with an increased anion gap. Increased mortality occurs in patients with high serum lactic acid level, and metabolic acidosis, however normal serum lactate and pH do not rule out MVT. D-dimer testing is not helpful as it can also be increased in other infectious or inflammatory pathologic processes.

After confirmation of MVT, testing for hypercoagulability is recommended in patients without known provoking conditions. Those tests will help predict associated complications of the underlying hypercoagulable state and determine the duration of anticoagulant therapy.

Treatment

The treatment of MVT involves nonoperative management alone or in combination with interventional radiological options and surgery.

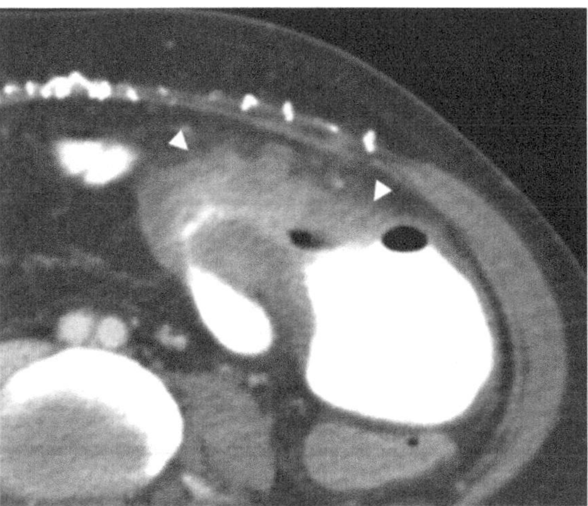

Fig. 24.4 Axial CT image from the same patient shows additional thrombus within the main portal vein extending into the right portal vein (arrowheads). Inflammatory bowel disease confers an in increased risk of venous thrombosis. With permission from Dr. Abraham Dachman and Dr. Justin Ramirez

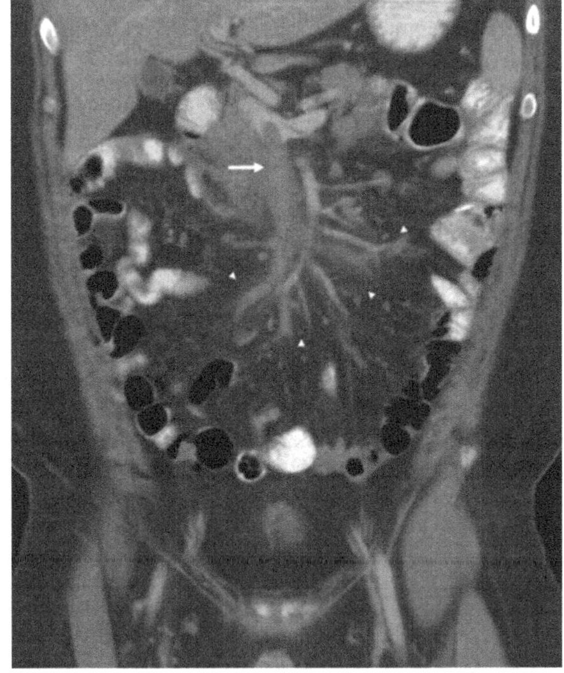

Fig. 24.5 Coronal CT image in a different patient again demonstrates a thrombus filling defect within the SMV (arrow) and ill-defined peri-visceral fat stranding (arrowheads). With permission from Dr. Abraham Dachman and Dr. Justin Ramirez

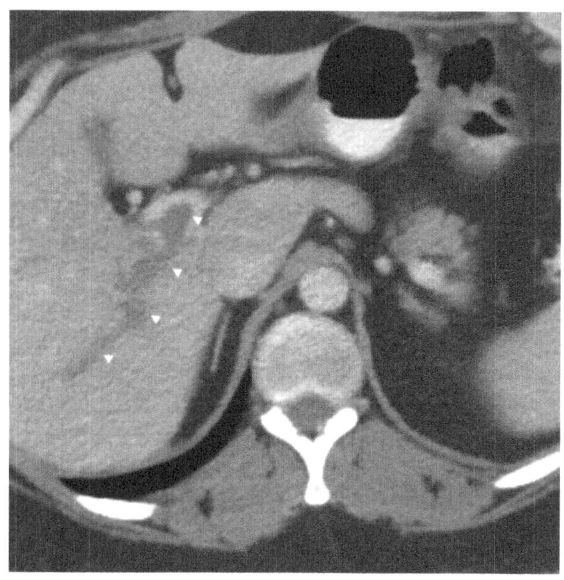

Fig. 24.6 Axial CT image from the same patient demonstrates bowel-wall edema with associated fat-stranding (arrowheads) suspicious for bowel ischemia. With permission from Dr. Abraham Dachman and Dr. Justin Ramirez

Nonoperative Management

In patients with MVT (acute or subacute) but no evidence of bowel infarction, the mainstay of treatment is systemic anticoagulation combined with supportive treatment. Supportive measures include bowel rest, nasogastric tube insertion for bowel decompression, intravenous fluid administration and pain control. The benefit of routine prophylactic antibiotics usage is not clear though they are required if the suspicion is high for bowel ischemia, infarction, and peritonitis.

Anticoagulant therapy is administered without delay and is the cornerstone of treatment. Anticoagulation limits thrombus progression, allows for venous recanalization, promotes bowel perfusion and improves survival. Anticoagulation therapy is initiated with either intravenous infusion of unfractionated heparin or subcutaneous injection with low molecular weight heparin (LMWH). The latter should not be used in patients who may undergo operative interventions or in those with renal dysfunction. After stabilization of patient's condition, anticoagulation is transitioned to oral agents (vitamin K antagonist such as warfarin or non-vitamin K oral anticoagulant, such as rivaroxaban). Anticoagulation for at least 3–6 months is recommended for cases with provoked and reversible causes. Idiopathic cases and those with hypercoagulable state usually require extended-duration anticoagulation.

Patients with chronic MVT may also benefit from anticoagulation assuming that the benefit of long-term anticoagulation outweighs its risk. In cases of chronic MVT presenting with gastrointestinal bleeding, anticoagulation is often withheld until control of bleeding is achieved.

Interventional Radiology Procedures

In patients with inadequate response to anticoagulation but without evidence of bowel infarction, interventional radiology is indicated. These procedures are generally performed in specialized centers with available technical expertise. However, these procedures are not a substitute for anticoagulation. According to some literature, catheter-directed fibrinolysis results in improvement of symptoms as well as reduces the rate of bowel resection. However, bleeding is a serious complication of catheter-directed fibrinolysis, and its reported rate is as high as 60%. Therefore, candidacy for fibrinolysis needs to be carefully determined depending on potential benefits and risks.

Catheter-assisted thrombectomy is considered as an option in cases of sizeable venous thromboses for patients in whom use of thrombolytic agents is contraindicated. Thrombectomy is most effective in the setting of an acute thrombus.

Surgery

Surgery should be performed without delay when there is evidence of bowel infarction or perforation. Surgical resection of necrotic bowel and anastomosis is the standard procedure with open laparotomy preferred over a laparoscopic procedure. Determining the extent of bowel resection can be challenging as the differentiation between viable and ischemic bowel is often difficult in the acute stage of MVT. Some surgeons advocate the use of intraoperative Doppler ultrasound or fluorescein infusion to assess bowel viability. Some authors recommend a conservative strategy to limit the extent of initial resection when there are areas of bowel with questionable viability, followed by second-look surgery within 24–48 h to reassess bowel viability as well as to determine the need of additional resection. The intention of a conservative strategy is to minimize unnecessary bowel resection, short bowel syndrome and its long-term consequences.

Prognosis

Compared to other forms of acute mesenteric ischemia, acute mesenteric venous thrombosis has a better prognosis. According to an extensive systematic review of acute mesenteric ischemia cases, the overall mortality rate of MVT was 44%, compared with 66–89% for arterial occlusive or nonocclusive ischemia. Early recognition and treatment have led to improved survival in MVT cases. In rapidly diagnosed and treated cases, some recent studies report mortality rates between 10 and 20% for acute MVT. Patients with intestinal infarction have a higher mortality rate that is greater than 75%. Important prognostic factors for survival from MVT also include age, the comorbidities predisposing to MVT and timing of diagnosis and treatment.

Table 24.2 Prognosis and mortality of MVT

• Prognostic factors
- Age - Co-morbidities (e.g. malignancy, cirrhosis, portal hypertension, gastrointestinal bleeding) - Time to diagnosis and treatment - Complications (e.g. intestinal infarction or perforation)
• The overall mortality rate of acute MVT = 44% • Cases of early diagnosis and treatment = 10 and 20% mortality • Cases complicated with intestinal infarction >75% • Prognosis of chronic MVT is variable depending on nature and severity of underlying etiology. 5-year survival rates are 78 to 82%

The prognosis of chronic MVT is related to the nature and severity of underlying illness. 5-year survival rates have been reported as high as 78 to 82% (Table 24.2).

Suggested Reading

1. Harnik IG, Brandt LJ. Mesenteric venous thrombosis. Vasc Med. 2010;15:407–18.
2. Russell CE, Wadhera RK, Piazza G. Mesenteric venous thrombosis. Circulation. 2015;131:1599–1603.
3. Hmoud B, Singal AK, Kamath PS. Mesenteric venous thrombosis. J Clin Exp Hepatol. 2014; 4(3):257–263.
4. Acosta S, Alhadad A, Svensson P, Ekberg O. Epidemiology, risk and prognostic factors in mesenteric venous thrombosis. Br J Surg. 2008;95:1245–51. https://doi.org/10.1002/bjs.6319.
5. Sulger E, Gonzalez L. Mesenteric venous thrombosis. StatPearls [Internet]. Treasure Island (FL): StatPearls Publishing; 2018 Jan-Oct 27. PMID: 29083683.
7. Acosta S, Ogren M, Sternby NH, et al. Mesenteric venous thrombosis with transmural intestinal infarction: a population-based study. J Vasc Surg. 2005;41:59.
8. Kanasaki S, Furukawa A, Fumoto K, Hamanaka Y, Ota S, Hirose T, Inoue A, Shirakawa T, Hung Nguyen LD, Tulyeubai S. Acute mesenteric ischemia: multidetector ct findings and endovascular management. Radiographics. 2018;38(3):945–961.
9. Zhang J, Duan ZQ, Song QB, et al. Acute mesenteric venous thrombosis: a better outcome achieved through improved imaging techniques and a changed policy of clinical management. Eur J Vasc Endovasc Surg. 2004;28:329.
10. Singal AK, Kamath PS, Tefferi A. Mesenteric venous thrombosis. Mayo Clin Proc. 2013;88 (3):285–294.
11. Orr DW, Harrison PM, Devlin J, et al. Chronic mesenteric venous thrombosis: evaluation and determinants of survival during long-term follow-up. Clin Gastroenterol Hepatol. 2007;5:80.
12. Kim HS, Patra A, Khan J, Arepally A, Streiff MB. Transhepatic catheter-directed thrombectomy and thrombolysis of acute superior mesenteric venous thrombosis. J Vasc Interv Radiol. 2005;16:1685–91.
13. Bala M, Kashuk J, Moore EE, Kluger Y, Biffl W, Gomes CA, Ben-Ishay O, Rubinstein C, Balogh ZJ, Civil I, Coccolini F, Leppaniemi A, Peitzman A, Ansaloni L, Sugrue M, Sartelli M, Di Saverio S, Fraga GP, Catena F. Acute mesenteric ischemia: guidelines of the World Society of Emergency Surgery. World J Emerg Surg. 2017;7(12):38.

14. Schoots IG, Koffeman GI, Legemate DA, et al. Systematic review of survival after acute mesenteric ischaemia according to disease aetiology. Br J Surg. 2004;91:17.
15. Management of the Diseases of Mesenteric Arteries and Veins: Clinical Practice Guidelines of the European Society of Vascular Surgery (ESVS).
16. Björck M, Koelemay M, Acosta S, Bastos Goncalves F, Kölbel T, Kolkman JJ, Lees T, Lefevre JH, Menyhei G, Oderich G, Esvs Guidelines Committee, Kolh P, de Borst GJ, Chakfe N, Debus S, Hinchliffe R, Kakkos S, Koncar I, Sanddal Lindholt J, Vega de Ceniga M, Vermassen F, Verzini F, Document Reviewers, Geelkerken B, Gloviczki P, Huber T, Naylor R. Management of the diseases of mesenteric arteries and veins: clinical practice guidelines of the European Society of Vascular Surgery (ESVS). Eur J Vasc Endovasc Surg. 2017;53(4):460–510.

Mesenteric Arterial Occlusion

25

Adib Chaus, Khaja M. Siraj, and Eli D. Ehrenpreis

Definition and Description of the Disease

Acute mesenteric arterial occlusion refers to the sudden onset of hypoperfusion to the small intestine due to reduction or cessation in arterial inflow. Acute arterial occlusion can be caused by embolic or thrombotic obstruction in the mesenteric vessels. Abnormal blood flow from the Superior Mesenteric Artery (SMA) and its branches are the most common cause of mesenteric arterial occlusion. It rarely occurs as a result of abnormal blood flow in the Celiac Artery (CE) or Inferior Mesenteric Artery (IMA) since both receive extensive arterial collateral circulation from the SMA. There are two primary etiologies of mesenteric arterial occlusion—embolism and thrombosis.

Epidemiology

Mesenteric arterial occlusion accounts for 0.01–0.1% of all hospital admissions in the United States. The incidence has been increasing incidence with aging of the US population. There is a slight predominance in women compared to men. The median age at diagnosis is 70 years old. In a prospective study in 2003 on acute

A. Chaus
Department of Internal Medicine – Cardiology, Advocate Lutheran General Hospital, 1775 Dempster St, 6 South, Park Ridge, IL 60068, USA

K. M. Siraj
Department of Internal Medicine, NorthShore Medical Group, Evanston, IL, USA

E. D. Ehrenpreis (✉)
Department of Medicine, Advocate Lutheran General Hospital, 1775 Dempster St, 6 South, Park Ridge, IL 60068, USA
e-mail: e2bioconsultants@gmail.com

E. D. Ehrenpreis et al. (eds.), *The Mesenteric Organ in Health and Disease*,
https://doi.org/10.1007/978-3-030-71963-0_25

thrombo-embolic occlusion of the SMA, the estimated annual incidence was 5.3 per 100,000 persons in the general population. However, this was likely an underestimate. A study that investigated a population with a high autopsy rate of 87% in Malmö, Sweden revealed the incidence of MAT to be 8.6 per 100,000 persons. This population consisted of 23,446 clinical and 7569 forensic autopsies. The study specifically looked at 997 patients in the clinical autopsy group and 9 patients in the forensic autopsy group that were coded for bowel ischemia. 211 of 997 and 2 of 9 autopsies demonstrated thrombo-embolic occlusion of the SMA. Additionally, the study also demonstrated increasing mortality with age.

Patients At Risk (See Table 25.1).

Any condition that increases the risk of arterial thrombosis or embolism from the heart to the arteries can lead to acute mesenteric arterial occlusion. The risk of embolism is higher in patients with aortic atherosclerosis, cardiac arrhythmias (notably atrial fibrillation and atrial flutter), cardiac valvular disease, infective endocarditis with systemic embolism, recent myocardial infarction, ventricular aneurysm, and aortic aneurysm. Among those etiologies, a dislodged thrombus from the aorta, left ventricle, left atrium, or cardiac valves are the most common causes of mesenteric arterial occlusion.

Thrombotic occlusion occurs in patients with general risk factors for atherosclerosis. Major risk factors for atherosclerosis include hyperlipidemia, hypertension, smoking, insulin resistance, diabetes, obesity, lack of physical activity, unhealthy diet, older age, and family history of early heart disease, chronic inflammation. These patients may have prior history of atherosclerotic disease, particularly coronary, cerebrovascular, and peripheral arteries. In these patients, any clinical scenario that leads to low flow or hypotension can result in acute-on-chronic arterial thrombosis.

Table 25.1 Risk factors for mesenteric arterial embolism and mesenteric arterial thrombosis

Risk factors	
Mesenteric arterial embolism	**Mesenteric arterial thrombosis**
Cardiac arrhythmias, especially atrial fibrillation	Atherosclerosis
Severe cardiac valvular disease	Advanced age
Recent cardiac surgery	Aortic dissection
Infective endocarditis	Severe dehydration
Vascular aortic prosthetic grafts proximal to SMA	Recent MI
Advanced age	Recent cardiac surgery
Low cardiac output	Vasculitis
	Fibromuscular dysplasia

Adapted from: Tryforos M. Acute Mesenteric Ischemia. In: Ferri FF, ed. Ferri's Clinical Advisor. Elsevier, 2019;53–55

Anatomy

The celiac artery (CA), superior mesenteric artery (SMA), and the inferior mesenteric artery (IMA) provide circulation to the foregut, midgut, and hindgut, respectively. Acute thromboembolic occlusion most commonly affects the SMA. The SMA supplies the distal duodenum, small intestine, and the large intestine to the mid-transverse colon. Additionally, it supplies collaterals to both the CA and IMA. The SMA originates approximately 1 cm distal to the CA and at an acute angle compared to the CA. The branches of the SMA include inferior, anterior, and posterior pancreaticoduodenal arteries, middle colic artery, right colic artery, ileocolic artery, jejunal and ileal branches. For additional details, see Chap. 4 Vascular Anatomy of the Mesentery.

Pathophysiology

Arterial Embolism—An embolus in the mesenteric artery migrates from the left atrium or ventricle, particularly in the setting of recent myocardial infarction (MI), atrial fibrillation, or atrial flutter. An embolism may also originate from artificial cardiac valves or the aorta. In mechanical valves, thromboembolic complications, including systemic emboli, occur at a rate of 0.7–6% patient years. Thrombosis or embolization from bioprosthetic valves is rare, except in the early postoperative period due to lack of endothelialization of the suture zone, which requires a few weeks to complete.

The gastrointestinal tract is the second most common destination for systemic embolization from endocarditis, the most common site being the central nervous system. The movement of a thrombus that has undergone arterial embolization will terminate at points of previous stenosis, narrow anatomical diameter and branch points. The SMA is highly susceptible because of its large diameter, acute takeoff angle from the abdominal aorta and rapid taper that begins only a few centimeters distal to its origin (Fig. 25.1). Due to this reason, the more proximal jejunal branches are spared. Therefore, the more distant region from the collateral circulation of the celiac and inferior mesenteric arteries, the middle segment of the jejunum, is most often involved in ischemia (Also see Chap. 26 Ischemic Enteropathy).

Arterial thrombosis—Atherosclerosis causes narrowing of the blood vessels, ultimately leading to thrombosis in areas of severe narrowing. The most common site of arterial thrombus in the mesentery is where the SMA and celiac artery originate from the aorta. The acuity of this process determines the level of collateral formation. Many of these of patients have a history consistent with chronic mesenteric ischemia. In those who may have developed progressive atherosclerosis over many years, collateral circulation from the celiac artery and inferior mesenteric artery may have resulted.

Fig. 25.1 Differing locations of SMA atherosclerotic thrombosis versus embolic occlusion

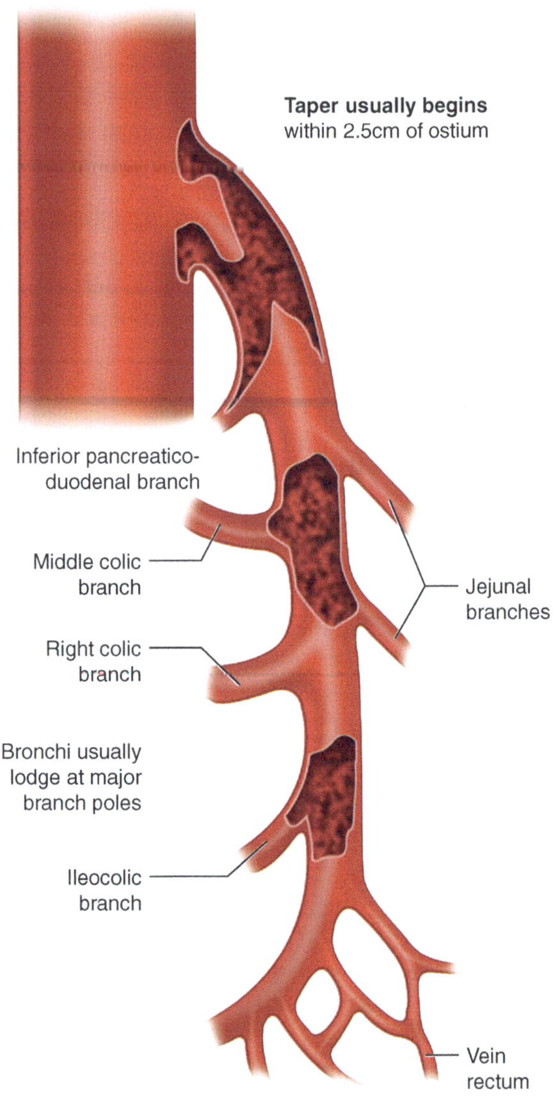

Arterial embolization has been shown to be a more common cause of MAT in studies comparing arterial embolization to arterial thrombosis. In the previously discussed Swedish autopsy study, the ratio of superior mesenteric embolus to thrombus was 1.4:1. With thrombotic occlusions, location was more proximal than embolic occlusions and intestinal infarction was more extensive.

Many factors are involved in leading to the atherosclerotic plaque being disrupted (Fig. 25.2). Turbulent blood flow impacting the plaque and vessel wall stress are some external factors affecting plaque stability. Inflammatory cell activity inside the plaque also has an impact on plaque stability. These include macrophages secreting metalloproteinases, which can weaken the fibrous cap, secretion of inflammatory cytokines that stimulate production of reactive oxygen species and proteolytic enzymes, and recruitment of monocytes and T-lymphocytes by chemotactic factors. Rupture of the fibrous cap due to these internal and external forces exposes thombogenic material to the blood in the vessel lumen. This leads to platelet accumulation then fibrin formation.

Signs and Symptoms

Arterial embolism—Patients with an acute embolic SMA occlusion classically present with pain out of proportion to their physical examination, particularly of the abdomen. The clinical triad associated with acute arterial embolism of the mesentery includes the sudden onset of abdominal pain associated with nausea, vomiting, and diarrhea, bowel emptying, with a clearly embolic source (such as atrial fibrillation or infective endocarditis). Over 20% of embolic events can involve multiple organs, therefore, a complete vascular examination in suspected acute

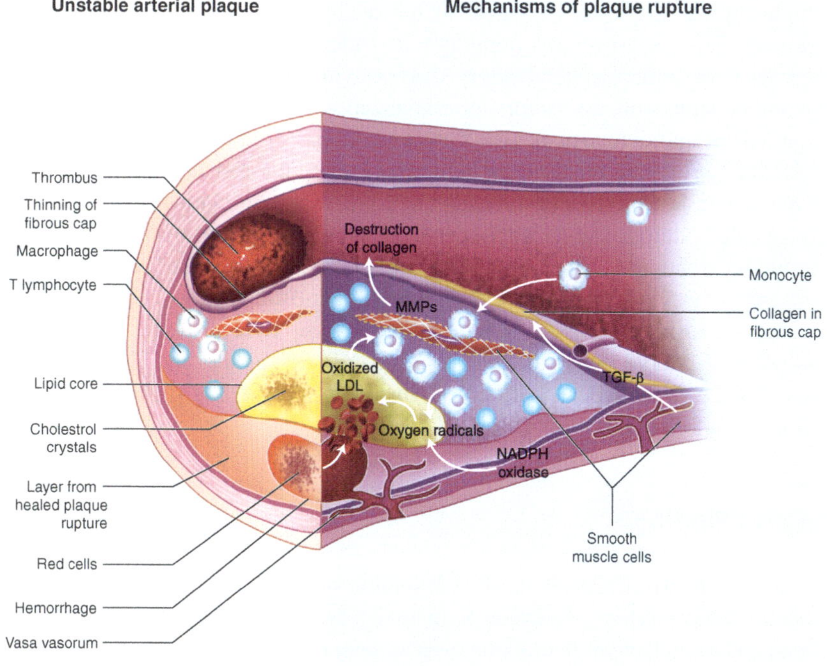

Fig. 25.2 Pathogenesis of atherosclerosis and plague rupture

mesenteric embolism should include an evaluation of the carotid arteries, as well as the upper and lower extremities to look for evidence of synchronous embolism.

Arterial thrombosis—Patients that present with arterial thrombosis usually have a prior history of atherosclerotic diseases involving the coronary, cerebrovascular, or peripheral arteries. When patients with these risk factors present with acute abdominal pain, the diagnosis of acute SMA thrombosis should be considered. Acute superimposed on chronic occlusion of a preexisting atherosclerotic lesion often results in a more insidious onset of pain, with collateral flow, which mitigates the severity of ischemia. Symptoms of chronic mesenteric artery thrombosis include postprandial abdominal pain, aversion to food, diarrhea, and weight loss.

In many patients, progressive atherosclerosis occurring over several to many years may have resulted in collateral circulation to the SMA. Therefore, these patients may or may not have symptoms of chronic mesenteric ischemia such as chronic postprandial abdominal pain. Prior history of vascular disease or cardiac diseases are important clues for differentiating thrombotic from embolic ischemia, possibly influencing the choice of initial treatment.

Diagnosis

A high clinical suspicion is critical to the diagnosis of acute mesenteric arterial occlusion. This is especially important in patients with known risk factors for atherosclerosis and peripheral artery disease with or without chronic abdominal pain and patients with risk factors for embolization (atrial fibrillation, valve disease, recent MI). Rapid diagnosis is vital to prevent intestinal infarction. Imaging studies are needed for definitive diagnosis. Abdominal computerized angiography (CTA) is the diagnostic test of choice in these patients with a high clinical suspicion. CTA may be able to delineate differences between an embolic versus thrombotic cause of mesenteric ischemia. Thrombotic occlusion generally appears as a thrombus overlying a heavily calcified lesion at the SMA ostium (Figs. 25.3 and 25.4). Embolic occlusion is usually characterized by an oval-shaped thrombus in a non-calcified artery, surrounded by contrast, in the middle and distal proximal SMA. If CTA is equivocal, but clinical suspicion remains high, angiogram of the mesenteric arteries may be required to confirm the diagnosis.

Physical Findings

The classically described physical examination in a patient with acute mesenteric ischemia is occurrence of severe abdominal exam in a patient with a relatively normal abdominal exam. In embolic events, there are differences in early versus late presentations of mesenteric ischemia. Early in the course of presentation, the presence of acute abdominal pain may represent reversible ischemia. This pain may then decrease in intensity, followed by an increase in abdominal pain with signs of

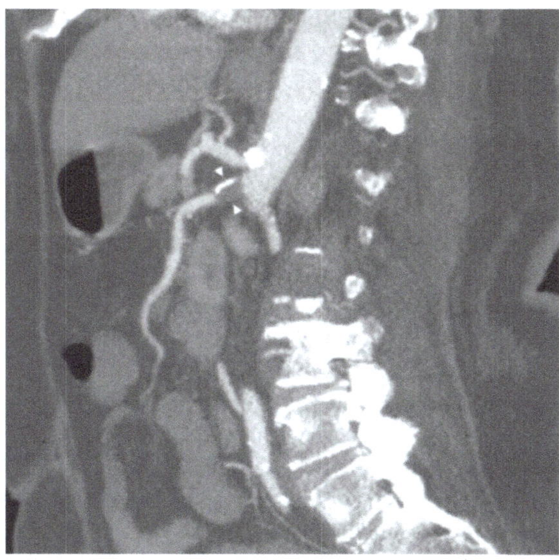

Fig. 25.3 Superior mesenteric artery thrombus. Sagittal CT angiogram image through the superior mesenteric artery shows an occlusive thrombus (arrowheads) at the SMA origin with eccentric calcification (high-density component). Contrast within the distal SMA is likely from celiac axis collateralization

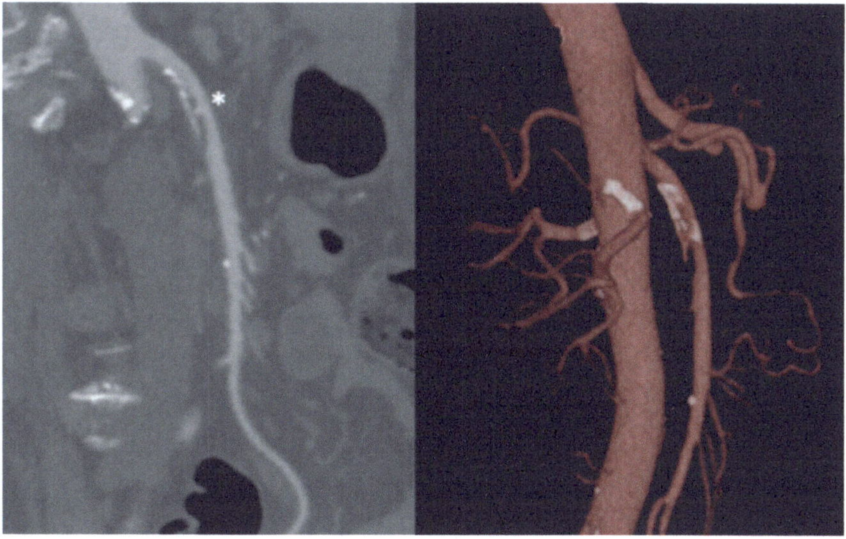

Fig. 25.4 Sagittal CT angiogram and 3D volume-rendered images from a different patient show an eccentric, non-occlusive SMA thrombus (asterisk) distal to the SMA origin. Appropriate window selection allows for clear differentiation between intraluminal contrast and calcified thrombus. With permission from Dr. Abraham Dachman and Dr. Justin Ramirez

peritonitis. The latter symptom represents irreversible ischemia. In a study of mesenteric ischemia, 95% of patients presented with abdominal pain, 44% with nausea, 35% with vomiting, 35% with diarrhea, and 16% with blood per rectum. If the etiology is thrombotic, patients may also present with acute abdominal pain, but progressive atherosclerosis in these patients may have resulted in collateral circulation to the SMA leading to a less dramatic presentation. This population could have long-standing postprandial abdominal pain, diarrhea, a fear of eating (sitophobia), and weight loss.

Laboratory Findings

Laboratory findings of mesenteric arterial thrombosis are neither specific or sensitive; however, certain lab values may be helpful in diagnosis. More than 90% of patients have an elevated leukocyte count and 88% have metabolic acidosis due to elevated lactate levels, indicating systemic versus local hypoperfusion. Serum lactate levels have a high sensitivity (91–100%), but a low specificity (42%), therefore, normal lactate levels do not rule out early mesenteric ischemia. Lab studies have shown that lactate produced in the porto-mesenteric venous circulation can be effectively metabolized by the liver, leading to normal lactate levels. D-dimer levels reflect ongoing clot formation and degradation via fibrinolysis. D-dimer levels have also been found to be a highly sensitive early marker (sensitivity approaching 100%), but D-dimer levels have a low specificity for mesenteric arterial occlusion, since many conditions are associated with elevated D-dimer levels. A normal D-dimer may exclude the diagnosis of SMA occlusion. A prospective study that included 47 patients with the clinical suspicion for acute mesenteric ischemia (AMI), was performed to determine if plasma D-dimer alone could replace a biphasic CT with mesenteric CTA examination in these patients. AMI was diagnosed in 28 of the 47 patients. The mean value of D-dimer testing was significantly higher in those with AMI, with sensitivity of 94.7% and specificity of 78.6%, The sensitivity and specificity values of biphasic CT with mesenteric CTA were 92.9% and 89.5% respectively.

Other nonspecific laboratory tests for MAT include intestinal fatty acid binding protein (I-FABP), serum alpha-glutathione S-transferase (alpha–GST), and cobalt-albumin binding assay (CABA). I-FABP is a small water-soluble protein that is present in mature enterocytes, and abundant in the mucosa of the small bowel from the duodenum to the ileum. I-FABP release into the bloodstream occurs during intestinal injury. In a prospective study evaluating 208 patients with a clinical suspicion of acute intestinal ischemia, vascular intestinal ischemia was diagnosed in 24 patients and non-vascular ischemia in 62 patients. The levels of most biomarkers (CRP, lactate, LDH, AST, D-dimer) were higher in those with vascular ischemia than the other groups. However, I-FABP was best at diagnosing vascular intestinal ischemia.

Imaging

CT Angiography—Multidetector CTA, a form of high-resolution CT scan, which allows for CT scanners to acquire multiple sections or slices with a thickness of 1 mm or smaller and increase the speed of the CT image acquisition. This scan, performed with and without contrast in the arterial and portal venous phases (triphasic protocol) is the recommended imaging technique. Three-dimensional reconstruction of CTA images may be performed. CTA is useful for demonstrating the presence arterial emboli and thromboses (Figs. 25.3 and 25.4). In advanced irreversible ischemia, CTA finding include intestinal dilation and thickness, decreased or absent visceral enhancement, pneumatosis intestinalis, and free intraperitoneal air. (See Chap. 26 Ischemic Enteropathy).

Conventional Angiography—In patients with mesenteric embolization, angiogram would demonstrate filling of the proximal SMA with no visualization of the distal vessels. Patients with thrombosis may have collateral circulation due to long-standing ischemia. Conventional angiography has the advantage of offering therapeutic and diagnostic options. However, it is highly invasive and unsuitable for critically ill patients. Additionally, it is not readily available and may delay surgical management.

MRI/MRA—In AMI, MRI findings are similar to those on CT. In comparing MRAs to CTAs, CTA has higher spatial resolution and faster acquisition times. This allows for better assessment of peripheral visceral branches of the IMA with better accuracy. MRA with contrast does allow the advantage of no radiation and lack of iodinated contrast agents and therefore, can be used in children and in those with renal dysfunction.

Duplex ultrasound—Ultrasound is not considered an appropriate imaging method for diagnosis of mesenteric arterial occlusion. It may show a thrombus or absent flow in a particular vessel, but is also operator-dependent. It is able to identify proximal vessel occlusions, but will not be able to detect distal occlusions.

Plain Abdominal Radiography—Signs on conventional radiography are late and nonspecific, but include ileus, thickened bowel with or without edema, pneumatosis intestinalis, and gas within the portal vein. Plain films were found to be completely normal in 25% of patients in one study with those with confirmed mesenteric ischemia. In this retrospective study, 23 proven cases of mesenteric infarction were reviewed with comparison of abdominal CT and plain film findings. Criteria considered specific for bowel infarction on plain films were identified in seven (30%) of 23 patients and comprised focally edematous bowel in six patients (26%) and pneumatosis intestinalis in one patient (4%).

Treatment

Overview

For patients with mesenteric arterial thrombosis leading to mesenteric ischemia, initial management should focus on active resuscitation and efforts to reduce vasospasm, and prevent propagation of intravascular clot, and minimize reperfusion injury. Intravenous fluid resuscitation with crystalloids should be started promptly to maintain intravascular volume and correct volume deficits. If there is no contraindication to anticoagulation, such as an active bleed as in ischemic colitis, therapeutic IV heparin should be administered to limit thrombus propagation. Broad-spectrum antibiotics should be started with focus on covering anaerobes and gram-negative organisms. Common antibiotics include cefepime with metronidazole, pipercillin-tazobactam, or meropenem. If vasopressors are required, vasoconstricting agents should be avoided to prevent exacerbation of mesenteric ischemia. Vasopressors such as dobutamine, low-dose dopamine, or milrinone should be considered to decrease the effect on mesenteric perfusion compared to other vasopressors.

The goal of treatment is to rapidly restore intestinal blood flow. The specific approach to treatment depends on whether the occlusion is embolic or thrombotic (Fig. 25.5a, b). Other important clinical factors to take into account include patient's risk factors for intervention and clinical stability. Options for treatment include surgical, endovascular, or a combined approach.

For patients who are poor surgical candidates with extensive infarction and no meaningful recovery, a palliative approach may be appropriate.

In patients that are hemodynamically stable with no signs of advanced ischemia, it may be appropriate to observe on anticoagulation if good collateral circulation is seen on imaging. Frequent lab studies and serial physical exams should be performed with low threshold for repeat imaging or intervention if symptoms progress.

Patients that are good surgical candidates with signs of advanced ischemia such as peritonitis, free air, or pneumatosis, should be immediately taken for laparotomy. While undergoing surgical exploration, bowel viability should be assessed. If the bowel is viable, attempts at reperfusion with methods such as thromboendarterectomy, mesenteric bypass, mesenteric stent should precede bowel resection.

For an embolic source for mesenteric occlusion, open surgical embolectomy is traditionally performed with removal of the thrombus, direct visualization and assessment for bowel viability. In mesenteric thrombosis, mesenteric bypass may be the preferred option because a thrombectomy may not provide a long-term solution for advanced atherosclerosis. Other surgical options for mesenteric thrombosis include retrograde angioplasty and arterial stenting. For hemodynamically stable patients with no signs of advanced ischemia, an endovascular approach may be appropriate. For patients with acute embolism, options include percutaneous clot aspiration or catheter-directed thrombolytic therapy. In patients with acute thrombosis, angioplasty/stenting is another reasonable option (see next section).

Surgical

Laparotomy—For patients with findings consistent with peritonitis, a laparotomy is indicated to evaluate the extent of intestinal infarction. During the laparotomy, appearance, pulsations, bleeding from cut surfaces, peristalsis is evaluated. The intestinal tract is also assessed for impending or gross perforation. If bowel

Management of acute embolic mesenteric occlusion

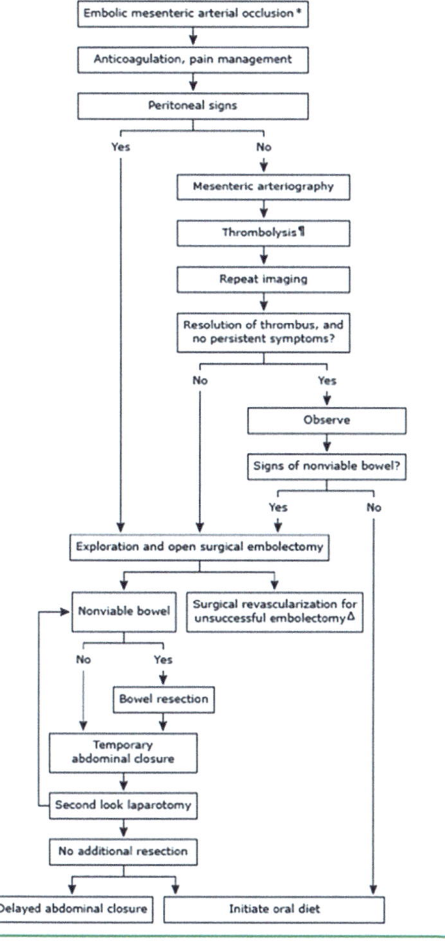

CT: computed tomography.
* Diagnosis typically made on CT angiography of the abdomen (refer to associated UpToDate algorithm on intestinal ischemia).
¶ In the absence of contraindications to thrombolytic therapy (eg, active bleeding).
Δ Includes thromboendarterectomy, mesenteric bypass, retrograde mesenteric stent, or a combination of these.

Fig. 25.5 Algorithms for management of acute embolic (**a**) and thrombotic mesenteric arterial (**b**) occlusion. Reproduced with permission from: Morgan J, Raut CP, Duensing A, Keedy VL. Epidemiology, classification, clinical presentation, prognostic features, and diagnostic work-up of gastrointestinal stromal tumors (GIST). In: UpToDate, Post TW (Ed), UpToDate, Waltham, MA. (Accessed on [Date].) Copyright © 2020 UpToDate, Inc. For more information visit www.uptodate.com

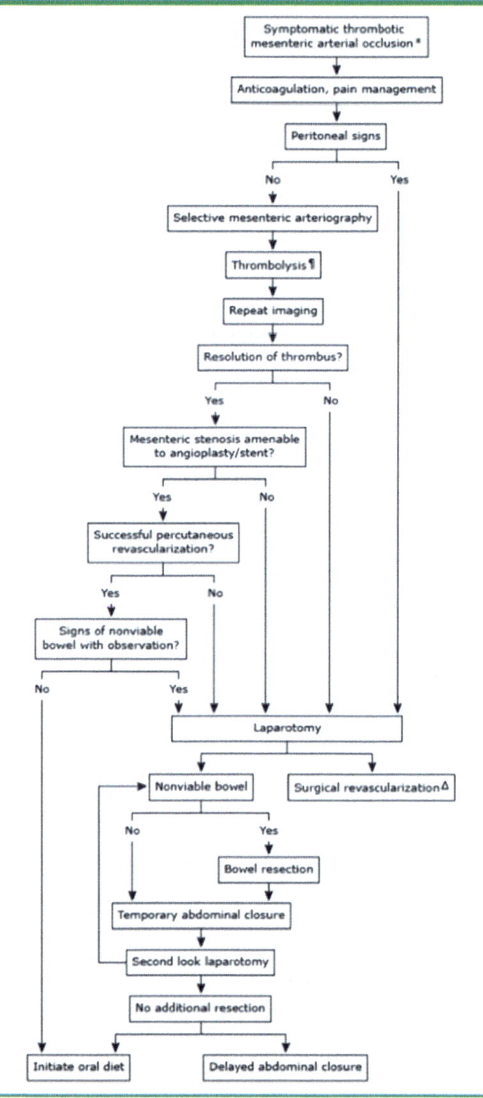

Fig. 25.5 (continued)

perforation is evident, the segment that is affected is resected with reconstruction or stoma formation until a second look laparotomy. Arterial revascularization is preferred and should precede bowel-resection in almost all patients, except when a segment of bowel is clearly non-viable. If there is question of bowel viability, perfusion should be restored and bowel viability should then be reassessed. For a majority of patients, a second-look laparotomy after mesenteric revascularization is performed, usually 24–48 hours after the initial laparotomy.

Surgical Revascularization

Open SMA Embolectomy—Arterial revascularization is preferred over resection. When a laparotomy is performed, the SMA is exposed. A transverse arteriotomy is then performed, a 3 or 4 Fr Fogarty catheter is introduced to embolectomy catheters are then introduced to remove the clot. Restoration of pulsatile blood flow confirms clearance of the thrombus after multiple passes are made and there is confirmation of absence of additional thrombi. Confirmation can be done by angiography of the SMA or transit time flow measurement. If neither are available, palpation of the pulse distally in the mesentery can be performed.

SMA Bypass—Typically used in patients with arterial thrombosis due to atherosclerotic disease. This is usually a vascular bypass graft that can originate from the infrarenal or supraceliac aorta. Autologous reversed saphenous vein is the preferred conduit to avoid intra-abdominal contamination from perforation or infarction of the bowel. Pre-operative CTA can be used to guide decision-making and can be used to determine sites with extensive atherosclerotic lesions. Bypass from the infrarenal aorta to the SMA is the simplest procedure that can be performed; however, the aorta is often heavily calcified and is therefore rarely performed. The supraceliac aorta or the anterior common iliac artery may be used as the inflow vessel.

Endovascular Procedures

An endovascular approach is considered in patients who are hemodynamically stable with no signs of advanced ischemia. Options include aspiration embolectomy, balloon angioplasty with stent placement, local pharmacologic thrombolysis.

Aspiration embolectomy—A catheter is passed into the ileocolic branch of the SMA. An introducer is placed proximal to the embolus in the SMA with aspiration with a smaller guiding catheter. Angiography is then performed and repeated aspirations done to remove clots (Fig. 25.6).

Angioplasty with stenting—Stenting can be performed in an anterograde vs retrograde fashion. In the antegrade approach, the catheter is introduced through the aorta into the SMA. A balloon expandable stent is placed at the level of the stenosis. Results are conformed by angiography and pressure measurements. A pressure gradient of more that 12 mmHg would indicate the need for additional angioplasty and/or stenting.

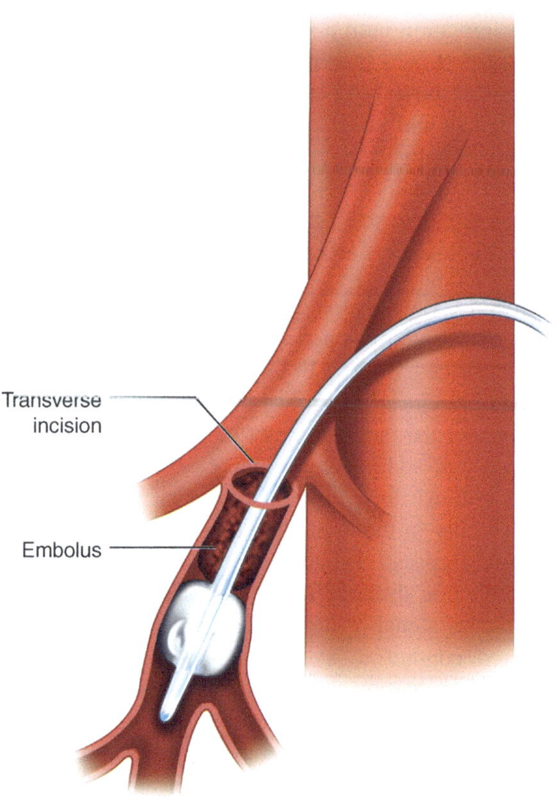

Fig. 25.6 Aspiration embolectoy of the SMA

If percutaneous access fails, laparotomy is performed with exposure of the SMA. A retrograde approach with recanalization and stenting is an option for reperfusion. The retrograde approach is performed with the device introduced distal to the obstruction via the SMA. Periodic stent surveillance should be done using duplex ultrasound or CT angiography, however, there are no specific guidelines for post-intervention follow-up.

Local SMA thrombolysis—In patients with incomplete aspiration embolectomy or distal mesenteric embolization, catheter-directed thrombolysis is an alternative. Low-dose heparin (500 units/hour) is administered after an end-hole catheter is advanced up to the clot or a multiple sidehole catheter is advanced through the clot. Infusion should be stopped if progressive signs of ischemia occur. Arteriography is repeated within the first four hours to assess for clot lysis. If this is not accomplished, abdominal exploration in the OR is indicated.

Medical Therapy

Medical treatment for those who survive after acute mesenteric arterial occlusion is focused on improving long-term survival. For patients with an embolic source for the occlusion, treatment should focus on the prevention of future embolic events. For atrial fibrillation, therapy with novel oral anticoagulants or a vitamin k antagonist is indicated. For patients with mesenteric arterial occlusion due to atherosclerosis, antiplatelet therapy and stain is indicated, in addition to decreasing risk factors such as tobacco use. A meta-analysis that studied the efficacy and safety of intensive lowering of LDL cholesterol, looked at 170,000 participants in 26 randomized trials. This meta-analysis concluded that there a 10% reduction in total mortality per 1.0 mmol/L reduction in LDL levels.

Prognosis

Mortality from acute mesenteric occlusion is high, even with aggressive treatment approaches. In patients undergoing intervention for acute mesenteric arterial occlusion, with open and endovascular treatment, perioperative mortality was 22% with complications occurring in 68%. Additionally, mortality was similar for open and endovascular approaches. Causes of death included, but not limited to, intra-abdominal sepsis with multiorgan dysfunction, hemorrhage, and cardiac causes. Complications for those who survived bowel resection develop short-gut syndrome, requiring long-term parenteral nutrition.

Suggested Reading

1. Björck M, et al. Management of the diseases of mesenteric arteries and veins. Eur J Vasc Endovasc Surg. 2017;53(4):460–510. https://doi.org/10.1016/j.ejvs.2017.01.010.
2. Bala M, et al. Acute mesenteric ischemia: guidelines of the world society of emergency surgery. World J Emerg Surg. 2017. https://doi.org/10.1186/s13017-017-0150-5.
3. Tilsed JV, et al. ESTES guidelines: acute mesenteric ischaemia. Eur J Trauma Emerg Surg: Off Publ Eur Trauma Soc. 2016;42(2):253–70. https://doi.org/10.1007/s00068-016-0634-0.
4. Clair D, Beach J. Mesenteric ischemia. N Engl J Med. 2016;374(10):959–68. https://doi.org/10.1056/NEJMra1503884.
5. Acosta S. Mesenteric ischemia. Curr Opin Crit Care. 2015;21:171–8.
6. Ha C, Magowan S, Accortt NA, Chen J, Stone CD. Risk of arterial thrombotic events in inflammatory bowel disease. Am J Gastroenterol. 2009;104(6):1445–51. [Medline].
7. Vokurka J, Olejnik J, Jedlicka V, Vesely M, Ciernik J, Paseka T. Acute mesenteric ischemia. Hepatogastroenterology. 2008;55(85):1349–52. [Medline].
8. Tallarita T, Oderich GS, Macedo TA, Gloviczki P, Misra S, Duncan AA, et al. Reinterventions for stent restenosis in patients treated for atherosclerotic mesenteric artery disease. J Vasc Surg. 2011;54(5):1422–9.e1. [Medline].
9. Cardin F, Fratta S, Perissinotto E, Casarrubea G, Inelmen EM, Terranova C, et al. Clinical correlation of mesenteric vascular disease in older patients. Aging Clin Exp Res. 2012;24(3 Suppl):43–6. [Medline].

10. Acosta S, Ogren M, Sternby NH, Bergqvist D, Björck M. Clinical implications for the management of acute thromboembolic occlusion of the superior mesenteric artery: autopsy findings in 213 patients. Ann Surg. 2005;241(3):516–22.
11. Bjornsson S, Resch T, Acosta S. Symptomatic mesenteric atherosclerotic disease-lessons learned from the diagnostic workup. J Gastrointest Surg. 2013;17:973–80.
12. Kougias P, Lau D, El Sayed HF, Zhou W, Huynh TT, Lin PH. Determinants of mortality and treatment outcome following surgical interventions for acute mesenteric ischemia. J Vasc Surg. 2007;46:467–74.
13. Matsumoto S, Sekine K, Funaoka H, Yamazaki M, Shimizu M, Hayashida K, Kitano M. Diagnostic performance of plasma biomarkers in patients with acute intestinal ischaemia. Br J Surg. 2014;101:232–8.
14. Menke J. Diagnostic accuracy of multidetector CT in acute mesenteric ischemia: systematic review and meta-analysis. Radiology. 2010;256:93–101.
15. Oliva IB, Davarpanah AH, Rybicki FJ, Desjardins B, Flamm SD, Francois CJ, et al. ACR appropriateness criteria (R) imaging of mesenteric ischemia. Abdom Imaging. 2013;38:714–9.
16. Ryer EJ, Kalra M, Oderich GS, Duncan AA, Gloviczki P, Cha S, Bower TC. Revascularization for acute mesenteric ischemia. J Vasc Surg. 2012;55(6):1682–9.
17. Block TA, Acosta S, Bjorck M. Endovascular and open surgery for acute occlusion of the superior mesenteric artery. J Vasc Surg. 2010;52:959–66.
18. Kim HK, Hwang D, Park S, Huh S, Lee JM, Yun WS, Kim YW. Effect of clinical suspicion by referral physician and early outcomes in patients with acute superior mesenteric artery embolism. Vasc Spec Int. 2017;33(3):99–107. https://doi.org/10.5758/vsi.2017.33.3.99.
19. Gutstein DE, Fuster V. Pathophysiology and clinical significance of atherosclerotic plaque rupture. Cardiovasc Res. 1999;41(2):323–33. https://doi.org/10.1016/S0008-6363(98)00322-8.
20. Akyildiz H, et al. The correlation of the D-dimer test and biphasic computed tomography with mesenteric computed tomography angiography in the diagnosis of acute mesenteric ischemia. Am J Surg. 2009;197(4):429–33.
21. Thuijls G, van Wijck K, Grootjans J, Derikx JP, van Bijnen AA, Heineman E, et al. Early diagnosis of intestinal ischemia using urinary and plasma fatty acid binding proteins. Ann Surg. 2011;253:303–8.
22. Rafieian-Kopaei M, Setorki M, Doudi M, Baradaran A, Nasri H. Atherosclerosis: process, indicators, risk factors and new hopes. Int J Prev Med. 2014;5(8):927–46.
23. Tryforos M. Acute mesenteric ischemia. Ferri's Clin Advis. 2019:53–55.
24. Gargiulo S, Gramanzini M, Mancini M. Molecular imaging of vulnerable atherosclerotic plaques in animal models. Int J Mol Sci. 2016;17:1511. https://doi.org/10.3390/ijms17091511.
25. Pearl G, Nieuwsma J. Acute mesenteric arterial occlusion. UpToDate. 2018; https://www.uptodate.com/contents/acute-mesenteric-arterial-occlusion.

26 Ischemic Enteropathy (Also Called Mesenteric Ischemia)

Dorsa Samsami and Eli D. Ehrenpreis

Definition and Description of the Disease

Diminished small intestinal blood flow and its associated consequence are termed either ischemic enteropathy or small intestinal mesenteric ischemia. This condition can arise from a decrease in arterial or venous supply to the small intestine.

Classification and Terminology

There are a variety of classifications systems used to characterize decreased blood flow of mesenteric vascular origin. For the purpose of this chapter, mesenteric ischemia is divided by temporal occurrence (acute versus chronic) and the site of diminished mesenteric blood flow (arterial versus venous). A variety of etiologies of acute and chronic mesenteric ischemia are present. According to most classification systems, mesenteric ischemia affecting the colon alone is termed ischemic colitis (Table 26.1). Because of contrasting etiologies, presentations and treatments for ischemic colitis, this chapter will focus on small intestinal disease caused by diminished mesenteric blood flow. Further details regarding colonic disease are presented in Chap. 27 Ischemic Colitis.

Acute mesenteric ischemia is characterized by the sudden disruption of blood flow in the mesenteric vasculature and is associated with devastating clinical consequences. Chronic mesenteric ischemia is characterized by a more insidious

D. Samsami
Department of Internal Medicine, Advocate Lutheran General Hospital, 1775 Dempster St, 6 South, Park Ridge, IL 60068, USA

E. D. Ehrenpreis (✉)
Department of Medicine, Advocate Lutheran General Hospital, 1775 Dempster St, 6 South, Park Ridge, IL 60068, USA
e-mail: e2bioconsultants@gmail.com

E. D. Ehrenpreis et al. (eds.), *The Mesenteric Organ in Health and Disease*,
https://doi.org/10.1007/978-3-030-71963-0_26

Table 26.1 American gastroenterological association classification of mesenteric ischemia

Acute mesenteric ischemia
Superior mesenteric artery thrombosis
Superior mesenteric artery embolus
Nonocclusive mesenteric ischemia
Superior mesenteric vein thrombosis
Focal segmental ischemia

process due to the slow development of mesenteric arterial obstruction and the abundance of collateral blood flow to the intestines. Late complications of mesenteric ischemia include intestinal stricture and short bowel syndrome.

Acute ischemic enteropathy is a serious medical condition that may lead to bowel infarction and sepsis with an overall morbidity and mortality approaching 50%. Because of its severity and potential for rapid and fatal consequences, early diagnosis of ischemic enteropathy is of vital importance. Acute intestinal ischemia may be result of either arterial or venous obstruction, while chronic mesenteric ischemia is primarily due to atherosclerotic narrowing of and diminished blood flow through the mesenteric arterial circulation.

Intestinal Vascular Anatomy

The arterial supply to the intestines is primarily from the superior mesenteric artery and the inferior mesenteric artery. The superior mesenteric artery supplies the entire small intestine except for the proximal duodenum. Both the superior and inferior mesenteric arteries supply the colon. There is also an extensive collateral circulation to the intestines. (See Chap. 4, Mesenteric Vascular Anatomy).

Epidemiology

Acute mesenteric ischemia is estimated to be the primary diagnosis for 0.1% of hospital admissions. Mesenteric arterial occlusion is the cause in 1% of patients presenting with an acute abdomen. Mesenteric artery stenosis occurs in as many as 10% of the population over 65 years of age. The majority of these patients are asymptomatic, with chronic mesenteric ischemia accounting for about 5% of intestinal ischemic events. The disparity between the occurrence of mesenteric artery stenosis and the development of chronic mesenteric ischemia is due in part to the extensive network of arterial collaterals. This network provides compensation for diminished blood flow from stenotic mesenteric arteries.

Patients at Risk

Risk factors for acute mesenteric arterial occlusion include any processes that are associated with diminished mesenteric blood flow, intrinsic arterial or venous thrombosis or increase the potential for mesenteric arterial embolism. These include congestive heart failure, cardiac arrhythmia, valvular heart disease, infective endocarditis, aortic atherosclerosis, acute myocardial infarction, aortic aneurysm, and ventricular aneurysm. Patients with peripheral arterial disease, advanced age, and low cardiac output are at increased risk of acute mesenteric ischemia. Other risk factors include the use of vasoconstrictive medications and drugs of abuse, primary and secondary causes of increased platelet concentration, and vasculitis, most commonly polyarteritis nodosa.

Chronic mesenteric ischemia occurs in patients with mesenteric arterial atherosclerosis. Risk factors for chronic intestinal ischemia include hyperlipidemia, hypertension, smoking and diabetes mellitis. Symptoms present most commonly between the ages of 50 and 70 in patients with pre-existing atherosclerotic disease. Chronic mesenteric ischemia is more common in females. Specific etiologies of acute mesenteric ischemia and their frequency of occurrence are shown in in Table 26.2.

Pathophysiology

Between 25 and 35% of cardiac output is directed toward splanchnic blood flow, the majority of which supplies the mucosal and submucosal layers of the intestine. Multiple regulatory pathways are involved in the maintenance of adequate blood flow to the intestines. These pathways may be viewed as having intrinsic components involving local metabolic and myogenic functions and extrinsic components involving neural and humoral regulatory systems, Decreased splanchnic blood flow resulting from arterial or venous obstruction or dysregulation related to the aforementioned pathways, all may result in the development of ischemic damage to the intestine.

Table 26.2 Causes and approximate frequencies of acute mesenteric ischemia

Etiology	Cause	Frequency (%)
Arterial	Superior mesenteric artery thrombosis	54–68
	Superior mesenteric artery embolus	26–32
	Nonocclusive mesenteric ischemia	10
Venous	Superior mesenteric vein thrombosis	5
	Focal segmental ischemia	5

Adapted from Brandt LJ, Feuerstadt P. Intestinal Ischemia. In: Gastrointestinal and Liver Disease. 10th ed. p. 2076–2101

Table 26.3 Physiologic and pharmacologic factors regulating mesenteric blood flow

	Decrease blood flow by splanchnic vasoconstriction	Increase blood flow by splanchnic vasodilation
Humoral (endogenous and exogenous) factors	Epinephrine (high dose)	Epinephrine (low dose)
	Norepinephrine (moderate to high dose)	Norepinephrine (low dose)
	Dopamine (high dose)	Dopamine (low dose)
	Phenylephrine	Dobutamine
	Vasopressin	Sodium nitroprusside
	Angiotensin II	Papaverine
	Digoxin	Nitric oxide
Neural factors	α-Adrenergic receptors	β-Adrenergic receptors
	Dopaminergic receptors	

Adapted from Oldenburg WA, Lau LL, Rodenberg TJ, Edmonds HJ, Burger CD. Acute Mesenteric Ischemia. A Clinical Review. *Arch Intern Med.* 2004;164(10):1054–1062

The majority of cases of acute intestinal ischemia result from blockage of blood flow in the superior mesenteric artery. Pharmacologic causes of mesenteric ischemia are also shown in Table 26.3. Reduction in either arterial or venous blood flow may result in the development of acute mesenteric ischemia, while chronic mesenteric ischemia is generally caused by arterial narrowing. (See Chap. 25 Mesenteric Arterial Thrombosis and Chap. 24 Mesenteric Venous Thrombosis).

Acute Mesenteric Ischemia of Arterial Origen

There are several mechanisms resulting in the development of an acute decrease in mesenteric arterial blood flow.

SMA embolus (SMAE): Although controversial, it is currently believed that SMAE is the most common cause of acute mesenteric ischemia and is responsible for 50% of cases. The origins of SMAE in these patients are thromboses in the left atrium, left ventricle, cardiac valves and the abdominal aorta. Atrial fibrillation and other arrythmias predispose to atrial emboli. Ventricular stasis and poor cardiac output are associated with venous thrombi. Valvular heart disease, especially when complicated by endocarditis, may be accompanied by valvular emboli. The SMA is a relatively large artery and low takeoff angle (25–60 degrees), from the aorta. This creates ideal condition for embolic obstruction, as emboli tend to become wedged in areas of anatomic narrowing (Fig. 26.1). In general, embolic obstruction occurs about 6–8 cm distal to the origin of the SMA, at a narrowed location near the beginning of the middle colic artery. Emboli to other arterial systems are seen in 20% of cases and diminished circulation to other organs including the spleen, or kidney may be seen.

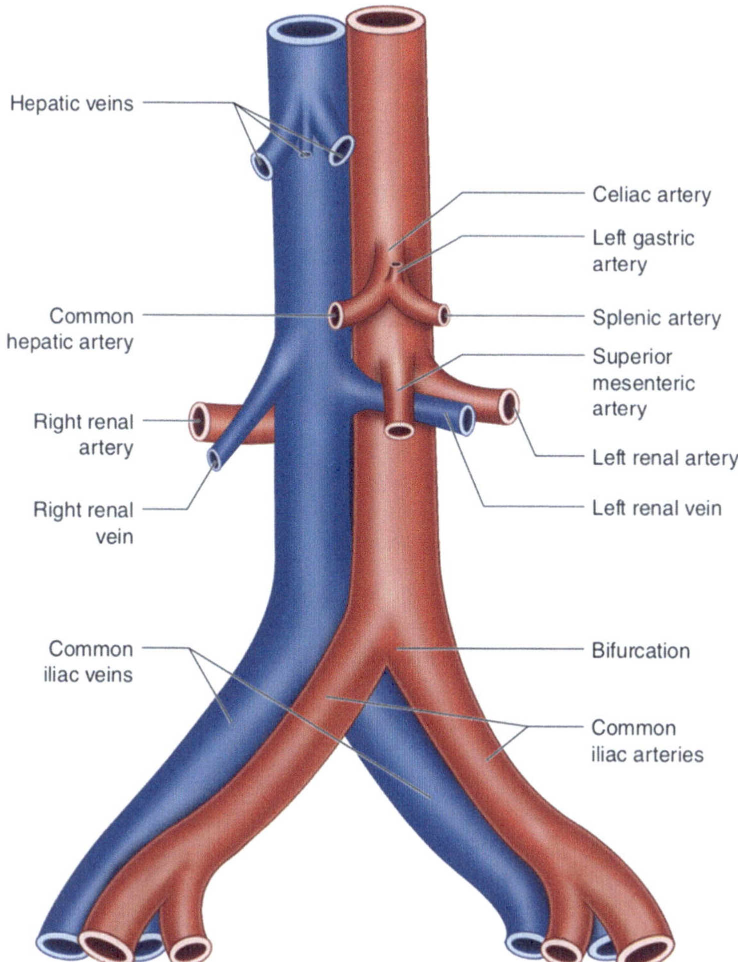

Fig. 26.1 Schematic representation of the abdominal aorta and inferior vena cava. The superior mesenteric artery is shown having a very sharply angulated takeoff from the abdominal aorta

SMA Thrombosis (SMAT)

Approximately 25% of acute intestinal ischemia of arterial origin are due to SMAT. Thrombosis of the SMA occurs in patients with longstanding atherosclerotic involvement and narrowing of the SMA. These patients may have chronic symptoms including sitophobia (fear of eating) and post- prandial abdominal pain, although relative compensation for SMA narrowing takes place by the formation of collateral blood flow. Decreased blood circulation in the SMA after cardiac decompensation from an acute myocardial infarction, congestive heart failure or vessel dissection results in acute thrombus formation. In general, symptomatic acute mesenteric ischemia develops when the celiac trunk and SMA are involved.

Nonocclusive Mesenteric Ischemia (NOMI)

20% of cases of ischemic enteropathy of arterial origin are due to NOMI. NOMI occurs when constriction of the splanchnic bed occurs after an acute cardiovascular event without the presence of an anatomic obstruction. It is most commonly seen in critically ill patients with sepsis and markedly diminished cardiac output, in which failure of normal autoregulation of mesenteric blood flow is present. A variety of pharmacologic agents that reduce the reginal blood flow within the splanchnic circulation may result in the development of NOMI. These include vasopressors such as norepinephrine, epinephrine, vasopressin and phenylephrine, digitalis (which diminishes splanchnic blood flow), vasoconstrictors such as cocaine and sumatriptan as well as diuretic medications.

Focal Segmental Ischemia (FSI)

FSI refers to the development of damage to short segments of the small intestine as a result of vascular compromise. Causes of FSI may be thrombotic, embolic, traumatic and vasculitic. FSI may occur secondary to medications such as oral contraceptives. It may also be seen with strangulated intestinal hernias.

Acute Mesenteric Ischemia of Venous Origen (MVT)

Approximately 10% of all cases of ischemic enteropathy result from mesenteric venous thrombosis. The superior mesenteric vein is the most common site of occurrence of MVT. A variety of conditions that are associated with hypercoagulable states predispose to the development of MVT. These include neoplastic disorders, cirrhosis of the liver, acute pancreatitis, and hereditary hypercoagulable states including Factor V Leiden, prothrombin gene mutation, deficiencies of anticlotting clotting (such as antithrombin, protein C and protein S), and elevated levels of homocysteine or fibrinogen or dysfibrinogenemia. Acute MVT has an expected 30% survival at 30 days and a poor longt-erm prognosis. For additional information see Chap. 24, Mesenteric Vein Thrombosis.

Table 26.4 Causes of chronic mesenteric ischemia

Causes of chronic mesenteric ischemia
Occlusion from atherosclerosis
Dissection of mesenteric vasculature
Inflammation of the vessels (vasculitis)
Fibromuscular dysplasia of the mesenteric vessels
Radiation-induced vascular injury
Chronic cocaine abuse

Chronic mesenteric ischemia: Chronic mesenteric ischemia is an insidious process and is usually related to arterial occlusion from atherosclerosis. Additional causes of chronic mesenteric ischemia include dissection of mesenteric vasculature, inflammation of the vessels (vasculitis), fibromuscular dysplasia of the mesenteric vessels, radiation-induced vascular injury, and chronic cocaine abuse (Table 26.4).

Since the development of mesenteric atherosclerosis is a slow process, collateralization of mesenteric vessels occurs. This process results in clinical compensation for loss of flow in vessels affected by atherosclerosis. Because of this phenomenon, signs and symptoms of intestinal ischemia typically occur when at least two and sometimes three arteries are affected.

Signs and Symptoms of Ischemic Enteropathy

The medical history usually does not provide enough information to secure the diagnosis of acute ischemic enteropathy. Acute mesenteric ischemia and nonocclusive mesenteric ischemia should always be considered in patients with rapidly developing and diffuse abdominal pain. This is particularly true in patients with risk factors for acute mesenteric ischemia including patients with cardiovascular disease such as congestive heart failure or recent myocardial infarction, hereditary hypercoagulable states, vasculitis or the use of vasoconstricting medications including cocaine. Patients with acute ischemic enteropathy present with abdominal pain that is out of proportion to the physical exam. Associated symptoms may include nausea, vomiting, fever, and gastrointestinal hemorrhage. As ischemia persists, symptoms may evolve and the pain may temporarily disappear indicating that significant bowel necrosis has occurred. The classic presentation of chronic ischemic enteropathy is recurrent post-prandial pain. Signs and symptoms typically appear when at least two arteries are affected. The pain is often described as dull, crampy, located in the epigastric region, and occurring approximately one hour after eating. Symptoms often resolve within a couple of hours. Patients often develop food aversion (sitophobia) due to anticipation of post-prandial pain, thus leading to weight loss.

Diagnosis

The diagnostic approach is largely dependent on the clinical presentation. If a patient is hemodynamically stable, without signs of sepsis, the first recommended imaging study is a contrast enhanced computerized tomogram (CT) or CT angiography (CTA). Although a CT scan is a sensitive test, it can be negative in early stages of bowel ischemia. If the CT scan is non-diagnostic and suspicion for mesenteric ischemia remains high, the next diagnostic test would be a mesenteric angiogram with arterial and venous phases. On the other hand, if a patient presents with hemodynamic instability, plain abdominal films can be of utility to show free air and other signs of advanced ischemia. If these signs are present, a laparotomy is indicated.

Physical Findings

In early acute mesenteric ischemia, even in the presence of severe pain, the abdominal examination may be normal or only mildly abnormal with minimal abdominal distention and tenderness. This is why the term "pain out of proportion to exam" has become a classic description of acute mesenteric ischemia. As intestinal ischemia progresses and transmural bowel infarction develops, the abdomen can become grossly distended, bowel sounds are diminished or absent, and peritoneal signs can develop. There are a few other pertinent exam findings. For instance, occult blood may also be present in the stool. A feculent odor of the breath can also be observed with bowel ischemia. In addition, there can be non-specific signs consistent with dehydration and shock that indicate a deteriorating clinical course. On the other hand, physical exam findings for chronic mesenteric ischemia are usually nonspecific. Due to the fear of eating from postprandial pain, these patients can be undernourished and cachectic appearing. The abdominal exam is soft and non-tender however may be slightly distended. An abdominal bruit is commonly present but this is also a nonspecific finding.

Laboratory Findings

Laboratory values are typically non-specific. While abnormal laboratory findings can support the diagnosis of acute ischemic enteropathy, normal laboratory values should not exclude the diagnosis or delay further workup. The reason for this is because most abnormalities only arise once ischemic enteropathy has progressed to bowel necrosis. The most common findings for mesenteric ischemia include leukocytosis, elevated amylase, phosphorus, and metabolic acidosis. Leukocytosis can be moderate to markedly elevated, however 10% of patient have a normal white blood cell count. Elevated lactic acid is 86% sensitive and 44% specific for acute mesenteric ischemia. The specificity increases once other causes of lactic acidosis are ruled out such as shock and diabetic ketoacidosis. Elevated serum amylase has been observed in 50% of patients. Studies have shown that the bowel contains high concentrations of both organic and inorganic phosphate. When there is ischemia, phosphate is released from the bowel, and elevated phosphorus has been observed in 80% of patients. D-dimer can also be checked, as it is an early marker for ischemia, increasing as early as 30 minutes after an ischemic event. While a normal d-dimer can help exclude acute mesenteric ischemia, an elevated level is not diagnostic as there are many other conditions that increase d-dimer levels.

Imaging

A plain abdominal radiograph is often the initial test ordered in patients with acute abdominal pain however it has very little diagnostic role in mesenteric ischemia. Its primary purpose is to rule out other diagnoses such as perforated viscus or a bowel obstruction. Findings suggestive of mesenteric ischemia include presence of an ileus, bowel wall thickening, as well and pneumatosis intestinalis. A plain radiograph can also show free intraperitoneal air, a late sign of ischemia indicating that intestinal perforation has already occurred and that urgent surgical exploration is required (Fig. 26.2).

For patients with suspected acute mesenteric ischemia who are hemodynamically stable, a contrast enhanced CT scan, especially a CTA is performed. Findings on CT that are associated with mesenteric ischemia include bowel wall thickening, submucosal edema or hemorrhage, pneumatosis intestinalis, portal vein gas, porto-mesenteric thrombosis, mesenteric arterial calcification, and mesenteric artery occlusion. Findings consistent with intestinal infarction include discontinuous mucosal enhancement and adjacent mesenteric stranding (Fig. 26.3).

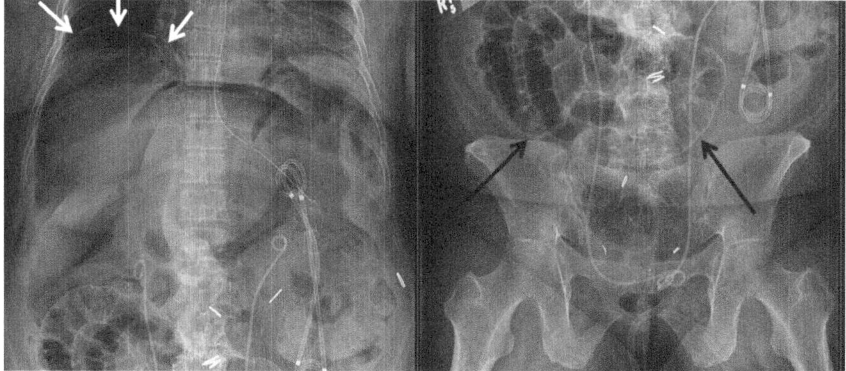

Fig. 26.2 Anterior–posterior radiographic images of the abdomen show massive pneumoperitoneum with air under the right hemidiaphragm (white arrows) and lucent air outlining the small bowel walls (black arrows). The latter finding is call Rigler's sign

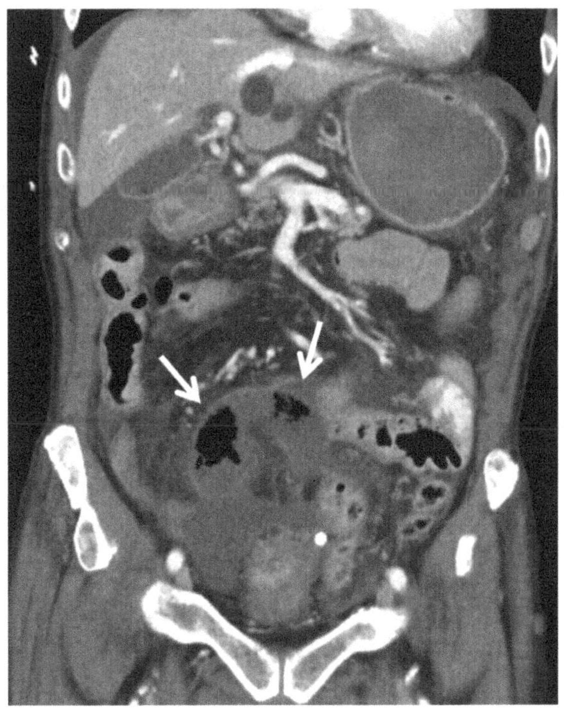

Fig. 26.3 Coronal, contrast-enhanced computed tomography image shows a dilated small bowel loop with discontinuous mucosal enhancement (arrows) and adjacent mesenteric fat-stranding suggestive of small bowel infarction in this patient with SMA thrombosis (not shown). With permission from Dr. Abraham Dachman and Dr. Justin Ramirez

CTA is the imaging modality of choice and should be performed without delay. A meta-analysis of six studies published between 1996 and 2009 on the diagnostic accuracy of multidetector CT in acute mesenteric ischemia showed a pooled sensitivity of 93% and specificity of 96%. It is important to note that the CT angiogram should not be performed with oral contrast. This is because oral contrast can obscure the mesenteric vessels and bowel wall enhancement. Disadvantages of CT include radiation exposure, the risk of contrast nephropathy, and allergic reactions to iodinated contrast. CT angiography should still be performed in patients despite renal failure as the consequences of delayed diagnosis outweigh detrimental renal affects. If CT angiography findings are equivocal, catheter directed angiography may be necessary.

MRA is a good option in in patients to avoid radiation exposure and for those who cannot have iodinated contrast, however it takes significantly longer to perform. MRA additionally appears to be more sensitive at identifying venous occlusions. By contrast, CT demonstrates better visualization of the IMA, the peripheral splanchnic vessels, calcified plaques, and previously placed stents.

Duplex ultrasonography is another imaging modality used in the appropriate clinical setting. It can identify occlusions in the celiac or superior mesenteric arteries or arterial stenosis. Duplex ultrasonography has limitations based on body habitus, presence of bowel gas, effects of respiration, and operator expertise.

Histologic Findings

Histological specimens demonstrate the pathophysiologic processes associated with mesenteric ischemia, and may provide useful information regarding the extent and severity of intestinal injury. Findings differ by temporal occurrence ischemia as well as the site of diminished mesenteric blood flow. In the small intestine, ischemic changes due to arterial insufficiency initially occur at the tips of the intestinal villi. Early stages of injury show minor surface epithelial degeneration and mild hyalinization of the lamina propria caused by leakage of plasma proteins from injured capillaries. With further ischemic insult, the necrosis extends through the entire length of the villi and starts to involve the bases of the crypts. "Crypt ghosts" are empty spaces within the mucosa and represent remnants of intestinal crypts at this stage. The submucosa is typically edematous and erythematous with engorged veins. It is described as friable, and may have associated hemorrhage. The muscularis propria is the last layer to be affected. As this layer undergoes necrosis, the intestinal wall becomes thin, friable, and easily disrupted. Organizing thrombi may be detected within blood vessels (Figs. 26.4 and 26.5).

Acute intestinal ischemia due to venous obstruction ultimately acts in a manner similar to arterial insufficiency. However, it is differentiated by an initial phase that is characterized by severe vascular congestion, hemorrhage, and marked tissue edema. With progressive tissue insult, arterial flow ceases.

Chronic mesenteric ischemia has a histologically distinct appearance when compared to acute causes of mesenteric ischemia. Chronic mesenteric ischemia is characterized by fibrosis that usually involves all layers of the intestinal wall, including the mucosa. Hemosiderin deposits may also be present. These indicate previous episodes of hemorrhage. Chronic intestinal ischemia may also be characterized by the presence of linear ulcerations and strictures.

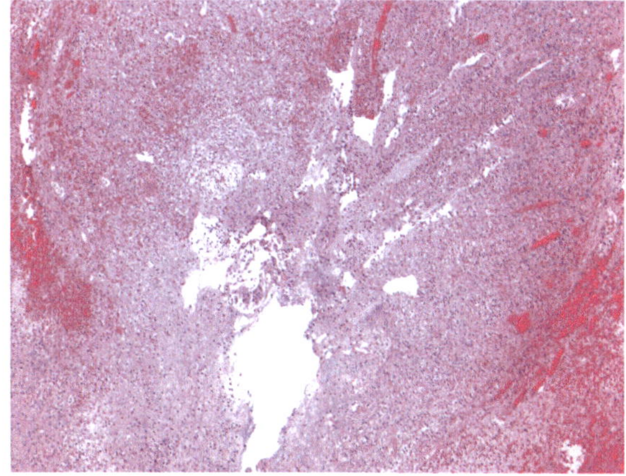

Fig. 26.4 Granulation tissue, ulcerations, and necrosis extend into submucosa and surrounding smooth muscle fibers of muscularis mucosa. There is a surface exudate of neutrophils, fibrin, and mucosal necrosis (H&E, 4x)

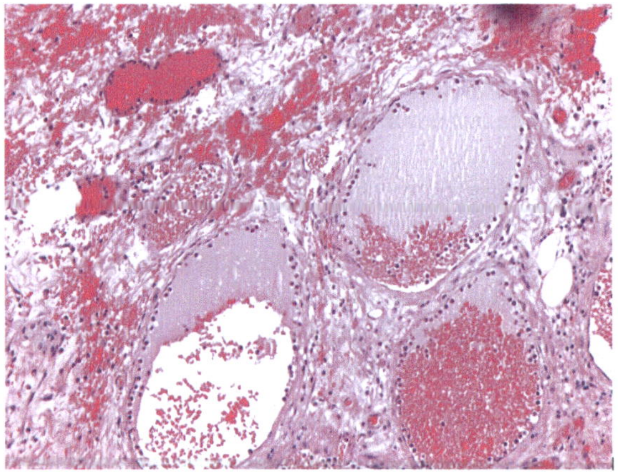

Fig. 26.5 Organizing thrombi in the blood vessels (H&E, 20x) (Histology). With permission from Dr. Ismail Mohamed Ahmed Elbaz Younes

Treatments for Acute Mesenteric Ischemia

Acute mesenteric ischemia is a medical emergency. Once the diagnosis is made, intervention should be initiated without delay.

Medical

Initial management is focused on resuscitation and stabilization. This consists of intravenous fluid management with crystalloids to correct volume deficits and metabolic derangement and blood products. Fluid volume requirements may be high due to significant capillary leakage. Early intubation should be considered in unstable patients to improve oxygenation and allow for more aggressive fluids resuscitation. Vasopressors should be avoided if possible as they worsen ischemia by producing vasoconstriction. However, if necessary, dobutamine, low dose dopamine, and milrinone all improve cardiac function and have been shown to have less impact on mesenteric blood flow. Additionally, all patients with possible intestinal ischemia should be started on broad spectrum antibiotics as early as possible. This is because intestinal ischemia leads to loss of the mucosal barrier, facilitating bacterial translocation.

Pharmacologic therapy for acute ischemic enteropathy typically consists of papaverine, thrombolytics, and anticoagulants. Papaverine is a vasodilator that is indicated to treat arterial spasm. Therefore, it is useful for all arterial forms of acute mesenteric ischemia, and it is the only treatment for NOMI. Papaverine is usually infused directly into the SMA resulting in an improvement of intestinal blood flow.

Thrombolytic agents have been shown to be beneficial in SMAE but they have shown no benefit in SMAT. The infusion must be started within 8 hours of symptom onset. If symptoms do not improve within 4 hours, or if peritonitis develops, then the infusion is stopped and surgical treatment should be initiated.

Heparin anticoagulation is the first-line treatment for MVT and early use has been associated with improved survival. There are no studies that explore the use of enoxaparin or another low molecular weight heparin as a substitute for heparin.

Surgical

If there are peritoneal signs on physical examination, this usually indicates bowel infarction rather than ischemia alone. There is a significant reduction in survival in these patients and emergency laparotomy is required. Goals for surgical intervention in acute mesenteric ischemia include re-establishment of the blood supply to the ischemic intestine, preservation of viable intestine, and resection of non-viable organs.

Treatments for Chronic Mesenteric Ischemia

Medical

Medical therapy alone is usually reserved for patients who are asymptomatic or those who are not good surgical candidates. In these patients, treatment consists of long-term anticoagulation therapy with medications such as warfarin. Novel oral anticoagulants have not been studied extensively for this indication.

Surgical

Traditionally, surgical repair has been the standard method of treatment for chronic mesenteric ischemia. However, published success rates as well as decrease in morbidity and mortality have shifted practices toward endovascular therapy as primary treatment for chronic mesenteric ischemia.

Open surgical repair consists of several different procedures including transaortic endarterectomy, direct reimplantation of the aorta, and anterograde or retrograde bypass grafting. Two of the most favored techniques include anterograde or retrograde bypass grafting. The anterograde bypass originates from the supraceliac aorta and should revascularize multiple visceral arteries. Benefits of this technique include reduction in kinking and turbulent blood flow, along with greater ease of multivessel revascularization. However, this technique does increase the risk of renal ischemic injury due to brief cross-clamping of the aorta above the renal vessels. Retrograde bypass originates from either the infrarenal aorta or the iliac arteries and it usually only revascularizes the SMA.

Surgical revascularization has been shown to be successful in 84% to 94% of patients with recurrence rates of 7–14%. Major complications include acute renal failure (11%), bowel infarction requiring resection (8%), cardiac arrest (6%), respiratory complications (5%), myocardial infarction (5%), and gastrointestinal hemorrhage (3%).

Endovascular repair includes angioplasty and/or stent placement within the mesenteric vasculature. Technical success rates are high and reported to be from 90 to 100%. Relapse rates from restenosis are reported to be as high as 28%. These require repeat endovascular intervention. Major complication rates of endovascular repair are acute renal failure (6%), infarction requiring bowel resection (3%), and acute myocardial infarction (3%). Review of the literature shows a better patency rate with mesenteric artery stenting when compared to angioplasty alone. Stenting eliminates the possibility of elastic recoil following conventional angioplasty.

Prognosis

Acute ischemic enteropathy has an overall mortality approaching 50%. The prognosis is largely dependent on the mechanism of ischemia as well as treatment, which explains the heterogeneity in numbers reported in the literature. Surgical repair is considered to have high rate of technical success and can produce symptomatic improvement in patients with chronic mesenteric ischemia. However, these surgeries are associated with the significant morbidity and mortality rates of 38% and 13%, respectively. Weight loss and malnutrition, including low albumin levels, are predictors of increased morbidity and mortality after major surgeries, and these are common problems in patients with chronic mesenteric ischemia that undergo surgical management. Endovascular repair has lower morbidity and mortality rates of 20% and 3%, respectively.

Late deaths, defined as greater than 30 days following mesenteric revascularization, are generally not related to ongoing mesenteric ischemia. A retrospective review conducted by Tallarita et al. in 2013 evaluated long-term patient survival and causes of death after open or endovascular revascularization for atherosclerotic chronic mesenteric ischemia. A total of 343 patients were reviewed and there were 144 late deaths. Reported causes of late death included cardiac conditions (35%), cancer (15%), pulmonary complications (13%), and mesenteric ischemia (11%).

Recurrent stenosis is a concern following mesenteric revascularization and occurs in 5–15% of patients. For this reason, routine follow-up is important to monitor for recurrence of symptomatic disease. It is recommended to have an initial follow-up imaging study to establish a baseline for future comparison. The imaging study of choice is a duplex ultrasound. It is repeated at three to six-month intervals thereafter.

Suggested Reading

1. Fidelman N, Aburahma AF, Cash BD, Kapoor BS, Knuttinen M, Minocha, et al. ACR appropriateness criteria ® radiologic management of mesenteric ischemia. J Am Coll Radiol. 2017;14(5). https://doi.org/10.1016/j.jacr.2017.02.014.
2. Lo RC, Schermerhorn ML. Mesenteric arterial disease: epidemiology, pathophysiology, and clinical evaluation. In: Rutherford's vascular surgery and endovascular therapy, 9th ed. n.d. Elsevier. p. 1725–34.
3. Danczyk RC, Moneta GL. Clinical evaluation and treatment of mesenteric vascular disease. In: Vascular medicine: a companion to Braunwalds heart disease, 2nd ed. Philadelphia, PA: Elsevier/Saunders. p. 328–39.
4. Galandiuk S, Jorden JR, Rice J, Deveaux PG. Acute and chronic mesenteric ischemia. In: Current therapy in colon and rectal surgery, 3rd ed. n.d. Elsevier. p. 412–7.
5. Brandt LJ, Feuerstadt P. Acute intestinal ischemia. In: Sleisinger and Fordtran's gastrointestinal and liver disease.
6. Hohenwalter EJ. Chronic mesenteric ischemia: diagnosis and treatment. Semin Intervent Radiol. 2009;26(4):345–51.
7. Sreenarasimhaiah J. Chronic mesenteric ischemia. Best Pract Res Clin Gastroenterol. 2005;19 (2):283–95.
8. Brandt LJ, Boley SJ. AGA technical review on intestinal ischemia. Gastroenterology. 2000;118:954–68.
9. Management of the diseases of mesenteric arteries and veins clinical practice guidelines of the European society of vascular disease.
10. Patel R, Costanza M. Mesenteric ischemia, chronic. [Updated 2018 Oct 27]. In: StatPearls [Internet]. Treasure Island (FL): StatPearls Publishing; 2018 Jan.
11. Bala M, Kashuk J, Moore EE, et al. Acute mesenteric ischemia: guidelines of the world society of emergency surgery. World J Emerg Surg. 2017;12:38. Published 2017 Aug 7. https://doi.org/10.1186/s13017-017-0150-5.
12. Dang CV, Su M. Acute mesenteric ischemia. https://emedicine.medscape.com/article/189146-overview. Accessed 3 January 19.
13. Oldenburg WA, Lau LL, Rodenberg TJ, Edmonds HJ, Burger CD. Acute mesenteric ischemia. A clinical review Arch Intern Med. 2004;164(10):1054–62.
14. Mitchell KA, Brian West A. Vascular disorders of the gastrointestinal tract. In: Odze RD, Goldblum JR, editors. Surgical pathology of the GI tract, liver, biliary tract and pancreas. Chapter 10, p. 215–255.e3.
15. Harris JW, Mark Evers B. Small intestine. In: Sabiston textbook of surgery. Chapter 49, p. 1237–95.
16. Tallarita T, Oderich GS, Gloviczki P, Duncan AA, Kalra M, Cha S, Misra S, Bower TC. Patient survival after open and endovascular mesenteric revascularization for chronic mesenteric ischemia. J Vasc Surg. 2013;57(3):747–55.

27 Colonic Ischemia (Also Known as Ischemic Colitis)

Chloe Lee and Eli D. Ehrenpreis

Introduction

Colonic ischemia is a condition that results when blood flow to the colon is compromised, causing an inability to maintain adequate colonic tissue perfusion and nutrition. This physiologic dysfunction can lead to ischemic damage and reperfusion injury, resulting in either reversible or irreversible damage. Although most cases are self-limited, treatment for each patient should be tailored to severity of presentation and associated risk factors.

Epidemiology

CI is the most frequent form of intestinal ischemia. The incidence of CI based on a national insurance claims-based survey was estimated to be 17.7 cases per 100,000 person-years. CI occurs in adults of all ages but it is more common in women and in the older population of patients in their sixties and seventies. However, younger individuals may develop CI in settings of hypercoagulable states, drug-induced conditions and collagen vascular diseases.

C. Lee
Department of Internal Medicine, Advocate Lutheran General Hospital, 1775 Dempster St, 6 South, Park Ridge, IL 60068, USA

E. D. Ehrenpreis (✉)
Department of Medicine, Advocate Lutheran General Hospital, 1775 Dempster St, 6 South, Park Ridge, IL 60068, USA
e-mail: e2bioconsultants@gmail.com

E. D. Ehrenpreis et al. (eds.), *The Mesenteric Organ in Health and Disease*,
https://doi.org/10.1007/978-3-030-71963-0_27

Risk Factors

A variety of conditions are associated the development of CI (Table 27.1). Patients of advanced age who have accompanying comorbid conditions are at especially high risk of developing CI. Hypertension, diabetes mellitus, coronary artery disease, dyslipidemia, chronic obstructive pulmonary disease, congestive heart failure, atrial fibrillation, and renal disease are comorbidities that are associated with the development of CI. Notably, coronary artery disease and atrial fibrillation were approximately twice as common in patients with right colon ischemia compared with ischemia in other segments of colon. A variety of surgeries can result in compromised arterial blood flow to the colon, including abdominal aortic aneurysm repair, cardiopulmonary bypass and renal transplantation. Colonoscopy and barium enema as well as a variety of medications and illicit drugs are other iatrogenic factors that are associated with the development of CI. Patients with irritable bowel syndrome (IBS) have been determined to have an increased risk for CI, possibly as a consequence of excessive sympathetic activity in the mesenteric vasculature. Severe constipation is a risk factor for CI and is also associated with a segmental form of colonic ischemia termed stercoral ulcers. These are due to pressure on the colonic wall and decreasing blood flow that is obstructed by hard stool. Thrombophilia is another risk factor for CI. Younger patients with CI should be evaluated for hereditary coagulopathies such as Factor V Leiden mutation and deficiencies of protein C, protein S and antithrombin III. Of interest, hematochezia and endoscopically diagnosed right sided IC has been described in long distance runners, particularly in females. Pharmacologic agents including vasoconstrictors, psychotropics, oral contraceptives, Type 1 interferon, 5-hydroxytrptamine-1 (5-HT1) agonists, and illicit drugs are known causes of nonocclusive CI (Table 27.2). Despite a long list of possible causes of CI, many patients with CI will have no identifiable risk factors for the disease.

Pathophysiology

CI is caused by a sudden reduction in blood flow to the mesenteric vasculature and subsequent decrease in oxygen to regions of the colon. The large intestine receives its blood supply from the superior mesenteric artery and inferior mesenteric artery, while the rectum has a dual blood supply that includes the inferior mesenteric artery and the internal iliac arteries. The "watershed" regions of the colon include the splenic flexure (Griffith's point), the joining point where the blood supplies of the superior mesenteric artery and inferior mesenteric artery meet and part of sigmoid colon (Sudeck's point), where the blood supplies of the inferior mesenteric arteries and rectal arteries connect. These areas are at increased risk for ischemia due to limited collateral blood flow (Fig. 27.1). Nonocclusive ischemia is the most common mechanisms for the development of CI. The degree of systemic perfusion and collateral circulation determine the overall development of hypoxia and subsequent

Table 27.1 Medical risk factors for colonic ischemia

Occlusive disease	Nonocclusive disease
Arterial	**Hypoperfusion state**
Thrombosis	Congestive heart failure
Cardiac emboli	Pancreatitis
Cholesterol emboli	Hypotension
Small vessel disease	Shock
Diabetes	Hemodialysis
Vasculitis	**Pharmacologic**
Atherosclerosis	Medications
Amyloidosis	Illicit drugs
Rheumatoid arthritis	**Gastrointestinal**
Radiation	Irritable bowel syndrome (IBS)
Venous	Constipation
Mesenteric venous thrombosis	
Portal hypertension	
Hypercoagulable state	
Lymphocytic phlebitis	
Surgical	
Aortoiliac reconstruction	
Cardiopulmonary bypass	
Colectomy	
Colonoscopy	
Barium enema	
Renal transplant	

Modified from Washington, Christopher, and Joseph C. Carmichael. "Management of Ischemic Colitis." Clinics in Colon and Rectal Surgery 25.4 (2012): 228–235. Reused with permission Copyright © Thieme

reperfusion. The most common etiologies of nonocclusive CI include sepsis, hemorrhage and heart failure. During shock states, there is redistribution of blood flow to the systemic circulation and away from the splanchnic circulation. Hypoperfusion of the colon enhances the risk of ischemic injury due to increased demand in setting of preexisting diminished circulation. It is thought that renin-angiotensin axis is involved in the splanchnic vasoconstriction that occurs in response to shock. Various medications are also known to cause vasospasms leading to nonocclusive CI.

Traditionally, CI is also divided into gangrenous and non-gangrenous forms with the latter accounting for 80–85% of cases. Gangrenous CI requires surgical intervention. Further classification defines CI into Type I or II disease. Type I disease refers to idiopathic, nonocclusive ischemia likely due to small vessel disease. These patients are generally managed with supportive care. Type II disease is associated

Table 27.2 Pharmacologic risk factors for ischemic colitis

Drugs
Moderate evidence
Constipation inducing drugs
Immunomodulator drugs (Antitumor necrosis factor-alpha, Type 1 interferon)
Illicit drugs (amphetamines, cocaine)
Low evidence
Antibiotics
Appetite suppressants
Chemotherapeutic drugs
Decongestants
Diuretics
Ergot alkaloids
Hormonal therapies
Laxatives
Psychotropic drugs
Serotoniergic drugs
Very low evidence
Digitalis
Kayexalate
Non-steroidal anti-inflammatory drugs (NSAIDs)
Statins
Vasopressors

Adapted from: Brandt, Lawrence. ACG Clinical Guideline: Epidemiology, Risk Factors, Patterns of Presentation, Diagnosis, and Management of Colon Ischemia. AM J Gastroenterol 2015; 110:18–44. Copyright © Wolters Kluwer 2015

with an identified etiology and usually followed by systemic hypotension, decreased cardiac output or aortic surgery.

The left colon is the segment most affected with CI, and accounts for 32.6% of patients with the disease. CI occurs in the distal colon 24.6%, right colon 25.2% and entire colon 7.3% of the time. It has been shown that patients with right sided CI have a poorer outcome than those with CI occurring in any other segments of the colon. Figure 27.1 shows the watershed regions of the colon.

CI is usually a transient, self-limited condition. However, stricturing and prolonged colonic inflammation can occur as long-term complications of the disease.

Clinical Presentations

Clinical manifestations of CI are highly variable. In general, patients with CI develop sudden lower abdominal cramping and the urge to defecate. This is often accompanied by the passage of bloody diarrhea, usually within 24 hours of the

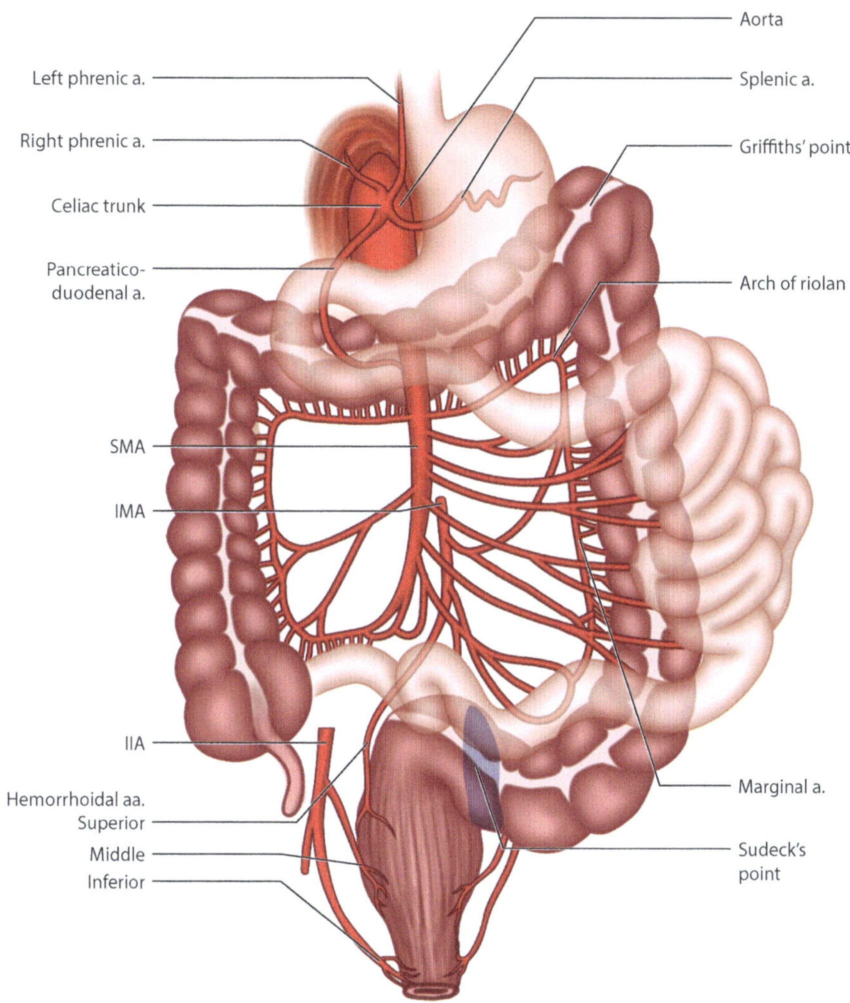

Fig. 27.1 Watershed regions of the colon

onset of abdominal pain. Although blood loss in CI is usually mild, severe anemia due to excessive hemorrhage can occur and are often associated with gangrenous CI and/or fulminant pancolitis. Other nonspecific symptoms occurring in patients with CI include fever, nausea, vomiting, anorexia and presyncope.

Relatively mild tenderness of the abdomen, sometimes at the region of the affected segment of the colon is found on physical examination. Severe tenderness with peritoneal signs suggests the presence of advanced disease, including peritonitis. In addition, severe cases of CI can demonstrate signs of overwhelming infection including high fever, tachycardia and hypotension.

In most patients, signs and symptoms of the disease are transient and self-limited. More than half of the patients with CI have resolution of their symptoms within 2–3 days. Later recurrences of CI are relatively uncommon and are seen in 6.8–16% of patients. The rates of complicated CI are 14.3% for stricture formation, 9.9% for gangrenous colitis and 2.5% for fulminant colitis.

Diagnosis

Because of the nonspecific nature of symptoms of CI, the large differential diagnosis for the common presentations of CI and the fact that CI is often idiopathic, a high index of suspicion for mesenteric ischemia, including CI, should be entertained in patients with abdominal pain and rectal bleeding. Nonetheless, establishing the diagnosis and severity of CI often requires an assessment of findings from the medical history (including risk factors), physical examination, laboratory testing, imaging and endoscopic evaluation of the colon. Computerized tomography (CT) of the abdomen is frequently performed in patients with abdominal pain and/or gastrointestinal bleeding. Colonoscopy is generally performed within 48 hours and can confirm colonic ischemia based on its typical appearance and biopsy confirmation.

Laboratory Findings

Although laboratory studies are neither sensitive nor specific for CI, they can help predict the severity of CI. Elevated white blood cell count, lactate dehydrogenase (LDH) and decreased hemoglobin (Hgb) concentration occur commonly in patients with CI. Serum levels of D-lactate may be a marker for CI. Severe cases of CI have characteristic laboratory findings. These include elevated serum lactate levels, the presence of metabolic acidosis, and a significant base deficit usually accompanied by profound anemia. Since hematochezia with or without abdominal pain may be infectious in origin, stool studies for culture, ova and parasite exam, and Clostridium toxin assay are included in patient evaluation.

Imaging

Plain films of the abdomen can show “thumbprinting,” representing edema or thickening of the bowel in early ischemia (Fig. 27.2). CT of the abdomen with intravenous contrast is usually the imaging of choice with high suspicion for CI. Findings may include colon wall thickening and luminal narrowing, although these findings are nonspecific (Fig. 27.3).

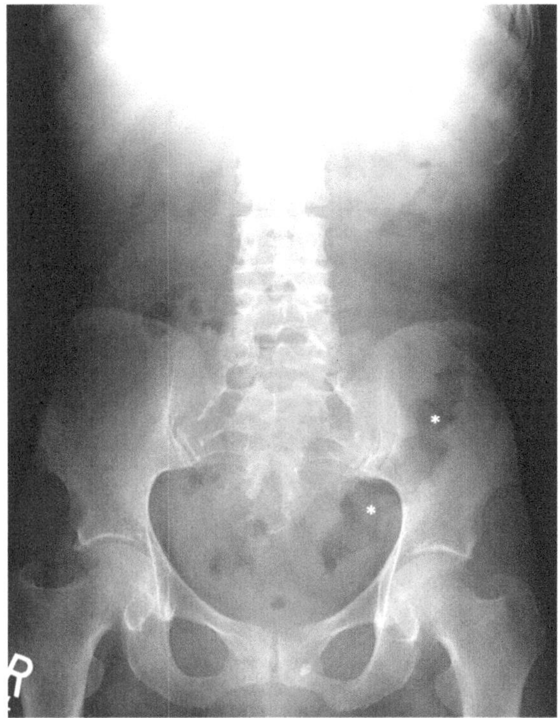

Fig. 27.2 AP abdominal radiograph of a different patient shows the plain film appearance of “thumbprinting” involving the descending colon (asterisks)

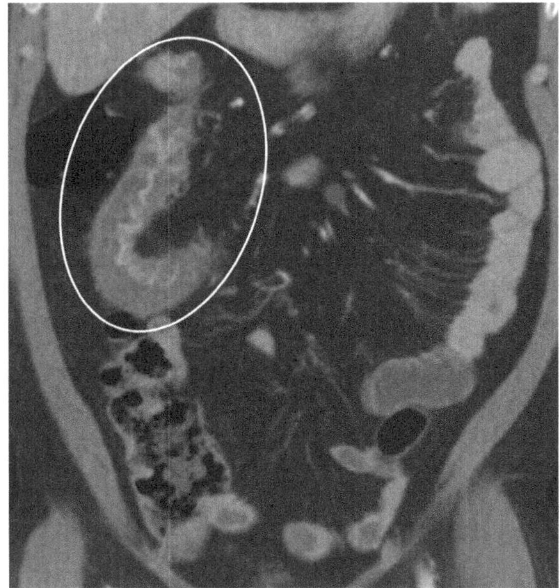

Fig. 27.3 Coronal and axial contrast-enhanced computed tomography images with wall thickening and irregular luminal narrowing of the transverse colon (oval), indicative of a nonspecific focal colitis. Colonoscopy and pathology confirmed the diagnosis of ischemic colitis. With permission from Dr. Abraham Dachman and Dr. Justin Ramirez

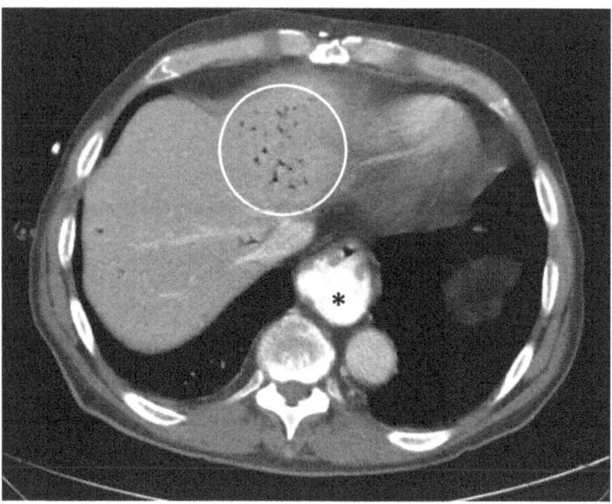

Fig. 27.4 Axial computed tomography image demonstrates branching radiolucencies (circle) within the hepatic parenchyma compatible with portal venous gas. Additional foci are present at the periphery of the right hepatic lobe and adjacent to the middle hepatic vein and IVC. Incidental note of contrast filled hiatal hernia (asterisk)

Contrast enhanced CT and CTA have high sensitivity and specificity for assessing vascular occlusion as an etiology of CI, although it is generally not required since most cases are due to non-occlusive disease. Presence of pneumoperitoneum and hepatic portal venous gas indicate bowel necrosis although surgical manipulation can also cause such finding (Figs. 27.4 and 27.5). Pneumatosis intestinalis is a radiological finding of gas bubbles within the mucosa or submucosa of the intestinal wall most commonly associated with bowel ischemia. Mucosal disruption from bowel ischemia, systemic infection, or inflammation may lead to the development of pneumatosis intestinalis. US and MRI are less used imaging modalities in diagnosis of CI.

Colonoscopy

Colonoscopy with biopsy is the most accurate means of diagnosing CI. It provides the means to directly visualize the abnormal colonic mucosa and allow tissue collection for histology. Necrotic mucosa can be identified by the presence of a dusky and/or cyanotic discoloration in affected segments of colon. If necrosis is identified, immediate removal of the colonoscope, and consideration of rapid surgery is required. Early colonoscopy (within 48 hours of presentation) is recommended in most patients who are suspected to have CI.

The location of abnormal mucosa can also be determined. Severe and gangrenous forms of the disease may be identified and require immediate withdrawal of the colonoscope. The presence of submucosal hemorrhage in the splenic flexure is a

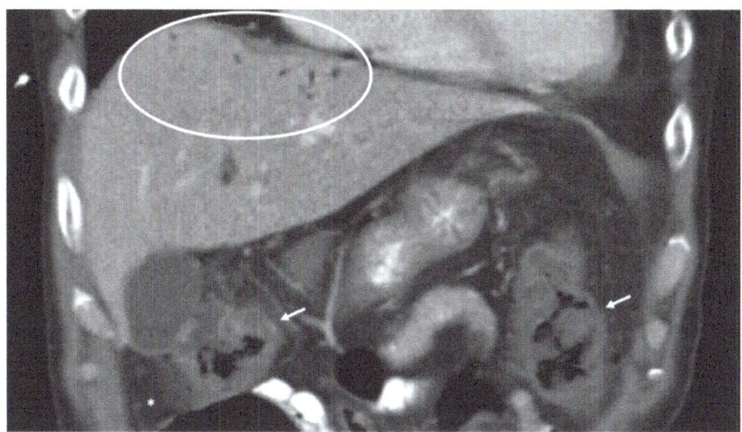

Fig. 27.5 Coronal CT image of the same patient demonstrates the peripheral location of the portal venous gas (oval). Imaged portions of hepatic flexure and descending colon (arrows) show marked colonic wall and haustral thickening giving the appearance of "thumbprinting". Note adjacent fat stranding (asterisk) involving the hepatic flexure. With permission from Dr. Abraham Dachman and Dr. Justin Ramirez

characteristic finding on colonoscopy in patients with ischemic colitis (Fig. 27.6). Other mucosal lesions including edema, erythema, pale regions and bluish hemorrhagic nodules. Frank bleeding may also be present. Biopsies demonstrate mucosal and submucosal hemorrhage, edema, crypt destruction, and granulation tissue.

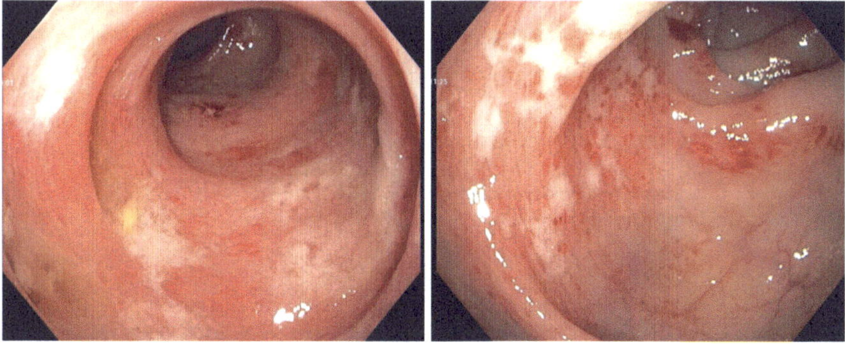

Fig. 27.6 Colonoscopy shows submucosal hemorrhage and mucosal edema in a patient with colonic ischemia

Histologic Findings

Histologic findings include mucosal and submucosal hemorrhage, edema, crypt destruction, and granulation tissue. These changes can mimic histologic findings of Crohn disease. Ghost cells, preserved cellular outlines without cell content, can be pathognomonic for infarction in CI.

Treatments

Medical

Treatments of CI vary according to the severity of the disease (Table 27.3).

Based on previous studies, severity of CI is generally divided into mild, moderate and severe forms. Patients with mild disease have typical symptoms of CI

Table 27.3 Ischemic colitis: classification of disease severity and management

Disease severity	Criteria	Treatment
Mild	Typical symptoms of CI with a segmental colitis not isolated to the right colon and with none of the commonly associated risk factors for poorer outcome that are seen in moderate disease	Observation Supportive care
Moderate	Any patient with CI and up to 3 of the following factors: Male gender Hypotension (systolic BP <90 mm Hg) Tachycardia (HR >100 beats/min) Abdominal pain without rectal bleeding BUN >20 mg/dl Hgb <12 g/dl LDH >350 U/l Serum sodium <136 mEq/l (mmol/l) WBC >15 cells/cmm (×109/l) Colonic mucosal ulceration identified colonoscopically	Correction of cardiovascular abnormalities Broad-spectrum antibiotic therapy Surgical consultation
Severe	Any patient with CI and more than 3 of the criteria for moderate disease or any of the following: Peritoneal signs on physical examination Pneumatosis or portal venous gas on radiographic imaging Gangrene on colonoscopic examination Pancolonic distribution or isolated right colonic ischemia (IRCI) on imaging or colonoscopy	Emergent surgical consultation Transfer to intensive care unit Correction of cardiovascular abnormalities Broad-spectrum antibiotic therapy

Adapted from: Brandt, Lawrence. ACG Clinical Guideline: Epidemiology, Risk Factors, Patterns of Presentation, Diagnosis, and Management of Colon Ischemia. *AM J Gastroenterol* 2015; 110:18–44. Copyright © Wolters Kluwer 2015

without isolated right colonic disease. Moderate disease has other associated factors (male patient, hypotension, tachycardia, abdominal pain without rectal bleeding, blood urea nitrogen >20 mg/dl, Hgb <12 g/dL, LDH >350 U/l, serum sodium <136 mEq/l, WBC >15 × 109/l or mucosal ulcerations. Severe CI is defined by more than three of the above factors, peritoneal signs, pneumatosis on CT or gangrene on colonoscopy.

Most cases of CI are mild and self-limited. If hospitalized, patients can be treated with supportive measures with intravenous hydration, electrolyte correction and bowel rest. There is currently no strong evidence to support the administration of antibiotics in patients with mild CI, although they may be indicated in moderate to severe disease. Animal models do demonstrate that antibiotics decrease bacterial translocation during acute ischemia and reperfusion injury. Currently, the choice of antimicrobial agents and optimal duration of therapy have not been established. Of interest, some antibiotics are also associated with the development of CI. In severe cases of CI, surgical intervention is required as mortality from necrotic bowel can be as high as 100%. Acute indications for immediate surgery include peritoneal signs, massive bleeding, fulminant colitis, and portal venous gas and pneumatosis intestinalis on imaging. In addition, failure to respond to treatment within 2–3 weeks with continued symptoms or a protein-losing colopathy may require segmental colectomy. The type of surgical procedure depends on the affected region of colon. Risk stratification should be considered prior to surgical intervention as mortality post operation for CI may be as high as 37–47%.

Complications

CI is usually a self-limited condition which requires supportive treatment in most cases. However, severe ischemia can cause inflammation which can develop into a stricture or segmental ulcerating colitis. Ischemic strictures can be asymptomatic or cause obstructive symptoms. For patients with symptoms, a repeat colonoscopy may be required to visualize strictures or segmental colitis. Some strictures resolve in one to two years but symptomatic strictures usually require segmental bowel resection. Recurrent episodes of colonic ischemia associated with segmental colitis may also require elective segmental resection. Unfavorable outcomes were more frequent in isolated right colon ischemia compared with other regions of the colon (40.9 vs. 10.3%).

Suggested Readings

1. Alpern MB, Glazer GM, Francis IR. Ischemic or infarcted bowel: CT findings. Radiology. 1988;166:149.
2. Bailey RW, Bulkley GB, Hamilton SR, Morris JB, Smith GW. Pathogenesis of nonocclusive ischemic colitis. Ann Surg. 1986;203(6):590–9.

3. Salk A, Ehrenpreis ED. Ischemic colitis with Type I interferons used in the treatment of hepatitis C and multiple sclerosis: an evaluation from the food and drug administration adverse event reporting system and review of the literature. Ann Pharmacother. 2013;47:537–42.
4. Brandt L. ACG clinical guideline: epidemiology, risk factors, patterns of presentation, diagnosis, and management of colon ischemia. AM J Gastroenterol. 2015;110:18–44.
5. Kontogianni A, Delakidis S, Basioukas P, Tsantoulas D, Tamvakis N, Demonakou M. Ischemic colitis: clinical, endoscopic and histologic spectrum of 254 cases. Ann Gastroenterol. 2003;16.
6. Longstreth GF, Yao JF. Diseases and drugs that increase risk of acute large bowel ischemia. Clin Gastroenterol Hepatol. 2010;8:49.
7. Luo CC, Shih HH, Chiu CH, et al. Translocation of coagulase-negative bacterial staphylococci in rats following intestinal ischemia-reperfusion injury. Biol Neonate. 2004;85:151–4.
8. Midian-Singh R, Polen A, Durishin C, et al. Ischemic colitis revisited: a prospective study identifying hypercoagulability as a risk factor. South Med J. 2004;97:120–3.
9. Mitsudo S, Brandt LJ. Pathology of intestinal ischemia. Surg Clin North Am. 1992;72:43.
10. Montoro M, Brandt L, Santolaria S, et al. Clinical patterns and outcomes of ischaemic colitis: results of the working group for the study of ischaemic colitis in Spain (CIE study). Scand J Gastroenterol. 2011;46:2:236–46.
11. Mosele M, Cardin F, Inelmen EM, et al. Ischemic colitis in the elderly: pre-dictors of the disease and prognostic factors to negative outcome. Scand J Gastroenterol. 2010;45:428–33.
12. Smerud MJ, Johnson CD, Stephens DH. Diagnosis of bowel infarction: a comparison of plain films and CT scans in 23 cases. AJR Am J Roentgenol. 1990;154:99.
13. Theodoropoulou A, Koutroubakis IE. Ischemic colitis: clinical practice in diagnosis and treatment. World J Gastroenterol. 2008;14:7302.
14. Washington C, Carmichael JC. Management of ischemic colitis. Clin Colon Rectal Surg. (2012);25(4):228–35. PMC. Web. 6 Oct. 2018.

Mesenteric Lymphangioma

28

Eli D. Ehrenpreis and Siddhartha Guru

Definition

Lymphangiomas are very rare, benign lesions that are a result of an abnormality of thin-walled lymphatic structures. Lymphangiomas are characterized by an increase in the number or size of lymphatic channels and lymphatic spaces, or the occurrence of large, partially organized cystic lesions. A very small number of cases involve the portion of the mesentery that is proximal to the small intestine. They appear as mass-like or cystic lesions on abdominal imaging.

Epidemiology

Lymphangiomas are rare lesions with fewer than 900 cases reported in the medical literature. Of these, only 72 cases involving the mesentery adjacent to the small intestine have been described. Lymphangiomas are most commonly found in the soft tissues of the head and neck. These are most often diagnosed in children, with 50% occurring at birth and 90% before the age of 2. A 3:1 male to female ration has been suggested for adult forms of mesenteric lymphangioma.

E. D. Ehrenpreis (✉) · S. Guru
Department of Medicine, Advocate Lutheran General Hospital, 1775 Dempster St, 6 South, Park Ridge, IL 60068, USA
e-mail: e2bioconsultants@gmail.com

E. D. Ehrenpreis et al. (eds.), *The Mesenteric Organ in Health and Disease*,
https://doi.org/10.1007/978-3-030-71963-0_28

Patients at Risk

Pediatric cases of lymphangioma are congenital, with no stated relationship to other congenital disorders. Abdominal trauma, abdominal surgery, lymphatic obstruction and radiation injury appear to be risk factors for the development of the acquired form of mesenteric lymphangioma in adults.

Pathophysiology

Congenital lymphangiomas arise from abnormal embryonic development of lymphatic tissue. Lymphangioma occurrence may result blockage or arrest of normal growth of the primitive lymph channels, failure of the primitive lymphatic sac to reach the venous system, or inappropriate location of lymphatic vessels during embryogenesis. In adult patients, lymphangiomas may occur after abdominal surgery trauma, inflammation, or radiation injury.

Clinical Presentation

A variety of presentations of mesenteric lymphangioma have been described, ranging from asymptomatic patients to an acute abdomen and bowel obstruction from volvulus or intussusception. In a recently published series of six patients with mesenteric lymphangiomas, all were diagnosed after acute presentations, but one patient had a two-year history of abdominal bloating. Patients may develop symptoms from a mass effect from compression of adjacent small bowel or other organs. Fever, nausea, diarrhea and gastrointestinal bleeding have also been described.

Physical Findings

In the head and neck, lymphangiomas present as painless, soft, and compressible masses that have a slow rate of growth. In patients with mesenteric lymphangiomas, abdominal tenderness and a palpable mass may be present on examination. Signs and symptoms of intestinal obstruction can be seen in cases with volvulus or intussusception. The acute abdomen presentation of lymphangioma can be confused with appendicitis and a myriad of other causes.

Laboratory

Complications of mesenteric lymphangiomas may be manifested by anemia, leukocytosis, and a variety of biochemical abnormalities.

Histology

Lymphangiomas may be divided into three histologic subtypes: capillary (simple), cavernous and cystic. The capillary lymphangioma are composed of small thin walled lymphatic vessels and are usually located in the skin. Cavernous lymphangiomas are connected to nearby lymphatic spaces and are composed of various sized lymphatic vessels and lymphoid stroma. Cystic lymphangiomas are characterized by the formation of large cystic lymphatic spaces with collagen and smooth muscle bundles lined by endothelial layers. These cystic lymphangiomas with lack connections to adjacent normal lymphatics and major lymphatic vessels of the small intestine.

Immunohistochemical testing with D2-40 on cystic wall endothelium confirms the presence of lymphatic endothelial cells. Differentiation from Lymphangiomyoma and multicystic mesothelioma is performed by testing for HMB-45 immunoreactivity and staining for calretinin. Lymphangiomyoma has HMB-45 immunoreactivity and mesothelioma will stain positive for calretinin, while both of these are negative in lymphangioma. Cell that line the cystic walls of lymphangiomas may be immunoreactive for factor VIII-related antigen.

Imaging

Abdominal radiography may be used to diagnose bowel obstruction, volvulus and intussusception. Abdominal ultrasound and CT are the most commonly imaging studies used to diagnose mesenteric lymphangioma and to differentiate it from other disorders. Nonetheless, due to the rarity of mesenteric lymphangiomas, they are often misdiagnosed when seen on standard imaging studies. Ultrasound may reveal a multiloculated cystic mass or masses with thin internal septae. On CT scan, mesenteric lymphangiomas present as unilocular or multilocular masses with contrast enhancement of the cyst walls and septae. Cystic lesions are most often homogeneous, but appear inhomogeneous if they contain proteinaceous fluid, blood or fat components. Lymphatic vessels may be seen coursing between intralesional locules (Figs. 28.1, 28.2, 28.3, 28.4 and 28.5).

Retroperitoneal lymphangiomas ae very uncommon and may present as non-specific masses or cystic lesions within the retroperitoneal space (Figs. 28.4 and 28.5).

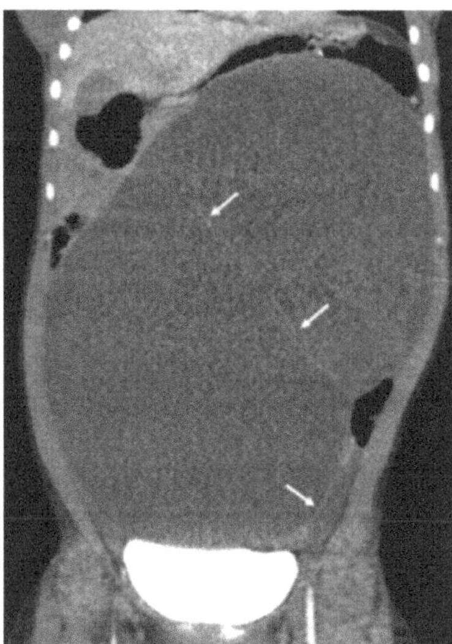

Fig. 28.1 Coronal CT image from a pediatric patient demonstrates mesenteric lymphangioma. An expansile intraperitoneal cystic lesion with thin enhancing septations (arrows) is seen. Note the displaced gas filled bowel at the periphery of the lesion

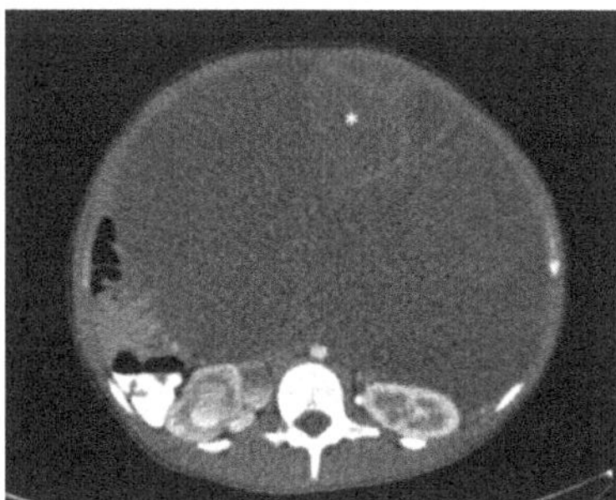

Fig. 28.2 CT image from the same patient demonstrates discrete, large intraperitoneal cystic structures displacing adjacent organs. Subtle increased attenuation within an anterior cystic structure (asterisk) suggests a more complex fluid component such as protein or blood products

MRI may demonstrate hemorrhagic lesions with increased signal intensity on the T2 images. Imaging studies cannot in themselves be used to diagnosis mesenteric lymphangioma as their appearance may mimic other cystic lesions such as teratomas.

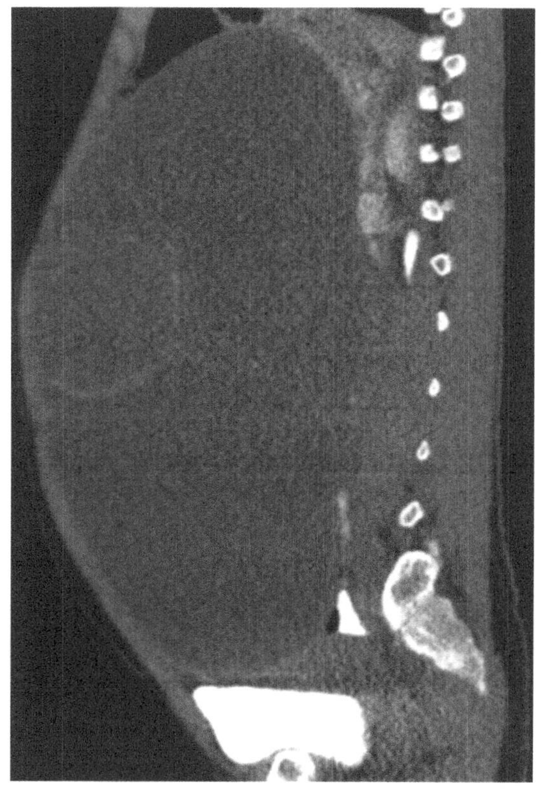

Fig. 28.3 Sagittal CT image better demonstrates the significant mass effect, nearly filling the entire

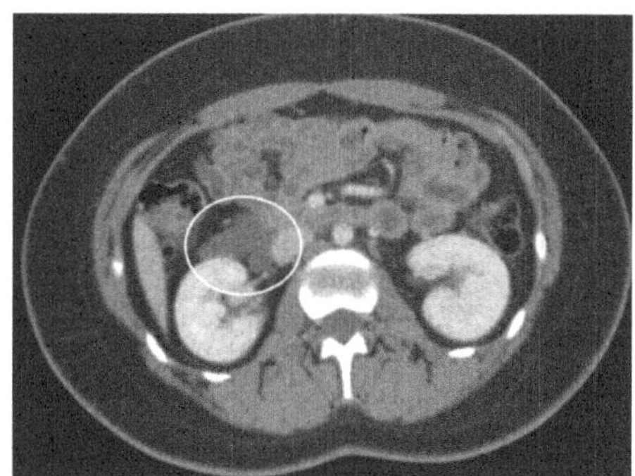

Fig. 28.4 Axial CT image from a patient showing a retroperitoneal lymphatic malformation in an adult (oval)

Diagnosis

The diagnosis of mesenteric lymphangioma is diagnosed on histological examination and staining of tissue from excised lesions, (see Histology section above).

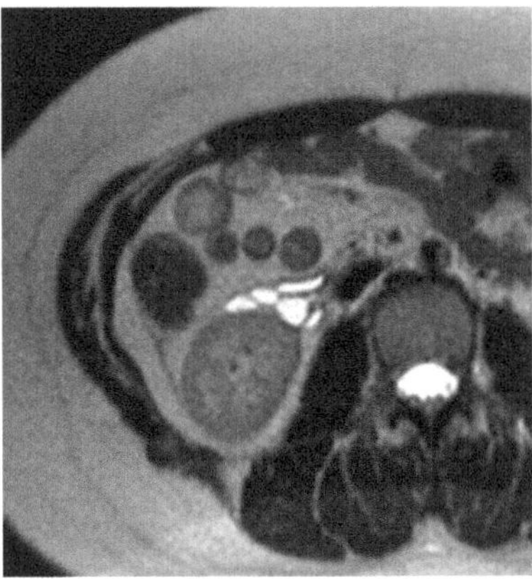

Fig. 28.5 Axial T2-weighted MR image from the same patient demonstrates the septated, cystic nature of the lesion. When discovered incidentally, as in this patient, the lesion is often asymptomatic with little or no mass effect on adjacent structures. With permission from Dr. Abraham Dachman and Dr. Justin Ramirez

Immunohistochemical testing with D2-40 on cystic wall endothelium is done to determine the presence of lymphatic endothelial cells. Which would be positive in both lymphangiomyoma and multicystic mesothelioma hence are further differentiated by testing using HMB-45 immunoreactivity and staining for calretinin. While lymphangiomyoma will be positive with HMB-45 immunoreactivity and mesothelioma will stain positive for calretinin. They are also immune reactive for factor VIII related antigen.

Treatment

Medical

Although surgery is the mainstay of treatment for lymphangiomas, broad spectrum antibiotics with symptomatic treatment with acetaminophen may be used as adjunctive therapies. Intralesional injection therapy with OK-432, a lyophilized biological preparation containing cells of Streptococcus Pyogenes su-strain treated with Benzylpenicillin has demonstrated efficacy in children with head and neck lesions.

Surgery

Surgical excision is the treatment of choice for mesenteric lymphangioma. The rationale for surgery is that lesions can increase in size and impact adjacent organs. Complications of unresected mesenteric lymphangiomas include bowel obstruction, infection, hemorrhage and possibly malignant transformation. Following resection, ongoing imaging is performed to exclude recurrences or residual lesions. Percutaneous catheter drainage and sclerotherapy has been performed as an alternative to surgery in some cases.

Suggested Reading

1. Chen J, Du L, Wang DR. Experience in the diagnosis and treatment of mesenteric lymphangioma in adults: a case report and review of literature. World J Gastrointest Oncol. 2018;10(12):522–7.
2. Grasso DL, Pelizzo G, Zocconi E, Schleef J. Lymphangiomas of the head and neck in children. Acta Otorhinolaryngol Ital. 2008;28(1):17–20.
3. Johnson PT, Horton KM, Fishman EK. Nonvascular Mesenteric disease: utility of multidetector CT with 3D volume rendering. RadioGraphics. 2009;29(3):721–40.
4. Suthiwartnarueput W, Kiatipunsodsai S, Kwankua A, Chaumrattanakul U. Lymphangioma of the small bowel mesentery: a case report and review of the literature. World J Gastroenterol. 2012;18(43):6328–32.
5. Rebuffini E, Zuccarino L, Grecchi E, Carinci F, Merulla VE. Picibanil (OK-432) in the treatment of head and neck lymphangiomas in children. Dent Res J (Isfahan). 2012;9(Suppl 2): S192–6.

Radiation-Induced Mesenteric Injury

29

Eli D. Ehrenpreis and Charles Broy

Pathophysiology of Radiation-Induced Tissue Injury

It is reasonable to expect that some damage to mesenteric tissue occurs during the course of radiation therapy for solid tumors of abdominal organs and during radiation therapy for lymphoma. Radiation deposits energy by changing neutral atoms to ions. Adjacent anatomic structures that are exposed to therapeutically administered radiation become innocent bystanders to these effects. In a biologic environment, this causes activated oxygen species as well as irreparable damage to DNA. The subsequent host response includes rapid inflammation around the area of exposure. These insults can cause long-term injury and possibly cell death. Unfortunately, ionizing radiation effects all cells within the area of exposure and not just the area of desired treatment. Routine imaging and radiation therapy for cancer most often uses both X-Rays (photons) and electrons. Other heavy particles (including protons and neutrons) are also used. Although the repair process is more efficient in normal cells than malignant cells, acute and chronic injurious affects to normal tissue may occur.

The primary site of radiation-induced cellular injury is damage to double stranded DNA. These effects are most pronounced in cells that exhibit rapid proliferation and cellular regeneration. In the abdomen, the rapid turnover of epithelial cells that form the inner lining and glands within the intestinal organs are most sensitive to the damaging effects of radiation. Other cell types, including the mesenchymal-derived cells contained within the mesentery, are more radiation resistant. The mesentery is a continuous structure to which the abdominal digestive

E. D. Ehrenpreis (✉)
Department of Medicine, Advocate Lutheran General Hospital, 1775 Dempster St, 6 South, Park Ridge, IL 60068, USA
e-mail: e2bioconsultants@gmail.com

C. Broy
Department of Gastroenterology, Edward Hines Jr VA Hospital, Hines, IL, USA
e-mail: Charles.Broy@va.gov

E. D. Ehrenpreis et al. (eds.), *The Mesenteric Organ in Health and Disease*,
https://doi.org/10.1007/978-3-030-71963-0_29

organs are directly attached. The mesentery then collectively connects the abdominal organs to the posterior abdominal wall. Within the mesenteric structure are vascular, lymphatic and biliary systems. The surfaces of the mesentery are lined with mesothelial cells. The undersurface of the mesothelium of the mesentery is made up of connective tissue and also contains some mesenchymal cells. The components of connective tissue include fibrocytes, adipose cells, collagen fibers, and elastin fibers. Lymphatic channels and capillaries coalesce into larger structures within the mesentery. The mesentery also contains macrophages, lymphocytes and polymorphonuclear neutrophils (PMNs). (See Chap. 3, General Anatomy of the Mesentery and Chap. 7, Cellular Anatomy of the Mesentery). Most of these cell types have a low rate of mitotic activity and are relatively resistant to radiation damage.

Radiation exposure causes increased bacterial translocation, inhibition of blood vessel hormones and alters mast cell function in the rat/mouse model. All irradiated blood vessels show loss of smooth muscle, intimal fibrosis, accumulation of foam cells and luminal narrowing. This causes vascular insufficiency that may lead to ulcerations, injury-prone vessels and bleeding. Irradiation of connective tissue can cause a loss of flexibility and increased stiffness.

The most important factors in the development of radiation injury to normal tissue are total radiation dose, fraction size, duration and the volume of organs irradiated. The use of combination chemotherapy and radiation therapy results in improved response and survival for a variety of cancers and is a standard approach in modern oncology. Chemotherapies are often specifically applied as radiosensitizing agents in patients receiving radiation therapy. Concomitant chemotherapy also enhances the risk of radiation injury. Although data is most robust for 5-fluorouracil-based chemotherapy, other agents including oxaliplatin and irinotecan have been implicated in worsening radiation damage to the abdominal organs. Others risk factors include vascular disease, inflammatory bowel disease, collagen vascular diseases and previous injury to the mesentery (Table 29.1). The relative effects of these risk factors on the development of radiation-induced injury to the mesentery have not been studied.

Table 29.1 Risk factors for the development of radiation injury to normal abdominal tissues

Risk factors	
Cell type	Rapidly dividing cells (epithelial cells, glands)
Radiation	Total radiation dose, fraction size, duration and the volume of irradiated organs
Chemotherapy	5-fluorouracil, oxaliplatin, irinotecan
Medical diseases	Vascular disease, inflammatory bowel disease, collagen vascular diseases and previous injury to the mesentery

Radiation-Induced Intestinal Injury-Enteropathy and Colopathy

Radiation-induced injury to the small and large intestine occurs in acute and chronic forms. Acute intestinal toxicity is a result of destruction and delayed regeneration of the intestinal epithelium. These changes are characterized by dilation of intestinal crypts and mild inflammation. Epithelial atrophy and mucosal ulceration are subsequently followed by cellular regeneration and restoration of normal mucosa. Clinical presentation of acute radiation injury of the intestine includes diarrhea, nausea, and abdominal pain. The mechanism of chronic radiation enteropathy and colopathy is complex and involves vascular inflammation, mucosal atrophy, intestinal wall fibrosis, telangiectasia formation and vascular sclerosis. This can present with deep ulcerations, gastrointestinal bleeding, altered motility, obstructive symptoms, fistula formation, constipation, diarrhea from small intestinal bacterial overgrowth, nutrient deficiencies from small intestinal malabsorption and intestinal perforation.

Measures to prevent abdominal organ damage from radiation therapy include displacement of the intestine away from the field of radiation, antioxidants, free radical scavengers, dietary supplementation with glutamine. Research into cytokine manipulation and modulators of endothelial function is ongoing.

The degree of involvement of the mesentery in the pathophysiology and clinical manifestations of acute and chronic radiation-induced intestinal injury has not been explored to date.

Medical Literature on Radiation and the Mesentery

Basic Research

Early studies in mice demonstrated that radiation of mesenteric vascular structures produces increased cellularity of the arterioles and vascular permeability within 3 months changes, with a gradual recovery within a year. Vascular fibrosis, and occasional luminal occlusion occurred after 18 months. Venous and capillary stasis were described using a rat model. Acute vascular changes occur following conventionally fractionated RT with mitotic cell death without apoptosis occurring in endothelial cells. Larger fractional radiation doses may cause apoptotic endothelial cell death. Rapid hydrolysis of water stimulated by radiation exposure, induces the production of hydroxide, superoxide, and hydroxyl radicals. These, in turn have been implicated in the development of endothelial vascular injury. A single study showed that the production of free radicals in the mesenteric circulation occurred following leukocyte activation and adherence. Using a macaque model, a recent study demonstrated that mesenteric lymph nodes in radiated animals showed reduced number of lymphoid cells immediately after radiation exposure. Subsequently, mesenteric lymph nodes also developed fibrotic changes, that were attributed to abnormal immunomodulatory cells and/or signaling molecules that originated from portions of the intestine with radiation-induced injury.

Of interest, various studies have shown that mesenchymal stem cells can attenuate the injurious effects of radiation on tissues of the lung, periosteum, heart and brain.

A study using a rat model demonstrated that an immune-enhanced diet containing arginine, omega-3-fatty acids and RNA fragments reduced bacterial counts in mesenteric lymph nodes after radiation exposure.

Clinical Publications

There is a single published case of a desmoid tumor of the mesentery that developed nineteen years after abdominal radiation therapy for Hodgkins disease. Mesenteric stranding was found to occur in a study of four patients with radiation enteropathy that underwent magnetic resonance enterography. A single case of inadvertent injection of Y90 microspheres throughout the mesentery that occurred during a radioembolization (RE) procedure was reported. The patient was treated with amofostine and survived the exposure.

Summary

There is a true paucity of information currently available on the overall effects of abdominal radiation on mesenteric physiology and function. The relationship between radiation-induced mesenteric injury and acute and chronic entropathy, colopathy and proctopathy have yet to be explored.

Suggested Reading

1. Hall CR. Pathology of radiation effects on healthy tissues in the pelvis. In: Ehrenpreis ED, Marsh R de M, Small W, editors. Radiation therapy for pelvic malignancy and its consequences. N.Y.: Springer;2015.
2. Gunnlaugsson A, Kjellén E, Nilsson P, et al. Dose-volume relationships between enteritis and irradiated bowel volumes during 5-fluorouracil and oxaliplatin based chemoradiotherapy in locally advanced rectal cancer. Acta Oncol. 2007;46:937.
3. Sehdev A, Marsh R de W. Contribution of chemotherapy to the toxicity of pelvic irradiation. In: Ehrenpreis ED, Marsh R de M, Small W, editors. Radiation therapy for pelvic malignancy and its consequences. N.Y.: Springer;2015.
4. Song DY, Lawrie WT, Abrams RA, et al. Acute and late radiotherapy toxicity in patients with inflammatory bowel disease. Int J Radiat Oncol Biol Phys. 2001;51:455.
5. Lin A, Abu-Isa E, Griffith KA, Ben-Josef E. Toxicity of radiotherapy in patients with collagen vascular disease. Cancer. 2008;113:648.
6. Panés J, Granger DN. Neutrophils generate oxygen free radicals in rat mesenteric microcirculation after abdominal irradiation. Gastroenterology. 1996;111(4):981–9.

7. Parakkal D, Ehrenpreis ED. Medical management of radiation effects on the intestines. In: Ehrenpreis ED, Marsh R de M, Small W, editors. Radiation therapy for pelvic malignancy and its consequences. N.Y.: Springer;2015.
8. Atasoy BM, Deniz M, Dane F, Özen Z, Turan P, Ercan F, Çerikçioğlu N, Aral C, Akgün Z, Abacioğlu U, Berrak Yeğen Ç. Prophylactic feeding with immune-enhanced diet ameliorates chemoradiation-induced gastrointestinal injury in rats. Int J Radiat Biol. 2010;86(10):867–79.
9. Sabet A, Ahmadzadehfar H, et al. Survival after accidental extrahepatic distribution of Y90 microspheres to the mesentery during a radioembolization procedure. Cardiovasc Intervent Radiol. 2012;35(4):954–7.
10. Feldmeier JJ, Hampson NB. A systematic review of the literature reporting the application of hyperbaric oxygen prevention and treatment of delayed radiation injuries: an evidence based approach. Undersea Hyperb Med. 2002;29:4.
11. Wegner HE, Fleige B, Dieckmann KP. Mesenteric desmoid tumor 19 years after radiation therapy for testicular seminoma. Urol Int. 1994;53(1):48–9.

Drug Induced Mesenteric and Retroperitoneal Diseases

30

Sarah Burroughs and Eli D. Ehrenpreis

Definition

Nonocclusive mesenteric ischemia occurs as a sudden form of hypoperfusion of the intestine that is not due to an arterial or venous mechanical blockage. Nonocclusive arterial hypoperfusion is most commonly due to primary splanchnic arterial vasoconstriction, with the majority of cases involving vasospasm of branches of the superior mesenteric artery supplying the small intestine and proximal colon. Several vasoconstricting medications and drugs of abuse have been associated with the development of nonocclusive mesenteric ischemia (Table 30.1).

The pathophysiology of the various types of mesenteric ischemia, including non-occlusive mesenteric ischemia are shown in Fig. 30.1.

Epidemiology

While nonocclusive mesenteric ischemia accounts for 5 to 15 percent of patients with acute mesenteric ischemia overall, the incidence of drug-induced nonocclusive mesenteric ischemia has not been determined. The occurrence of nonocclusive mesenteric ischemia in hospitalized patients has declined due to the to the widespread use of invasive hemodynamic monitoring, early correction of hypotension, removal of pharmacologic and drugs of abuse, and the use of systemic vasodilators in cardiac failure. Nonetheless the mortality for nonocclusive mesenteric ischemia remains high, in part due to the complexity of establishing its diagnosis.

S. Burroughs · E. D. Ehrenpreis (✉)
Department of Medicine, Advocate Lutheran General Hospital, 1775 Dempster St, 6 South, Park Ridge, IL 60068, USA
e-mail: e2bioconsultants@gmail.com

E. D. Ehrenpreis et al. (eds.), *The Mesenteric Organ in Health and Disease*,
https://doi.org/10.1007/978-3-030-71963-0_30

Table 30.1 Medications and drugs of abuse associated with nonocclusive mesenteric ischemia

Medications and drugs of abuse associated with nonocclusive mesenteric ischemia
Digoxin
Alpha-adrenergic agonists
Cocaine
Methamphetamine

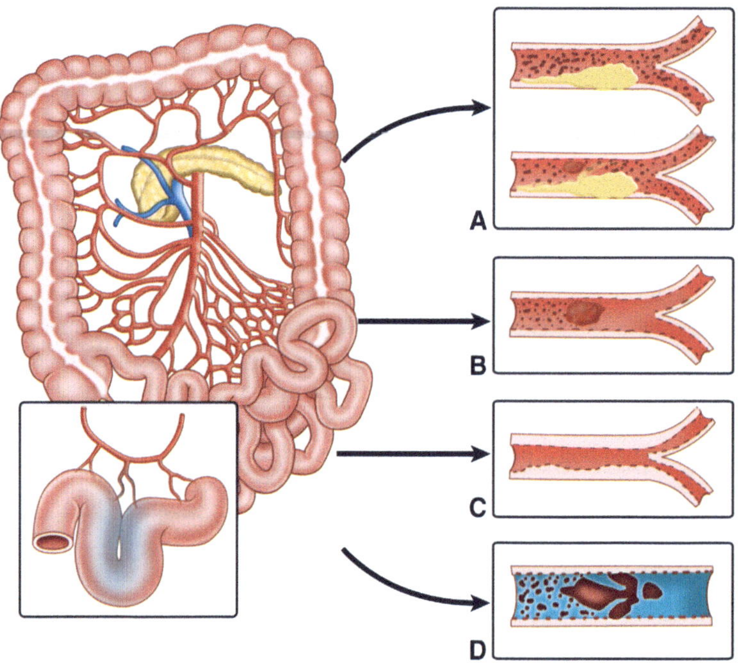

Fig. 30.1 Pathophysiology associated with mesenteric ischemia. **a** Arterial thrombosis, **b** arterial embolus, **c** nonocclusive mesenteric ischemia, **d** Venous thrombosis. Nonocclusive mesenteric ischemia as shown in option C occurs via vasoconstriction

Patients at Risk

Patients who are at risk of developing nonocclusive mesenteric ischemia include those who are critically ill, such as those with cardiogenic shock or sepsis and patients with a history of cardiovascular disease such as congestive heart failure, coronary artery disease. Drug-induced causes of nonocclusive mesenteric ischemia is associated with medications, or drugs of abuse that are known to produce intestinal vasoconstriction, such as digoxin, alpha-adrenergic agonists, cocaine, or methamphetamine.

Pathophysiology

The pathogenesis of nonocclusive mesenteric ischemia is related to the regulation of cardiac and cerebral blood flow at the expense of the splanchnic and peripheral circulation. The neurohormonal mediators, vasopressin and angiotensin, are often putative agents in the occurrence of nonocclusive mesenteric ischemia. When nonocclusive mesenteric ischemia is caused by medications or drugs of abuse, such as cocaine and methamphetamine, the primary instigating factor is drug-induced sympathomimetic mesenteric vasoconstriction, leading to a spectrum of secondary effects ranging from mild intestinal injury to frank bowel infarction. A number of mechanisms have been described in the development of mesenteric ischemia from cocaine abuse. These include (1) direct action of cocaine on the endothelium causing vasoconstriction, (2) increased norepinephrine and activation of noradrenergic receptors causing vasoconstriction, (3) accelerated atherosclerosis, and (4) platelet aggregation leading to mesenteric thrombosis. Similar to cocaine and methamphetamine, digoxin can also induce a vasoconstrictive effect on the splanchnic circulation, thereby increasing risk of mesenteric ischemia.

Diagnosis, Physical Findings and Laboratory Findings

The diagnosis of nonocclusive mesenteric ischemia is based on clinical suspicion comprised of patient risk factors and clinical signs and symptoms of mesenteric ischemic injury. Both the severity and location of abdominal pain in the setting of nonocclusive mesenteric ischemia are highly variable. Patients often report having a rapid onset of severe abdominal pain out of proportion to their accompanying physical examination. In about twenty five percent of patients with nonocclusive ischemia, abdominal pain may be absent or may start with nonspecific symptoms such as bloating, nausea, and vomiting. Progressive abdominal pain may then occur. Initial abdominal examination may be normal or nonspecific. The presence of peritoneal signs such as rebound tenderness and guarding occur when ischemia develops and transmural bowel infarction ensues. Nonocclusive mesenteric ischemia is often concomitant with other disorders and clinical presentation resulting in the masking of its symptoms. Because of this, necrosis and perforation may develop before a definitive diagnosis can be established, adding to the urgency of making a prompt diagnosis.

Laboratory studies can be nonspecific. Elevated white cell count, elevated serum lactate, metabolic acidosis, and elevated hematocrit have all been shown to occur in acute mesenteric ischemia.

Imaging

Plain abdominal radiographs have a limited role in diagnosing mesenteric ischemia and these are normal in 25% of cases. However, findings suggestive of all forms of mesenteric ischemia include the presence of an ileus, bowel wall thickening, or pneumatosis intestinalis. Computerized tomographic (CT) imaging of the abdomen may also be nonspecific, but abdominal CT can be used rule out other causes of acute abdominal pain. Findings consistent with acute mesenteric ischemia on abdominal CT include focal or segmental bowel wall thickening, bowel dilation, or mesenteric stranding (Also see Chap. 26, Ischemic Enteropathy).

Definitive imaging of nonocclusive mesenteric ischemia relies upon the demonstration of narrowing or spasm of mesenteric vasculature. This can be demonstrated on CT angiography or selective mesenteric arteriography as reduced intramural vessel filling, reduction in mesenteric vessels, and irregularity of the arterial branches of the mesenteric vasculature on vascular imaging.

Treatment

The goal of treatment of patients with nonocclusive mesenteric ischemia is to restore intestinal blood flow. Removal of vasoconstrictive medications is a critical aspect of this treatment. If necessary, other medical agents with limited inhibitory effects on mesenteric perfusion may be indicated. These agents include dobutamine, dopamine, and milrinone. Hemodynamic support, treatment of additional underlying causes, and intra-arterial infusion of vasodilators including papverine, prostaglandins, and nitroglycerine are also used. For patients with signs of advanced bowel ischemia, abdominal exploration and possible bowel resection are indicated. For patients suspected of having intestinal perforation surgical intervention should not be delayed.

Prognosis

When compared with other forms of mesenteric ischemia, nonocclusive mesenteric ischemia has the lowest survival with a mortality rates ranging from 70 to 90%. This high mortality has been attributed to the severity of comorbid conditions that contribute to the reduction of mesenteric perfusion, as well delay in diagnosis. Several case studies have focused on the prognosis of drug- induced nonocclusive mesenteric ischemia. Drugs such as methamphetamine and cocaine are associated with a high risk of significant microvascular compromise and mesenteric ischemia. Despite having fewer comorbidities, patients presenting with cocaine related mesenteric ischemia have a relatively poor prognosis. In one study, 5 of 19 (26%)

patients with cocaine-induced nonocclusive mesenteric ischemia died, with mortality in these patients resulting from septic shock caused by extensive bowel injury.

Angiotensin Converting Enzyme (ACE)-Inhibitor Induced Visceral Edema

Definition

ACE-inhibitors are widely prescribed medications for a hypertension, coronary artery disease, heart failure, diabetes, and chronic kidney disease. Common side effects of these medications include dizziness, headache, diarrhea, and dry cough. ACE-inhibitors are also known to be the leading cause of drug-induced angioedema in the United States and in some cases intermittent abdominal pain secondary to intestinal angioedema.

Epidemiology

Angioedema from ACE-inhibitor therapy is relatively rare and occurs in 0.1 to 0.7% of recipients. However, because they are so commonly prescribed, ACE-inhibitors account for 20 to 40% of all emergency department visits for angioedema per year. Compared to the general population, patients of African descent have a five times higher rate of occurrence of angioedema from ACE-inhibitor therapy. The Omapatrilat Cardiovascular Treatment vs. Enalapril (OCTAVE) trial was a prospective study which evaluated over 12,500 patients with hypertension who were treated with enalapril, an ACE-inhibitor or omapatrilat, an inhibitor of both neprilysin (neutral endopeptidase, NEP) and ACE. The rate of the development of angioedema secondary to enalapril over a period of six months was 0.68%.

Patients at Risk

Potential risk factors for developing ACE-inhibitor angioedema include a previous history of episodes of angioedema, (specifically those related to NSAID use), age greater than 65 years, intake of aspirin or other NSAIDs, and smoking.

Pathophysiology

ACE-inhibitors prevent the enzyme ACE from metabolizing Angiotensin I to Angiotensin II. Angiotensin II is responsible for inactivating bradykinin and acts as a vasoconstrictor by stimulation of angiotensin I and II receptors. ACE inhibitors

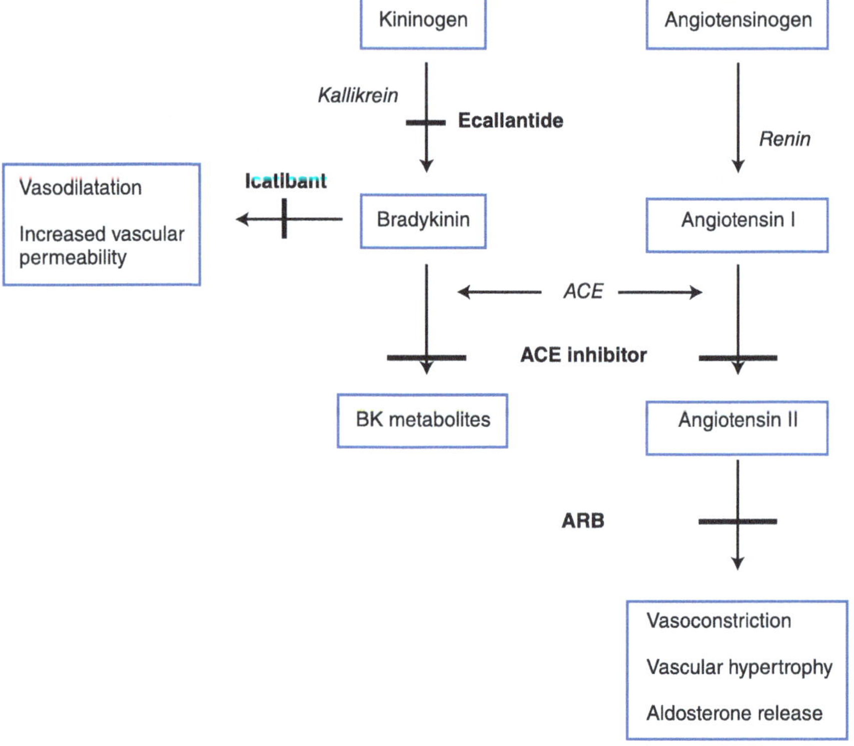

Fig. 30.2 Relationship between angiotensin II, ACE-inhibitors, and Bradykinin. Bold text: drugs. *Italicized text*: enzymes. ACE: angiotensin-converting enzyme; BK: bradykinin; ARB: angiotensin II receptor blocker. Modified from: Hurst M, Empson M. Oral angioedema secondary to ACE inhibitors, a frequently overlooked association: case report and review. N Z Med J 2006; 119: U1930

affect the signaling and the degradation of bradykinin, as they inhibit Angiotensin II from inactivating bradykinin. This in turn stimulates the release of nitric oxide and prostaglandins resulting in vasodilation as bradykinin levels increase. Increased capillary permeability of the blood vessels may also occur (Fig. 30.2).

Signs and Symptoms

Angioedema is often described at asymmetric, nonpitting swelling most commonly affecting nondependent areas of subcutaneous or submucosal tissues. Commonly affected areas are the lips, tongues and face. Visceral angioedema involving the mesentery that is associated with ACE-inhibitor use has also been described in a few case reports and reviews. Signs and symptoms described in these cases include diffuse abdominal pain, diarrhea, vomiting, anorexia and ascites. Angioedema

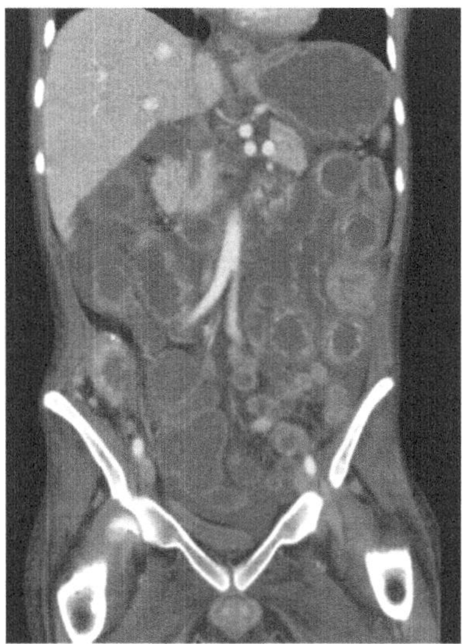

Fig. 30.3 Angiotensin converting enzyme inhibitor (ACE)-induced angioedema: coronal view. With permission from Dr. Abraham Dachman and Dr. Justin Ramirez

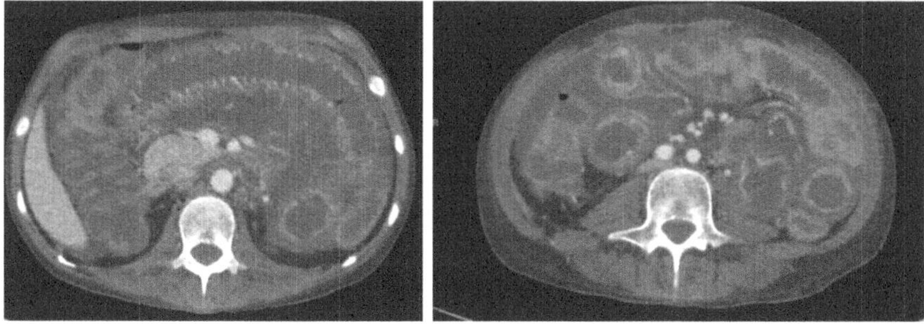

Fig. 30.4 **a** and **b** Angiotensin converting enzyme inhibitor (ACE)-induced angioedema: axial views. With permission from Dr. Abraham Dachman and Dr. Justin Ramirez

associated with ACE-inhibitors may appear within a week of starting or increasing medications, or after years of use.

Laboratory Findings

No definitive laboratory tests exist to diagnose ACE-inhibitor induced angioedema. Obtaining a complement protein 4 level may be reasonable if there is clinical suspicion that another cause of bradykinin-mediated angioedema could be present. Monitoring for resolution of symptoms after discontinuation of ACE inhibitor confirms the diagnosis.

Imaging

Visceral angioedema can be appreciated on computed tomography. Findings include dilated bowel loops, thickened mucosal folds, mesenteric edema, perihepatic fluid, ascites, or small intestinal "doughnut" or "stacked coin" appearance. Most reported cases of ACE inhibitor-induced intestinal angioedema involve the small bowel with circumferential thickening of the small bowel wall, ascites and incomplete small bowel obstruction (see Figs. 30.3 and 30.4a, b).

Discontinuation of ACE-inhibitors is the primary treatment of ACE-inhibitor induced angioedema. After discontinuation, angioedema usually resolves within 24 to 72 hours. H1 and H2 histamine blocker and, corticosteroids are commonly used treatments for angioedema. In severe cases, intramuscular or subcutaneous epinephrine is administered.

Drug Induced Retroperitoneal Fibrosis

Definition, Description, and Epidemiology

The retroperitoneum is the anatomical space that lies between the posterior parietal peritoneum and anterior to the transversalis fascia. The organs within the retroperitoneal space include the adrenal glands, kidneys, ureters, aorta, inferior vena cava, pancreas, esophagus, ascending and descending colon, rectum, and anal canal. The primary drug-induced disease in this region is retroperitoneal fibrosis. Retroperitoneal fibrosis is a disease process characterized by progressive inflammation and fibrosis of connective and adipose tissue in the retroperitoneal space. This inflammatory process generally originates inferior to the level of the aortic bifurcation near the sacral promontory and then covers the retroperitoneum with extension bilaterally along the aorta and inferior vena cava. It was first known as Ormond's disease, in the idiopathic, likely autoimmune form, and is uncommon, with an occurrence of 1 in 200,000 patients. With an overall male predominance of 2–3 to 1, retroperitoneal fibrosis usually presents in the fifth and sixth decades of life.

Pathophysiology

Retroperitoneal fibrosis has been linked to paraneoplastic syndromes and may occur as the result of severe retroperitoneal inflammation in acute pancreatitis. At least two thirds of retroperitoneal fibrosis are considered idiopathic. However, several medications have also been implicated as causes of retroperitoneal fibrosis (Table 30.2).

Methysergide, an ergot derived prescription medication used for the prophylaxis of migraine and cluster headaches, is a serotonin antagonist and has clearly been

Table 30.2 Medications associated with drug induced retroperitoneal fibrosis

Medications
Methysergide
Ergotamine
Hydralazine
Beta-adrenergic blocking agents

shown to be linked to the development of retroperitoneal fibrosis with lengthy use. The estimated incidence of methysergide-induced retroperitoneal fibrosis is 1 in 5,000 treated patients. While the main mechanism is unknown, it has been hypothesized that methysergide has profibrotic-reactive activities. In these patients, methysergide enhances myofibroblast proliferation and overproduction of extracellular matrix components such as collagen, fibronectin, tenascin, glycosaminoglycans.

Other pharmacologic agents such as ergotamine, hydralazine, and Beta-adrenergic blocking agents, (both nonselective and Beta-1 selective), have also been implicated in the development of retroperitoneal fibrosis by similar mechanisms.

Diagnosis, Physical Findings and Laboratory Findings

Presenting symptoms of retroperitoneal fibrosis are usually the result of gradual compression of structures within the retroperitoneal space, especially the ureter. The most common individual symptom of retroperitoneal fibrosis is a dull, constant pain localized to the back or flank, radiating to the lower abdomen. Other symptoms which may be present are nausea, diarrhea, fever, and weight loss, as well as hypertension. Laboratory abnormalities may include leukocytosis, elevated erythrocytic sedimentation rate, and elevated alkaline phosphatase levels. As the disease progresses, and there is further compression of the retroperitoneal structures, most commonly the ureters, azotemia, renal insufficiency, and possible renal failure can be seen. Lymphatics and venous obstruction can also occur secondary to compression and result in edema, specifically of the lower extremities. Arterial insufficiency can be seen with symptoms such as leg claudication secondary to aorta involvement. Rarely this disease process involves the duodenum, common bile duct, or colon resulting in obstruction.

Imaging

CT scan is the imaging of choice for the diagnosis of retroperitoneal fibrosis. Retroperitoneal fibrosis is seen as a periaortic soft tissue mass of variable thickness that envelops the retroperitoneal structures with obliteration of the plane between these structures and the adjacent psoas musculature (Figs. 30.5, 30.6 and 30.7).

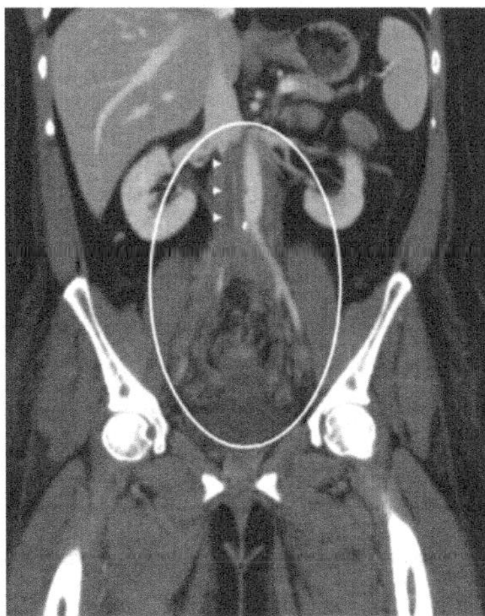

Fig. 30.5 Coronal contrast-enhanced CT image demonstrates confluent soft tissue encasing the infrarenal abdominal aorta and bilateral common iliac arteries (oval). The infrarenal IVC is thrombosed (arrowheads)

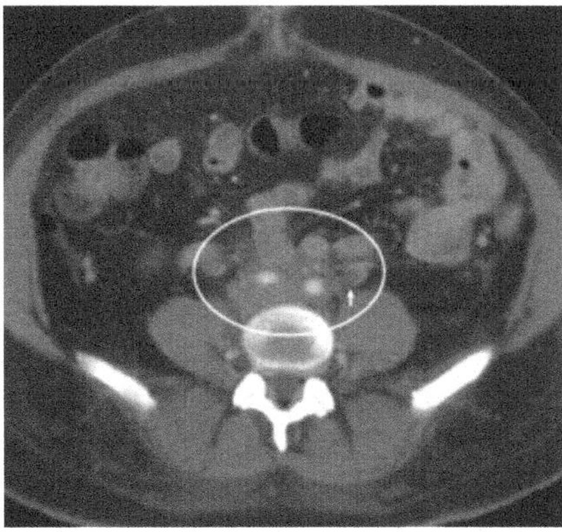

Fig. 30.6 Axial CT image at the level of the bilateral common iliac arteries shows the retroperitoneum completely replaced by abnormal soft tissue (oval) with additional thrombus in the left gonadal vein (arrow)

Magnetic resonance imaging (MRI) is also being more frequently used to define the staging clearly, particularly using T2-weighted imaging.

Intravenous pyelogram is also used to define retroperitoneal fibrosis. Classic findings that have been described include delayed excretion of contrast material with hydronephrosis, ureteral narrowing and medial deviation of the ureters at the L4–L5 level, (as opposed to lateral deviation found in retroperitoneal neoplasms). It is important to note, however, that medial deviation of the ureters is not a constant finding in patients with retroperitoneal fibrosis.

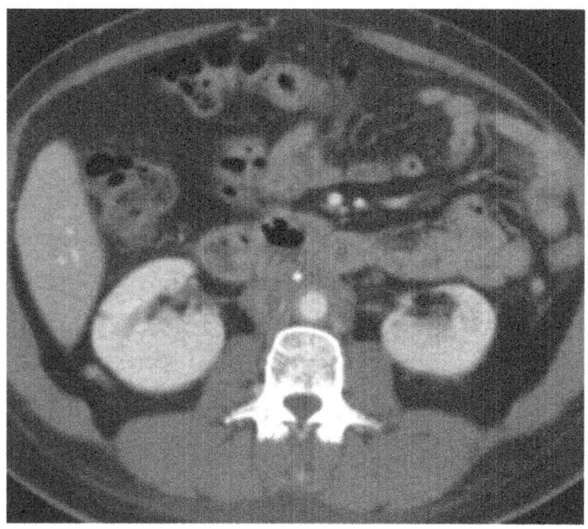

Fig. 30.7 Axial CT image at the level of the kidneys again shows abnormal retroperitoneal soft tissue encasing the aorta and IVC (Asterix). Images courtesy of Dr. Abraham Dachman

Histologic Findings

As the disease progresses, abundant collagen disposition and fibrous scarring occurs. Grossly, it will appear as woody, white, and fibrous plaque. Microscopically, features include abundant lymphocytes, plasma cells, and macrophages interspersed with fibroblasts and collagen bundles.

Treatments

The goal of treatment for patients with retroperitoneal fibrosis includes establishing the diagnosis, relieving the obstruction, and preventing progression of the disease.

Medications that are causative for retroperitoneal fibrosis are discontinued. Corticosteroids, implemented in early, active inflammatory stage of retroperitoneal fibrosis, have been shown to be effective. Other immunosuppressive drugs, such as azathioprine and cyclophosphamide, also have been used with success. Symptoms caused by venous obstruction are best treated nonoperatively, with elevation and elastic support of the leg eventually resulting in the development of collateral circulation.

Surgical

Surgical management of retroperitoneal fibrosis involves removing obstruction, often involving the ureter utilizing ureterolysis with intraperitoneal transposition of the ureters. Other maneuvers described to prevent recurrence of the ureteral obstruction include rapping the ureters in omental fat or interposing retroperitoneal fat between the ureters and the fibrosis.

Prognosis

In patients with idiopathic retroperitoneal fibrosis without renal compromise and effective ureterolysis, the prognosis is excellent with long-term success rates exceeding 90%.

Drug Induced Pancreatitis

Introduction

Acute pancreatitis is a condition in which the pancreas becomes inflamed and is often clinically characterized as epigastric abdominal pain with elevated levels of amylase and lipase. In some cases, inflammation of the mesentery may occur as a consequence of acute pancreatitis (See Chap. 16, Pancreatitis). A number of conditions are known to cause acute pancreatitis, with gallstones and alcohol abuse accounting for most cases. Drug-induced pancreatitis is the cause of up to 2% of patients with acute pancreatitis.

A number of medications have been associated with acute pancreatitis with timing of onset varying, from immediately upon drug administration, to several months. Since causative medications are usually identified in case reports and small cases studies, the likelihood of evidence for causation is classified based on timing of exposure and relationship to use and discontinuation of the drug (Table 30.3). A listing of drugs demonstrating the greatest potential for causing acute pancreatitis are shown in Table 30.4.

Varying mechanisms have been proposed for drug-induced pancreatitis. These include an immunologic reaction as seen with sulfonamides, 6-mercaptopurine, and aminosalicylates; ischemia as seen with diuretics such as furosemide and azathioprine; intravascular thrombosis as seen with estrogen; direct toxic effect as seen

Table 30.3 Classification system for drug-induced acute pancreatitis

Class Ia drugs
At least 1 case report with positive rechallenge, excluding all other causes, such as alcohol, hypertriglyceridemia, gallstones, and other drugs
Class Ib drugs
At least 1 case report with positive rechallenge; however, other causes, such as alcohol, hypertriglyceridemia, gallstones, and other drugs were not ruled out
Class II drugs
At least 4 cases in the literature
Consistent latency (>75 percent of cases)
Class III drugs
At least 2 cases in the literature
No consistent latency among cases
No rechallenge
Class IV drugs
Drugs not fitting into the earlier-described classes, single case report published in medical literature, without rechallenge

Reused with permission from Badalov N, Baradarian R, Kadirawel I, et al. Drug-Induced Acute Pancreatitis: An Evidence-Based Review. Clin Gastroenterol Hepatol 2007; 5:648. Copyright © 2007 AGA Institute

Table 30.4 Summary of drug-induced acute pancreatitis based on drug class

Class Ia	Class Ib	Class II	Class III	Class IV
α-methyldopa	All-trans-retinoic acid	Acetaminophen	Alendronate	Adrenocorticotrophic hormone
Azodisalicylate	Amiodarone	Chlorohiazide	Atorva statin	Ampicillin
Bezafibrate	Azathioprine	Clozapine	Carbamazepine	Bendroflumethiazide
Cannabis	Clomiphene	Didanosine	Captopril	Benazepril
Carbimazole	Dexamethasone	Erythromycin	Ceftriaxone	Betamethasone
Codeine	Ifosfamide	Estrogen	Chlorthalidone	Capecitabine
Cytosine	Lamivudine	L-asparaginase	Cimetidine	Cisplatin
Arabinoside	Losartan	Pegaspargase	Clarithromycin	Colchicine
Dapsone	Lynestrenol/methoxyethinylestradiol	Propofol	Cyclosporin	Cyclophosphamide
Enalapril	6-mercaptopurine	Tamoxifen	Gold	Cyproheptadine
Furosemide	Meglumine		Hydrochlorothiazide	Danazol
Isoniazid	Methimazole		Indomethacin	Diazoxide
Mesalamine	Nelfinavir		Interferon/ribavirin	Diclofenac
Metronidazole	Norethindronate/mestranol		Irbesartan	Diphenoxylate
Pentamidine	Omeprazole		Isotretinoin	Doxorubicin
Pravastatin	Premarin		Ketorolac	Ethacrynic acid
Procainamide	Trimethoprimsulfamethazole		Lisinopril	Famciclovir
Pyritonol			Metolazone	Finasteride
Simvastatin			Metformin	5-fluorouracil
Stibogluconate			Minocycline	Fluvastatin
Sulfamethoxazole			Mirtazapine	Gemfibrozil
Sulindac			Naproxen	Interleukin-2
Tetracycline			Paclitaxel	Ketoprofen
Valproic acid			Ponatinib	Lovastatin
			Prednisone	Mefenamic acid
			Prednisolone	Nitrofurantoin
				Octreotide
				Oxyphenbutazone
				Penicillin

(continued)

Table 30.4 (continued)

Class Ia	Class Ib	Class II	Class III	Class IV
				Phenophthalein Propoxyphene Ramipril Ranitidine Rifampin Risperidone Ritonavir Roxithromycin Rosuvastatin Sertraline Strychnine Tacrolimus Vigabatrin/lamotrigine Vincristine

Adapted from Badalov N, Baradarian R, Kadirawel I, et al. Drug-Induced Acute Pancreatitis: An Evidence-Based Review. Clin Gastroenterol Hepatol 2007; 5:648.

with diuretics and sulfonamides; and accumulation of toxic metabolites such as valproic acid, didanosine, pentamidine, and tetracycline data has shown that the use of prophylactic antibiotics plays a role in the management of acute pancreatitis.

Local complications that can arise from acute pancreatitis include necrosis, pancreatic pseudocysts, and fluid collection. In patients who develop necrotizing acute pancreatitis or pseudocysts, complications such as venous thrombosis, although rare, can occur due to extrinsic compression or secondary inflammation. This leads to mesenteric ischemia. Mesenteric edema and inflammation may occur in the setting of acute pancreatitis as well.

Suggested Reading

1. Boley SJ, Brandt LJ, Sammartano RJ. History of mesenteric ischemia. The evolution of a diagnosis and management. Surg Clin North Am. 1997;77:275.
2. Clair DG, Beach JM. Mesenteric Ischemia. N Engl J Med. 2016;374:959.
3. Acosta S. Epidemiology of mesenteric vascular disease: clinical implications. Semin Vasc Surg. 2010;23:4.
4. Bobadilla JL. Mesenteric ischemia. Surg Clin North Am. 2013;93:925.
5. Anderson JE, Brown IE, Olson KA, Iverson K, Cocanour C, Galante JM. Nonocclusive mesenteric ischemia in patients with methamphetamine use. 2017 WTA Podium Paper; Trauma Acute Care Surg. 2017; 84(6).
6. Ofer A, Abadi S, Nitecki S, et al. Multidetector CT angiography in the evaluation of acute mesenteric ischemia. Eur Radiol. 2009;19:24.
7. Mazzei MA, Mazzei FG, Marrelli D, et al. Computed tomographic evaluation of mesentery: diagnostic value in acute mesenteric ischemia. J Comput Assist Tomogr. 2012;36:1.
8. Banerji A, Clark S, Blanda M, et al. Multicenter study of patients with angiotensin-converting enzyme inhibitor-induced angioedema who present to the emergency department. Ann Allergy Asthma Immunol. 2008;100:327.
9. Gibbs CR, Lip GY, Beevers DG. Angioedema due to ACE inhibitors: increased risk in patients of African origin. Br J Clin Pharmacol. 1999;48:861.
10. Hoover T, Lippmann M, Grouzmann E, et al. Angiotensin converting enzyme inhibitor induced angio-oedema: a review of the pathophysiology and risk factors. Clin Exp Allergy. 2010;40:50.
11. Tenner S, Baillie J, DeWitt J, et al. American College of Gastroenterology guideline: management of acute pancreatitis. Am J Gastroenterol. 2013;108:1400.
12. Tadataka Y, Larson SD, Mark Evers B. Diseases of the peritoneum, retroperitoneum, mesentery, and omentum. In: Textbook of gastroenterology, vol. 2, 5th ed. 2009. p. 2514–6.
13. Geoghegan T, Byrne AT, Benfayed W, et al. Imaging and intervention of retroperitoneal fibrosis. Australas Radiol. 2007;51:26.
14. Alberti C. Drug-induced retroperitoneal fibrosis: short aetiopathogenetic note, from the past times of ergot-derivatives large use to currently applied bio-pharmacology. G Chir. 2015;36 (4):187–91.
15. Elramah M, Einstein M, Mori N, Vakikl N. High mortality of cocaine-related ischemic colitis: a hybrid cohort/case-control study. Gastrointest Endosc. 2012;75(6):1226–32.

Part VI
Neoplasms of the Mesentery

31 Primary Solid Neoplasms

Jasper B. van Praag, Robert C. Keskey, Eli D. Ehrenpreis, and John C. Alverdy

Introduction

Primary tumors arising in the mesentery are relatively rare and are most often detected incidentally by computerized tomography (CT). Because the mobility of the mesentery permits these tumors to occupy a large anatomic space, mesenteric tumors tend to grow to substantially large sizes. The most common symptoms and signs of mesenteric tumors are nonspecific and include abdominal pain, weight loss, diarrhea, and a palpable abdominal mass. Most primary neoplasms of the mesentery are mesenchymal in origin and, in the majority of cases, are histologically benign. There is, however, a wide variety of different types of mesenteric masses. This is due to the many different cellular structures that make up the composition of the mesentery.

J. B. van Praag
Department of Surgery, University Medical Center Groningen, Groningen, The Netherlands

R. C. Keskey
Department of Surgery, University of Chicago, Chicago, IL, USA

E. D. Ehrenpreis
Department of Medicine, Advocate Lutheran General Hospital, Park Ridge, IL, USA

J. C. Alverdy (✉)
Department of Surgery, University of Chicago, 5841 S. Maryland, Chicago, IL 60637, USA
e-mail: jalverdy@surgery.bsd.uchicago.edu

E. D. Ehrenpreis et al. (eds.), *The Mesenteric Organ in Health and Disease*,
https://doi.org/10.1007/978-3-030-71963-0_31

Desmoid Tumor of the Mesentery

Symptoms and Signs

Symptoms and signs of desmoid tumors are abdominal pain, a palpable abdominal mass or complications of the mass such as gastrointestinal bleeding, small bowel obstruction, fistula formation, or bowel perforation.

Pathophysiology

Desmoid tumors are uncommon and make up approximately 3% of all soft tissue tumors. Mesenteric desmoid tumors are locally aggressive proliferations of fibroblasts and can be infiltrative, but do not metastasize. They are often large in in size when discovered and commonly obstruct or damage local bowel loops, vascular structures and/or the urinary tract. Their etiology is unknown, although mesenteric desmoid tumors are often associated with Familiary Adenomateous Polyposis (FAP) (Figs. 31.1 and 31.2).

Diagnosis

CT shows tend to show a well-circumscribed soft tissue mass with variable attenuation and enhancement, it can demonstrate heterogeneous low attenuation areas due to necrosis (Fig. 31.1).

Immunohistochemistry is needed for differentiation from other neoplasms. Additionally, histologic diagnosis confirms the diagnosis of a desmoid tumor after biopsy or removal. Histologically, desmoids are characterized by monoclonal spindle-shaped proliferations in fibrous stroma (Fig. 31.2).

Treatment

The management of patients with mesenteric desmoid tumors is controversial as current therapies have varying success. Surgical resection can be curative but is also associated with significant morbidity and is associated with a high recurrence rate. Therefore, surgery should be reserved for patients with intestinal obstruction or for desmoid tumors that are symptomatic. These issues notwithstanding, complete excision is often difficult. Adjuvant aggressive medical therapy with non-steroidal anti-inflammatory agents (NSAIDs, e.g. sulindac), hormonal (tamoxifen) or cytotoxic chemotherapy (anthracyclines in combination with methotrexate and vinblastine) are most often used in lieu of surgery. Given the varying degree of response rates and high level of recurrence, watchful waiting has been employed for the initial management of desmoid tumors. Watchful waiting has been shown to

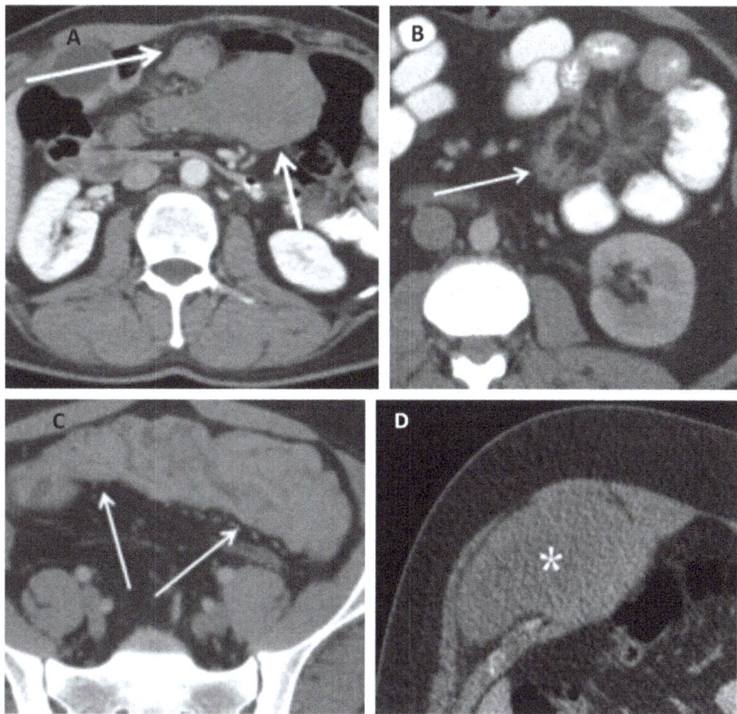

Fig. 31.1 Contrast-enhanced axial CT images from a single patient with previously undiagnosed Gardener syndrome. **a** shows show a well-circumscribed soft tissue mass with variable attenuation and enhancement, it can demonstrate heterogeneous low attenuation areas due to necrosis (arrows). **b** shows soft tissue attenuation in a stellate pattern within the mesenteric fat that is intimately associated with small bowel loops (arrow). **c** shows innumerable polyps throughout the imaged colon (arrows). **d** Most desmoid tumors will appear as well-circumscribed enhancing masses in the mesentery or anterior abdominal wall, as it is a benign, non-inflammatory fibroblastic tumor (asterix)

Fig. 31.2 Histology showing fibromatosis—desmoid type https://www.pathologyoutlines.com/topic/softtissuefibromatosisdeep.html Accessed December 3rd, 2019. © Copyright PathologyOutlines.com, Inc

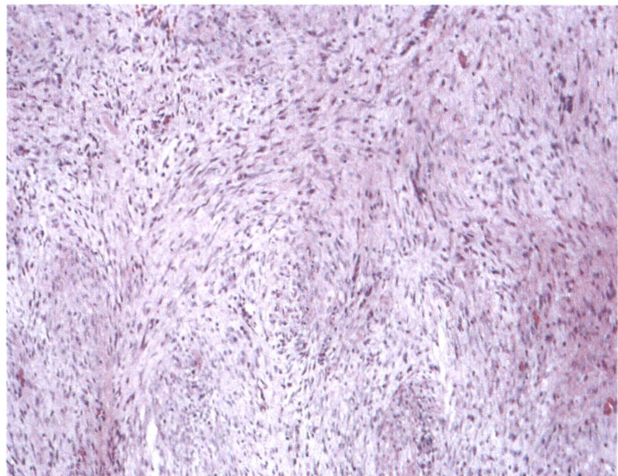

have a similar 5-year progression free survival as those treated with medical therapy (49.9 vs. 58.6%). Given these findings, the European Organization for Research and Treatment of Cancer and the Soft Tissue and Bone Sarcoma Group recommend watchful waiting for 1–2 years with frequent MRIs to elucidate the biology of the tumor.

Prognosis

Mesenteric desmoids are potentially life threatening. Because desmoid tumors are rare, little is known about prognosis or management options for individual cases. They have a fairly high recurrence rate of between 25 and 60% at 5 years following treatment. Progressive disease tends to occur when tumor size is greater than 7 cm, has an extra-abdominal location, and in patient that is more than 37 years old.

Primary Leiomyosarcoma of the Mesentery

Symptoms and Signs

Mesenteric leiomyosarcomas are mainly asymptomatic; however, when symptoms and signs are present, they include abdominal distension, pain and a palpable abdominal mass. When tumors infiltrate the gastrointestinal tract, they can cause bleeding, obstruction and pain.

Pathophysiology

Leiomyosarcoma is a mesenchymal tumor arising from smooth muscle. Leiomyosarcoma represents 5–10% of all newly diagnosed soft tissue sarcomas. Most often these tumors arise from the smooth muscle within the uterus, gastrointestinal tract, or blood vessels. It is believed that mesenteric leiomyosarcomas are derived from the smooth muscle of blood vessels within the mesentery. They have a reported incidence of only 1:350,000, most commonly occurring in middle-aged individuals. Infrequently these tumors produce neuron-specific enolase.

Diagnosis

Due to its frequent asymptomatic character, these neoplasms are often detected after metastases have developed. A solid mass can be found on ultrasonography and CT, but findings are non-specific (Fig. 31.3). Diagnosis can only be confirmed by histopathology and immunological staining. Pathologically, leiomyosarcoma

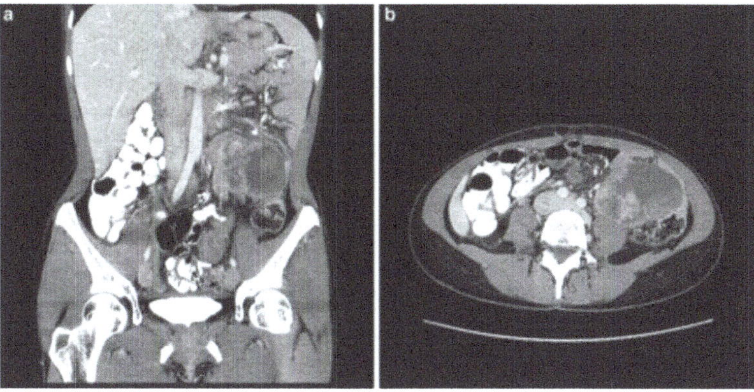

Fig. 31.3 CT scna showing a leiomyosarcoma of the mesentery located in the left upper quadrant. *Source* Varghese M et al. Metastatic mesenteric dedifferentiated leiomyosarcoma: a case report and a review of literature. Clin Sarcoma Res. 2016;6:2.

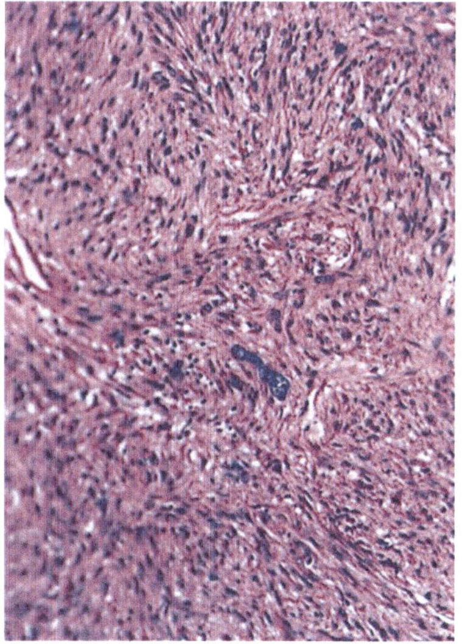

Fig. 31.4 Histology showing a leiomyosarcoma. https://www.pathologyoutlines.com/topic/softtissuefibromatosisdeep.html. Accessed December 3rd, 2019. © Copyright PathologyOutlines.com, Inc

exhibits high mitotic activity, usually atypical. Immunohistochemistry is positive for α-smooth muscle actin, vimentin and desmin; and stains negative for CD117 and CD34 (Fig. 31.4).

Treatment

Earlier detection using ultrasonography and CT, as well as complete surgical resection, may improve long-term prognosis in patients with primary leiomyosarcoma of mesenteric origin. Surgical excision with a wide margin of normal tissue is most effective since adjuvant chemotherapy or radiotherapy have poor responses.

Prognosis

Patients with leiomyosarcoma of the mesentery generally have a poor prognosis, also due to late diagnosis. The overall 5-year survival rate for this tumor is only 20–30%, and complete primary surgical resection is critical for achieving the best outcome. Conversely, recurrence can occur within 5 years, so that close and long-term follow-up of such patients for 5 years or more, with particular attention to the gastrointestinal tract, liver, and lung as sites of metastasis, is required. Prognostic factors associated with leiomyosarcomas include size >5 cm, age >62 years, high grade lesions and incomplete excision.

Primary Carcinoid Tumors of the Mesentery

Symptoms and Signs

Cases of primary mesenteric carcinoid tumors are extremely rare and symptoms before diagnosis are palpable abdominal mass and non-specific abdominal pain. These tumors can exhibit the so-called carcinoid syndrome, diarrhea and flushing are the most prominent symptoms.

Pathophysiology

Carcinoid is used to refer to well-differentiated neuro-endocrine tumors of the digestive tract. Mesenteric carcinoid tumors are slow-growing neoplasms that display neuroendocrine properties. They have varying degrees of fibrosis, calcification, focal or diffuse neurovascular bundle invasion, and necrosis or lymph node metastasis. In addition, liver metastases have been reported.

Diagnosis

Usually diagnosed by CT, mesenteric carcinoid tumors appear as a smooth-contoured soft tissue mass surrounded by radial bands (Fig. 31.5). Primary mesenteric tumors are diagnosed by exclusion of other conditions. Mesenteric carcinoids often (70%) have

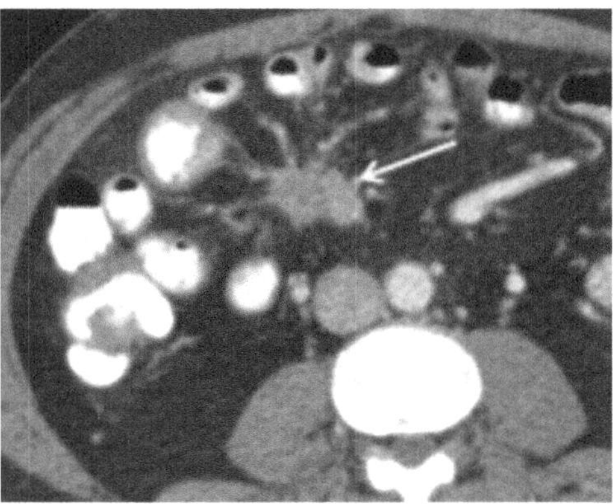

Fig. 31.5 A mesenteric soft tissue mass (arrows) with linear bands radiating in the surrounding mesenteric fat is consistent with neuroendocrine tumor shown on contrast-enhanced axial and coronal CT images

calcifications, in varying degrees. Metastatic Immunohistochemistry can show that the tumor demonstrates positive staining for neuroendocrine markers. Biochemical markers serum/urine 5-hydroxy indoleacteic acid (5-HIAA) and serum chromogranin A are elevated in functioning carcinoid tumors and can be used for functional imaging with somatostatin scan (Fig. 31.6a, b).

Treatment

No specific treatments for primary mesenteric carcinoid tumors are known. In general, surgery of carcinoid tumors and their metastases is considered the only curative treatment. Debulking might give relieve of symptoms due to secreted endocrines or (intestinal) obstruction. Long-acting somatostatin analogues can also be useful in the relief of symptoms. Other forms of treatment have not been proven successful for these neoplasms.

Prognosis

Due to the infrequency of mesenteric carcinoid tumors, prediction of prognosis is difficult. However, surgical resection of the tumor and/or its metastases are known to prolong survival.

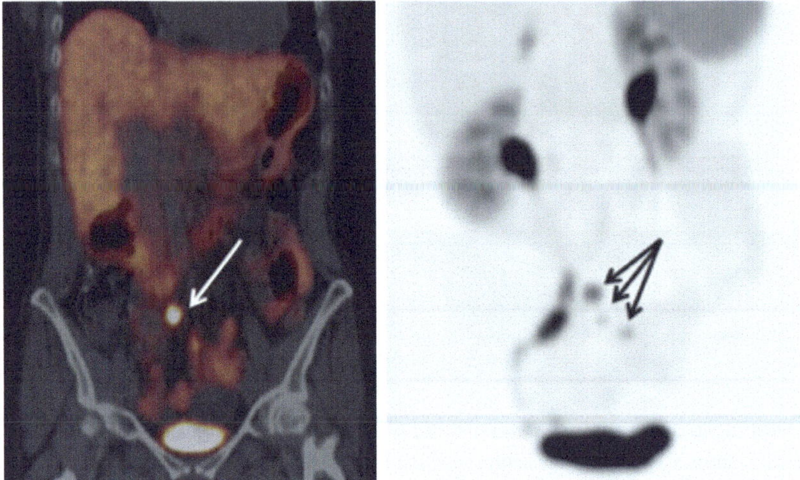

Fig. 31.6 Coronal fused CT (**a**)/PET (**b**) image demonstrates a small markedly avid lesion in the pelvis just to the right of midline (arrow). Maximum intensity projection (MIP) image demonstrates this and two other smaller foci located more inferiorly(arrows). The radiopharmaceutical (Dotatate) is a somatostatin receptor antagonist, and it is therefore highly sensitive for detection of neuroendocrine tumors (i.e., carcinoid). This agent is eliminated really and physiologic activity is present in the kidneys, collecting systems, and urinary bladder on the coronal maximum-intensity projection (MIP) image

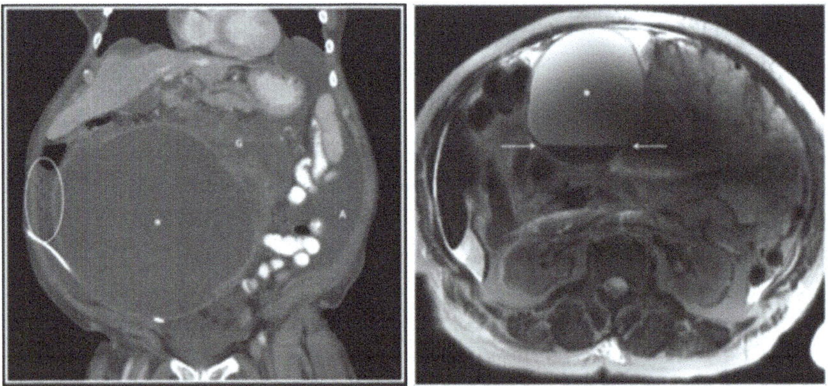

Fig. 31.7 **a** Coronal contrast-enhanced computed tomography and axial T2-weighted magnetic resonance imaging demonstrate an ill-defined, heterogeneously enhancing GIST (G) with a large cystic component (asterisk). Additional heterogeneous soft tissue along the periphery of the lesion is suggestive of peritoneal sarcomatosis and associated ascites (**b**). Layering T2-hypointense signal (arrows) is indicative of necrosis with layering blood products. With permission from Dr. Abraham Dachman, Dr. Scott Sorensen and Dr. Justin Ramirez

Primary Liposarcoma of the Mesentery

Symptoms and Signs

Patients may present with complaints of gradual abdominal distention, palpable abdominal mass, abdominal pain and weight loss.

Pathophysiology

As discussed above, soft tissue neoplasms are classified based on their tissue of origin. The mesenteric variant is rare, but liposarcoma is a common mesenchymal malignancy usually occurring within the retroperitoneum. These tumors are locally aggressive, are composed of lipoblasts and can secrete substances such as granulocyte-colony-stimulating factor (G-CSF) There are several histologic subtypes that vary by histologic grade. The most common variant is the well-differentiated (low-grade) liposarcoma. The second most common variant is dedifferentiated liposarcoma. This tumor contains regions of non-lipogenic sarcomatous tissue within a well-differentiated tumor. The dedifferentiated subtype of liposarcoma can be associated with distant metastasis. Liposarcomas are usually avascular to moderately vascular and can cause displacement of the major vessels.

Diagnosis

CT images show inhomogeneity, infiltration or poor margination and contrast enhancement. The density of the lesion exceeds that of subcutaneous fat. The enhancement on CT changes vary with the degree of histological grade. Well-differentiated liposarcomas are hyperintense on T2-weighted MRI with minimal or no enhancement.

Treatment

Surgical resection with sufficient surgical margin is the treatment of choice. Although the role of additional treatment is not clear, adjuvant therapy with chemotherapy consisting of doxorubicin, cisplatin and ifosfamide might cause shrinkage of tumor size.

Prognosis

The prognosis of mesenteric liposarcomas is unclear but appears to depend on the degree of cellular differentiation. In general, prognostication and stage are based on both tumor size and its histologic grade. Obtaining clear margins during resection is essential as these tumors have a high risk of local recurrence even after apparently complete resection.

Primary Stromal Tumor of the Mesentery

Symptoms and Signs

Symptoms are generally non-specific. Signs of extra-gastrointestinal stromal tumors (EGIST) of the mesentery include abdominal distension that gradually increases with tumor development.

Pathophysiology

Gastrointestinal stromal (GIST) tumours are mesenchymal tumors that arise from stromal cells, particularly interstitial cells of Cajal. EGIST of the mesentery are very rare. EGIST generally originate from the omentum, peritoneum, and mesentery. Based on histology and immunophenotyped stromal tumors of the mesentery are identical to gastrointestinal stromal tumors. The tumors are CD117 (C Kit) and CD34 positive mesenchymal neoplasms and are suggested to derive from pluripotential precursor cells.

Diagnosis

On CT scan, mesenteric GIST is a well-defined lobular mass showing heterogeneous contrast enhancement with areas of hypodensity (See Fig. 31.7a, b). On histological examination, spindle cells with eosinophilic cytoplasm are observed and are CD117 positive, CD34 positive, vimentin positive, S100 negative, and desmin negative (Fig. 31.8).

Treatment

The treatment of mesenteric stromal tumors is the performance of an *en bloc* resection, especially when tumors are >2 cm in size. Given the high level of KIT mutations, tyrosine kinase inhibitor drug treatment has been proven to prolong survival. Imatanib can be given in both the adjuvant and neoadjuvant setting. In the adjuvant setting, trials have shown a significant disease-free survival for GISTs <2 cm. Adjuvant treatment has been shown to improve disease free survival and should be administered for three years post resection in high risk GISTs. Whether treatment should be continued longer than 3 years has not been determined. In the

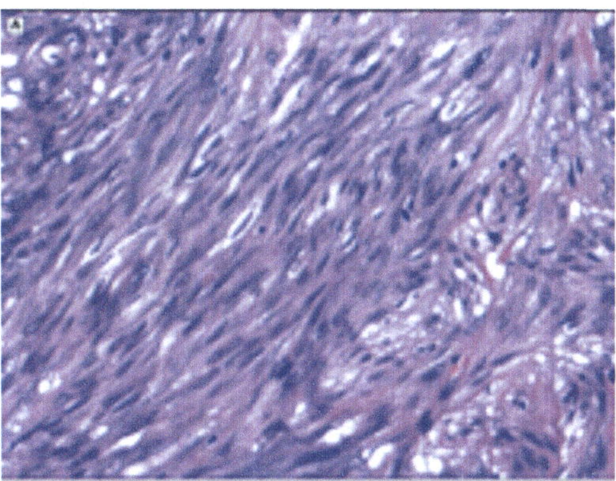

Fig. 31.8 H&E staining of a GIST tumor demonstrating uniform spindle cells with elongated nuclei and eosinophilic cytoplasm. Reused with permission Reproduced with permission from: Morgan J, Raut CP, Duensing A, Keedy VL. Epidemiology, classification, clinical presentation, prognostic features, and diagnostic work-up of gastrointestinal stromal tumors (GIST). In: UpToDate, Post TW (Ed), UpToDate, Waltham, MA. (Accessed on [Date].) Copyright © 2020 UpToDate, Inc. For more information visit www.uptodate.com

neoadjuvant setting, some studies have shown imatinib can be used to decrease tumor size. This in turn can improve the ability to obtain negative margins and decrease the morbidity of the surgery. In the setting of metastatic GISTs, imatinib has become first line treatment. There is some question whether or not resection has some benefit in the setting of metastatic GISTs.

Prognosis

For GIST tumors, prognosis is largely determined by location, size of the tumor, and the mitotic rate (uniquely measured in GISTs). GISTs tend to have a high recurrence rate, and EGIST arising from the mesentery tend to have less favorable outcomes. Predictive models exist that are largely based on tumor size and mitotic rate. These prediction models can be used to predict the need for adjuvant therapy for imatinib. There is limited data on the prognosis of EGIST, given its low prevalence. However, models may also be used to estimate risk. These include the general understanding that mesenteric GISTs tend to have a poor prognosis.

Summary

Primary tumors of the mesentery are rare and often present as incidental findings on CT scan. All tumors should be aggressively worked up with biopsy if indicated. Prognosis for tumors are dependent on the tumor biology, extent of disease and completeness of resection. Adjuvant chemotherapy is needed in most cases.

Suggested Reading

1. Yannopoulos K, Stout AP. Primary solid tumors of the mesentery. Cancer. 1963;16:914–27.
2. Sheth S, Horton KM, Garland MR, Fishman EK. Mesenteric neoplasms: CT appearances of primary and secondary tumors and differential diagnosis. RadioGraphics. Radiological Society of North America; 2003;23(2):457–73.
3. Dufay C, Abdelli A, Le Pennec V, Chiche L. Mesenteric tumors: diagnosis and treatment. J Visc Surg. 2012;149(4):e239–51.
4. Brooks AP, Reznek RH, Nugent K, Farmer KCR, Thomson JPS, Phillips RKS. CT appearances of desmoid tumours in familial adenomatous polyposis: further observations. Clin Radiol. 1994;49(9):601–7.
5. Healy JC, Reznek RH. The peritoneum, mesenteries and omenta: normal anatomy and pathological processes. Eur Radiol. 1998;8(6):886–900.
6. Li M, Cordon-Cardo C, Gerald WL, Rosai J. Desmoid fibromatosis is a clonal process. Hum Pathol. 1996;(9):939.
7. Lotfi AM, Dozois RR, Gordon H, Hruska LS, Weiland LH, Carryer PW, et al. Mesenteric fibromatosis complicating familial adenomatous polyposis: predisposing factors and results of treatment. Int J Colorectal Dis. 1989;4(1):30–6.
8. Jalini L, Hemming D, Bhattacharya V. Intraabdominal desmoid tumour presenting with perforation. Surgeon. 2006;4(2):114–6.
9. Burtenshaw SM, et al. Toward observation as first-line management in abdominal desmoid tumors. Ann Surg Oncol. 2016;23(7):2212–9.
10. Bonvalot S, et al. Extra-abdominal primary fibromatosis: aggressive management could be avoided in a subgroup of patients. Eur J Surg Oncol. 2008;34(4):462–8.
11. Salas S, et al. Gene expression profiling of desmoid tumors by cDNA microarrays and correlation with progression-free survival. Clin Cancer Res. 2015;21(18):4194–200.
12. Clark SK, Phillips RKS. Desmoids in familial adenomatous polyposis. Br J Surg. 1996;83(11):1494–504.
13. Mizoe A, Takebe K, Kanematsu T. Primary leiomyosarcoma of the jejunal mesentery: report of a case. Surg Today. Springer; 1998;28(1):87–90.
14. Fukunaga M. Neuron-specific enolase-producing leiomyosarcoma of the mesentery. Case report. APMIS. Wiley/Blackwell (10.1111); 2004;112(2):105–8.
15. Hashimoto H, Tsuneyoshi M, Enjoji M. Malignant smooth muscle tumors of the retroperitoneum and mesentery: a clinicopathologic analysis of 44 cases. J Surg Oncol. 1985;28(3):177–86.
16. Koczkowska M, Lipska BS, Grzeszewska J, Limon J, Biernat W, Jassem J. Primary leiomyosarcoma of the mesentery in two sisters: clinical and molecular characteristics. Pol J Pathol. 2013;64(1):59–63.
17. de Vries H, Verschueren RCJ, Willemse PHB, Kema IP, de Vries EGE. Diagnostic, surgical and medical aspect of the midgut carcinoids. Cancer Treat Rev. 2002;28(1):11–25.
18. Pantongrag-Brown L, Buetow PC, Carr NJ, Lichtenstein JE, Buck JL. Calcification and fibrosis in mesenteric carcinoid tumor: CT findings and pathologic correlation. Am J Roentgenol. American Public Health Association; 1995;164(2):387–91.

19. Karahan OI, Kahriman G, Yikilmaz A, Ozkan M, Bayram F. Gastrointestinal carcinoid tumors in rare locations: imaging findings. Clin Imaging. 2006;30(4):278–82.
20. Park I-S, Kye B-H, Kim H-S, Kim H-J, Cho H-M, Yoo C, et al. Primary mesenteric carcinoid tumor. J Korean Surg Soc. 2013;84(2):114–7.
21. Wasnik AP, Maturen KE, Kaza RK, Al-Hawary MM, Francis IR. Primary and secondary disease of the peritoneum and mesentery: review of anatomy and imaging features. Abdom Imaging. Springer, US; 2014;40(3):626–42.
22. Norton JA. Surgical management of carcinoid tumors: role of debulking and surgery for patients with advanced disease. Digestion. 1994;55 Suppl 3(3):98–103.
23. Horiguchi H, Matsui M, Yamamoto T, Mochizuki R, Uematsu T, Fujiwara M, et al. A case of liposarcoma with peritonitis due to jejunal perforation. Sarcoma. Hindawi; 2003;7(1):29–33.
24. Nakamura A, Tanaka S, Takayama H, Sakamoto M, Ishii H, Kusano M, et al. A mesenteric liposarcoma with production of granulocyte colony-stimulating factor. Internal medicine. Jpn Soc Intern Med. 1998;37(10):884–90.
25. Fletcher CDM, Bridge JA, Hogendoorn PCW, Mertens F. World health organization classification of tumours of soft tissue and bone. 4th ed. Lyon: IARC Press; 2013.
26. Jain SK, Mitra A, Kaza RCM, Malagi S. Primary mesenteric liposarcoma: an unusual presentation of a rare condition. J Gastrointest Oncol. 2012;3(2):147–50.
27. Friedman AC, Hartman DS, Sherman J, Lautin EM, Goldman M. Computed tomography of abdominal fatty masses. Radiology. 1981;139(2):415–29.
28. Ishiguro S, Yamamoto S, Chuman H, Moriya Y. A case of resected huge ileocolonic mesenteric liposarcoma which responded to pre-operative chemotherapy using doxorubicin, cisplatin and ifosfamide. Jpn J Clin Oncol. 2006;36(11):735–8.
29. Lucas DR, Nascimento AG, Sanjay BK, Rock MG. Well-differentiated liposarcoma. The Mayo Clinic experience with 58 cases. Am J Clin Pathol. 1994;102(5):677–83.
30. Sinha R, Verma R, Kong A. Mesenteric gastrointestinal stromal tumor in a patient with neurofibromatosis. Am J Roentgenol. American Roentgen Ray Society; 2004;183(6):1844–6.
31. Miettinen M, Monihan JM, Sarlomo-Rikala M, Kovatich AJ, Carr NJ, Emory TS, et al. Gastrointestinal stromal tumors/smooth muscle tumors (GISTs) primary in the omentum and mesentery: clinicopathologic and immunohistochemical study of 26 cases. Am J Surg Pathol. 1999;23(9):1109–18.
32. Lai ECH, Lau SHY, Lau WY. Current management of gastrointestinal stromal tumors–a comprehensive review. Int J Surg. 2012;10(7):334–40.
33. Morgan K, et al. Adjuvant and neoadjuvant imatinib for gastrointestinal stromal tumors. UpToDate Nov 2019. Accessed 12 May 2019.

Metastatic Diseases of the Mesentery

32

John C. Alverdy and Robert C. Keskey

Symptoms

As with other mesenteric masses, metastasis to the mesentery are initially indolent and may be asymptomatic while they are small. When symptoms develop from mesenteric masses, they may include the development of abdominal pain from compression of vascular structures or intestinal obstruction. Other symptoms and signs of obstructive symptoms may include nausea, emesis, abdominal distension, constipation, and obstipation. As with any malignancy, there is also the propensity to develop constitutional symptoms including fatigue and weight loss. In special circumstances, the presence of a neuroendocrine tumor may result in the development of carcinoid syndrome (represented by flushing and diarrhea). (See Chap. 31, Mesenteric Neoplasms).

Pathophysiology

Metastasis to the mesentery occur most commonly from a small intestinal neoplasms that spread along mesenteric lymphatic channels. Mesenteric metastases are seen in small bowel adenocarcinoma and more commonly in small bowel neuroendocrine tumors. Bulky mesenteric nodal disease can cause compression of mesenteric vessels and result in the development of chronic intestinal ischemia or bowel obstruction.

J. C. Alverdy (✉) · R. C. Keskey
Department of Surgery, University of Chicago, 5841 S. Maryland, Chicago, IL 60637, USA
e-mail: jalverdy@surgery.bsd.uchicago.edu

E. D. Ehrenpreis et al. (eds.), *The Mesenteric Organ in Health and Disease*,
https://doi.org/10.1007/978-3-030-71963-0_32

Metastatic invasion of the mesentery can also occur in conjunction with abdominal carcinomatosis. Peritoneal carcinomatosis occurs when either gastrointestinal (typically colorectal or gastric cancer), gynecological (ovarian cancer), or primary peritoneal malignancies (mesothelioma, pseudomyxoma peritonei) disseminate throughout the peritoneal cavity. Carcinomatosis occurring from gynecological or gastrointestinal malignancies, occurs as a result of tumor penetration through the serosa of the GI tract or cortex of the ovary. Carcinomatosis can involve any surface throughout the peritoneal cavity including the mesentery.

Diagnosis

Metastatic disease from lymphatic spread of small bowel tumors is diagnosed in a similar manner as other mesenteric masses using cross sectional imaging with computerized tomography (CT), or magnetic resonance imaging (MRI) with or without positron emission tomography (PET). Depending on the degree of lymphatic invasion, there may be some associated imaging signs of mesenteric ischemia associated with the metastatic lesions. The diagnosis of ischemia however cannot be definitively until tissue is visualized, and resected or biopsied. Carcinomatosis can be difficult to detect on imaging and may require diagnostic laparoscopy for definitive diagnosis. Some signs of carcinomatosis that may be visualized on imaging studies include scalloping of the liver and spleen and the presence of ascites. Occasionally, omental thickening can be detected as evidence of disease localizing to the omentum (Fig. 32.1a, b).

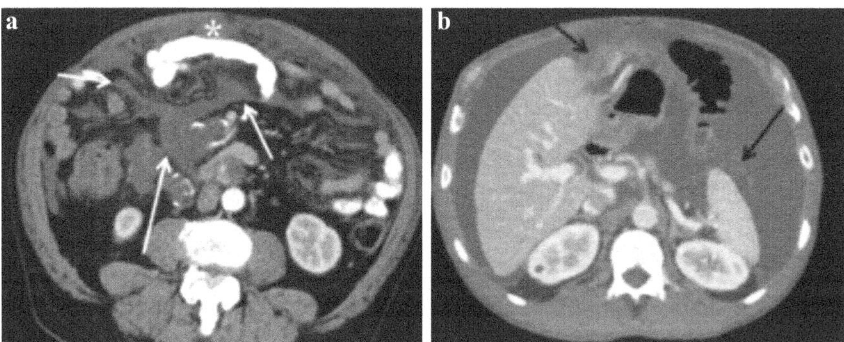

Fig. 32.1 **a** CT showing peritoneal carcinomatosis. Irregular soft tissue of the omentum (*) and nodular thickening of the mesenteric root (arrows). **b** CT demonstrating enhancing of soft tissue nodules anterior to the liver and spleen (black arrows). With permission from Dr. Abraham Dachman, Dr. Scott Sorensen and Dr. Justin Ramirez

Treatment

For metastases to the mesenteric lymph nodes, resection of the mass along with adjacent segment of bowel should be performed. In the setting of peritoneal carcinomatosis, cytoreductive surgery and HIPEC (intraperitoneal hypothermic chemotherapy) have been found to provide survival benefit for some cases depending on the extent of the disease and type of malignancy.

Malignant Bowel Obstructions and Palliative Care

When a patient is diagnosed with an intra-abdominal malignancy, goals of care should be established with the patient and their primary decision makers. In this setting, the main goals of treatment are to obtain symptomatic relief, especially in the case of peritoneal carcinomatosis as the overall prognosis in these patients is generally very poor. For mesenteric neoplasms, extra-luminal compression of the intestine can result in bowel obstruction. Malignant bowel obstruction manifests as nausea, vomiting, and abdominal pain. When a patient is diagnosed with malignant bowel obstruction, the overall predicted survival is about 125 days, with 42% patients dying within 90 days of surgical consultation.

Initial management of a malignant bowel obstruction from a mesenteric neoplasm should be nasogastric tube decompression to alleviate gastric distension from a distal obstruction. By decreasing gastric distension, patients tend to have rapid improvement in nausea. Placement of an endoscopic gastrostomy tube can be considered for long-term management in these patients, especially when medical therapy of nausea fails. Placement of a gastrostomy tube allows for periodic venting and relief of gastric distension. Retrospective studies have demonstrated that patients who have venting gastrostomy tubes have fewer hospital readmissions and a lower rate of in-hospital deaths. Another consideration for palliative care in this patient group is the performance of colonic stenting to help alleviate obstruction. The performance of colonic stenting in individual patients is largely depend on the anatomic location of the obstruction and the availability of skilled endoscopists. Stenting offers a short-term solution to malignant obstructions; however, up to one third of patients with colonic stents will develop recurrent obstruction and require repeated intervention. Both endoscopic gastrostomy tubes and colonic stents have significant potential complications but may provide a means for achieving end of life goals.

The mainstay of pharmacologic management of malignant bowel obstruction is anti-secretory therapy. Octreotide, a somatostatin analog is generally the first line treatment for decreasing intestinal secretions. Randomized controlled trials have shown that somatostatin analogs are more effective at relieving symptoms in malignant bowel obstructions than anticholinergic medications. Long acting depot injections of somatostatin analogs can be considered for outpatient treatment. Some anticholinergic medications that are also used include hyoscine, butylbromide, and glycopyrrolate. These are thought to have fewer central nervous system side effects than other anticholinergic agents. Glucocorticoids represents another adjunctive measure for patients with mesenteric metastatic disease. However, there is limited evidence regarding the efficacy of glucocorticoid therapy in this clinical setting.

Conclusion

Metastatic disease to the mesentery should be strongly considered in patients who develop a mesenteric mass. Metastases to the mesentery occur from lymph node drainage from adjacent small bowel malignancies or from peritoneal carcinomatosis. In either case, depending on the extent of disease, surgical resection has a significant potential role in patient management. Finally, palliative options are available for patients with malignant bowel obstruction from mesenteric metastases. These treatments should be considered to improve their quality of life.

Suggested Reading

1. Freedman AS, Friedberg JW, Aster, JC. Clinical presentation and diagnosis of non-Hodgkin lymphoma, in UptoDate. 2018.
2. Coccoloni F, Gheza F, Lotti M, et al. Peritoneal carcinomatosis. World J Gastroenterol. 2013;19(41):6979–94.
3. Dufay C, Abdelli A, Le Pennec V, Chiche L. Mesenteric tumors: diagnosis and treatment. J Visc Surg. 2012;149(4):239–51.
4. Pujara D, et al. Selective approach for patients with advanced malignancy and gastrointestinal obstruction. J Am Coll Surg. 2017;225(1):53–9.
5. Lilley EJ, et al. Survival, healthcare utilization, and end-of-life care among older adults with malignancy-associated bowel obstruction: comparative study of surgery, venting gastrostomy, or medical management. Ann Surg. 2018;267(4):692–9.
6. Faraz S, et al. Predictors of clinical outcome of colonic stents in patients with malignant large-bowel obstruction because of extracolonic malignancy. Gastrointest Endosc. 2018;87(5):1310–7.
7. Arthur J. Constipation and bowel obstruction. In: Yennurajalingam S, Bruera E, editors. Oxford American handbook of hospice and palliative medicine and supportive care, 2nd ed. Oxford University Press; 2016.

Mesenteric Lymphoma

33

Emily Papazian, Robert C. Keskey, and John C. Alverdy

Symptoms and Signs

Mesenteric lymphoma commonly presents with abdominal pain and presence of a palpable abdominal mass. Advanced stage disease presents with systemic symptoms including fever, night sweats and weight loss and may rarely result in complications such as intestinal obstruction, perforation and hemorrhage. Mesenteric lymphoma may also present with hepatosplenomegaly. Thus, evaluation of the size and consistency of the liver and spleen should be evaluated along with the presence of peripheral lymph nodes in staging of the disease.

Pathophysiology

Lymphoma is the most common solid mesenteric tumor. Lymphomas are broadly categorized into two main subtypes: Non-Hodgkin's lymphoma and Hodgkin's lymphoma. Non-Hodgkin's lymphoma is usually of B cell origin. The disease is often diagnosed with a peripheral lymph node biopsy and disease in the abdomen is presumed to be due to systemic disease. Disease may be detected in mesenteric lymph nodes in approximately 30–50% of patients with non-Hodgkin's lymphoma, of which 40–67% are large B cell tumors. Spread to the mesentery should especially be considered in patients with para-aortic disease.

E. Papazian · R. C. Keskey · J. C. Alverdy (✉)
Department of Surgery, University of Chicago, 5841 S. Maryland, Chicago, IL 60637, USA
e-mail: jalverdy@surgery.bsd.uchicago.edu

E. D. Ehrenpreis et al. (eds.), *The Mesenteric Organ in Health and Disease*,
https://doi.org/10.1007/978-3-030-71963-0_33

Diagnosis

Computerized tomography (CT) is used to identify the presence of mesenteric masses. A typical radiologic finding in abdominal lymphoma is the "sandwich sign", a confluence of enlarged lymph nodes surrounding mesenteric fat and vessels (Fig. 33.1a). On imaging, mesenteric lymphoma presents as a lobulated mass of homogenously enhancing nodes that displaces the small bowel. Mesenteric lymphoma may more rarely present as segmental "misty mesentery" on CT with hypermetabolism in adjacent nodes on positron emission tomography (PET) (Fig. 33.1b). Mesenteric disease is often accompanied by extensive retroperitoneal lymphadenopathy, which may aid in diagnosis. Since imaging alone is not adequate for diagnosing mesenteric lymphoma, tissue diagnosis is essential to obtain prior to treatment. Surgical biopsy using a laparoscopic technique, or open laparotomy, may be used to differentiate mesenteric lymphoma from other mesenteric masses, including mesenteric panniculitis a benign, inflammatory disorder. CT guided biopsy of the mesenteric mass is another consideration.

Treatment

Chemotherapy is the primary treatment indicated for mesenteric lymphoma. The chemotherapy regiment usually consists of CHOP (cyclophosphamide, doxorubicin, vincristine, and prednisone) with or without rituximab, (known as R-CHOP).

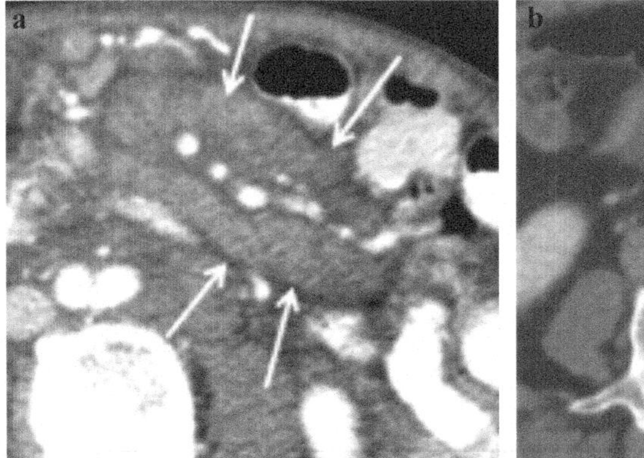

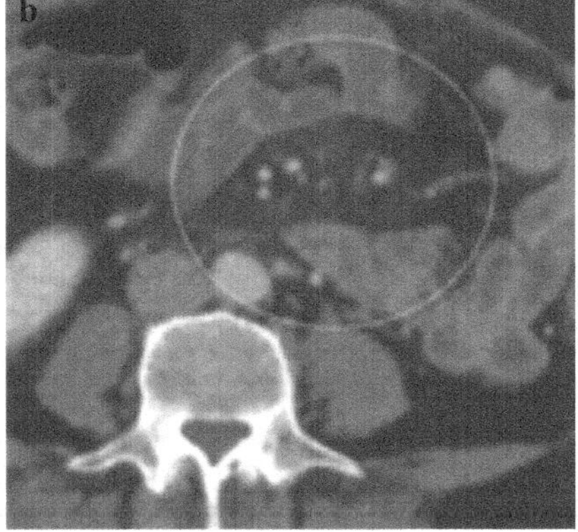

Fig. 33.1 CT cross sectional imaging revealing a confluence of enlarged lymph nodes referred to as the 'sandwich sign' (**a**). The image on the right is a CT scan showing the 'misty mesentery' which can be associated with lymphoma (**b**). With permission from Dr. Abraham Dachman, Dr. Scott Sorensen and Dr. Justin Ramirez

Prognosis

The revised International Prognostic Index may be used to stratify patients into risk groups to predict survival. High tumor bulk is a poor prognostic indicator in follicular lymphoma, the most common subtype of mesenteric lymphoma. A study of early-presenting patients with mesenteric lymphoma demonstrated a ten-year survival of 54% in patients with bulky disease and of 76% in patients who did not have bulky disease.

Summary

Diagnosis of mesenteric lymphoma may require a surgical biopsy. Treatment of mesenteric lymphoma consists of systemic chemotherapy. High tumor bulk is a negative prognostic indicator in mesenteric lymphoma.

Suggested Reading

1. Salemis NS, Gourgiotis S, Tsiambas E, Karagkiouzis G, Nakos G, Karathanasis V. Diffuse large B cell lymphoma of the mesentery: an unusual presentation and review of the literature. J Gastrointest Canc. 2009;40(3):79.
2. Dufay C, Abdelli A, Le Pennec V, Chiche L. Mesenteric tumors: diagnosis and treatment. J Visc Surg. 2012;149(4):e239–51.
3. Sheth S, Horton KM, Garland MR, Fishman EK. Mesenteric neoplasms: CT appearances of primary and secondary tumors and differential diagnosis. RadioGraphics. 2003;23(2):457–73.
4. Karaosmanoglu D, Karcaaltincaba M, Oguz B, Akata D, Özmen M, Akhan O. CT findings of lymphoma with peritoneal, omental and mesenteric involvement: peritoneal lymphomatosis. Eur J Radiol. 2009;71(2):313–7.
5. Mueller PR, Ferrucci JT, Harbin WP, Kirkpatrick RH, Simeone JF, Wittenberg J. Appearance of lymphomatous involvement of the mesentery by ultrasonography and body computed tomography: the "sandwich sign." Radiology. 1980;134(2):467–73.
6. Taffel MT, Khati NJ, Hai N, Yaghmai V, Nikolaidis P. De-misty-fying the mesentery: an algorithmic approach to neoplastic and non-neoplastic mesenteric abnormalities. Abdom Imaging. 2014;39(4):892–907.
7. Romaguera JE, Velasquez WS, Silvermintz KB, Fuller LB, Hagemeister FB, McLaughlin P, et al. Surgical debulking is associated with improved survival in stage I–II diffuse large cell lymphoma. Cancer. 1990;66(2):267–72.
8. Daly SC, et al. Laparoscopy has a superior diagnostic yield than percutaneous image-guided biopsy for suspected intra-abdominal lymphoma. Surg Endosc. 2015;29(9):2496–9.

Castleman Disease with Mesenteric Involvement

34

Eli D. Ehrenpreis

Definition and Description of the Disease

Castleman disease (CMD) is the name given to a several types of benign, non-neoplastic lymphoproliferative disorders. CMD most commonly involves a single nodal group in the mediastinum. However, lymphoid involvement may also occur individually in other sites or in multiple locations. When a single nodal group is involved, the condition is called Unicentric CMD. If the disease assumes a form that involves multiple nodal locations it is called Multicentric CMD. CMD with mesenteric involvement is very rare.

Epidemiology

CMD in all forms is a very rare disorder. Unicentric disease restricted to one lymph node grouping is usually seen in young and middle-aged patients with no gender differences. To the present, 55 cases of CMD with mesenteric involvement has been described in the medical literature.

Patients at Risk

Several cases of CMD have been described in patients with Sjogren's Syndrome. Other autoimmune diseases occurring in patients with CMD include rheumatoid arthritis, systemic lupus erythematosus, myasthenia gravis, Evans' syndrome,

E. D. Ehrenpreis (✉)
Department of Medicine, Advocate Lutheran General Hospital, 1775 Dempster St, 6 South, Park Ridge, IL 60068, USA
e-mail: e2bioconsultants@gmail.com

E. D. Ehrenpreis et al. (eds.), *The Mesenteric Organ in Health and Disease*,
https://doi.org/10.1007/978-3-030-71963-0_34

vitiligo, Graves' disease, ulcerative colitis, celiac disease, immune-mediated thrombocytopenia. Two cases of CMD have been reported in patients with Crohn's disease. POEMS syndrome (polyneuropathy, organomegaly, endocrinopathy, M proteins, skin changes) has occurred in a number of patients with CMD. Some patients with CMD and TAFRO syndrome (thrombocytopenia, anasarca, fever, reticulin fibrosis, and organomegaly) have been described.

Cases of CMD have been reported as a complication of infection with the human immunodeficiency virus (HIV). Other patients have developed CMD de novo without the presence of underlying autoimmune diseases. It is presumed that the risk of CMD with mesenteric involvement is similar to other patients with CMD.

Pathophysiology

Because a number of cases of CMD occurring in patients with autoimmune and immunologically-mediated diseases, CMD is postulated to originate as an autoimmune disorder. Reports of CMD in patients with HIV infection, suggests a possible of a role of immunodeficiency in the development of CMD. It has now been determined that symptoms experienced by patients with the plasma cell histologic subtype of CMD are related to excessive production of IL-6 by affected lymph nodes resulting in direct systemic overactivity of IL-6.

Signs and Symptoms

Patients with symptoms generally fall into two categories; those that develop symptoms at the site of the lesion due to local mass effect, and those that develop systemic symptoms. The severity and types of symptoms occurring in patients with CMD depend on the location, muticentricity and histologic subtype of the disease. In general, patients with unicentric disease, (especially if it is of the hyaline vascular histologic subtype, see below) are asymptomatic or have mild symptoms, while multicentric disease and plasma cell histologic subtype disease are associated with more severe systemic symptoms. Typical systemic symptoms of CMD include fever, night sweats, weight loss, fatigue and anemia. Signs and symptoms in patients with CMD and mesenteric involvement that have been described in the literature have included lack of symptoms, vague abdominal pain, nausea, acute abdominal pain, obstructive symptoms and tenderness, the presence of a palpable abdominal mass, and the signs and symptoms associated with systemic disease.

Diagnosis

In asymptomatic patients, the diagnosis of CMD is generally based on an incidental finding on imaging. The presentation of patients with CMD are dependent on disease location, the presence of multicentricity and the histologic subtype.

Physical Findings

The physical examination in patients with CMD requires a thorough palpation of the lymph node chains to assess the extent of muticentricity of the disease. Specific physical findings are dependent on disease location, and the presence of multicentricity. Cases of CMD with mesenteric involvement that are described in the literature have demonstrated normal examinations, abdominal tenderness and the presence of a palpable abdominal mass.

Laboratory Findings

These are nonspecific. Anemia is a cardinal finding in multicentric CMD. Hypoalbuminemia has been linked to a poorer prognosis of the disease.

Imaging

Some reported cases of CMD with mesenteric involvement have initially been diagnosed on abdominal ultrasound. Small intestinal radiography with barium contrast has been used to demonstrate extrinsic compression and displacement of small bowel loops. Abdominal CT and MRI are the radiographic tests of choice to identify and characterize CMD with mesenteric involvement. Findings on abdominal CT and/or MRI that have been described in the literature include a localized solid mass (some with calcification and various degrees of contrast enhancement), a mass effect with small intestinal displacement, mesenteric nodal enlargement, fascial thickening around the mass and satellite lesions. Pelvic and retroperitoneal lesions have also been described. Hepatosplenomegaly may also be present (Fig. 34.1a, b).

Histologic Findings

CMD can be classified based on two specific histologic subtypes. The most common of these is the hyaline vascular form of the disease, which occurs in 80–90% of cases. Most of these cases are unicentric disease and present as a single mass on imaging. On microscopic examination, excised specimens of the hyaline vascular subtype show a lymphoid tissue mass containing very large lymphoid follicles, capillary proliferation and hyalinization. The plasma cell histologic subtype occurs in the remaining 10–20% of cases. The plasma cell variant of CMD demonstrates architecturally recognizable lymph node containing solid sheets of plasma cells in various forms of differentiation. These patients generally have multicentric disease.

Treatments

Most treatments used for patients with CDM are based on individual case reports and case series. Some patients with asymptomatic unicentric CMD may be monitored without treatment. However, surgical resection is usually performed for solitary mesenteric lesions, as enlarging and increasingly symptomatic mesenteric masses have been described. Complete surgical excision is generally curative of the CDA including cases with mesenteric involvement. Since multicentric CDA is representative of a systemic lymphoproliferative disorder and carries a poorer prognosis, treatment consists of anti-inflammatory and antineoplastic medications. Reported treatments include rituximab, combination chemotherapy, autologous stem cell transplantation, bortezomib, thalidomide, anakinra (an IL1-antagonist), interferon-a, and all-trans retinoic acid.

Improvement of CMD symptoms and abnormal laboratory results following administration of anti-IL-6 antagonists has been recently demonstrated. A recent randomized, placebo-controlled trial of siltuximab, a chimeric monoclonal neutralizing antibody against IL6, was performed in 79 patients with multicentric CMD. In that study, sustained tumor and symptomatic responses occurred in 18 (34%) of 53 patients in the siltuximab group and none of 26 in the placebo group (difference = 34.0%, 95% CI 11.1–54.8%, $p = 0.0012$).

Prognosis

Surgical resection of solitary lesions is curative of the disease. Recently, a retrospective analysis of 113 patients with CMD evaluated at the Mayo Clinic and University of Nebraska was performed. This study demonstrated a 65% 5-year overall survival in patients with multicentric disease and a 91% overall survival for patients with unicentric disease. Other cofactors that influenced prognosis included the presence of POEMS. Patients with multicentric CMD with POEMS syndrome and no osteosclerotic lesions had the lowest overall survival (27%). The presence of HIV infection is an additional confounding factor affecting prognosis of CMD. It is assumed that similar prognostic indicators occur patients with CMD with mesenteric involvement.

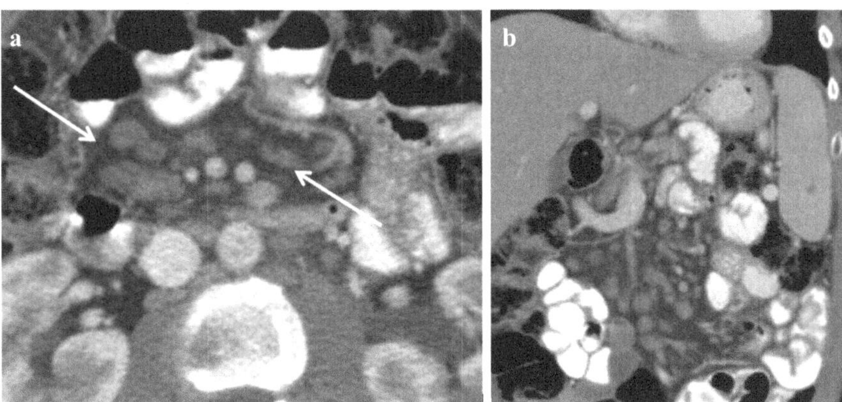

Fig. 34.1 Castleman disease with mesenteric involvement. Contrast-enhanced axial (**a**) and coronal (**b**) CT images showing numerous enlarged mesenteric lymph nodes (arrows), hepatomegaly and mild (14 cm) splenomegaly. In themselves, these are non-specific findings. This case was histologically-proven multicentric Castleman disease. With permission from Dr. Scott Sorensen and Dr. Abraham Dachman

Suggested Reading

1. Bracale U, Pacelli F, Milone M, et al. Laparoscopic treatment of abdominal unicentric castleman's disease: a case report and literature review. BMC Surg. 2017;17(1):38.
2. Lv A, Hao C, Qian H, Leng J, Liu W. Castleman disease of the mesentery as the great mimic: incidental finding of one case and the literature review. BioScience Trends. 2015;9(3):198–202.
3. Dei-Adomakoh YA, Quarcoopome L, Abrahams AD, Segbefia CI, Dey DI. Sjögren's and plasma cell variant Castleman disease: a case report. Ghana Med J. 2018;52(1):61–5.
4. Papaziogas B, Chatzimavroudis G, Koutelidakis I, Grigoriou M, Atmatzidis K. A rare form of isolated mesenteric Castleman's disease presenting as an abdominal mass (isolated mesenteric Castleman's disease). J Gastrointestin Liver Dis. 2006;15(2):171–4.
5. Caroline Ribeiro Sales A, Romão de Souza Junior V, Iglis de Oliveira M, Azevedo Braga Albuquerque C, de Barros Campelo Júnior E, Sérgio Ramos de Araújo P. Multicentric Castleman's disease in human immunodeficiency virus infection: two case reports. J Med Case Rep. 2018;12(1):117.
6. Yoshizaki K, Murayama S, Ito H, Koga T. The role of interleukin-6 in Castleman disease. Hematol Oncol Clin North Am. 2018;32(1):23–36.
7. van Rhee F, Greenway A, Stone K. Treatment of idiopathic Castleman disease. Hematol Oncol Clin North Am. 2018;32(1):89–106.
8. van Rhee F, Wong RS, Munshi N, Rossi JF, Ke XY, Fosså A, et al. Siltuximab for multicentric Castleman's disease: a randomized, double-blind, placebo-controlled trial. Lancet Oncol. 2014;15(9):966–74.

Part VII
Surgical Diseases of the Mesentery

Mesenteric Resection in Upper Abdominal Surgery

35

Sara Gaines and John C. Alverdy

Indications for Surgery

The mesenteric resection in upper abdominal surgery is directed toward the primary pathology of the accompanying organ. The most surgical common procedure that includes resection of the mesentery is small bowel resection. Mesenteric resection is performed in these cases as an extension of treatment for ischemia, trauma, cancer, ulceration, bleeding, obstruction, or Crohn's disease (Table 35.1).

In acute mesenteric ischemia, the mesentery may show signs of edema, hemorrhage, and venous congestion. The presence of these findings is an indication for exploration and possible resection of the mesentery. Trauma with suspected bowel injury includes a thorough intraoperative exploration and a "running" of the small bowel and the mesentery. Full thickness or multi segment small bowel injuries are resected and repaired primarily. A mesenteric injury may be seen without an accompanying small bowel injury. In this case, bowel viability dictates whether the bleeding vessel in the mesentery is suture ligated or if bowel and mesenteric resection is needed. Small, non-expanding mesenteric hematomas are not generally explored.

Tumors of the small bowel require resection of the adjacent lymph nodes that are embedded within the mesentery. In this setting, resection of the mesentery is required for adequate oncologic staging. Patients with Crohn's disease often develop stricturing disease of the small bowel, segments of which may eventually require resection. The mesentery in these patients is often thickened leading to difficulties with traditional resection and ligation techniques. Also see Chaps. 39 and 40, Crohn's Disease). Another common reason for resection of the mesentery is for the creation of the roux limb in gastric bypass, gastrectomy, or pancreatico-duodenectomy operations. The roux limb is the middle portion of the jejunum,

S. Gaines · J. C. Alverdy (✉)
Department of Surgery, University of Chicago, 5841 S. Maryland, Chicago, IL 60637, USA
e-mail: jalverdy@surgery.bsd.uchicago.edu

E. D. Ehrenpreis et al. (eds.), *The Mesenteric Organ in Health and Disease*,
https://doi.org/10.1007/978-3-030-71963-0_35

Table 35.1 Indications for mesenteric resection

1. **Removing adjacent small bowel**: obstruction, ischemia, trauma, bleeding, cancer, etc.
2. **Removing a primary mesenteric neoplasm** (associated small bowel must be resected due to loss of blood supply): Lymphoma, desmoid, GIST, etc.
3. **Dividing the mesentery for creation of a roux limb**: gastric bypass, gastrectomy, whipple, etc.

usually 40–75 cm from the ligamental of Trietz, that is used to reestablish small intestinal continuity after resection. This portion of the intestine is used due to the long length of the associated mesentery which allows the limb to be proximally attached without significant tension (Fig. 35.1a, b). Resection of the mesentery is also required for the removal of mesenteric cysts and tumors. A full review of mesenteric cysts and tumors can be found in earlier chapters of this book.

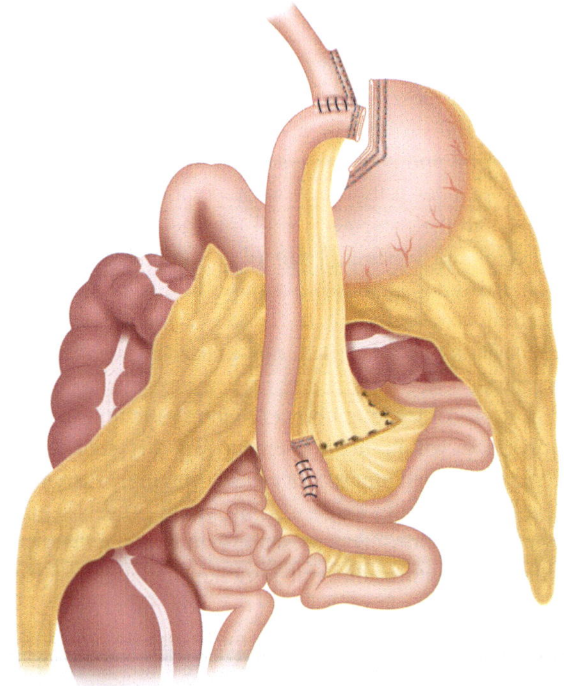

Fig. 35.1 Roux limb demonstrating resected mesentery to allow release of the limb for anastomosis

Contraindications

Contraindications to mesenteric resection relate to contraindications for small bowel resection. Absolute contraindications include poor blood supply to the ends of the bowel for the anastomosis or unclear bowel viability after attempted revascularization. Relative contraindications include peritoneal sepsis and hemodynamic instability.

Description of Surgery

Small Bowel Resection with Adjacent Mesentery (Open Approach, Hand Sewn Anastomosis)

The abdomen is entered through a standard midline incision. The abdomen is explored in all four quadrants, and lysis of adhesions is performed in the area of pathology only. The margins of resection are determined. The mesentery is scored using electrocautery at the planned margins. A surgeon must ensure that the mesentery encompasses only the vessels and lymph nodes (for oncologic staging if required) related to the section of bowel that is to be removed (Fig. 35.2). A window is then made in the mesentery next the bowel at the margins. This can be done with a right angle. A GIA stapler is then passed through the windows on either side of the bowel. A blue (3.8 mm) load is typically used to divide the bowel. The mesentery is then divided using electrocautery, suture ligation, or a harmonic scalpel along the score lines. Division of mesentery in patients with Crohn's disease can be difficult due to its thickening from chronic inflammation. Vessel sealing devices may not adequately provide hemostasis; therefore, overlapping clamps with suture ligations should be used (Fig. 35.3). For a handsewn anastomosis, a 3-0 silk is then used in an interrupted fashion to close the posterior layer (Lembert stitch). The staple line is excised. The inner layer is then closed starting in the middle using a 3-0 absorbable, double arm suture. Each arm closes the posterior inner layer in a continuous fashion. After the corner, a transition stitch is performed from suturing inside to outside to allow closure of the anterior layer. The outer anterior layer is closed with a Lembert stitch. The mesentery is closed with 3-0 interrupted or continuous silk to prevent internal herniation. The patency of the anastomosis is confirmed by manual exploration. The anastomosis and mesentery is observed for hemostasis. The abdomen is closed in a standard fashion.

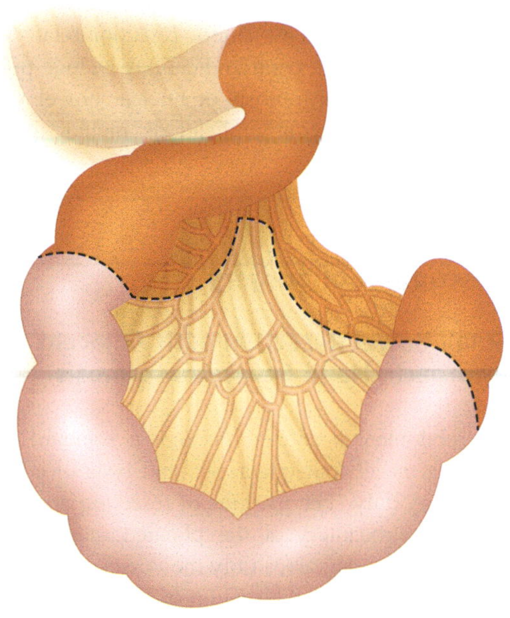

Fig. 35.2 Margins of small bowel resection with adjacent mesentery and accompanying vasculature

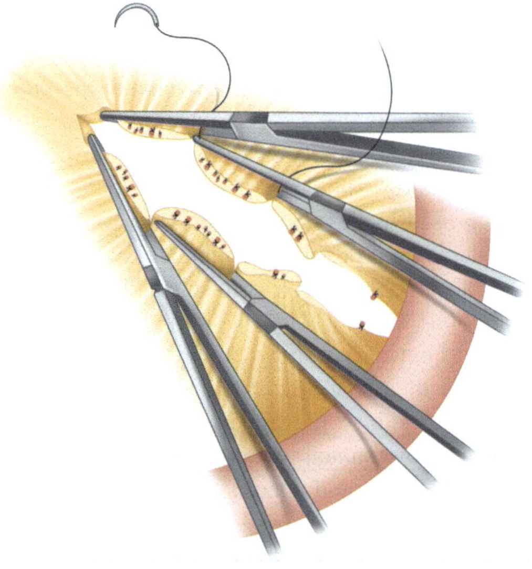

Fig. 35.3 Small bowel resection with thickened mesentery performed with suture ligation

Complications

Early complications include mesenteric hematoma and anastomotic bleeding. Late complications include bowel ischemia, prolonged ileus, anastomotic leak, mechanical obstruction, wound infections, and enterocutaneous fistula.

Summary

Margins of a small bowel resection and mesentery must encompass relevant vasculature and lymph nodes. Status of the mesentery dictates bowel viability. It is critical to ensure mesenteric hemostasis.

Suggested Reading

1. Minter RM, Doherty GM, editors. Current procedures: surgery. New York, NY: McGraw-Hill.
2. Cameron JL, Cameron AM, editors. Current surgical therapy. Philadelphia, PA: Elsevier.
3. Mulholland MW, Lillemoe KD, Doherty GM, Maier RV, Simeone DM, Upchurch GR, editors. Greenfield's surgery. Philadelphia, PA: Lippincott Williams & Wilkins.

Mesenteric Considerations in Surgery of the Colon and Rectum

36

Hermann Kessler, Mariane G. M. Camargo, and Kristen T. Crowell

Introduction

The mesentery is contiguous along the intestines from the esophagus to the distal rectum. It has a different shape along the GI tract and provides suspension of the intestines and encases the feeding blood vessels and draining lymphatics. In this chapter we will describe the anatomic and functional aspects of the mesentery which are important in surgical decision making in malignant and benign colon and rectal resections.

Mesenteric Role in Diseases of the Colon and Rectum

Malignancy

The mesorectal plane, or the "Holy Plane", was first described by R. J. Heald in 1984 and defined as the optimal plane for rectal cancer resection. Sharp dissection along embryologic planes that separate the visceral (mesorectal) fascia off of the surrounding parietal peritoneum provides a resected specimen containing the rectum, lymph nodes, lymphatic vessels, and surrounding fat within the mesorectum

H. Kessler (✉)
Digestive Disease and Surgery Institute, Department of Colorectal Surgery, Cleveland Clinic, 9500 Euclid Avenue, A30, Cleveland, OH 44195, USA
e-mail: kessleh@ccf.org

M. G. M. Camargo
Department of Gastroenterology and Nutrology, Service of Colorectal Surgery, Hospital das Clínicas da Faculdade de Medicina da Universidade de São Paulo, São Paulo, Brazil

K. T. Crowell
Department of Surgery, Division of Colon and Rectal Surgery, Beth Israel Deaconess Medical Center, Boston, MA, USA

E. D. Ehrenpreis et al. (eds.), *The Mesenteric Organ in Health and Disease*,
https://doi.org/10.1007/978-3-030-71963-0_36

en bloc. This technique revolutionized rectal cancer operations and began the investigation into complete mesenteric resections.

Malignant lymphatic spread to mesenteric lymph nodes is an important mechanism for systemic spread. Lymph node drainage from colorectal adenocarcinoma follows the supplying arteries. Therefore, in addition to adequate intestinal margins, resection of the mesenteric lymph nodes en bloc with the tumor is a key factor in principles of oncologic resections. Lymph node involvement in colorectal cancer is a significant and even an independent prognostic indicator. When lymph nodes are found to be positive, adjuvant treatment is recommended. An increased number of resected lymph nodes more precisely defines pathologic staging and improves survival. Guidelines recommend resection and pathologic evaluation of at least 12 lymph nodes in all colon and rectal cancer specimens.

In the 1990s W. Hohenberger transferred the principles of TME to the colon in an attempt to improve colon cancer outcomes. He introduced central vascular ligation (CVL) and dissection along embryologic planes with preservation of the anterior and posterior peritoneum of the mesocolon and named the procedure "Complete Mesocolic Excision" (CME). With this technique, outcomes in colon cancer improved stepwise at his home institution, the University of Erlangen, Germany. The 5-year local recurrence rate decreased from 6.5 to 3.6%, and the 5-year disease related survival increased from 82.1 to 89.1% Since that time, multiple studies succeeded in reaffirming the relationship between anatomic-based colon and rectal cancer surgery and improved outcomes.

Although TME dissection plane in rectal cancer is globally accepted, the implementation of CME for colon cancer is far less prevalent. It is critical for the surgeon to understand the anatomic planes to achieve the optimal resected specimen. Achieving this goal requires preoperative discussion with radiologists as well as postoperative evaluation with the pathologist to provide surgical quality control and feedback to improve the surgical techniques.

Benign Disease

Complete resection of the mesentery is not required in colon and rectal resections for benign disease since there is no concern for systemic spread. Thus, in a prophylactic colectomy or proctocolectomy for hereditary colorectal cancer, such as familial polyposis syndrome or Lynch syndrome, performed to prevent cancer occurrence, CME or TME is not required as long as the presence of cancer has been ruled out prior to resection (Fig. 36.1). The resection margin may violate the mesentery, and the resection plane can proceed close to the bowel wall, leaving behind some of the mesentery.

Colon and rectal resections for inflammatory bowel disease (IBD) also do not require an oncologic CME or TME. Ulcerative colitis is a disease of the intestinal mucosa, and the mesentery has not been implicated as a factor in disease severity or outcomes, thus complete resection of the mesentery is not indicated. Although not ontologically required, during a proctectomy for IBD, the authors recommend

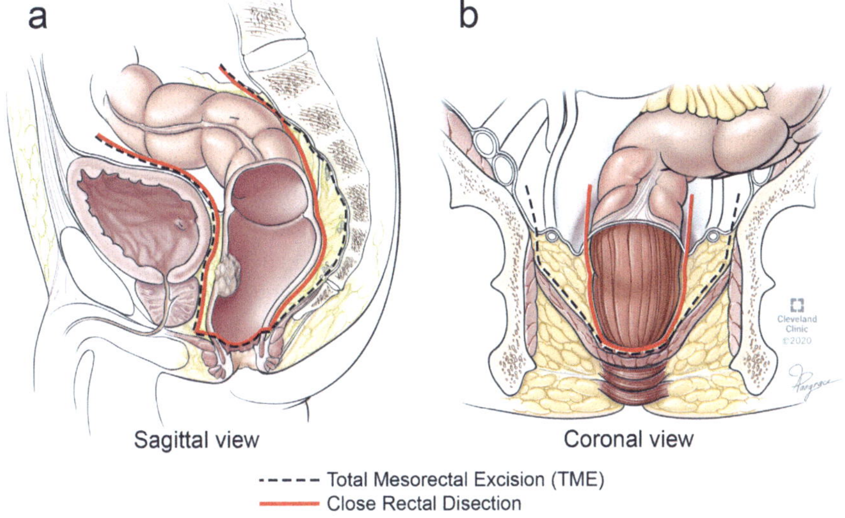

Fig. 36.1 Sagittal (**a**) and coronal (**b**) views comparing the dissection planes of TME versus close rectal dissection. Reprinted with permission, Cleveland Clinic Center for Medical Art & Photography ©2020. All Rights Reserved

Fig. 36.2 A tumor thrombus is seen within a mesenteric vessel that is only 0.1 mm from the mesenteric surface

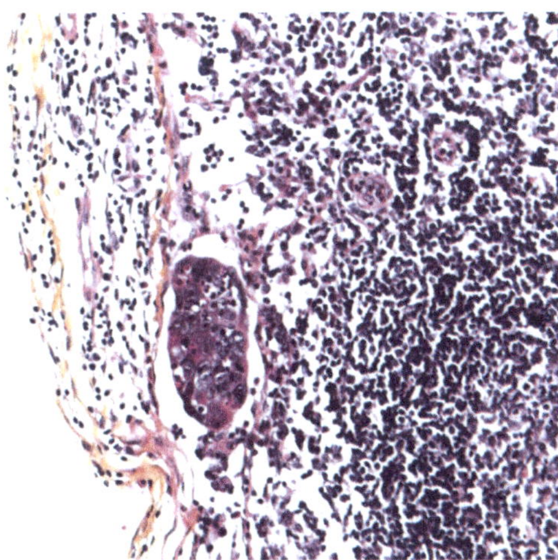

dissecting in the avascular TME plane to avoid bleeding associated with dissecting through the mesorectum. Crohn's disease has a characteristic "fat wrapping" of the mesentery onto the bowel serosa and thickening of the mesentery. When thickening and foreshortening of the mesentery is extensive, vascular ligation can become difficult and increase the blood loss, thus resections are classically performed close to the bowel wall or at a central area of the mesentery apart from the diseased bowel where inflammation and thickening have not yet arrived. Suture ligation may be required instead of vascular sealing devices, which more easily ligate vessels in normal mesentery. Although, Crohn's disease is conventionally defined a disorder of the intestine itself, the contribution of hypertrophic mesenteric adipose tissue in the disease process is under debate. Coffey and colleagues found that ileocolic resections with a mesenteric excision had a lower rate of recurrence requiring surgery compared to conventional resection next to the bowel wall (10% vs. 30%). More patients were smokers in the conventional resection group, however, even in multivariate analysis, mesenteric excision was an independent protective factor. A retrospective study by de Groof et al. investigating proctectomy without restoration of intestinal continuity in Crohn's disease, found a significant increase in perineal wound complications when the mesentery was resected close to the rectum compared to a TME resection (59.5% vs. 17.6%). A secondary resection of the remaining mesorectum in patients with nonhealing wounds resulted in a 75% healing rate. Additionally, the re-resected mesorectum still expressed higher levels of the proinflammatory cytokine TNFα mRNA, suggesting the remaining mesorectum created a persistent proinflammatory environment even after resection of the diseased bowel. Further studies are required to determine the optimal mesenteric resection for Crohn's disease. These issues are discussed in other chapters in this textbook.

Colon Cancer Principles

In many Western Countries, the survival in rectal cancer has escalated to the extent that it now exceeds that of colon cancer, due to the improvement in rectal cancer treatment. There is a wide variation in colon cancer outcomes based upon location, with concern that the surgical quality may be an important variable. One consistent feature of colon cancer surgical treatment is removal of lymph nodes within the mesentery along the feeding vessels. The choice of surgical resection is based upon the location of the tumor. Each tumor has a unique anatomy and drainage, and therefore the operation needs to be individually tailored. Adjuvant therapy can improve outcome based upon lymph node involvement and tumor pathology, such that an ideal mesocolic resection with an increased excision and pathologic evaluation of mesenteric lymph nodes can correctly guides which patients should receive adjuvant therapy.

In Europe, CME and central vascular ligation (CVL) as proposed by Hohenberger have been introduced first. Sharp and careful dissection of the mesocolon's visceral plane from the retroperitoneal plane is mandatory to prevent spread of potentially involved lymph nodes, tumor deposits, or micrometastases in the mesocolon into the peritoneal cavity. Along with CME, ligating the named vessels through CVL, also referred to as high vascular ligation, promotes maximal harvest of regional lymph nodes including those at the root of the mesentery. If the peritoneal surface of the mesentery is penetrated, tumor cells may be disseminated (Fig. 36.2). Achieving CME and CVL offer the most accurate staging along with significant survival advantage with decreased local recurrence rates and improved 5-year cancer-related survival rates.

The Stockholm colon cancer project was initiated from 2004–2007 focusing on interdisciplinary exchange of knowledge and creating multidisciplinary teams. Improvements targeted more appropriate staging by screening, daytime specialists performing emergency surgeries, improved surgery by a more complete resection, maximizing lymph nodes by increasing extent of resection and pathologic evaluation of nodes in the specimen, and decreasing hospital and surgeon variability through training and centralization. This project lead to an increase in specimens with >12 lymph nodes from 42.2 to 81.4%, and an increase in the mean number of lymph nodes from 11 to 18. Overall survival was increased in all stages, and disease-free survival also improved, which was most pronounced in stage III disease. Thus, standardization of technique along with CME should be considered for all colon cancer resections. Examples of resected specimens are shown in Fig. 36.3a and b.

a

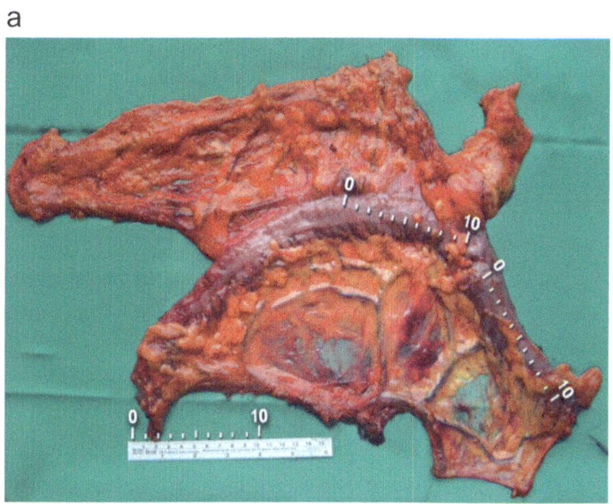

b

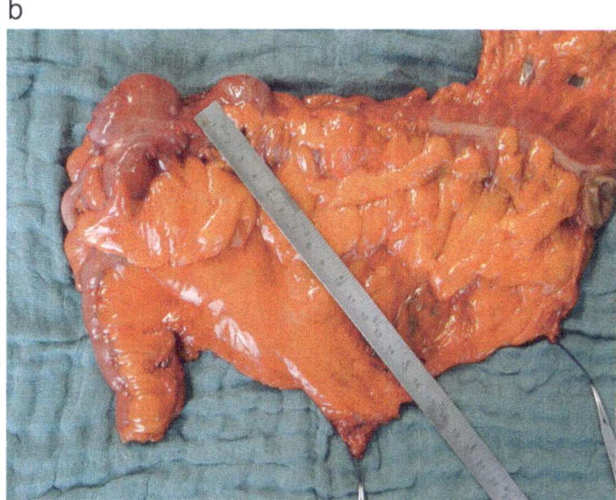

Fig. 36.3 CME specimens for carcinoma at the splenic flexure (**a**) and ascending colon (**b**)

In Japan, the "D3 lymphadenectomy" was described in 1977 and is now the standard approach there and also in, China, Korea, and Taiwan. The lymph nodes are described as epicolic and paracolic, D1, intermediate, D2, and central lymph nodes along trunks of named arteries and veins, D3. This approach focuses more on anatomic location of the lymph nodes rather than the embryologic dissection planes, and the D3 lymphadenectomy is recommended in T2–T4 tumors and in presence of lymph node metastases. Another difference between D3 lymphadenectomy in Asia and CME and CVL in the Western World is a shorter length of resected intestine.

The length of resected bowel is largely based upon the goal to obtain adequate lymph node resection. Initially published in 1983, Jinnai recommended that lymph nodes 10 cm laterally from the tumor should be resected. Additional studies supported this finding and further characterized that the distance of metastatic lymph node spread from the tumor increases with an increasing T stage. In a study of pathological evaluation of resected specimens, in T1 tumors, lymph node metastases were found up to 3 cm from the tumor, while in T3 and T4 tumors, metastatic lymph nodes were found up to 7 cm from the tumor. Since T stage is difficult to predict preoperatively, a proximal and distal margin of 10 cm is recommended to resect all potentially positive lymph nodes.

Studies of Anatomy and Embryology

Surgical embryology and anatomy have a key function for optimized colon cancer surgery. The primitive gastrointestinal tract in an embryo at 4–6 weeks (Fig. 36.4a and b) is reduced to a single somatic tube with the visceral peritoneum directly overlying the gastrointestinal tract and a second external layer, the parietal peritoneum which lines the body wall and associated organs including the retroperitoneal structures. The visceral peritoneum surrounds the somatic tube including the mesentery containing the vessels from the aorta to the intestines. As the organs grow, the primitive GI tube is divided into the foregut including stomach and duodenum, the midgut with jejunum, ileum, ascending and transverse colon, and the hindgut including left colon and rectum. The intestines grow and twist to elongate. In some regions of the intestine, the visceral peritoneum fuses with the parietal peritoneum to become fixed in specific areas. The transverse colon lies over the duodenum in close proximity to the stomach, and the anatomy in this area becomes more complex. The greater omentum attaches to the anterior transverse colon creating the lesser sac. The mesentery in all of these connections is covered by the visceral peritoneum which can be separated from the adjacent parietal peritoneum to recreate the embryologic planes (Fig. 36.5). The ascending and descending colon become fused to the parietal peritoneum lining the body wall laterally and posteriorly the kidneys and ureters. Toldt's fascia is a fusion of the visceral peritoneum of the bowel mesentery and the parietal peritoneum of the retroperitoneal structures. Gerota's fascia is a distinct layer overlying the kidney

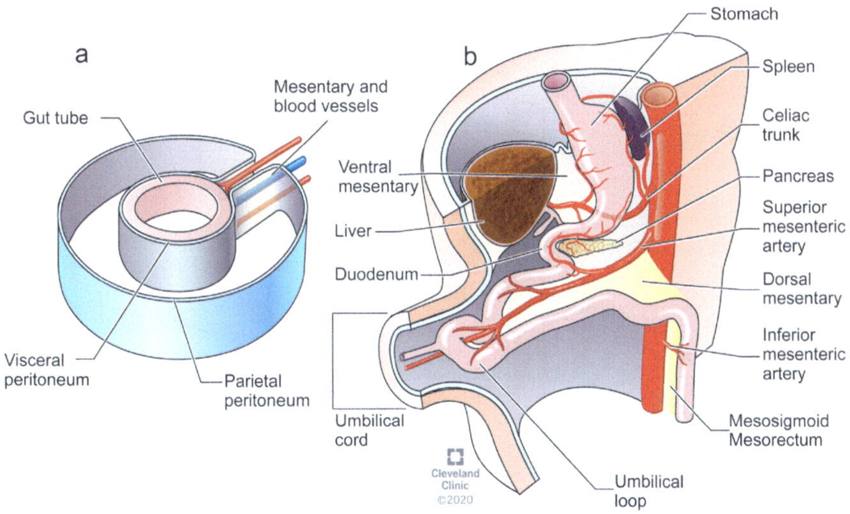

Fig. 36.4 Embryologic gastrointestinal (GI) tube at 4–6 weeks of life, shown in cross section (**a**), and sagittal view after further development (**b**). The GI tube is a central tube lined by visceral peritoneum which wraps directly around the visceral organs. The parietal peritoneum is the outer layer that lines the organs embedded in the body wall. There is a double layer of visceral peritoneum allowing blood vessels to enter the GI tube from the aorta which will become the mesentery. The parietal and visceral peritoneum join at the origin of the visceral arteries. Reprinted with permission, Cleveland Clinic Center for Medical Art & Photography ©2020. All Rights Reserved

and adrenal glands within the retroperitoneum. The correct dissection planes discussed later in this chapter should be between the visceral and parietal peritoneum, rather than extending into the retroperitoneum to expose Gerota's fascia.

Surgical Principles

Preoperative Vascular Anatomical Considerations

A part of the preoperative evaluation of patients with colon and rectal cancer is staging of the disease in the context of involvement of the mesocolon and mesorectum. Staging CT scans of the abdomen and pelvis with IV and PO contrast are indicated for both colon and rectal cancer to identify distant metastases but can also help identify vascular anatomy as well as potentially enlarged lymph nodes. MRI has become the recommended modality to stage rectal cancer, which is most useful in classifying T and N stages, and if adjacent structures are involved by the tumor. MRI also helps determine whether preoperative radiation is indicated prior

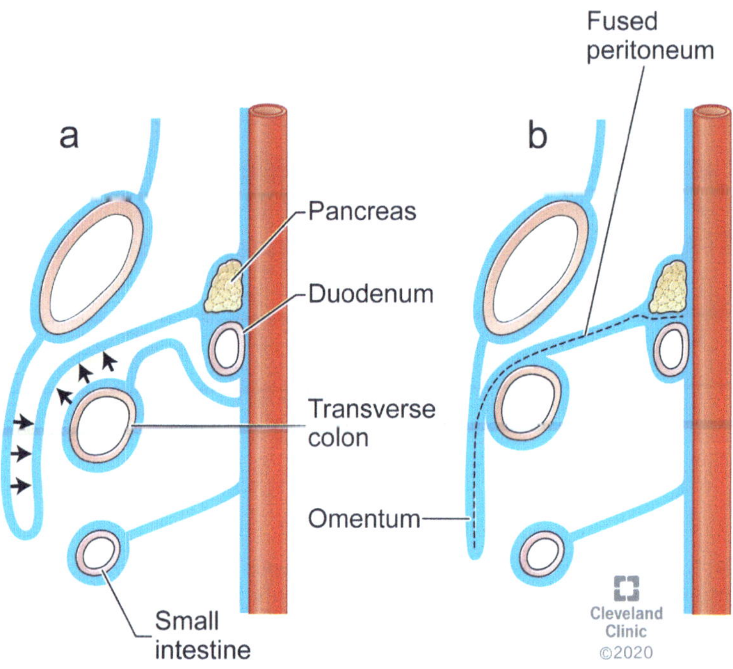

Fig. 36.5 The fusion planes of the peritoneum during development. The mesentery of the transverse colon fuses with the greater omentum. There is a continuously running plane from the pancreas across the duodenum towards the transverse mesocolon which can be identified during surgery to 'turn back' embryology. The ligament of Treitz is shown here as a duplication of the mesentery on the aorta. Reprinted with permission, Cleveland Clinic Center for Medical Art & Photography ©2020. All Rights Reserved

to surgical resection, which is recommended for T3–T4 cancers, any lymph node involvement, and in cases where the tumor is close to the mesorectal fascia.

The preoperative identification of vascular anatomical variations can be helpful at the time of the surgery. The anatomy of the left sided colonic arteries is relatively simple although variations can occur (Fig. 36.6). The left colic artery (LCA) is absent in only 5% of individuals. The origin of the first sigmoid artery is variable, as it arises from the inferior mesenteric artery (IMA) as a separate artery (Type 1) in 41–58% of individuals, from the LCA (Type 2) in 27–45%, and at the angle between the LCA and IMA (Type 3) in 9–15%. The number of sigmoid arteries varies usually between one and five, with two (21–40%), three (32–50%), and four (7–25%) as the most common numbers.

The right side of the colon derives from the midgut and is supplied by the superior mesenteric artery (SMA). The ileocolic artery (ICA) is consistent, and supplies the cecum, ascending colon and, through anastomoses with the SMA, the terminal ileum (Fig. 36.7a–c). The anatomy of the middle colic artery (MCA) and the right colic artery (RCA) varies. The MCA has more than one separate branch

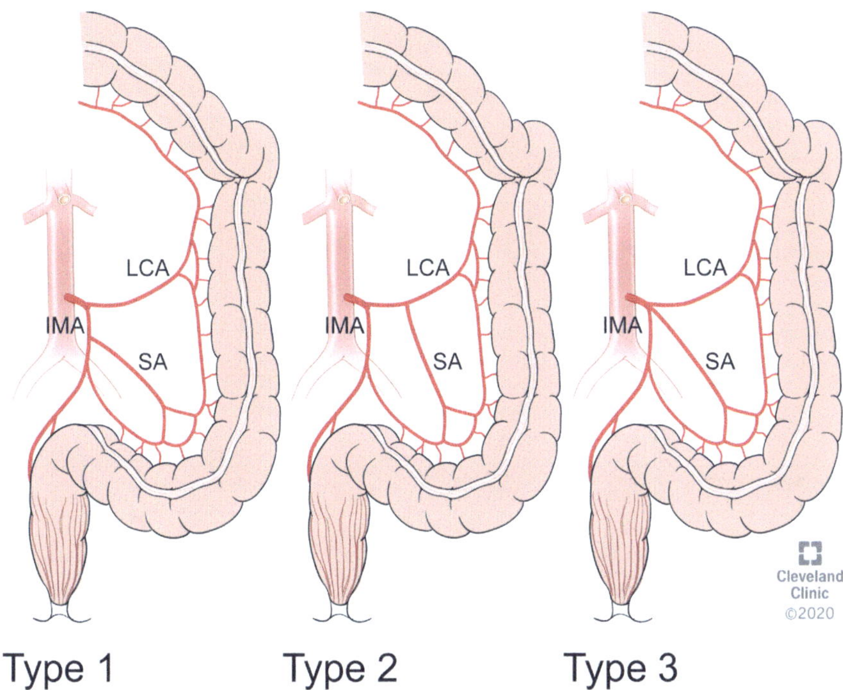

Fig. 36.6 Variations in anatomy of inferior mesenteric artery (IMA). In type 1, the first sigmoid artery (SA) branch arises directly from the IMA. In type 2, the SA arises from the left colic artery (LCA). In type 3, the SA and LCA arise together at an angle from the IMA. Reprinted with permission, Cleveland Clinic Center for Medical Art & Photography ©2020. All Rights Reserved

arising directly from the SMA in 4–36% of cases. In a few cases a left branch of the MCA might arise from the dorsal pancreatic artery or the IMA. The RCA is an artery supplying the middle part of the ascending colon, arising from the SMA, and running inferior to the avascular mesocolic window covering the duodenum. However, the RCA is present as a separate artery in only 11–20% of cases.

Surgical Approach

The classic open approach utilizes a lateral-to-medial dissection for benign and malignant disease, although a posterior and medial-to-lateral approach have also been described. With the integration of minimally invasive surgery in colon and rectal cancer, the development of the medial-to-lateral approach has become widely adopted, but the lateral-to-medial approach is equally acceptable. There are a variety of minimally invasive options for colon and rectal resections. Laparoscopy enables an intracorporeal anastomosis while hand assisted surgery

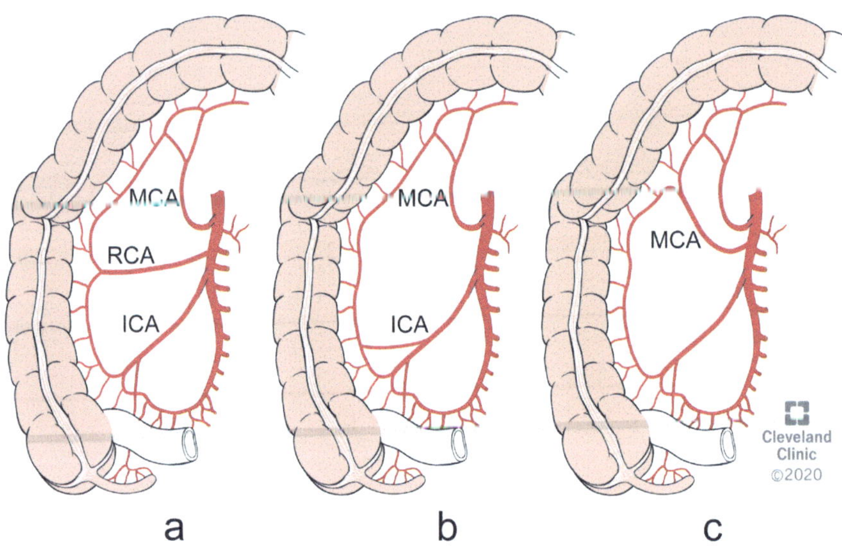

Fig. 36.7 Variations of arterial supply to the right colon. **a** The ileocolic (ICA), the right colic (RCA) and the middle colic (MCA) arteries originate independently from the superior mesenteric artery (SMA). **b** The RCA originates from the ICA, and a common trunk of the MCA supplies the hepatic flexure and the right two-thirds of the transverse colon. **c** The two main branches of the MCA originate separately from the SMA, with the right branch of the MCA supplying the distal part of the ascending colon and the hepatic flexure. Reprinted with permission, Cleveland Clinic Center for Medical Art & Photography ©2020. All Rights Reserved

facilitates entrance of a hand into the dissection field to assist with retraction and additionally provides a potential extraction site for assistance with the anastomosis. In laparoscopic assisted surgery, the dissection is laparoscopically performed, while the anastomosis is performed through a mini-laparotomy under direct visualization. More recently, robotic assisted surgery is becoming integrated into colon and rectal resections. The surgical approach should not alter the surgical goals, to obtain a complete resection including a complete lymph node resection, negative circumferential margins, while preserving surrounding structures and preventing unnecessary complications. For simplicity, this chapter will focus on the description of the open approach.

Surgical Management of Colon Cancer

Surgical principles of colon cancer resection are based upon lymph nodes in the colonic mesentery and as described above, the technical details of lymph node dissection within the specimen vary based upon location. It is globally accepted that an increased number of lymph nodes improves survival, and the main relevant vessels should be centrally ligated. The techniques currently practiced vary from a traditional high ligation of the vessels and dissection of at least 12 lymph nodes without specific considerations regarding the mesocolon, to CME and CVL as refined techniques, and the D3 lymphadenectomy in mainly Japan. We describe the CME and CVL technique in resection of colon cancer which focuses on sharp separation of the colon and mesentery following the embryologic planes and respects the completeness and quality of the preservation of surfaces of the mesocolon.

Cecal and Ascending Colon Carcinoma

For right-sided cancer, or tumors located in the cecum and ascending colon below the hepatic flexure, a right colectomy is the treatment of choice (Fig. 36.8). The resection should include ligation at the origin of the ileocolic artery and right branch of the middle colic artery, with lymph node removal along both vessels down to the origin. If the right colic artery is present, the ligation should also be at the origin. To enter the correct surgical plane between the retroperitoneal and visceral peritoneum, retraction of the lateral peritoneal reflection allows for identification of the white line of Toldt where parietal and visceral peritoneum are fused, and the initial entry point should be away from the tumor to preserve the circumferential surgical

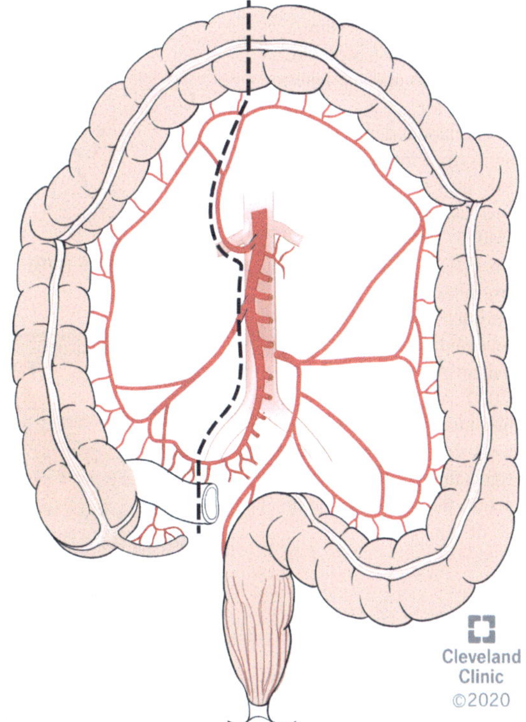

Fig. 36.8 Right hemicolectomy. Reprinted with permission, Cleveland Clinic Center for Medical Art & Photography ©2020. All Rights Reserved

margins. With careful traction on the ascending colon using a gauze, the mesocolon is sharply dissected from the parietal peritoneum of the retroperitoneum, with care to preserve the mesocolon. This dissection continues to mobilize the entire right colon from the lateral and posterior attachments following the embryologic planes cephalad. Medially, areolar tissue is centrally encountered and the duodenum and pancreatic head are sharply mobilized off of the mesocolon and remain in the retroperitoneum. To completely expose the central nodes, a Kocher maneuver is recommended. The anterior surface of the superior mesenteric vein (SMV) is a landmark and is exposed and the superior mesenteric artery (SMA) remains to its left and further posterior.

Mobilization in this area should be performed with caution due to numerous anatomic variations of the veins forming the gastrocolic trunk of Henle (GTH) or the "bleeding point" (Fig. 36.9a–e). Proximally the anterior and superior

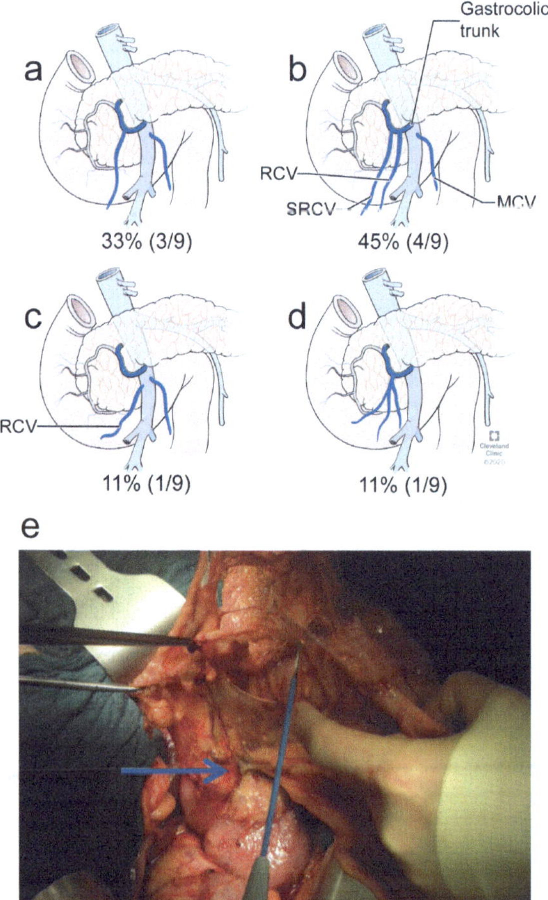

Fig. 36.9 Gastrocolic trunk of Henle (GTH). The venous drainage around the hepatic flexure is highly variable and bears high risk of bleeding during surgical dissection. Henle's trunk depicts the point of drainage of the superior right colic vein (SRCV), right colic vein (RCV), right gastroepicolic vein (RGEV) and the anterior and superior pancreaticoduodenal vein as they form a trunk to enter the SMV near the head of the pancreas, many variations are possible. The middle colic vein (MCV) can also be in close proximity (**a–d**). Reprinted with permission, Cleveland Clinic Center for Medical Art & Photography ©2020. All Rights Reserved This intraoperative picture shows the veins (arrow) and depicts that with inappropriate traction the veins can be at risk of injury (**e**)

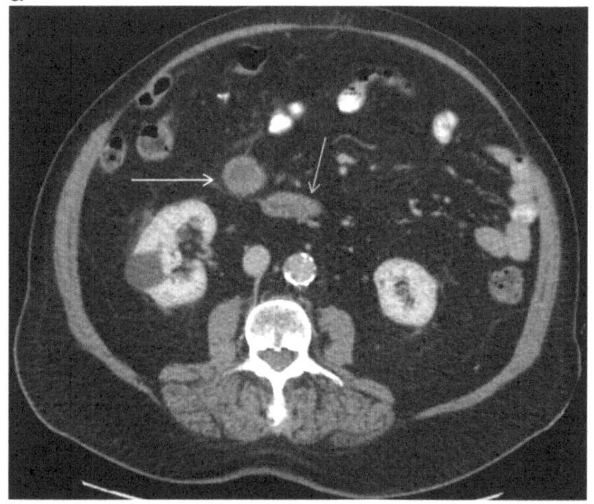

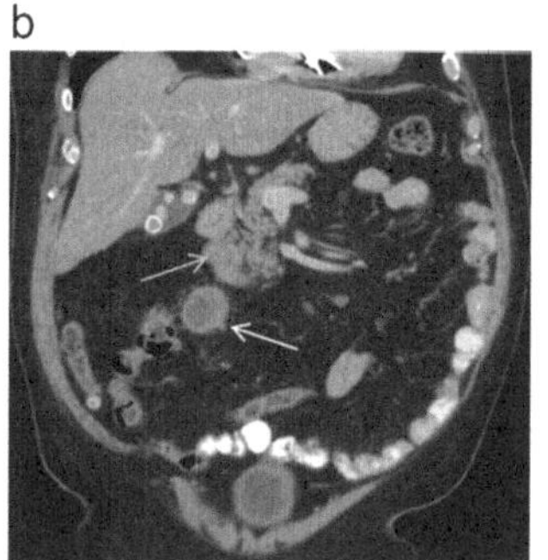

Fig. 36.10 CT scan of right colon cancer with tumor recurrence in the central mesentery (white arrow) in axial (**a**) and coronal (**b**) views at root of mesocolon likely from an incomplete mesenteric resection. The duodenum is identified by a yellow arrow showing the importance of complete central resection to prevent recurrence

pancreaticoduodenal veins, the right gastroepiploic vein, right colic vein, and superior right colic vein form the trunk of Henle; however, many variations are possible as also right colic vein and middle colic vein (MCV) may flow to the GTH. Most frequently, the middle colic vein has a separate mouth towards the SMV. Also, the SMV itself may be split up into two branches: a jejunal and an ileal one. Any transection of the GTH demands clarity of the anatomic situation in order to avoid complications. In order to prevent bleeding, traction of the mesocolon in this area must be done carefully during the separation of the mesocolon from the pancreas and the duodenum. At least the right colic and superior right colic veins should be ligated to gain full access to the SMV.

After full access, the vessels are ligated; the ileocolic artery and vein should be identified and individually secured. The ileocolic artery may pass posterior to the SMV in 40% of patients, thus the origin of the ileocolic vessels should be identified to verify ligation of the artery at its origin and to avoid injury to the jejunal trunk. All lymph nodes should be dissected with the specimen in this area including those overlying the SMV at the root of the right mesocolon (Fig. 36.10a, b). If the right colic artery is present (11–20%) it should be taken with high ligation. The middle colic artery is identified and lymph nodes over the base should be removed en bloc. After the right branch of the middle colic artery is ligated at the branching of the middle colic artery, the dissection should continue through the mesentery toward the proximal transverse colon. Omental resection beyond 10 cm from the tumor is not required; omentum within 10 cm from the tumor can be dissected from the proximal transverse colon. After the transverse colon is divided between the right and left branch of the middle colic artery, the ileum is divided 5–10 cm proximal from the ileocolic valve. In this manner, the entire lymphatic field is resected after which an ileocolic anastomosis is performed.

Hepatic Flexure Carcinoma

Additional lymph node clearance is required for cancers at the hepatic flexure and central transverse colon. Approximately 5% of lymph node metastases can be found over the head of the pancreas or along the lower edge of the pancreas, and 5–8% lymph nodes along the greater gastroepiploic arcade along the greater curvature of the stomach if lymph node involvement is present. An extended right hemicolectomy is required (Fig. 36.11). During the required resection, the greater curvature

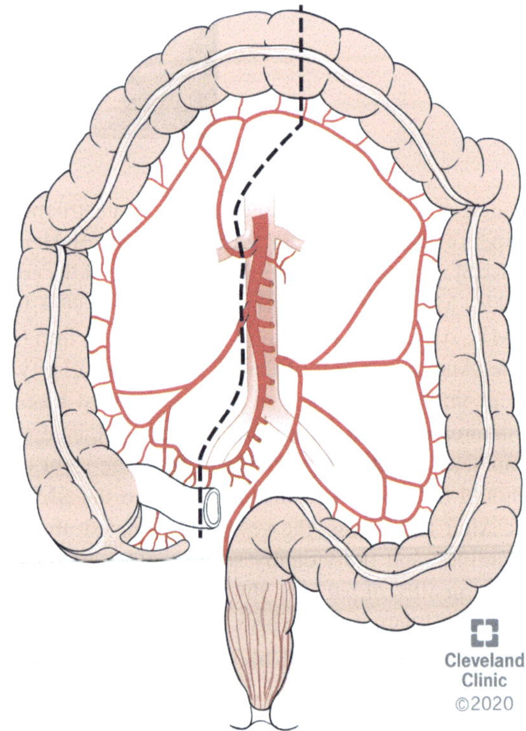

Fig. 36.11 Extended right hemicolectomy. Reprinted with permission, Cleveland Clinic Center for Medical Art & Photography ©2020. All Rights Reserved

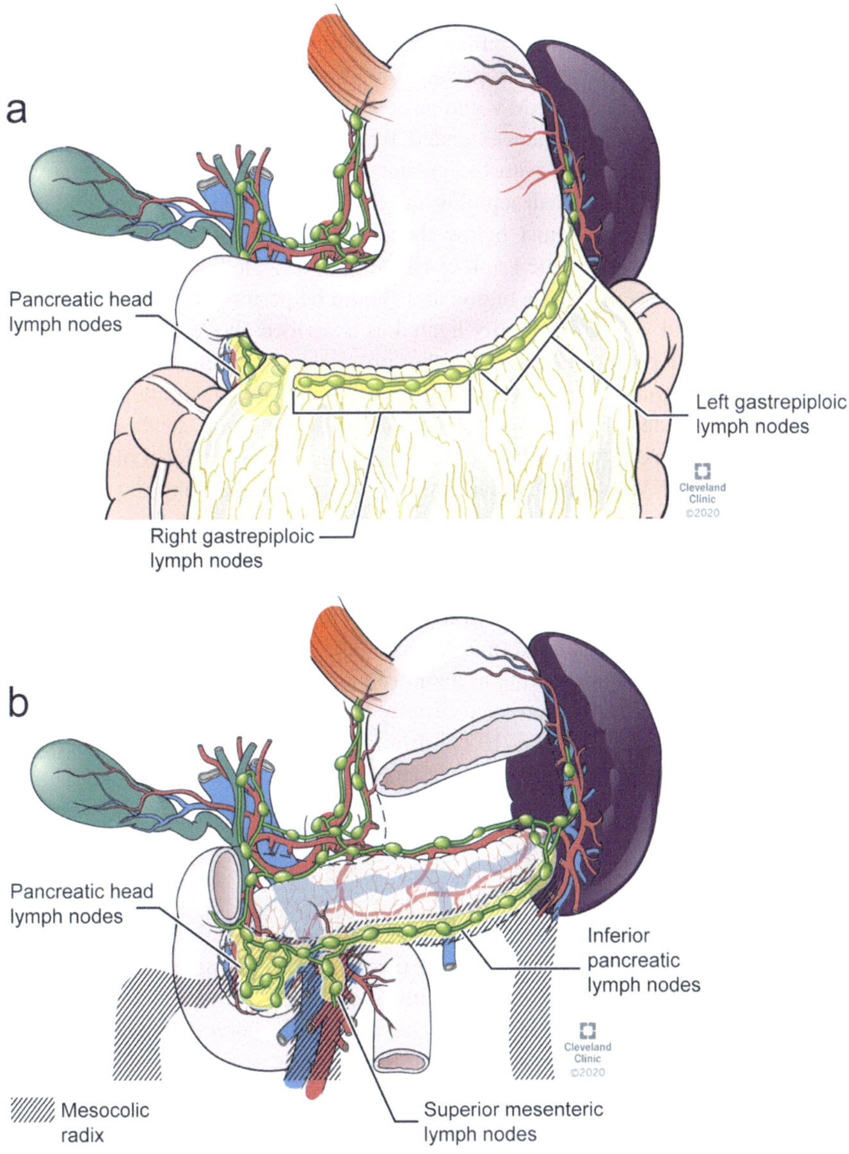

Fig. 36.12 Lymphatic drainage along the transverse colon which involves the gastroepiploic arcade (**a**) and lymph nodes overlying the head and lower border of the pancreas (**b**). Reprinted with permission, Cleveland Clinic Center for Medical Art & Photography ©2020. All Rights Reserved

should be freed of lymph nodes along the right gastroepiploic vessels to retrieve all possible lymph node metastases.

The initial dissection is similar to the right hemicolectomy above, initially incising into the lateral peritoneal reflection and continuing with sharp dissection of the ascending colon from the retroperitoneum, Kocher's maneuver, and identification of the ileocolic vessels, SMV and gastrocolic trunk of Henle which is ligated centrally. The dissection is then extended by entering the lesser sac. The right gastroepiploic artery is freed from the greater curvature of the stomach removing the nodes along the right gastroepiploic arcade (Fig. 36.12a, b). The right gastroepiploic artery is ligated just below the pylorus while preserving the gastroduodenal artery. The gastrocolic trunk of Henle is next ligated, again traction of the mesocolon may tear veins in this region and should be performed with caution. The ileocolic artery and vein are centrally ligated as described above. The middle colic vein and artery can be approached by further proximal dissection along the SMV as a landmark. The middle colic vein and artery are ligated near their respective origins and transected. Following the transection of the ileum at about 5–10 cm proximal to the ileocolic valve, the mesocolon is separated at the level of the ligated vessels along the SMV while preserving autonomic nerves along the SMV and SMA. The transverse mesocolon is then completely dissected from the lower margin of the pancreas. Lymph node metastases may be present on the lower border of the pancreas, therefore, these lymph nodes below the pancreatic edge should be removed and included when performing central transverse mesocolic dissection; these nodes should be left behind only for technical reasons if they are fixed or the patient has a history of pancreatitis to avoid risk of severe postoperative pancreatic complications. The colon is transected distally to the level of the left branch of the middle colic artery but at least 10 cm from the tumor edge, and an ileocolic anastomosis is created.

Transverse Colon Carcinoma

These tumors are treated by resecting at least the transverse colon, central ligation of the middle colic artery, and resection of the neighboring lymphatic vessels. The greater omentum with the gastrocolic ligament and the gastroepiploic arcade corresponding to the location of the tumor is resected again observing that a minimal clearance of 10 cm needs to be kept. As an anastomosis of the ascending colon to the descending colon is not recommended, the resection usually favors an extended right or left hemicolectomy or a subtotal colectomy.

Splenic Flexure Carcinoma

Splenic flexure carcinoma is unique because of its borderline localization between midgut and hindgut with bidirectional lymph node spread towards middle colic vessels and inferior mesenteric artery. Thus, it can be treated by a left or right sided approach, however, in most cases, an extended left colectomy is preferred. In order to avoid complications with a transverse-descending colonic anastomosis, another option is subtotal colectomy with an ileosigmoid anastomosis. The targeted arteries for resection are the middle colic and the left ascending colic artery along with the accompanying pericolic and central lymph nodes. Lymph nodes along the greater gastroepiploic arcade at the greater curvature of the stomach should be included as well requiring mobilization of the gastroepiploic arcade off the greater curvature of the stomach 10–15 cm opposite to the tumor site. This maneuver allows the omentum to accompany the specimen. There are potential communicating arteries along the inferior border of the pancreatic tail and splenic hilum, thus lymph nodes in this region should be removed.

The dissection starts with the skeletonization of the stomach; the transverse colon and gastroepiploic arcade are prepared and en-bloc resection of the gastrocolic ligament and the omentum is planned. Following full separation of the transverse mesocolon, the lesser sac is fully exposed, and the lymph nodes along the lower edge of the pancreatic corpus and tail are also dissected with central transection of the transverse mesocolon at its root. The ascending colon is mobilized but the ileocolic vessels and lymphatic dissection off the SMV and right colic vessels are not required. Central vascular ligation of the middle colic vessels is crucial as it allows dissection of the central lymph nodes. Up to 4–6% of lymph nodes at the root of the middle colic artery can be involved with carcinoma. Full mobilization of the splenic flexure and the descending colon is performed along the embryological planes. The descending colon is transected at least 10 cm distally to the tumor and the lymph nodes around the IMA are removed. The left colic artery is ligated at its origin along the IMA, and the IMV is ligated centrally at the lower border of the pancreas. The complete mesocolon, transverse and adjacent colon, omentum, and gastrocolic arcade are removed en-bloc as specimen.

Descending Colon Carcinoma

Carcinomas in the descending colon require a left hemicolectomy (Fig. 36.13) with high ligation of the inferior mesenteric artery. Lesions located 10 cm away from the splenic flexure do not necessitate removal of omentum or skeletonization. However, splenic flexure mobilization is crucial for a tension-free anastomosis. The lateral peritoneal attachment is sharply divided and the left mesocolon is dissected off the retroperitoneum. Fatty tissue around the kidney, left ureter and gonadal vessels remains covered by the parietal fascia. The omentum is mobilized from the left transverse colon and the lesser sac is fully exposed. Subsequently, both layers of the mesocolon are divided at the lower border of the pancreas and the integrity of the mesocolon should be strictly preserved. The peritoneum over the aorta is opened to achieve central access to perform central lymph node dissection around the IMA which is divided at about 1 cm distally to its origin at the aorta. Injury to the autonomic nerves in this area should be avoided. The IMV is transected centrally at the lower border of the pancreas facilitating a tension-free colorectal anastomosis. The colon is transected proximally at least 10 cm proximally and distally to the

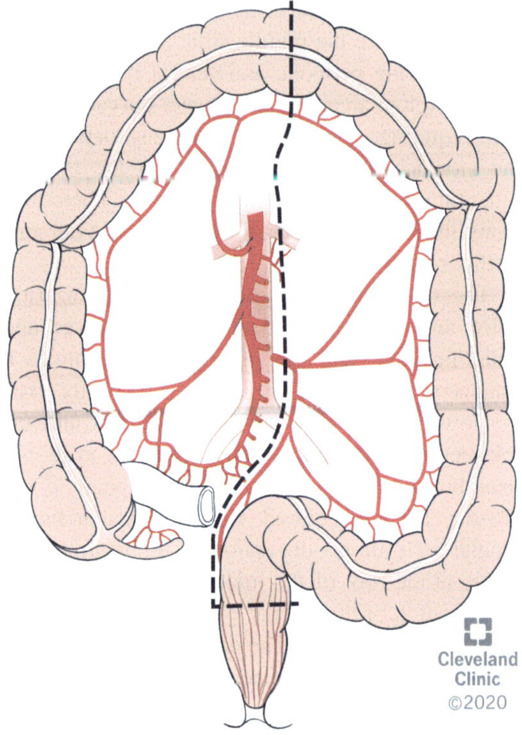

Fig. 36.13 Left hemicolectomy. Reprinted with permission, Cleveland Clinic Center for Medical Art & Photography ©2020. All Rights Reserved

tumor, but due to the central ligation of the IMA, the distal resection margin should be in the upper rectum for adequate perfusion of the colorectal anastomosis.

Sigmoid Carcinoma

The dissection planes are similar to those planes dissected when resecting a colon carcinoma but the proximal colonic transection is in the descending colon. Splenic flexure mobilization is again essential for a tension-free colorectal anastomosis. The left peritoneal reflection is divided, and sharp dissection of parietal and visceral planes is undertaken, leaving the left ureter, gonadal and iliac vessels safely below the parietal plane. The visceral fascia over the aorta is opened to allow for creation of a window at the base of the left mesocolon while preserving the autonomic nerves; through this window, the IMA is ligated. Further cephalad dissection allows us to ligate the IMV next to the duodenum. Proximal colonic transection occurs at least 10 cm proximally to the edge of the tumor. For reasons of oncology and blood supply, the distal transection is performed at the level of the upper rectum, at least 10 cm distally to the tumor. Partial mesorectal excision can be undertaken if indicated. For distal sigmoid carcinomas, 5 cm of distal margin may be sufficient if the dissection extends to the rectum, in which the original dissection plane between the visceral and parietal peritoneum is distally extended behind the mesorectum.

Total Mesorectal Excision Principles

The goals of the total mesorectal excision (TME) are to perform sharp dissection along the definable tissue planes around the mesorectal fascia and remove an en bloc specimen containing the lymph nodes along with the superior rectal and inferior mesenteric arteries; the first step is a high ligation of the IMA. Lateral attachments of the sigmoid are dissected along the white fusion plane of parietal and mesenteric fascia, and the sigmoid colon is retracted medially. The retroperitoneal structures including the left ureter and gonadal vessels are identified and preserved in the retroperitoneum. The dissection is taken further cephalad and medial to mobilize the left mesocolon off the parietal peritoneum. This mobilization permits identification of the IMA that is dissected free of the peritoneal attachments. A high ligation on the IMA is then performed within 1 cm of the origin off of the aorta. The dissection proceeds more cephalad to identify the origin of the IMV at the ligament of Treitz, and a central ligation of the IMV is performed at the lower border of the pancreas.

Next, the rectum and sigmoid are retracted cephalad and anterior to begin the mesorectal dissection. The correct dissection plane appears as an areolar tissue plane in between the mesorectal fascia and endopelvic fascia. The bulk of the mesorectum lies posterior to the rectum where the mesorectal fascia is most developed. Sharp dissection begins in this well-defined plane posteriorly and proceeds in a circumferential manner around the mesorectum, laterally on both sides, then finally anteriorly.

During the dissection, the surgeon should anticipate the structures at risk of injury. The superior hypogastric nerve plexus should be identified as these nerves enter the pelvis anterior and lateral to the sacral promontory as the right and left hypogastric nerves. Damage to nerves during proctectomy can lead to urinary and sexual dysfunction. Posteriorly the presacral fascia overlies the presacral veins, and the dissection should continue in the areolar plane between the mesorectum and presacral fascia to avoid injury to these presacral veins that can cause extensive bleeding if injured. Laterally as the dissection continues, the inferior hypogastric plexus curves around the mesorectum and should be preserved. If the middle rectal arteries are present, they should be divided with electrocautery; the middle rectal arteries are usually found in 20% of patients. The anterior dissection separates the rectum from Denonvilliers' fascia (fusion of the lower peritoneal sac), separating the prostate and seminal vesicles in men and the rectovaginal septum in women. Additionally, anteriorly the parasympathetic nerves (Walsh bundles) may be at risk if Denonvilliers' fascia is violated. In tumors of the mid and lower rectum, to complete the TME, the dissection must be continued to the level of the pelvic floor. The distal clearance of the rectum should ideally be 2 cm from the distal edge of the tumor. However, shorter margins may be acceptable depending on tumor factors and if neoadjuvant therapy was administered. If an anastomosis is planned, a colorectal or coloanal anastomosis then completes the operation if not a stapled rectal stump and end colostomy are created. If this distal margin cannot be achieved, an abdominoperineal resection may be required.

Surgical Quality

The quality of the resected specimen was introduced in rectal cancer with completeness of the TME. In addition to the surgical quality to follow the complete TME (Fig. 36.14), the pathologist must work in conjunction with the surgeon to correctly evaluate and report the specimen quality. This evaluation and feedback to the surgeon is key to assure optimal surgical treatment since the surgical quality is related to oncologic outcomes. TME evaluation and report by the pathologist are required to become accredited in the United States by the National Accreditation Program for Rectal Cancer.

The evaluation of CME specimens was investigated by Quirke and colleagues by a grading scale, and high quality specimens were associated with an overall survival advantage. The planes of mesocolic dissection were graded: mesocolic plane (mesocolon is resected entirely and intact), intramesocolic plane (mesocolon is disrupted and so the plane of dissection is within the mesentery), and muscularis propria plane (mesocolon is divided at the intestinal margin, with potential colonic

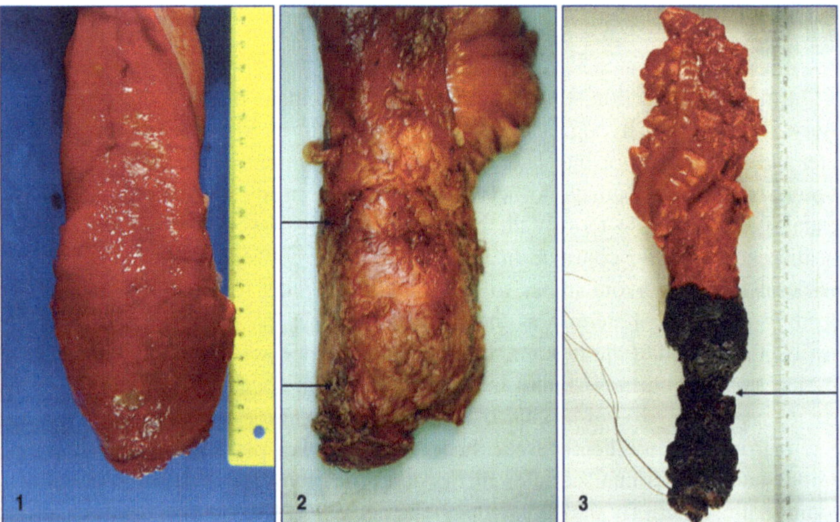

Fig. 36.14 Proctectomy specimen after resection. The shiny appearance of a complete TME is visualized with intact mesorectal fascia (**1**). Partial TME with minor disruption of the mesorectal fascia (**2**). Incomplete TME with disruption of the mesorectal fascia and narrowing or 'waisting' of the specimen where the embryologic planes were violated (**3**)

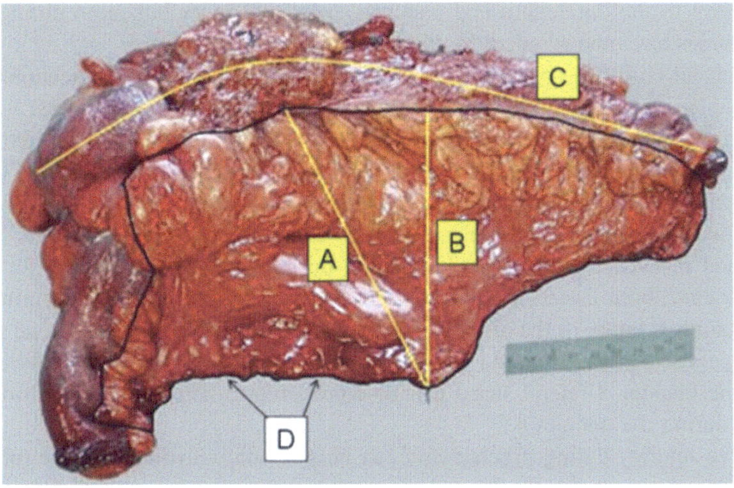

Fig. 36.15 Right colectomy specimen with CME dissection. The ileocolic vessel is taken with central ligation maximizing the distance from tumor to vascular ligation (A), and distance from vascular ligation to nearest bowel wall (B). The length of colon (C) and cross-sectional area of the mesentery (D) are measured as well

wall injuries); measurements were taken according to Fig. 36.15. The quality of the surgical resection is key to provide optimal care for colon and rectal malignancies.

Surgical Complications

Postoperative complications associated with complete mesenteric excisions and high ligations of vasculature may be due to dissection near important structures that can inadvertently become injured. Resections that are performed close to the bowel wall dissect through the mesentery, and the advantages include a lower bleeding risk during distal ligation of vessels and lower risk of inadvertent injury of nearby nerves or other organs. The evidence for complete mesocolic and mesorectal excision with high vascular ligation is convincing when resecting cancer to prevent recurrence and prolong survival. When these resections are indicated, the surgeon should understand the anatomy to provide adequate resection while preventing injuries.

Due to the increased extent of dissection for CVL and CME, the operation time on average is longer for a CME compared with standard surgery. During central dissection during CME, venous structures are at increased risk, especially at the trunk of Henle as well as the splenic and superior mesenteric veins. During ligation of the ileocolic vessels, the first jejunal branches can be inadvertently ligated, thus correct vascular identification is pertinent to preserve small bowel perfusion. Dissection of the duodenum and pancreatic head from the mesocolon is a crucial step, and injury to the duodenum and pancreas especially in mobilization during right colectomies has to be avoided. Mobilization of the splenic flexure can lead to injury of the distal tail of the pancreas or spleen itself, thus meticulous dissection in the correct plane and avoiding undue tension can help prevent these injuries.

During right colectomies, the right ureter should be identified if the retroperitoneal plane is violated. Injury to the right ureter should also be identified during mobilization of the distal ileum from the lateral retroperitoneal attachments. The left ureter is at risk during sigmoid colectomies and proctectomies and should be identified in every operation. Additionally, during the TME dissection for rectal mobilization, both ureters should be identified as they cross the pelvic brim. If a rectal tumor is bulky or the disease process has obliterated the normal lateral planes in the pelvis, the ureter should be identified laterally before they turn anteriorly to enter the bladder. Ureteral stents can be considered to help with ureteral identification during the operation.

Nerve injuries during proctectomy can be potentially avoided by careful identification of the sympathetic and parasympathetic nerves in the pelvis. The hypogastric nerves derive from the superior hypogastric plexus that enters the pelvis over the pelvic brim, identifiable as the right and left hypogastric nerves. Ligation of these sympathetic nerves can lead to retrograde ejaculation in a male and urinary incontinence. More distally in the pelvis, the inferior hypogastric nerves travel lateral and anterior outside of the mesorectum and can be injured when this dissection is too lateral, excessive traction is placed on the rectum, or during ligation of the lateral stalks. Injury to these sympathetic nerves leads to erectile dysfunction. Finally, during the anterior dissection, mixed sympathetic and parasympathetic nerves lie anterior to Denonvilliers' fascia, but if injured can result in neurogenic bladder requiring self-catheterization and erectile impotence.

Additional structures to identify include the internal iliac artery, especially on the left during the initial mobilization of the distal sigmoid colon and as the TME dissection continues over the pelvic brim. Organs anterior to Denonvilliers' fascia should also be avoided during the anterior dissection, including the prostate and seminal vesicles and the vagina.

Conclusion

When considering any colon or rectal resection, the surgical approach should be first based upon the diagnosis, especially considering whether there is any risk of malignancy. In benign diseases, an extensive mesenteric resection is not required and usually allows an easier dissection plane through the mesentery, which is associated with lower complication rate. Current studies are ongoing to determine the extent of mesenteric involvement in Crohn's disease and if a more extensive mesenteric resection could improve outcomes and decrease recurrence rates.

A more extensive resection of the bowel and complete resection of the mesentery improve survival in malignant disease. For rectal cancer, there is global agreement that the TME resection is the optimal surgical resection and includes the entire mesorectum. The ideal mesenteric resection for colon cancer are under debate, but a complete unviolated excision of the mesentery with the associated feeding vessels and central lymph nodes provides an oncologic advantage to incomplete resections through the mesentery. While clinical trials comparing mesentery to non-mesentery-based surgery for cancer are not ethical, current data comparing the outcomes of CME with CVL and D3 lymphadenectomy show similar favorable outcomes. These techniques offer standardization of good quality colon cancer surgery and may be the key to optimized survival in colon cancer surgery.

Suggested Readings

1. Heald RJ. The 'Holy Plane' of rectal surgery. J R Soc Med. 1988;81(9):503–8.
2. Goldstein NS. Lymph node recoveries from 2427 pT3 colorectal resection specimens spanning 45 years: recommendations for a minimum number of recovered lymph nodes based on predictive probabilities. Am J Surg Pathol. 2002;26(2):179–89.
3. Le Voyer TE, Sigurdson ER, Hanlon AL, Mayer RJ, Macdonald JS, Catalano PJ, Haller DG. Colon cancer survival is associated with increasing number of lymph nodes analyzed: a secondary survey of intergroup trial INT-0089. J Clin Oncol. 2003;21(15):2912–9.
4. Hohenberger W, Weber K, Matzel K, Papadopoulos T, Merkel S. Standardized surgery for colonic cancer: complete mesocolic excision and central ligation–technical notes and outcome. Colorectal Dis. 2009;11(4):354–64; discussion 364–5.
5. Merkel S, Weber K, Matzel KE, Agaimy A, Gohl J, Hohenberger W. Prognosis of patients with colonic carcinoma before, during and after implementation of complete mesocolic excision. Br J Surg. 2016;103(9):1220–9.
6. Nally DM, Kavanagh DO, Winter DC. Close rectal dissection in benign diseases of the rectum: a review. Surgeon. 2019;17(2):119–26.
7. Coffey CJ, Kiernan MG, Sahebally SM, Jarrar A, Burke JP, Kiely PA, Shen B, Waldron D, Peirce C, Moloney M, Skelly M, Tibbitts P, Hidayat H, Faul PN, Healy V, O'Leary PD, Walsh LG, Dockery P, O'Connell RP, Martin ST, Shanahan F, Fiocchi C, Dunne CP. Inclusion of the mesentery in ileocolic resection for Crohn's disease is associated with reduced surgical recurrence. J Crohns Colitis. 2018;12(10):1139–50.
8. de Groof EJ, van der Meer JHM, Tanis PJ, de Bruyn JR, van Ruler O, D'Haens G, van den Brink GR, Bemelman WA, Wildenberg ME, Buskens CJ. Persistent mesorectal inflammatory activity is associated with complications after proctectomy in Crohn's disease. J Crohns Colitis. 2019;13(3):285–93.
9. Bernhoff R, Martling A, Sjovall A, Granath F, Hohenberger W, Holm T. Improved survival after an educational project on colon cancer management in the county of Stockholm–a population based cohort study. Eur J Surg Oncol. 2015;41(11):1479–84.
10. Hashiguchi Y, Muro K, Saito Y, Ito Y, Ajioka Y, Hamaguchi T, Hasegawa K, Hotta K, Ishida H, Ishiguro M, Ishihara S, Kanemitsu Y, Kinugasa Y, Murofushi K, Nakajima TE, Oka S, Tanaka T, Taniguchi H, Tsuji A, Uehara K, Ueno H, Yamanaka T, Yamazaki K, Yoshida M, Yoshino T, Itabashi M, Sakamaki K, Sano K, Shimada Y, Tanaka S, Uetake H, Yamaguchi S, Yamaguchi N, Kobayashi H, Matsuda K, Kotake K, Sugihara K, Japanese Society for Cancer of the Colon and Rectum. Japanese Society for Cancer of the Colon and Rectum (JSCCR) guidelines 2019 for the treatment of colorectal cancer. Int J Clin Oncol. 2020;25(1):1–42.

11. Kotake K, Mizuguchi T, Moritani K, Wada O, Ozawa H, Oki I, Sugihara K. Impact of D3 lymph node dissection on survival for patients with T3 and T4 colon cancer. Int J Colorectal Dis. 2014;29(7):847–52.
12. West NP, Kobayashi H, Takahashi K, Perrakis A, Weber K, Hohenberger W, Sugihara K, Quirke P. Understanding optimal colonic cancer surgery: comparison of Japanese D3 resection and European complete mesocolic excision with central vascular ligation. J Clin Oncol. 2012;30(15):1763–9.
13. General rules for clinical and pathological studies on cancer of the colon, rectum and anus. Part I. Clinical classification. Japanese Research Society for Cancer of the Colon and Rectum. Jpn J Surg. 1983;13(6):557–73.
14. Hida J, Yasutomi M, Maruyama T, Fujimoto K, Uchida T, Okuno K. The extent of lymph node dissection for colon carcinoma: the potential impact on laparoscopic surgery. Cancer. 1997;80(2):188–92.
15. Yada H, Sawai K, Taniguchi H, Hoshima M, Katoh M, Takahashi T. Analysis of vascular anatomy and lymph node metastases warrants radical segmental bowel resection for colon cancer. World J Surg. 1997;21(1):109–15.
16. Murono K, Kawai K, Kazama S, Ishihara S, Yamaguchi H, Sunami E, Kitayama J, Watanabe T. Anatomy of the inferior mesenteric artery evaluated using 3-dimensional CT angiography. Dis Colon Rectum. 2015;58(2):214–9.
17. Garcia-Ruiz A, Milsom JW, Ludwig KA, Marchesa P. Right colonic arterial anatomy. Implications for laparoscopic surgery. Dis Colon Rectum. 1996;39(8):906–11.
18. Nesgaard JM, Stimec BV, Bakka AO, Edwin B, Ignjatovic D, RCC Study. Navigating the mesentery: a comparative pre- and per-operative visualization of the vascular anatomy. Colorectal Dis. 2015;17(9):810–8.
19. Nagtegaal ID, van de Velde CJ, van der Worp E, Kapiteijn E, Quirke P, van Krieken JH, Cooperative Clinical Investigators of the Dutch Colorectal Cancer Group. Macroscopic evaluation of rectal cancer resection specimen: clinical significance of the pathologist in quality control. J Clin Oncol. 2002;20(7):1729–34.
20. Leonard D, Penninckx F, Fieuws S, Jouret-Mourin A, Sempoux C, Jehaes C, Van Eycken E, PROCARE, A Multidisciplinary Belgian Project on Cancer of the Rectum. Factors predicting the quality of total mesorectal excision for rectal cancer. Ann Surg. 2010;252(6):982–8.
21. Bertelsen CA, Bols B, Ingeholm P, Jansen JE, Neuenschwander AU, Vilandt J. Can the quality of colonic surgery be improved by standardization of surgical technique with complete mesocolic excision? Colorectal Dis. 2011;13(10):1123–9.
22. Alhassan N, Yang M, Wong-Chong N, Liberman AS, Charlebois P, Stein B, Fried GM, Lee L. Comparison between conventional colectomy and complete mesocolic excision for colon cancer: a systematic review and pooled analysis: a review of CME versus conventional colectomies. Surg Endosc. 2019;33(1):8–18.

37 Mesocolic Resection in Colon Cancer

Felipe Quezada-Díaz, Winson Jianhong Tan, and J. Joshua Smith

Description of the Procedure

Resection of the mesocolon in colon cancer surgery requires dissecting the colon with its complete envelope of mesentery along well-defined anatomical planes. This type of mesocolon-oriented dissection has been expanded to the concept of complete mesocolic excision, similarly to total mesorectal excision for rectal cancer.

The principle underlying mesocolon-oriented dissection is to mobilize the mesocolon while keeping it intact, thus allowing central ligation of the vessels and lymph nodes contained within. This maneuver increases nodal yield, which may be associated with improved survival. To perform this dissection, the peritoneal attachments must be opened in a very specific manner. The mesocolic excision technique entails mobilization of the mesocolon and peritoneal attachments rather than just bowel division. In contrast, colonic resections for benign conditions do not require adherence to these principles, and the mesocolon can be resected without consideration of the adequacy of resection margins.

Interestingly, even though the concept of mesocolic excision makes sense from an oncological viewpoint, the technique was formally only relatively recently described by W. Hohenberger and colleagues. In their 2009 publication, the authors described their technique and reported outcomes based on data prospectively collected from 1329 patients over a 24-year period. The rate of locoregional recurrence was reduced from 6.5 to 3.6% and the 5-year rate of cancer-related survival improved from 82 to 89% after adoption of complete mesocolic excision. This report piqued the interest of the surgical community, and since then numerous case series have been published describing outcomes after complete mesocolic excision.

F. Quezada-Díaz · W. J. Tan · J. Joshua Smith (✉)
Department of Colorectal Surgery, Memorial Sloan Kettering Cancer Center, 1275 York Avenue, New York, NY 10065, USA
e-mail: smithj5@mskcc.org

E. D. Ehrenpreis et al. (eds.), *The Mesenteric Organ in Health and Disease*,
https://doi.org/10.1007/978-3-030-71963-0_37

Indications

The standard patient who will undergo a mesocolon-oriented resection corresponds to a pathological confirmed adenocarcinoma of the colon with no evidence of distant metastasis.

Contraindications

There are no absolute contraindications to mesocolon-oriented resection. Fibrosis or scarring from prior surgery involving the retroperitoneum may make the mesocolic plane of excision challenging to delineate. Inadequate surgical training is a relative contraindication, as the procedure entails close dissection along critical vascular structures, and lack of familiarity with the structures in the retroperitoneum could lead to surgical error. In addition, obesity increases the complexity of the procedure.

Preparation of the Patient

As with any preoperative planning for patients with colon cancer, proper clinical staging is required. Suspicious lymph nodes should be identified to ensure that they are incorporated within the extent of resection. In certain instances, the original resection field may need to be modified to achieve this goal.

Either open or minimally invasive approaches can be used to perform complete mesocolic excision. In open surgery, the mobilization of structures is aided by manual and/or instrumental retraction. With a laparoscopic approach, achieving an adequate amount of tension is more challenging due to the confined workspace. A combination of strategic patient positioning and adjustment of the assistant's retracting instrument is crucial. It is important to guard against musculoskeletal and/or peripheral neurological damage that may be potentiated by the extreme positions into which the patient is placed during the dissection. With the robotic platform, extreme positions can be avoided due the wider range of mobility of wristed instruments.

In anticipation of the use of extreme positioning, the surgeon should take every precaution to avoid causing injury. The use of anti-slip pads, padding of bony prominences and scheduled repositioning of extremities, especially for patients at high risk for compartment syndrome, should be part of standard intraoperative positioning protocols.

How the Procedure Is Performed

General Principles

Mobilization of the mesocolon is key regardless of the surgical approach. The surgeon must be able to perform an en bloc mobilization of the mesocolon and part of the small bowel mesentery without damaging the vascular structures and the visceral peritoneal layer that contains the envelope surrounding the tumor and suspected lymph nodes.

Opening the mesocolon is necessary to access the retroperitoneal plane (Fig. 37.1a and b), especially in minimally invasive surgery. In mobilizing the colon and mesentery, gentle traction should be used to avoid excessive bleeding, which can make identifying the planes challenging. The use of energy devices can be helpful, but dissection in the correct mesocolic plane is usually associated with minimal bleeding, as this plane is avascular.

In open surgery, care should be taken to adequately expose structures in order to avoid causing injury and dissecting in the wrong plane. Positioning of the patient is less important than in minimally invasive surgery because structures can be manually retracted for adequate tension and exposure, whereas laparoscopic and robotic approaches rely more on optimal positioning of the patient and proper traction with atraumatic instruments to avoid mesenteric tears or unrecognized enterotomies. Special care should be taken if the patient is obese or has previous intra-abdominal adhesions, because excessive traction may lead to mesenteric tears.

Control of the vascular pedicle can be achieved as the initial step (as initially described by Turnbull) or after the mesocolon is completely mobilized. If a minimally invasive procedure is used, we prefer to identify and isolate the vessels first using a medial-to-lateral approach. It is important to remove all the surrounding lymphatic tissue up to the origin of the supplying vessel (Fig. 37.2). In open surgery, simple ligation and division of the mesenteric vessels between clamps is preferred. In minimally invasive procedures, clips, energy devices or vascular staplers can be used to divide vessels.

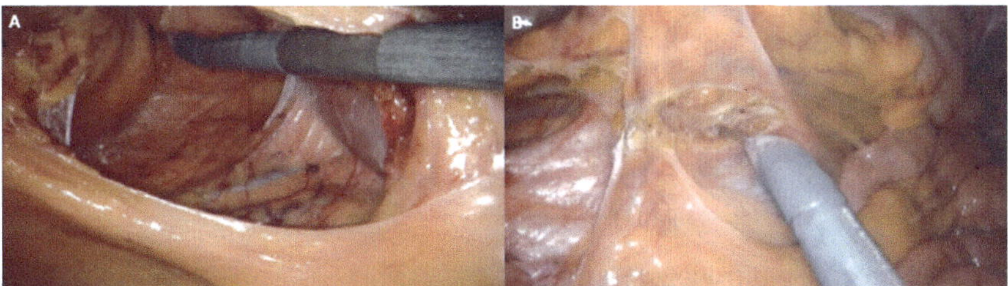

Fig. 37.1 Medial opening of the retroperitoneal tunnel to access the plane of mesocolon dissection. **a** right colon; **b** left colon

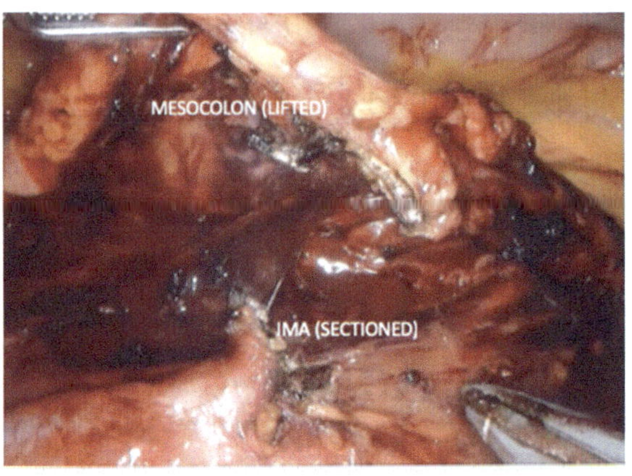

Fig. 37.2 Complete removal of lymphatic tissue during left mesocolic excision until the root of the inferior mesenteric artery (IMA) is reached

After the colon is fully mobilized from its peritoneal attachments, the margin of resection should be identified. Margins of at least 10 cm on either side of the tumor are recommended. It is crucial that the resected mesocolon include the principal vessel supplying the colon segment containing the tumor, as this is where lymph node metastasis occurs most commonly. The D3 dissection, developed in Japan, involves central vascular ligation at the root of either the superior or the inferior mesenteric vein, depending on the location of the tumor and uses the position of the main vascular supply of the tumor—rather than a predefined distance from the tumor—to define the limits of the bowel resection. This approach may result in shorter specimens in terms of bowel length but a similar amount of mesocolon, compared to using predefined resection limits.

For proper anastomosis, the mesocolon should be cleared from the resection margins to facilitate visualization of the bowel wall without compromising the blood supply. One way to assess blood supply is to rely on the identification of mucosal bleeding after bowel division with a scalpel. Visualization of pulsatile vessels at the remnant end of the transected bowel is another surrogate used to assess the adequacy of vascularity. More recently, indocyanine green fluorescence has been used to assess vascular perfusion.

The mesenteric defect can be closed using various techniques. In a complete mesocolic excision, the mesenteric defect is usually wide, and we opt to leave it open.

Specific Considerations

Right Colon

In open surgery, mobilization is frequently performed in a lateral-to-medial fashion. A key aspect in mobilizing the right colon is the opening of the peritoneum at the level of Toldt's fascia. The initial peritoneal incision should be made just medial to Toldt's fascia to avoid a plane of dissection that is excessively deep. Dissection in the correct plane ensures clean separation between the envelope of the mesocolon and the retroperitoneal fascia.

In a minimally invasive approach, a retrocolic tunnel is needed. This tunnel can be created using a medial-to-lateral approach, usually right below the tension area obtained from lateral traction of the ileocolic vascular pedicle. An alternative option is the bottom-up approach, starting from the peritoneal reflection of the cecum in the right lower quadrant. The duodenum and ureters should be identified before any vascular division is performed (Fig. 37.3). The vessels should be divided as centrally as possible using clips, energy devices or staples. For complete mesocolic excision, dissection over the axis of the superior mesenteric vein and artery allows accurate identification of the origin of the ileocolic vessels (Fig. 37.4). A central vascular ligation requires the division of the veins of the gastrocolic trunk, with possible preservation of the right gastroepiploic vein (Fig. 37.5).

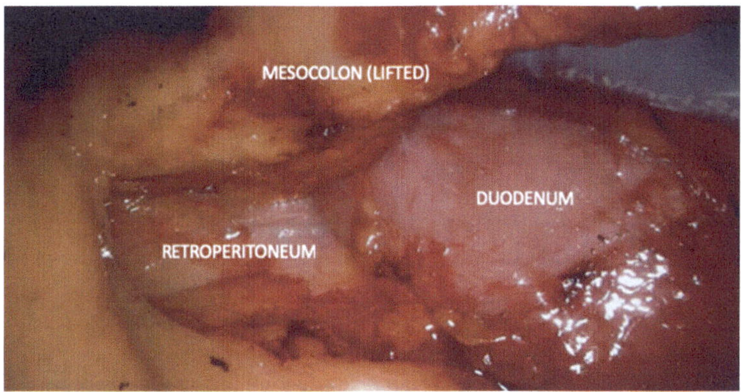

Fig. 37.3 Identification of the duodenum is required, with complete en bloc separation of the right mesocolon from the retroperitoneal plane

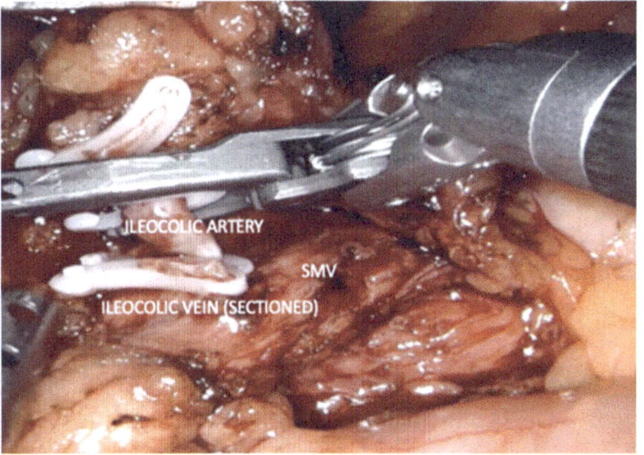

Fig. 37.4 Central ligation on the right colon requires identification of the ileocolic vessels at the level of the superior mesenteric vessels (SMV). Here, the ileocolic vessels run lateral to the vein

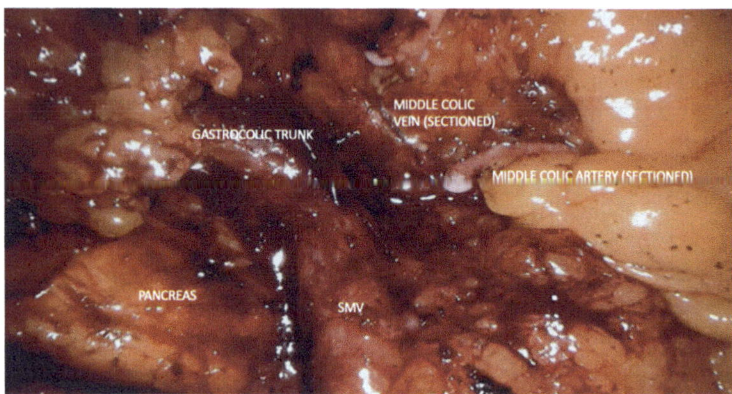

Fig. 37.5 Complete mesocolic excision in right-colon surgery requires complete dissection over the superior mesenteric vessels (SMV). The gastrocolic trunk runs directly above the pancreas

Transverse Colon and Flexures

The transverse mesocolon converges from the flexures to the center vascular pedicle defined by the middle colic vessels. In comparison with the right and left mesocolon, the transverse mesocolon is more freely mobile yet anchored by the middle colic vessels. The hepatic flexure and splenic flexure are sites of convergence between the transverse and right mesocolon and between the transverse and left mesocolon, respectively.

For proper mobilization of the flexures, the greater omentum needs to be mobilized or resected. A precise anatomical strategy is critical for the surgical approach, mainly due to the fact that the degree of adhesions to the anterior surface of the transverse colon varies between patients and is related to the peritoneal reflections of the flexures, which can make the dissection more challenging. The separation of the omentum from the transverse colon opens access to the lesser sac and enables central ligation of the middle colic vessels (Fig. 37.6). In open surgery, superior traction of the greater omentum reveals the dissection plane, allowing access to the lesser sac. In a minimally invasive approach, obtaining adequate tension to identify the plane of dissection may be more challenging but can still be achieved with appropriate retraction and countertraction of tissues.

Mobilization of the splenic flexure is one of the most challenging aspects of minimally invasive colorectal surgery. For proper mobilization of the mesocolon of the splenic flexure, identification of the inferior mesenteric vein at the level of the angle (ligament) of Treitz is required. The plane just below the inferior mesenteric vein is entered, and dissection proceeds superiorly as the mesocolon is lifted off the retroperitoneum. The dissection should proceed until the mesocolon is lifted away from the anterior aspect of the pancreas (Fig. 37.7). Once this phase of the dissection has been completed, the lateral peritoneal attachments of the left colon are then incised and mobilization is performed in the lateral-to-medial direction until

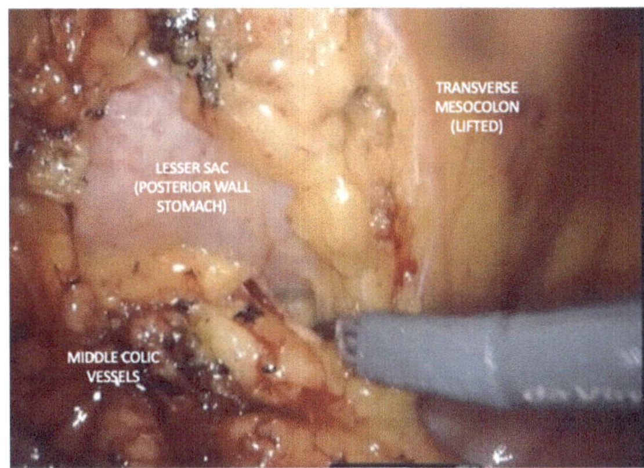

Fig. 37.6 Complete en bloc mobilization of the transverse mesocolon requires entering the lesser sac and visualizing the posterior aspect of the stomach

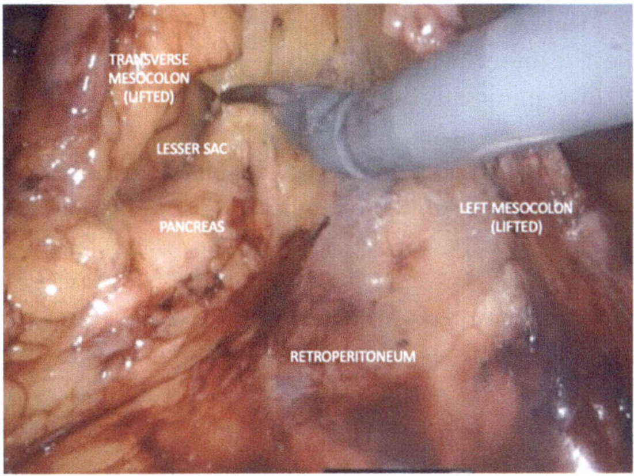

Fig. 37.7 Complete mobilization of the splenic flexure requires lifting the transverse and left mesocolon directly above the pancreas and retroperitoneum, respectively

the two planes meet. After mobilization is completed, vascular ligation can be performed. The left colic artery and the left branch of the middle colic vessels should be divided at their origin.

A complete oncological resection of the transverse mesocolon requires full mobilization of the flexures. The root of the transverse mesocolon can be identified by anterior traction of the mesocolon, which creates a fold separating the mesocolon from the small bowel mesentery (Fig. 37.8).

Left and Sigmoid Colon

The lateral-to-medial approach to resection of tumors of the left colon and sigmoid colon allows the surgeon to lift the mesocolon over the iliac vessels, the left ureter, the left gonadal vessels, Gerota's fascia covering the left kidney and the distal pancreas—important structures that need to be identified and preserved during dissection. The mesentery of the left side of the transverse colon is attached to the inferior border of the pancreas. During the dissection, the intersigmoid fossa should

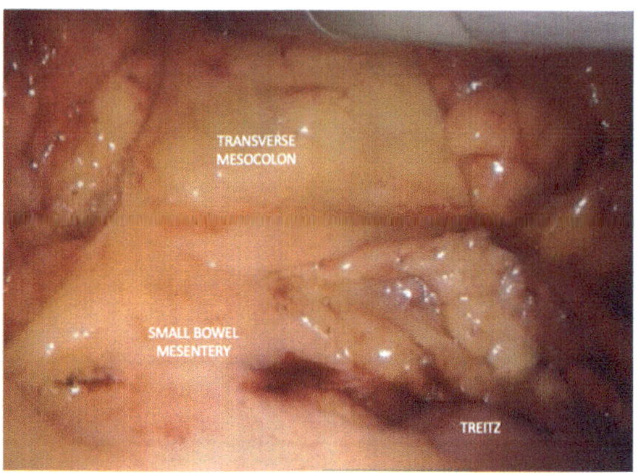

Fig. 37.8 A fold between the transverse mesocolon and small bowel mesentery at the level of the angle of Treitz highlights the surgical plane to follow for complete mobilization of the transverse mesocolon

be identified relative to the lateral peritoneal attachments in order to avoid injuring the left ureter, which is in close proximity.

In a minimally invasive approach, the resection of the mesocolon starts with identification of the inferior mesenteric vein at the level of the ligament of Treitz. The medial-to-lateral approach commences beneath the inferior mesenteric vein as the mesocolon is lifted over Gerota's fascia, gonadal vessels and the ureter. The left ureter is closely related to the inferior mesenteric vessels and should be identified prior to vascular ligation. For central ligation of the inferior mesenteric artery, the peritoneum is excised just above the left iliac artery until the identification of the aorta and the root of the vessel. The vessel can be controlled using clips, an energy device or a vascular stapler in case of calcified or large-diameter vessels.

Anatomical Variants and Typical Abnormal Findings

Anatomical variants in blood supply are often encountered during complete mesocolic excision. Surgeons should be familiar with these variations to avoid causing vascular injury during dissection.

The least-variable vessel in the right colon is the ileocolic artery. The right colic artery is the most variable vessel, arising in 40% of patients from the superior mesenteric artery. It is completely absent in 20% of patients. The middle colic artery usually arises from the superior mesenteric artery but is also absent in 20% of patients.

The inferior mesenteric artery is relatively constant, but the locations of its branches can vary. In 45% of patients, the left colic artery and sigmoid artery have a common trunk. The left colic artery is located lateral to the inferior mesenteric vein at the level of the origin of the inferior mesenteric artery in 73% of patients.

Excision of the mesocolon at the level of the splenic flexure is complex not only due to the surrounding organs but also due to variations in vascular supply. The three most important vessels for the splenic flexure are (i) the marginal artery, also known as artery of Drummond (present in all patients), which runs proximate and parallel to the edge of the colonic mesentery; (ii) the central arc of Riolan (present in 18% of patients), which joins the middle colic artery with the left colic artery; and (iii) the meandering mesenteric artery of Moskowitz (present in 11% of patients), which runs through the base of the mesocolon just above the ventral edge of the pancreas and connects the proximal segment of the middle colic artery with the ascending branch of the left colic artery. During medial-to-lateral dissection of the inferior mesenteric vein, injury of the arc of Riolan or the meandering mesenteric artery may compromise blood supply to the descending colon.

With respect to venous drainage, mesocolic resection of the right colon with high ligation encounters the most anatomical variations. Identification of the veins in the gastrocolic trunk right above the head of the pancreas is vital. Variations of drainage include a superior right colic vessel and a right colic vein that can drain to the trunk. In 20% of patients, the gastrocolic trunk is absent.

Complications

Complication rates after complete mesocolic excision range from 11.6 to 30.6%. A 2018 meta-analysis reported a pooled overall complication rate of 22.5%. Although most studies have reported equivalent complication rates for conventional excisions, the largest study thus far (University Hospital Erlangen) found higher morbidity after complete mesocolic excision (21.3% vs. 17.2%). The only study thus far to report the rate of intraoperative complications was a Danish population study that compared outcomes between a hospital where complete mesocolic excision was performed routinely and three hospitals where it was not performed routinely. The rate of intraoperative injury to intra-abdominal viscera was significantly higher in patients who underwent complete mesocolic excision (9.1% vs. 3.6%). Injuries to the spleen and the superior mesenteric vein accounted for the majority of the reported injuries. Postoperative respiratory failure (8.1% vs. 3.4%) and sepsis (6.6% vs. 3.2%) were also significantly higher in patients who underwent complete mesocolic excision.

Outcomes

Complete mesocolic excision results in superior pathology specimens and higher lymph node harvesting than does conventional surgery. However, oncological outcomes benefits remain unproven as most studies report higher disease-free or overall survival using only historical cohorts. In such studies, advancements in systemic therapy could explain the observed differences in survival. Several studies have also demonstrated superior oncological outcomes after complete mesocolic excision. However, the observed differences are confounded by variations in the quality of surgery and medical care. No randomized trials comparing oncological outcomes between complete mesocolic excision and conventional surgery have been published. A randomized trial currently in progress in China is expected to be completed in 2021.

Conclusion

Mesocolon-oriented cancer surgery, with its sound anatomical and oncological rationale, produces high quality resection specimens and may be associated with better outcomes than conventional colon cancer surgery. Since randomized trials face feasibility challenges, the best available evidence will likely come from well-designed prospective cohort studies.

Suggested Reading

1. Alhassan N, Yang M, Wong-Chong N, Liberman AS, Charlebois P, Stein B, Fried GM, Lee L. Comparison between conventional colectomy and complete mesocolic excision for colon cancer: a systematic review and pooled analysis: a review of CME versus conventional colectomies. Surg Endosc. 2018. https://doi.org/10.1007/s00464-018-6419-2.
2. Bertelsen CA, Neuenschwander AU, Jansen JE, Kirkegaard-Klitbo A, Tenma JR, Wilhelmsen M, Rasmussen LA, Jepsen LV, Kristensen B, Gögenur I, Copenhagen Complete Mesocolic Excision Study (COMES), Danish Colorectal Cancer Group (DCCG). Short-term outcomes after complete mesocolic excision compared with "conventional" colonic cancer surgery. Br J Surg. 2016;103:581–9. https://doi.org/10.1002/bjs.10083.
3. Coffey JC, Lavery IC, Sehgal R. Mesenteric principles of gastrointestinal surgery: basic and applied science. Abingdon, United Kingdom: Taylor & Francis; 2017.
4. Hohenberger W, Weber K, Matzel K, Papadopoulos T, Merkel S. Standardized surgery for colonic cancer: complete mesocolic excision and central ligation–technical notes and outcome. Colorectal Dis. 2009;11:354–64; discussion 364–65. https://doi.org/10.1111/j.1463-1318.2008.01735.x
5. Le Voyer TE, Sigurdson ER, Hanlon AL, Mayer RJ, Macdonald JS, Catalano PJ, Haller DG. Colon cancer survival is associated with increasing number of lymph nodes analyzed: a secondary survey of intergroup trial INT-0089. J Clin Oncol. 2003;21:2912–9. https://doi.org/10.1200/JCO.2003.05.062.
6. Merkel S, Weber K, Matzel KE, Agaimy A, Göhl J, Hohenberger W. Prognosis of patients with colonic carcinoma before, during and after implementation of complete mesocolic excision. Br J Surg. 2016;103:1220–9. https://doi.org/10.1002/bjs.10183.
7. Wang C, Gao Z, Shen K, Shen Z, Jiang K, Liang B, Yin M, Yang X, Wang S, Ye Y. Safety, quality and effect of complete mesocolic excision vs non-complete mesocolic excision in patients with colon cancer: a systemic review and meta-analysis. Colorectal Dis. 2017;19:962–72. https://doi.org/10.1111/codi.13900.

8. Watanabe T, Muro K, Ajioka Y, Hashiguchi Y, Ito Y, Saito Y, Hamaguchi T, Ishida H, Ishiguro M, Ishihara S, Kanemitsu Y, Kawano H, Kinugasa Y, Kokudo N, Murofushi K, Nakajima T, Oka S, Sakai Y, Tsuji A, Uehara K, Ueno H, Yamazaki K, Yoshida M, Yoshino T, Boku N, Fujimori T, Itabashi M, Koinuma N, Morita T, Nishimura G, Sakata Y, Shimada Y, Takahashi K, Tanaka S, Tsuruta O, Yamaguchi T, Yamaguchi N, Tanaka T, Kotake K, Sugihara K, Japanese Society for Cancer of the Colon and Rectum. Japanese Society for Cancer of the Colon and Rectum (JSCCR) guidelines 2016 for the treatment of colorectal cancer. Int J Clin Oncol. 2018;23:1–34. https://doi.org/10.1007/s10147-017-1101-6.

38 Mesenteric Resection in Rectal Cancer

Craig A. Messick

Description of Rectal Cancer and the Mesorectum

In the later part of the 1900s, rectal cancer gained national attention as outcomes in the U.S. were significantly worse compared to European and Asian countries. Local recurrence rates were roughly 25–30% following resection. As a result, increased attention by the American College of Surgeons (ACS) and the Commission on Cancer (CoC) in collaboration with the American Society of Colon and Rectal Surgery (ASCRS) Society of Gastrointestinal and Endoscopic Surgery (SAGES), Surgical Society of the Alimentary Tract (SSAT), the American College of Radiology (ACR) and the College of American Pathology (CAP) set out to improve outcomes in rectal cancer. This difficult task was assumed by numerous leaders in rectal surgery and continues to evolve. There have been landmark changes to every aspect of rectal cancer management including enhanced diagnostic modalities addressing circumferential resection margin, extratumoral deposits and extramural venous invasion (EMVI), imaging-based prediction of treatment response, standardized reporting systems using synoptic reporting methods, and tailored treatment sequences including strategic consideration of short course (5 × 5) radiation therapy and rejection (albeit with good intentions) of a one-size-fits-all radiation administration treatment algorithm. These adoptions have already shown improvement in local recurrence, long-term survival rates, and quality of life.

To understand rectal cancer, the magnitude of biological and environmental influences that cause accumulated mutational or methylated events that ultimately result in genetic change must be appreciated. But to understand rectal cancer treatment, as complicated as it has become, one must truly appreciate the role of the mesorectum. Its role is becoming increasingly clear, as decisions for wide local excision (WLE), radical resection of the mesorectum, induction chemotherapy,

C. A. Messick (✉)
Department of Surgical Oncology, The University of Texas MD Anderson Cancer Center, 1515 Holcombe Blvd., Houston, TX 77030, USA
e-mail: cmessick@mdanderson.org

E. D. Ehrenpreis et al. (eds.), *The Mesenteric Organ in Health and Disease*,
https://doi.org/10.1007/978-3-030-71963-0_38

multimodal chemoradiotherapy, or total neoadjuvant therapy with surgery last are at least in part based upon spatial relationships (extent of the T stage) between tumor location and the circumferential resection margin (CRM) at the mesorectal fascia.

Meso is the Latin term that means middle, as in a structure that lies between, or in the middle, of two structures. For the colon, the *meso*-colon lies between the bowel and the retroperitoneum or central vasculature (Fig. 38.1a and b). Regarding the rectum, the *meso*-rectum is the lymphatic and blood-vessel laden fatty tissue that is enveloped by its own fascia and lies between the rectal wall and the surrounding pelvic structures (Fig. 38.2). Those pelvic structures are as follows: anterior boundary, rectovaginal septum (females) or Denonvillier's fascia (males), lateral boundary, the endopelvic fascia or pelvic side walls, and posterior boundary, Waldeyer's fascia. Recognition of these structures, or boundaries, is essential when performing radical proctectomy for rectal cancer. Its complete, or enveloped, resection may not be necessary for a proctectomy when performed for benign disease like Crohn's disease (CD) or mucosal ulcerative colitis (MUC), but for malignancies including adenocarcinoma or neuroendocrine tumors, complete removal of all the fatty mesorectum containing all lymphatic tissue is the correct operation; nothing less is acceptable.

The mesorectum is an embryologic continuum from the mesocolon and its development has been proposed in several differing classical theories including the sliding and regression theories. The regression theory purports that the bowels outgrow the dorsal mesentery extending beyond its edges, and the sliding theory is based upon the notion that the bowels migrate to their final locations and pull their respective mesenteries with them. These and other theories were never adopted by main stream anatomists and embryologists. Understanding key vascular origins and suspension locations, a series of incomplete twists and the laying down of the mesentery on the retroperitoneum, has been much more widely accepted and reproduced. Recognition of these unique events affords rectal cancer surgeons an opportunity to improve upon rectal cancer management based upon embryological origins.

Indications for Surgery of the Rectum and Mesorectum

The American Joint Commission on Cancer (AJCC) and National Comprehensive Cancer Network (NCCN) have set clear indications for surgical resection and non-surgical management strategies for patients with rectal cancer. However, the decision for surgical approach is substantially different based upon the primary tumor's T (depth of wall invasion) stage. Tumors that are T0 or deemed muscularis mucosal invasion or carcinoma in situ (infrequent and better termed high-grade dysplasia) can safely be considered as pre-malignant (non-invasive) and surgically removed with polypectomy or endoscopic mucosal resection (EMR) or transanal excision (TAE) via an operating proctoscope, or minimally invasive (TAMIS). One consideration the surgeon must factor in the management strategy is the size of the

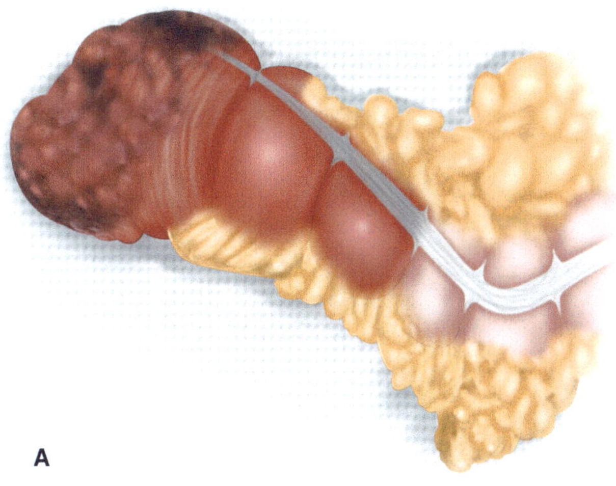

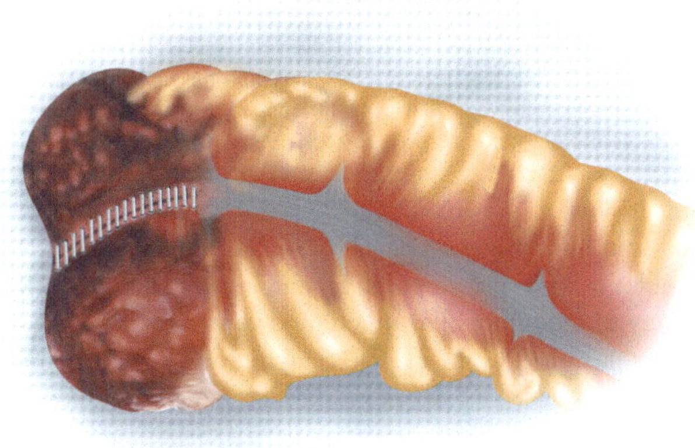

Fig. 38.1 **a** This is an image of the posterior surface of the mesorectum. The dissection off of Waldeyer's fascia, when performed correctly, yields a smooth, glistening surface as is shown in this image. The distal mesorectum (left side of the image) appears more lobular than smooth as this is the case when the mesorectum is transected at an appropriate margin. **b** This is an anterior view of the TsME resection. The taenia coli can be seen extending down onto the upper rectum, which then fans out to encompass the entirety of the remaining rectum. A staple line is shown here at the distal mesorectal resection margin

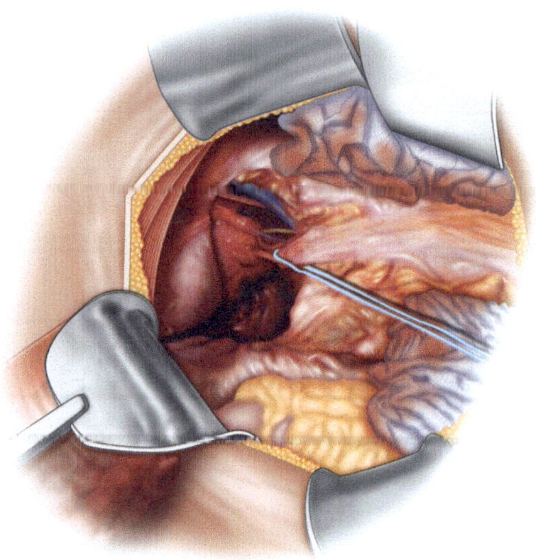

Fig. 38.2 This figure shows an anterior to posterior view from the patient's left side into the deep pelvis. The self-retaining retractor arms are noted and there are two towels used to retract the bowels. A blue vessel-loop is noted and encircles the right gonadal vein and ureter. The right pelvic endopelvic fascia (sidewall) has been resected and both the right iliac and obturator lymph nodes have been completely removed. The right internal iliac vein (blue) and artery branches (red) are seen as well as the obturator nerve (yellow)

lesion and location to the sphincter complex. Standard biopsies of the tumor when performed via colonoscopy are diminutive and may not necessarily reflect the complete pathology of the tumor. This likelihood is greater as the size of the tumor increases. Polyps that are 30 mm in size and return with high-grade dysplasia, harbor around a 30% chance of invasive disease not represented in the biopsy specimen.

It is imperative to ensure the local T stage prior to local excision to rule out the possibility of invasion beyond the submucosa, which should be avoided outright. Rectal cancers that are proven to be T1, invasive into but not beyond the submucosa, are amenable to full-thickness excision when surgically resected, and via an endoscopic submucosal dissection (ESD), when performed by an advanced endoscopist. Though it remains controversial, transrectal ultrasound (TRUS) may be most beneficial to delineate the tissue layers within the rectal wall for rectal cancers within a finger's reach due to the difficulties in differentiating tissue planes near the anus on MRI. For completeness sake of T1 rectal cancer (without high-risk [HR] features of LVI/PNI, poor differentiation or signet ring type histology) management following TAE or ESD, the concern for depth of SM invasion, Haggitt's levels, SM1, SM2, SM3 all carry very low, but inherent, risks for synchronous mesorectal lymph node (LN) involvement, ranging from 3 to 7%, representing historical radical proctectomy data. Lastly, where it was once thought that a 2 mm negative resection margin at the base was necessary, data has emerged that supports the same outcomes for negative margin only resections. Thus, for patients whose low rectal cancers are T1 and without high-risk features, wide local (transanal) excision (TAE) is an acceptable alternative to radical resection, which remains oncologically appropriate also given the risks disclosed to the patients.

Non-HR T2 rectal cancers are associated with a risk of LN involvement in excess of 25–30% and thus except in unique circumstances precludes a TAE as an appropriate surgical approach. The percentages for LN involvement in tumors with HRF is even greater extending up to 40% at diagnosis. For these stage Ib rectal cancers, resection extends beyond that of concern of the local tumor margins, but rather to the all-important mesorectal LNs. A radical resection, as initially and eloquently described by Dr. Bill Heald, involves precise attention to the details of the mesorectal fascia. Though LNs are not always visible on imaging, surgeons must understand that they may reside anywhere within the mesorectum, and dissection yielding a completely intact mesorectal fascia at the circumferential resection margin (CRM) with a high, central vascular pedicle ligation, must be performed (Fig. 38.3). Leaving a single LN behind serves only as an opportunity for local recurrences and spells certain disaster for rectal cancer patients. An intact CRM has become standard with a 5 cm distal mesorectal and mucosal margin for radical resections in patients who do not receive neoadjuvant chemoradiotherapy (CRT). For upper rectal tumors, the distal resection margin of 5 cm may not require a total mesorectal excision (TME) to the pelvic floor muscles (levator ani), but

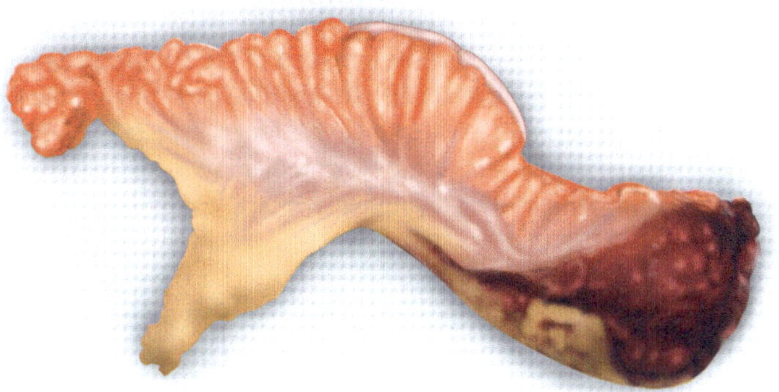

Fig. 38.3 This figure shows a complete and intact tumor specific mesorectal excision (TsME) and includes the sigmoid to the middle rectum. The central vascular resection with high-ligation of the IMA is noted with yellow mesenteric fat. The posterior surface of the mesorectum is smooth and extends to the level of the staple line which is seen at the most distal (right side of image) aspect

rather a modified, or tumor-specific mesorectal excision (TsME). Grading systems have been suggested to assess the "completeness" of mesorectal fascial excision but have been met with frustration as to precise definitions and leaves some degree of interpretation by either the pathologist or surgeon; often in disagreement.

The current standard of care for T3 lesions is concurrent neoadjuvant chemotherapy (oral administration) with long-course radiation in the US, though there has been an uptick in the use of short course (5 × 5). The discussion of long-course or short-course radiation is beyond the scope of this chapter, but suffice it to say, short-course has become more common than before and may have an increasing role in patients with oligo-metastatic disease who need immediate control of the pelvis with radiation and yet maintain the benefit of systemic therapy for their systemic disease, or as part of total neoadjuvant therapy (TNT). Still, it is paramount to evaluate the mesorectal fascia at the CRM in T3 lesions. The depth of tumor invasion for these tumors is completely through the wall of the rectum and into the mesorectal fat. Recognition and precise measurement of the tumor to the CRM is critical to the preoperative treatment plan and the operative strategy. The "early T3" or rectal cancers that are T3 but >5 mm from the CRM, have been studied on a national clinical trial [PROSPECT, NCT01515787] to evaluate the efficacy and outcomes of chemotherapy alone (3 months Folfox as neoadjuvant therapy without concurrent radiation) followed by surgery, and then 3 months of adjuvant chemotherapy. This point emphasizes the all-important mesorectum, its margin, and its relationship to the primary rectal tumor. As the US rectal cancer outcomes were deemed poor, the introduction of both the neoadjuvant, CRT and focus on an anatomic dissection of the mesorectum to circumferentially dissect free the mesorectum from the endopelvic fascia ensuring the complete resection of the mesorectal LN were instrumental in improved outcomes. The PROSPECT trial should shed some light on the notion of mesorectal dissection as being adequate for early T3 tumors; furthermore, delivery of CRT for T3 has been noted, but it has been standard for TxN1 disease as well. This practice may be questionable when the LN is free from the CRM. Careful consideration of CRT must be entertained for rectal cancer patients whose mesorectal LN are abutting or sitting on the CRM. For these LNs, CRT may make most sense, however, for mesorectal LN that are clear from the CRM, perhaps reconsideration of chemotherapy alone makes most sense as long as the surgeon can ensure a completely intact, circumferentially dissected mesorectum.

Never before has management of rectal cancer been so individually tailored and complicated to ensure the best and most oncologically sound operation. To say the least, these patients are best managed by a multidisciplinary tumor board where discussions for treatment are belabored and decided by all stakeholders involved in the treatment team. After CRT has been delivered, the distal margin requirement decreases to roughly 2 cm, but not universally agreed upon, and decisions for the extent of distal mesorectal dissection and transection should be carefully considered based upon post-treatment imaging. If LN are visible and remain abnormal, or suspicious on post-treatment MRI, they should be included in the radical dissection and TME/TsME. For tumors that are below the peritoneal reflection and T3 in nature, surgeons should strongly consider a TME to ensure the complete mesorectal resection, leaving no potentially positive LN behind. Numerous colorectal fellows have heard Dr. Ian Lavery say,

> You only get one chance to [manage rectal cancer patients correctly] do it right, or if you don't, you end up chasing the patient's recurrences the rest of their miserable lives.

When approaching patients with this understanding, surgeons absolutely *cannot* assume the role of technician, rather, maintain their status as 'captain of the ship' advocating on behalf of our patients at all times ensuring the best possible oncologic outcomes.

For the most locally advanced rectal cancers, T4, make no mistake about it, the involved organ identified at diagnosis must be included in the resection specimen and appropriate planning starts at their initial visit regardless of what the post-treatment images may suggest. T4 malignancies invade posteriorly into the sacrum or anteriorly into the bladder, seminal vesicles, vas deferens and prostate in males, and the uterus, cervix, or vaginal wall in females. Pelvic exentorative operations, either total or modified/posterior exentorations) require a great deal of local resources and an exhaustive conversation with these patients right from the start to set appropriate expectations. It seems illogical and improper to rely on either the medical or radiation therapies to "downstage" tumors that outright involve adjacent organs/structures. Moreover, it is inappropriate to rely on adjuvant chemotherapy to "clean up" what should have been surgically removed. With this approach, and following neoadjuvant CRT, the anticipated tumor treatment effect, regression, leaves a fibrotic scar in place of tumor. And whereas there may not be gross tumor in that scar or the previously involved organ, a radical resection that leaves even a single tumor cell behind is unacceptable and should never be considered when there is clearly a means to complete an R0 (margin negative by 2 mm) radical resection. This discussion should be held with the patient and any supporting family members who may need to assist with an extended recovery following such pelvic resections. When R1 resections are anticipated (or close margin), data from the University of Texas MD Anderson Cancer Center has shown that use of intraoperative radiation therapy (IORT) has similar results to that of an R0 resection, yielding this option for tumors whose resectability is feasible, but likely close on a margin: posterior along the sacrum, or lateral pelvis along the obturator space. All other locations should be surgically resected *en bloc* and reconstructed as appropriate. This extends to both simple coccygectomies and the more complex sacrectomies with/out internal hemipelvectomies extending to above S3 either unilaterally or bilaterally. Such operations often are extraordinarily challenging involving orthopedic/neurosurgery, a complex plastics reconstruction team, and a gifted anesthesia department as these cases often will pass through three or more anesthesiology providers during the surgery or be spread over two days. As these require such a magnitude of resources, such operations should be carefully considered and likely best served at institutions who perform them routinely.

Considering all these components to the rectal cancer patient, attention to the mesorectum is paramount and decisions for treatment and what surgery to be performed must be evaluated on case-by-case basis.

Contraindications of Surgery of the Rectum and Mesorectum

There are relatively few absolute contraindications to surgery on the rectum and mesorectum, generally related to surgeon and or resource availability. Facilities where use of intraoperative radiation therapy (IORT) is not available should not prevent surgeons from offering rectal cancer surgery, but when its use is anticipated but not available, referral to a more appropriate center should be performed. This is also the same for rectal surgeries that may require additional specialties to be involved. Lastly, whereas it may seem straight-forward to perform pelvic surgery, the careful and precise mesorectal dissection takes training and then, dedicated constant practice. There is not a consensus of how many mesorectal surgeries are required to be able to consistently perform TME surgery, but the fewer a surgeon performs typically results in inadequate mesorectal dissections yielding a greater percentage of incomplete mesorectums. That said, one can become proficient if the practice continues with higher volume and surgeons are held accountable to their dissections via photographs of their specimens. This notion of accountability has been recognized by the ACS and CoC. Practice accountability and standardization of rectal cancer surgery has been proposed and now administered by the National Accreditation Program for Rectal Cancer (NAPRC). With the utmost positive intentions for continuing to improve rectal cancer patients' outcomes through standardization and accountability, the program now offers accreditation for hospitals where rectal cancer surgery is performed including both high and low-volume centers. This patient-centric approach has already improved outcomes in other cancers including breast cancer, through the development of the National Accreditation Program for Breast Cancer (NAPBC). It is likely that such programs will be more largely adopted across most of the common cancers as the results speak for themselves.

There are far more relative than absolute contraindications, including patient factors such as additional medical comorbidities as well as surgeon and or resource availabilities. Patient medical comorbidities are real and should be heavily weighted at all stages of any surgical planning. Optimization of known cardiopulmonary disease is critical for minimizing 30- and 90-day perioperative mortality. As there is absolutely no rush to get patients to resection other than for the mental challenge of 'living with cancer' during treatment, all known medical disease should demand attention and optimization prior to surgery including cardiovascular, endocrine, diabetes, thyroid, nutrition, renal, coagulopathy, obesity, anemia, smoking, and cholesterol. There is a plethora of additional areas to focus, but clearly these need up front attention. When patients need neoadjuvant therapy, this allows the surgeon many months of work with patients' primary care providers (PCPs) to optimize their disease allowing for optimal outcomes for rectal cancer surgery.

Pelvic dissection can be very challenging not only based upon body habitus but also more challenging following radiation therapy. Waiting an extended period of 6–9 months to operate on the mesorectum following radiation can challenge even the most skilled, experienced surgeon. The extensive fibrosis that develops following radiation can make the distinction between benign fibrotic scar and malignant tumoral invasion very challenging. Posterior dissection beyond Waldeyer's fascia in an irradiated field can be more than unnervingly bloody. Injury to presacral veins or the internal iliac vessels in the lateral pelvic compartment can quickly produce life-threatening hemorrhage. One potential preemptive strategy is preoperative image review; another is the availability of blood and of sacral thumbtacks. Intentional lateral dissection beyond the mesorectal plane may be necessary to achieve and R0 resection and should be performed when necessary. Both urologists and the gynecologic oncologists have significant experience dissecting in the lateral pelvic sidewall as they routinely perform extended lymphadenectomies along the internal iliac and obturator lymph node basins. Lack of experience in this location should be only a relative contraindication and consultation of such services who can provide the complete lymphadenectomy in the lateral pelvis would be advised.

Technology has rapidly advanced and the robotics platform is purported to be beneficial especially in the narrow pelvis. It may offer superior views of the anterior mesorectal dissection in the low pelvis between the prostate in males and the rectovaginal septum in women when the dissection to the pelvic floor is required. Maneuverability under the pubis bone to see this extent of dissection is painstaking during a traditional open approach even with use of deep pelvic retractors such as the St. Mark's or Bright-track retractors. Still, when required for the appropriate distal dissection, the surgeon must achieve that level of dissection regardless of means. There are now approximately 10 years of data emerging on use of a top and bottom approach to total mesorectal surgery by using a transanal TME (TaTME) surgical approach. The most commonly employed approach may combine laparoscopy in the abdomen and a synchronous TaTME operation. Both access methods provide excellent distal mesorectal visualization. Learning curves are different for both and require mentoring from surgeons skilled in their use. However, before the advent of technology, rectal cancer surgery, including open TME resections were adequately performed and must be in the wheelhouse of every surgeon who offers rectal cancer surgery. Remember, the TME plane is the same no matter which approach is used and its complete resection must be performed regardless of the approach chosen for the operation. If and when the appropriate oncologic resection is prohibited by technology, the surgeon must resort to a different approach to ensure the oncologic appropriateness of the operation is completed.

Surgical Management: Technique

Approach to the posterior mesorectal fascial plane should be started only once the sigmoid has been freed from the left lateral pelvic side wall (open technique for lateral to medial mobilization) and the posterior mesocolon dissected free from the retroperitoneum (including left ureter and gonadal vein) up to and just proximal to the origin of the inferior mesenteric artery. Once the posterior plane has been correctly identified, the retroperitoneal structures protected and the vascular pedicles ligated, then the distal dissection of the mesorectum can commence. The initial steps previously outlined assist with delineation of the inferior hypogastric nerves as they course from medial on the anterior aorta to the sacral promontory where they begin to separate as part of the Posterior Waldeyer's fascia. At that level, as they begin to dive into the upper pelvis at the level of L5/S1 junction, they splay out coursing laterally into the lateral pelvic sidewall and become congruent with the endopelvic fascia. Once identified, the plane is just one cell layer onto the mesorectum, nothing more, which yields dissection into the mesorectum, and not one cell layer onto posterior Waldeyer's fascia, which will either injure the hypogastric nerves, or if deeper dissection is performed, invariably intense hemorrhage will ensue due to injury to the presacral veins. Extreme care must be taken not to injure these veins as they are a low-pressure system and will produce immediate and massive blood loss. They are extremely difficult to control as they tend to recoil into the intervertebral spaces or the foramina. Careful preoperative review of the imaging can serve as a useful guide to identifying these vessels and where they course so as not to encounter them. These vessels course often across midline of the sacrum posteriorly and wind around the posterior lateral aspects as the pelvic inlet is further dissected. This dissection is best visualized by use of the St. Mark's deep pelvic retractors, lighted or not, as per operator's preference. These retractors have an angled tip that has a lip at the end that facilitates the 'toe-in' motion provided at the handle. When this flattened lip is retracted distally into the mesorectum, the plane between the loose areolar connective tissue and the mesorectum becomes distinctly clear and proper dissection posteriorly can easily progress.

As further dissection continues, the loose areolar connective tissue often entices the operator, or often assistant, to dissect in the 'safe zone' in the middle of the tissue. However, this maneuver will invariably lead to dissection of the posterior Waldeyer's facia into the specimen as part of the mesorectum and ultimately lead to vascular injury. Deeper dissection requires advancement from the #2 to #3 retractor and facilitates posterior visualization to the deep pelvis where the sacrum begins to change direction and curve back anteriorly. When this is encountered, often use of the #4 is required to dissect completely to the pelvic floor, which is observed by muscle fasciculation when touched by the electrocautery. Dr. Victor Fazio is well-remembered for saying "the key to a great anterior dissection is a great posterior dissection." As much as possible, and as bearable, the posterior dissection should be completed before the anterior dissection begins. For the deep male pelvis, initial dissection of the lateral stalks may be necessary to facilitate exposure for the posterior deep pelvis. In order to appropriately dissect the lateral stalks, the posterior mesorectum is retracted with the St. Mark's retractor, and the rectum is withdrawn under tension up out of the pelvis. As this is performed, the lateral stalks can be seen and the dissection is easily completed using the cautery extension, and sometimes requiring the extender for the extender, from deep to anterior. The motion is directly anteriorly from deep to superficial and the mesorectum is dissected free from the endopelvic fascia. This is then repeated on the opposite side. When an adequate amount has allowed complete posterior dissection, then the remaining lateral peritoneum and mesorectal dissection can continue. After this portion is completed, the dissection ends anteriorly at the peritoneal reflection.

To begin the anterior dissection, the St. Mark's pelvic retractors are again used, but often, down-sizing to either the #2 or #3 is necessary as the tissues are no longer elbow deep in the pelvis. As the peritoneal reflection and bladder are lifted anteriorly with the retractors, there is a visible indentation that separates the typically much thinner mesorectum from the anterior Denonvillier's facia in men. The same retraction is applied for the dissection in women, however, the uterus is retracted anteriorly, or the vaginal cuff if the uterus is surgically absent, and the same indentation is appreciated. The peritoneum is scored with the cautery and then the mesorectal fascia and the anterior structures can be easily dissected, again, making full use of the St. Mark's retractors. As this dissection continues, it is vital to understand that the orientation of the mesorectum and rectum do not continue in a posterior dive, rather, begin to course more anteriorly, lending the dissection under the pelvis and symphysis pubis. Review of the sagittal pelvic imaging will confirm this direction. It is often that with use of a lap sponge, the rectum is firmly grasped or pressed down (key point) yet pulled out of the pelvis at the same time, to facilitate exposure. This retraction requires much practice and is easier in women than men, but positioning, dissecting, and repositioning will facilitate complete dissection to the pelvic floor and the operator's fingers can completely encircle the distal rectum at the bare area where the mesorectum obviously thins to absent and becomes intersphincteric. When the TME dissection continues too anteriorly or too laterally, the vaginal cuff will be easily encountered, and a typical sign of dissection too close to these structures is nuisance bleeding. When the bare area of the rectum (all mesorectum has been dissected distally) is encountered, this is usually met with great joy as this signifies the end of the dissection. It is worth noting that again, careful review of the imaging will lead the operator in the deep, lateral pelvis just prior to encountering the levator ani muscles of the pelvic floor, to a very wide dissection and ensures the mesorectum is not 'coned in' on as the dissection to the pelvis is completed.

Laparoscopic and robotic approaches have become more common place and their use has been boasted by optimal visualization in the deep pelvis. Laparoscopic approaches have been difficult to maneuver over the sacral promontory, it is possible and also requires careful consideration to port placement to achieve this. Through use of the robotics platform, the 3D camera and multiple operator arms including an additional assistant port facilitate this dissection with superior visualization during dissection. Robotics port placement is also of concern to facilitate dissection to the pelvic floor and they must be aligned in a semi-circular, pattern and often require "port burping" to dissection to the floor or intersphincteric when necessary. An angled port placement here will be prohibitive for complete dissection to the low pelvis and fully into the intersphincteric plane. Optimal port placement is beyond the scope of this chapter and should be based upon patient factors, tumor location, and comfort of the operator.

For TsME dissection, the mesorectal dissection should continue several centimeters beyond or distal to the level of rectal transection. This will allow for adequate mesorectal fat and lymph node inclusion and also facilitate a tension-free anastomosis, critical for ensuring low anastomotic leak rates. When the appropriate level is decided, the mesorectum can be safely dissected with an energy device if laparoscopically or robotic approaches are performed. During an open operation, or through a Pfannensteil incision, the mesorectum is scored and the dissection along the lateral rectal wall to underneath the rectum can be safely dissected. If the superior hemorrhoidal/rectal artery and vein are to be ligated prior to rectal transection, they may be ligated with an energy device, or between clamps with suture ligatures and ties, or hemolock clips as appropriate. Often the dissection of these vessels is facilitated by early rectal transection leaving only the remaining mesorectum for transection and vessel ligation, which can be easily performed as just described.

Surgical Complications

Early

There are many potential complications that may occur during surgery for rectal cancer or otherwise surgery that requires a complete mesocolic excision. During dissection, the inferior hypogastric nerves that run parallel to the aorta may be inadvertently encountered or even injured by energy devices or transected during standard dissection. Careful consideration of their anatomical location may improve injury rates, though it is often that they are not discovered to be injured until after recovery. Injury to these (both) will result in the male patient's inability to achieve and maintain an erection. Though there may be a delayed recovery, if gain of function has not occurred by 6–9 months, it is unlikely there will be further recovery of function. Proper counseling during obtaining informed consent should pay attention to this point especially in male patients; less is understood of the hypogastric nerve function in females but regardless, care should be taken during dissection.

In patients with central obesity, the visceral fat may be significant, and the IMA may be densely covered in mesenteric fat. Meticulous dissection is key to dissecting to the origin of the IMA and its complete skeletonization should be attempted to ensure its ligation with ties, clips, or stapler (operator preference) are able to occlude and stay steadfast on the transected vessel end. Ligation and transection of the major [named] vessels (specifically the IMA) by use of an energy device is certainly documented. However, caution should be maintained during this maneuver. Not all vessels are the same diameter, and a larger diameter vessel may not hold the cauterized edge for an extended time period very well. Smaller vessels may be ligated using such energy devices with impunity, but it only takes one patient that requires emergent reoperation for active hemorrhage for a surgeon to never ligate that vessel with such an energy device. Additional vessels that are easily injured resulting in major immediate complications are the presacral vessels. Though they are often not observed, careful review of the patient's images pre-operatively should steer the operator away from these low-pressure vessels. Injury to these, especially without adequate visualization, will result in audible hemorrhage that will stop under direct pressure, but continue upon immediate release. These are thin-walled, large sacral vessels that often retract into the sacral foramina when injured rendering them nearly impossible to control. Maneuvers to control bleeding include direct pressure or packing of the pelvis as an initial move to ensure the anesthesia providers can be ready for massive infusion/transfusion as needed. Use of multiple suction devices to provide visualization may be required and then use of the cautery on a maximal setting and using an "arcing" technique may slow or stop the bleeding. Another way to control this hemorrhage is suture ligation using 4–0 or 5–0 prolene suture. Wide bites may be necessary and extreme care must be taken not to put the suture under tension until the knot is tied down, or further vascular injury will ensue and lead to further hemorrhage. When these maneuvers are unsuccessful, a small plug of rectus muscles can be placed over the vessel/area of hemorrhage, and then again, using the cautery on max setting and as an arc without direct contact, the rectus plug is essentially welded to the site and halts hemorrhage due to direct pressure. If further hemorrhage is encountered, strong pelvic packing may be the last resort and close the abdomen to return to the operation in a 24–48-hour period with additional operator assistance.

During IMA dissection, if improper identification of the left gonadal vein and left ureter has not occurred, the left ureter may be incorporated into the IMA transection, especially when MIS approach is made. To prevent this injury at this location, the IMA must be completely skeletonized, especially on the posterior and left sides. When the vessel is completely isolated, then the operator can be sure the left ureter is maintained in the retroperitoneum and free from transection during IMA ligation. As the dissection continues for both MIS and traditional open approaches, the ureter is deeper in the retroperitoneum than the left gonadal vein, and visualization of both structures will further decrease the chance of left ureter injury. Often the gonadal vein is thin and look very similar to the ureter. As well, the ureter may not vermiculate, to clearly reveal it as the ureter. The ureter should be dissected off the posterior surface of the sigmoid mesentery as it crosses over the common iliac artery and vein and begins its pelvic trajectory to the bladder in the lateral pelvic compartment, outside the endopelvic fascia. When the ureter is injured along its course adjacent to the iliac vessels, the repair is much easier than its transection more proximally by the IMA. It should be clearly stated that when injury is identified, a urologist should be consulted for an optimal repair. If a urologist is unavailable, then full bladder mobilization should suffice for a tension free anastomosis (standard principle) with a primary re-implantation over a double-J stent is sufficient. It is prudent to place an adjacent drain incase of anastomotic leak. Suture here is critical and only an absorbable monofilament suture should be use, and never a braded suture, absorbable or permanent, due to risk of stone formation and chronic fibrosis leading to stricture. In the event a primary uretero-ureterostomy is required, the ends require full mobilization and should be "freshened up" by Potts scissors to remove any injured tissue. A spatulated, end-end anastomosis again is fashioned in the same way over a stent and a drain is placed adjacent to collect any urine leakage. Best outcomes are achieved when immediate identification and ureter repair are made as opposed to delayed identification.

Lastly, following any gastrointestinal surgery where bowel is resected continuity is restored, there is an inherent risk for an anastomotic leak. However, a leak following any rectal surgery has perhaps greater consequences. Ileocolic anastomotic leak rates vary in the literature but are currently reported around 7% according to NSQIP data. This seems tragically high, and for colorectal or coloanal anastomoses, the rates are even higher up to 10–12%. Consequent to a leak in the pelvis, overwhelming pelvic sepsis secondary to stool into the pelvis has disastrous consequences including death, but to a lesser degree and still importantly, pelvic floor dysfunction. It is well stated that a diverting stoma (loop colostomy or ileostomy) does not prevent a leak but diversion helps prevent overwhelming pelvic sepsis by minimizing the amount of stool in the pelvis. When leaks occur, reoperation is typically required for washout and control with diversion (if not previously diverted) and wide drainage for control.

To minimize the risk of leak, there are two fundamental principles: absence of anastomotic tension and pulsatile blood supply to both ends. One way to help ensure that this mandate is met is routine splenic flexure mobilization. An operative approach that starts with the plan to mobilize the flexure instead of one that requires it to be mobilized only after the anastomosis is perhaps under tension, provides a much smoother and likely faster operation. Additionally, the approach for a high ligation (open or MIS) with central mesenteric dissection should be performed for all rectal cancer surgeries. This necessitates a complete mesocolic lymphadenectomy onto the aorta and will facilitate adequate reach into the lower pelvis. To gain additional length, especially for low pelvic or anal anastomoses, the IMV must be ligated just lateral to the duodenum. Additionally, fully dissection of the transverse mesentery off the inferior border of the pancreas will be necessary to complete the splenic flexure mobilization. Lastly, complete dissection of the mesentery off the para-aortic retroperitoneum facilitates reach of the descending colon into the distal pelvis and even to be pulled through the anus for a hand-sewn anastomosis as indicated.

Upon completing a colorectal anastomosis, it is appropriate to investigate the integrity of the anastomosis and for any possible sign of a leak. Initial assessment begins with the anastomotic rings, or donuts as they are frequently termed. Once the safety has been put back on the stapler, the motion to remove the stapler from the rectum is critical, and often performed in such a way that puts additional radial tension on the anastomosis. Given the stapler is curved and not linear, when the hands are twisted, the functional end of the stapler will twist, moving the entire anastomosis in a circle creating tension at the anastomosis. This can be avoided quite easily, however. Instead of twisting the stapler using at the user's wrists, the hands of the user are moved around in a wide circle leaving the functional end of the stapler in-line with the colon and rectum and subsequently decreasing the unnecessary tension during stapler removal. The anvil is then separated from the stapler and donuts can be assessed; if the proximal anastomotic donut is unable to be removed without damage, the entire anvil and donut can be submitted to pathology as one specimen. To avoid damage to the distal donut and disrupt the rings, the surgeon should first withdraw the spike fully into the device and then remove the white ring from the end of the device using a hemostat or other instrument. Once removed, then the donut (tissue) can be safely removed from the stapler end. If during these maneuvers the donuts are not intact, the anastomotic leak rates have been shown to be higher than anastomoses with intact donuts.

Further evaluation of the anastomosis can then be performed using either flexible or rigid sigmoidoscopy. The pelvis is filled with fluid, typically isotonic solution (normal saline) but can also be sterile water (often used in cancer cases) and the proximal colon is clamped by either atraumatic bowel clamps (Glassman's) or simply an operator/assistant's fingers. The surgeon between the legs performs a digital anorectal exam feeling for the height of the anastomosis (when palpable), luminal patency, and any palpable defects if obvious. Following the digital exam, the sigmoidoscope is inserted just proximally to the anorectal ring (top of the levator ani) and gentle air insufflation can begin. Under direct visualization, the scope is advanced just beyond the anastomosis. During this maneuver, when using a rigid sigmoidoscope, it is prudent to always have suction available to assist with optimal visualization. When completely visualized, the entire circular anastomosis should be intact, with visible staples, and not hemorrhaging, though mild ooze may be present and self-limited. Careful attention to anastomotic hematoma formation is critical as the pressure and tension that develops can be enough to result in local ischemia leading to anastomotic failure and subsequent leak. Rectal dogears that were created if the distal rectal margin was transected using a stapler will also be visible, however, there will be no dogears if a double purse-string anastomosis was performed. Though this may be more difficult to perform, there are some data that support that technique over a double stapled anastomosis. If a steady hemorrhage is identified, the anastomosis can be revised if there is room, but conservative measures including an interrupted trans-anastomotic suture maybe used to control the bleeding. If unable to perform from outside the bowel, then endoscopic means should be considered. Endoscopic clips may be useful as well as injection with epinephrine, however injecting epinephrine into a fresh anastomosis seems somewhat illogical as it may cause ischemia in addition to control of hemorrhage. If it is only a mucosal hemorrhage, local intraluminal cautery can be easily used and very beneficial.

One last measure that has gained popularity to investigate the integrity of the anastomosis is use of fluorescence angiography. Fluorescent dye can be injected through a peripheral vein by the anesthesia providers. This travels to the gastrointestinal tract via arteries and then to capillaries. Sterile hand-held technology during traditional open approaches and during MIS (robotic and laparoscopic) surgeries will allow for visualization of the fluorescent dye in viable, or adequately perfused, bowel walls. This requires only a matter of seconds to be distributed to the entire bowel. Areas that are not fluorescent reveal inadequately perfused bowel and those areas should be considered non-viable, or ischemic. If the anastomosis appears to be ischemic, evidenced by lack of fluorescence, it should be taken down and a more viable conduit provided for the anastomosis. Routine use of technology may not be necessary if routine visualization of pulsatile bleeding from the marginal artery is observed and no tension is appreciated at the anastomosis, though recognition of their availability is critical to ensure the correct decisions are made. Attention to all these details provides optimal patient outcomes and better long-term function for patients.

Late

There are several very debilitating long-term, or delayed, complications that arise from rectal surgery when the complete mesorectal envelope is dissected from the endopelvic fascia. Injury to splanchnic nerves that reside around the anterior dissection planes, just below the seminal vesicles and prostate in males (too small to see typically) have been associated with sexual dysfunction but differently than the paired inferior hypogastric nerves. Injury to the splanchnic nerves at that level will result in a retrograde ejaculation. This is disturbing for men when they have an absent ejaculation, but it should be stated that from a functional standpoint, all other sexual functions remain normal. The retrograde ejaculation returns to the bladder instead of exiting the body through the urethra. It has been reported that recovery of normal antegrade ejaculation can be achieved through normal healing time up to 9 months. After 9 months it is unlikely that further recovery of baseline function will return.

A more significant late complication is bowel dysfunction. A great deal of attention has been turned to quality of life aspects following rectal surgery. It remains unclear as to when the expected post-operative bowel dysfunction that immediately follows organ preservation turns into what has been collectively termed the low anterior resection syndrome (LARS). A great deal of controversy exists regarding LARS, its full spectrum of symptoms, treatment, and prevention, but both clinicians and patients recognize it and patients suffer through it for many months before any improvement is reached. A recent Delphi analysis aimed at generating a consensus statement was completed and provided great insight into LARS. Symptoms included unpredictable bowel function and stool consistency, fecal urgency, fragmentation, stacking, difficulties emptying. However, when patients develop these symptoms, they must be coupled with at least a consequence, as not all patients are bothered by their symptoms to the same degree and thus have no difficulties with management. Consequences included significant time on the toilet, a preoccupation with bowel function, social and sexual activity, inability to appropriately perform work-related duties, and several others. Additionally, the observed symptoms that patients experience require a significant impact on their lives. Though the collective agreed upon symptoms are wide-ranging, not all patients will develop all of them, and some patients may even only develop a few. There is much to research regarding LARS and its development following organ preservation surgery including the mesorectum.

Summary

Review all imaging prior to an operation that involves mesorectal dissection. CT and MRI images provide meaningful data regarding metastatic disease, but they are pivotal for correct surgical technique to provide appropriate oncological margins. The images serve as a road map to getting in, successfully resecting the mesorectum and getting out of the pelvis without injury to vital structures.

Know where normal anatomical structures are supposed to reside and prepare for them during each move to ensure they are identified prior to injury. Injury prevention is key to a great outcome and continuing the proposed surgical plan.

Recognition of the local T and N stage of patients with rectal cancer is vital to ensuring they are treated with an appropriate use of neoadjuvant chemotherapy or radiation. Knowing the precise T and N stage may spare patients unnecessary treatment that is otherwise toxic and not beneficial. Recognition of the precise tumor location within the upper, middle, or lower rectum is just as critical to provide the patient with the best treatment.

Attention to surrounding anatomic structures is important to prevent organ injury such as presacral vessels and the ureters. Their anticipation during dissection will facilitate their being spared from injury during tedious dissection. But when injuries occur, consultation with a urologist will provide optimal outcomes for the patients.

The approach to rectal surgery should be individualized to each patient and their pelvises considered unique to the operator. This, combined with surgeons considering themselves as surgical anatomists, will lead them to be best prepared for the subtle, but important, anatomical variances that exist between patients and facilitate fewer complications and unanticipated inhibitors to the planned operative approach.

Suggested Readings

1. Agarwal A, et al. Quantified pathologic response assessed as residual tumor burden is a predictor of recurrence-free survival in patients with rectal cancer who undergo resection after neoadjuvant chemoradiotherapy. Cancer. 2013;119(24):4231–41.
2. Fleshman J, et al. Disease-free survival and local recurrence for laparoscopic resection compared with open resection of stage II to III rectal cancer: follow-up results of the ACOSOG Z6051 randomized controlled trial. Ann Surg. 2019;269(4):589–95.
3. Holliday EB, et al. Short course radiation as a component of definitive multidisciplinary treatment for select patients with metastatic rectal adenocarcinoma. J Gastrointest Oncol. 2017;8(6):990–7.
4. Hyngstrom JR, et al. Intraoperative radiation therapy for locally advanced primary and recurrent colorectal cancer: ten-year institutional experience. J Surg Oncol. 2014;109(7):652–8.
5. Ikoma N, et al. Impact of recurrence and salvage surgery on survival after multidisciplinary treatment of rectal cancer. J Clin Oncol. 2017;35(23):2631–8.
6. Malakorn S, et al. Who should get lateral pelvic lymph node dissection after neoadjuvant chemoradiation? Dis Colon Rectum. 2019;62(10):1158–66.
7. Rickles AS, et al. High rate of positive circumferential resection margins following rectal cancer surgery: a call to action. Ann Surg. 2015;262(6):891–8.
8. Massarweh NN, et al. Risk-adjusted pathologic margin positivity rate as a quality indicator in rectal cancer surgery. J Clin Oncol. 2014;32(27):2967–74.
9. Patel SH, et al. Circumferential resection margin as a hospital quality assessment tool for rectal cancer surgery. J Am Coll Surg. 2020;230(6):1008–18.e1005.
10. Paquette IM, et al. Impact of proximal vascular ligation on survival of patients with colon cancer. Ann Surg Oncol. 2018;25(1):38–45.
11. Rodriguez-Bigas MA, et al. Multidisciplinary approach to recurrent/unresectable rectal cancer: how to prepare for the extent of resection. Surg Oncol Clin N Am. 2010;19(4):847–59.
12. Sammour T, et al. Oncological outcomes after robotic proctectomy for rectal cancer: analysis of a prospective database. Ann Surg. 2018;267(3):521–6.
13. Silberfein EJ, et al. Long-term survival and recurrence outcomes following surgery for distal rectal cancer. Ann Surg Oncol. 2010;17(11):2863–9.
14. van der Valk MJM, et al. Erratum to "Compliance and tolerability of short-course radiotherapy followed by preoperative chemotherapy and surgery for high-risk rectal cancer—results of the international randomized RAPIDO-trial" [Radiother Oncol. 2020; 147:75–83]. Radiother Oncol. 2020;147:e1.
15. Keane C, et al. International consensus definition of low anterior resection syndrome. Dis Colon Rectum. 2020;63(3):274–84.

Mesenteric Resection in Crohn's Disease

39

Sherief Shawki and Steven D. Wexner

Introduction

While current armamentarium of medications play a substantial role in Crohn's disease (CD) remission and maintenance, still significant number of patients will require operative intervention when their disease is refractory to medical treatment, progress while on medications, develop complications or become end stage fistulizing and/or stricturing disease.

Although CD is mainly an intestinal disease. However, recent interest in the bowel mesentery in general, and especially in the extra-intestinal mesenteric manifestations that is tightly coupled with intestinal pathological features drew the attention to a potential role of the mesentery in CD. It has been proposed that the

S. Shawki (✉)
Department of Colon and Rectal Surgery, Cleveland Clinic, 9500 Euclid Ave, Desk A30, Cleveland, OH 44120, USA
e-mail: shawkis@ccf.org

S. D. Wexner
Department of Colorectal Surgery, Cleveland Clinic Florida, Weston, FL, USA

Cleveland Clinic Lerner College of Medicine of Case Western Reserve University, Cleveland, OH, USA

Charles E. Schmidt College of Medicine, Florida Atlantic University, Boca Raton, FL, USA

Herbert Wertheim College of Medicine, Florida International University, Miami, FL, USA

Ohio State University Wexner College of Medicine, Cleveland, OH, USA

Department of Surgery, University of South Florida Morsani College of Medicine, Tampa, FL, USA

Division of Surgery and Interventional Science, Department of Targeted Intervention, University College London, London, UK

E. D. Ehrenpreis et al. (eds.), *The Mesenteric Organ in Health and Disease*,
https://doi.org/10.1007/978-3-030-71963-0_39

bowel mesentery has its own substantive anatomic entity, as well as physiologic distinctiveness that it is no longer regarded as bystander tissue in human body. Being highly involved and active in local and systemic inflammatory response increased its recognition of centrality in homeostasis in disease process.

CD has a recurrent nature which results in a second surgical resection in about 40% of patients. It had been advocated that extended resection of the involved mesentery, regardless of its role primary or secondary, as opposed to limited resection may result in reduced recurrence of the disease. However, dealing with CD mesentery is challenging and associated with risk of bleeding. The recurrent nature also prompts bowel conservation resection. Some concerns exist related to extended mesenteric resection and associated more bowel resection than needed.

This chapter will review the main aspects evaluating the involvement of the intestinal mesentery in Crohn's disease and assess whether the mesentery has an independent net negative pathologic role in the disease process. Subsequently technical aspects of mesenteric resection in CD will be discussed.

The mesentery had traditionally been considered a "bystander" in CD. However, some authors have proposed that morphologic, and functional abnormalities in mesenteric structures may contribute to intestinal inflammation in CD. This bidirectional relationship could explain involvement of the mesentery in disease initiation and conceptually to disease recurrence.

Gross Features of the Bowel and Mesentery in Crohn's Disease

Gross examination of both diseased intestine and its associated mesentery provides increasing evidence that the mucosal and mesenteric diseases are topographically correlated in Crohn's disease. Where the disease is located in the intestine, there are associated changes in the respective mesentery including increase in visceral adipose tissues and fat wrapping. The creeping fat occurs when mesenteric fat extends onto the surface of the intestine. The mucosal ulcerations are usually confined to the mesenteric side of the intestinal lumen. These findings correlate with degree of disease severity. At the margin of the diseased segment the mesentery changes in a graduated manner back towards its normal pattern. A similar corresponding transition zone exists on the intestinal lumen where the mucosal disease ceases. These changes occur immediately opposite the mesenteric transition zone. Thus, the topographic coupling of mesenteric and intestinal disease points to a pathobiologic link.

Histopathologic Features of the Bowel and Mesentery in Crohn's Disease

Similar to the gross coupling of pathological changes in both mesentery and intestine, a similar correlation exists on the microscopic level where significant mesenchymal abnormalities in the mesentery closely resemble those observed in the submucosa. The histologic examination of the normal mesentery demonstrates a contiguous surface mesothelium overlying the mesentery. Underneath, there is a supportive connective tissue layer of variable thickness. This layer provides further connective tissue septations that compartmentalize the mesentery. At the interface between the mesentery and the intestines, the mesenteric connective tissue is contiguous with the intestinal serosa. In fact, there are additional connective tissue septations that further penetrate the underlying longitudinal muscular layer of the intestine.

In CD, the diseased mesentery is characterized by hyperplasia of both adepocytes and mesothelial cells as well as thickening of the connective tissue framework. These changes continue to occur at the intestinal-mesenteric interface and continues through the submucosa where there is a strikingly similar histologic features including adipocyte hyperplasia and connective tissue thickening. This bipolar pattern of mesenchymal abnormalities distribution might contribute to the transmural nature of the condition. This suggests that the transmural features of Crohn's disease are contributed to from both within at the submucosa level and from without at the mesentery level.

Mesenteric Adipose Tissue and Crohn's Disease

Mesenteric fat hypertrophy and fat wrapping have been recognized as indicators of severe Crohn's disease. The effect of visceral fat on the postoperative course in CD was studied. Mesenteric adipose tissue could contribute to intestinal and systemic inflammation by secretion of cytokines and adipokines. Increased mesenteric fat has been shown to correlate with CD activity and is associated with an increased risk of postoperative recurrence, suggesting that mesenteric fat is involved in the disease process of Crohn's disease.

It had been demonstrated that in the mesentery of patients with Crohn's disease, there is an imbalance between CD3+ T cells and CD14+ myeloid cells with significant deviation towards the myeloid population associated with an altered proinflammatory phenotype of macrophages. The presence of these changes irrespective of a defunctioning stoma, suggests a mesentery-specific rather than a reactive inflammation-induced phenotype.

Preliminary data extracted from CD patients, who underwent ileocolic resection or proctectomy, demonstrated a gradient in the mesentery where very few regulatory macrophages present surrounding the severely inflamed bowel, but increasing proportions of these cells exist towards the normal resection margin and central vasculature. This microscopic examination reflects a cutoff zone beyond which the advantages of further extended mesenteric resection does not outweigh the disadvantages of tissue damage.

Lymphatic System, Mesenteric Lymph Nodes Granulomas and Crohn's Disease

Mesenteric lymph nodes, present along the entire gastrointestinal tract, receive antigens from the bowel via lymphatic vessels. Another important function of this mesenteric lymphatic system is to regulate the pressure of the interstitial fluid by transporting excess fluid back into the circulation. This physiologic feature requires patent lymphatic vessels and functioning nodes. Lymphatics are heavily involved in immune cell–driven chronic intestinal inflammation by promoting immune cell egress which is consistent with the speculation that lymphangitis may be a cause or a consequence of CD.

The association between the presence of granulomas and the recurrence of CD remains controversial. Non caseating granuloma, identified in between 9 and 60% of patients with CD has long been considered as a histologic hallmark for CD. While the prognostic significance of the granuloma remains unknown, many investigators have attempted to link the presence of granulomas with postoperative recurrence and disease severity.

One study aimed to assess the predictive value of the presence of granulomas in MLN as well as in bowel wall for postoperative recurrence of CD. Surgical pathology reports of patients who underwent index ileocolic resection were reviewed and presence of granuloma in the bowel wall (29.4%) and in the mesenteric lymph nodes (11.9%) was confirmed. The study found that the presence of granulomas in the intestine per se was not associated with recurrence of CD. However, the presence of granulomas in MLN was an independent predictor for both endoscopic recurrence mesentery is thickened, edematous CD after the index ICR and ICA.

There had been interest in correlation between lymphatics and CD. One of the histologic features noted in CD is the obstructive lymphocytic lymphangitis and increased lymphatic vessel density. Decreased lymphatic vessel density had been also shown to be associated with a high risk of postoperative endoscopic recurrence suggesting that perhaps the dysfunction of lymphatic flow in the gut is somehow involved in recurrence of CD. The presence of granulomas in CD had been found to be associated with abnormal lymphatic vessels.

Based on these observations, granuloma formation within the mesenteric lymph nodes may interfere with lymphatic flow. This problem may lead to obstructed short segments of the regional lymphatics, which might be associated with segmental intestinal diseases such as CD. Mesenteric adipose tissue alterations in Crohn's disease are associated with lymphatic system dysfunction.

The mesenteric adipose tissue transformations in CD seems to be associated with the pathologic changes in the lymphatic system. This relation between mesenteric adipose tissue hyperplasia and lymphatic system in CD has been explored. Disruption of lymphatic system including lymphangiogenesis, lymphatic vessel dilation and obstructed lymphatic vessels and loss of tight junctions proteins have been demonstrated. This cascade results in a reduction in lymphatic drainage. Furthermore, the lymphatic vessels were discontinuous and/or ruptured with resultant leakage of antigens, digested lipids, and inflammatory cells into the mesentery promoting adipose tissue proliferation. The locally release cytokines from leaking lymphatics and irritated mesenteric adipose tissues further augments the inflammatory reaction in the mesentery.

Timing of Mesenteric Events in CD

Timing of mesenteric changes in CD in regards to the sequence of intestinal pathologic events is crucial. Based on the "mucosa first" theory, and the traditional believe that CD is mainly a luminal disease, then mesenteric alterations are secondary. However, there had been emerging radiographic data suggesting that mesenteric changes may occur early in the disease process. MRI-based imaging studies demonstrated some mesenteric abnormalities in the absence of endoscopic mucosal disease. The detected abnormal mesenteric hypervascularity was found to be predictive of subsequent development of CD.

Risk Factors for Recurrence and Extended Mesenteric Resection

After surgical resection, factors associated with postoperative recurrence include young age at diagnosis, penetrating disease type, tobacco smoking, extensive ileal disease/resection, short time interval between diagnosis and need of first surgery. Furthermore, the presence of family history of inflammatory bowel disease (IBD) and active smoking at the time of the index ileocolonic resection (ICR) are associated with an increased risk of the second ICR. However, the type and extent of mesenteric resection had not been recognized.

One study aimed to correlate limited versus inclusive mesenteric resection and rate of surgical recurrence during ileocolic resection in CD. Group A, n = 30, underwent conventional ileocolic resection with a limited mesenteric resection and the mesentery was divided flush with the intestine. Group B, n = 34, on the other side underwent corresponding mesenteric excision. Interestingly the cumulative reoperation rates were 40% and 2.9% respectively, P = 0.003. The lymph node yield was higher in group B, [12 vs. 2.4, P = 0.002], and indeed, surgical technique was an independent factor in determining outcome. Furthermore, advanced mesenteric disease predicted surgical recurrence. Based on the above discussion, it seems that endoscopic and surgical recurrence of CD can be reduced by mesenteric-based surgical strategies in this benign disease.

Mesenteric Resection in Crohn's Disease

The normal, non-inflamed, mesentery is soft and easy to handle surgically, in both the open and laparoscopic scenarios. Its vascular and avascular regions can be identified and differentiated. Hence a safe resection with secured control of the vascular pedicle and its arcades can generally be accomplished.

In CD, the mesentery is thickened, edematous, and stiff yet the tissues are very fragile. Due to the severe inflammation there is high arterial inflow and venous congestion associated with opening of the vascular collaterals. Collectively, is challenging if not impossible, to differentiate vascular and avascular mesenteric territories. Furthermore, in severely inflamed mesentery, the thickened tissues do not only interfere with proper vascular control but also with secured ligation. Because of these features, many surgeons tend to divide the mesentery flush with the intestine avoiding digression into the body of the mesentery to reduce the risk of intraoperative hemorrhage.

Occasionally, the mesentery will be foreshortened secondary to the chronicity of inflammation. This results in altering the native anatomy, mainly bringing the main blood vessels toward the specimen. Extreme caution should be exercised during vascular control to avoid inadvertent ligation of main blood vessels at the base of the mesentery. Should bleeding ensue, blind control by clamping mechanism may result in injuring important nearby structures such as duodenum, ureters, or if it is in small bowel mesentery the superior mesenteric bundle. In order to avoid such dreaded situation, proper mesenteric mobilization should be achieved, and anatomy of the specimen should be properly identified followed by deciding on the outline of mesenteric resection.

Currently, the majority of index CD surgeries are laparoscopically conducted. If the mesentery is not thickened, then mesenteric resection can be conducted intra-corporeally or extra-corporeally. This step may be undertaken using energy devices. However, avoiding usage of these devices in thick mesentery is advised for the above-mentioned reasons.

In the settings in which the mesentery laparoscopic full mobilization followed by extraction and mesenteric division may be preferred. The appropriate incision is crucial for accessibility and proper vascular control. Caution should be exercised to avoid traction on the specimen as force may easily cause bleeding at the base of the fragile easily torn mesentery, further complicating the situation.

In the bulky, fragile, and inflamed mesentery, where it is very difficult to dissect around the mesenteric vessels and bleeding could easily occur, the author prefers to use overlapping crushing technique using Kocher clamps. In this technique, a series of Kocher clamps are placed on the body side of the resection line in an overlapping fashion starting at the bowel transection point and heading to the base of mesenteric resection. Kelly clamps are placed on the specimen side. After placement of the first Kocher clamp, a larger size, and longer Kelly clamp is placed, on the specimen side, so that its tip surpasses the Kocher clamp. The mesentery is divided with a scissors to just before the tip of the Kocher. Subsequently, another Kocher is placed thus overlapping the first one and extend beyond the tip of the Kelly clamp. Then another Kelly clamp is placed. This process is continued until the mesenteric resection point is reached safely. Then, the mesentery is secured with overlapping suture ligature. These sutures include part of the tissues secured by the nearby Kocher clamp. In this way, it is guaranteed that all blood vessels (main and collaterals) are secured with the least manipulation of the mesentery and avoid blind penetration of the mesentery to place clamps.

Conclusion

It seems that there is a link between the mesentery and the intestine in CD. With the evolution of the intestinal disease there are associated concomitant mesenteric pathologic changes. These changes occur in a topographic relationship within the distribution of the disease in the bowel.

While it is believed that any mesenteric changes are primarily protective, emerging data are suggestive that mesenteric pathologic changes associated with CD could potentially be pathogenic. Specifically, leaving a residual diseased/inflamed mesentery could result in retained focus of inflammation acting as the culprit for early post-operative CD disease recurrence.

However, the current situation of mesenteric events in the context of pathologic evolution of CD and its recurrence remains under debate and further randomized studies are currently underway to investigate its role in recurrence and whether extended mesenteric resection in a way similar to oncologic resection is beneficial and associated with decreased recurrence.

Suggested Reading

1. Coffey JC, O'Leary DP. The mesentery: structure, function, and role in disease. Lancet Gastroenterol Hepatol. 2016;1:238–47.
2. Coffey JC, O'Leary DP, Kiernan MG, Faul P. The mesentery in Crohn's disease: friend or foe? Curr Opin Gastroenterol. 2016;32:267–73.

3. Desreumaux P, Ernst O, Geboes K, et al. Inflammatory alterations in mesenteric adipose tissue in Crohn's disease. Gastroenterology. 1999;117:73–81.
4. Coffey JC, O'leary DP. Defining the mesentery as an organ and what this means for understanding its roles in digestive disorders. Expert Rev Gastroenterol Hepatol. 2017;11 (8):703–5.
5. Li Y, Stocchi L, Liu X, Rui Y, et al. Presence of granulomas in mesenteric lymph nodes is associated with postoperative recurrence in Crohn's disease. Inflamm Bowel Dis. 2015;21:2613–8.

40 Mesenteric Resection in Crohn's Disease

Tara M. Connelly, Shoaib Ashfaq, and J. Calvin Coffey

Introduction

Until recently, the requirement for surgery in Crohn's disease had remained largely unchanged. In addition, the requirement for re-operation was unchanged. Mesenteric-based surgical treatment strategies emerged and became the standard approach in the management of colon and rectal cancer. The same has not occurred for the management of conditions such as Crohn's disease. The convention in Crohn's disease is to remove the intestine and leave the mesentery with which the intestine was continuous.

In tandem with the development of mesenteric based surgical treatment strategies in other areas of colorectal surgery, our understanding of the mesentery improved significantly. It is now recognized that the mesentery is one structure and that it connects all digestive organs with the rest of the abdomen. In turn the complex of mesentery and abdominal digestive organs, linked across the mesentery, determines the peritoneal landscape within the abdomen. This realization meant that the mesenteric-based principles of colon and rectal surgery could be applied at all levels of the abdominal digestive system. Together, these developments prompted the inclusion of the mesentery during resections for Crohn's disease, and an assessment of the associated outcomes.

One of the most reliable measures of the success or failure of a treatment in surgery is the requirement for re-operation. Rates of reoperation in Crohn's are up to 9% in the first three months after surgery. Lifetime postoperative recurrence rates are up to 80%. In general, most people do not wish to undergo an operation. This applies in particular to patients with Crohn's disease as these are young patients, in the prime of their lives, embarking on establishing independent careers. Crohn's

T. M. Connelly · S. Ashfaq · J. Calvin Coffey (✉)
School of Medicine and Department of Surgery, University of Limerick Hospital Group, Raheen, Dooradoyle, Limerick, Ireland
e-mail: calvin.coffey@ul.ie

E. D. Ehrenpreis et al. (eds.), *The Mesenteric Organ in Health and Disease*,
https://doi.org/10.1007/978-3-030-71963-0_40

disease generally tends to affect people in that particular age bracket and so the rate of surgery is a particularly reliable indicator of the stage of the disease.

A further point of importance is that surgery does not cure Crohn's disease. Following surgery, the pathobiologic process involved in the first instance recommences. Patients develop symptoms and ultimately may develop complications. This means that the disease appears to recommence and then progress. The best that we can hope for at the present is to slow or halt disease progression in the postoperative period.

Inclusion of the Mesentery During Ileocolic Resection

Long-Term Outcomes

A recent study by our group included the mesentery as part of resections in patients with Crohn's disease of the ileocolic region. Outcomes in this group were compared with patients in whom the mesentery had not been included in the resection. We found that re-operation was required in 2.5% of this cohort of patients. This contrasts with a re-operation rate of 40% in patients undergoing a conventional resection. This difference was statistically significant. Importantly, other factors such as smoking can also influence the requirement for re-operation. When we controlled for these factors in a multivariate analysis, the type of surgery (i.e. whether the mesentery was included or not) was retained as a factor affecting the rate of reoperation. This means that the surgical technique employed is an independent predictor of the requirement for re-operation.

Short Term Outcomes

The next question to ask is is it safe to remove the mesentery? One could claim that removal of the mesentery is a more radical operation than the conventional, because more tissue is removed. Alternatively, one could conclude that it is less radical, as one is detaching the mesentery along the embryological planes on which it was laid down in the first instance. During the conventional operation one is disrupting the intersection between the mesentery and the intestine, a zone that is highly vascular and full of nerve endings and lymphatic channels. This does not apply during controlled division of the fully mobilized mesentery.

We also examined the rate of short term complications after resection of the mesentery and found that this was not increased in comparison to the rate seen after conventional surgery. In addition, we noticed that some complications occurred at a reduced rate in the group in which the mesentery was removed, compared with the conventional cohort. This data suggests that the short term (i.e. postoperative) complications associated with inclusion of the mesentery may be less than those seen with conventional surgery. This is potentially explained by the more embryological and less disruptive nature of mesenteric resection.

Indications for Inclusion of the Mesentery During Surgery for Crohn's Disease

There are several phenotypes of Crohn's disease. To date, our group has investigated inclusion of the mesentery in most types (with exception of disease involving the upper gastrointestinal tract). We found that it is possible to conduct this type of surgery in patients with Crohn's colitis, in those with small intestinal Crohn's disease, in patients requiring removal of the rectum (proctectomy) and in patients who had an ileal pouch constructed that no requires excision.

These findings do not mean that the mesentery can be included in all surgical cases. We have a cohort of patients who underwent emergency surgery and were found to have disease so advanced it was not possible to remove the mesentery without placing the patient's life at risk. This is an unacceptable risk in a condition in which mortality occurs at a very low rate. In these patients we re-routed intestinal contents away from the diseased intestine, into an ileostomy. All patients recovered well and achieved an excellent quality of life. However, leaving diseased intestine is not recommended as symptoms will recur and worsen, and there is a risk of developing a malignancy with prolonged inflammation. On that basis we performed a staged repeat laparotomy in these patients and were able to remove the diseased intestine at the second operation. We were also able to restore intestinal continuity at the second operation. This means that mesenteric resection is not safe to conduct in a subgroup of patients. In this subgroup, a staged approach to resection may be more appropriate.

The Mesenteric Transition Zone

There are other means by which mesenteric manifestations of disease can be exploited to aid in the surgical management of Crohn's disease. For example, our study examining removal of the mesentery also highlighted the mesenteric transition zone, the zone at which the mesentery changes from normal to abnormal. If the resected specimen is opened and examined at this zone, one finds that mesenteric, mural and mucosal disease all start and increase in tandem at this zone (Fig. 40.1). This means that the surgeon can place the proximal intestinal incision just proximal to where mesenteric changes commence and be sure that all mucosal disease will be resected. The mesenteric transition zone is thus extremely important as it enables the surgeon conserve as much intestine as possible, whilst simultaneously being assured that all disease is being removed.

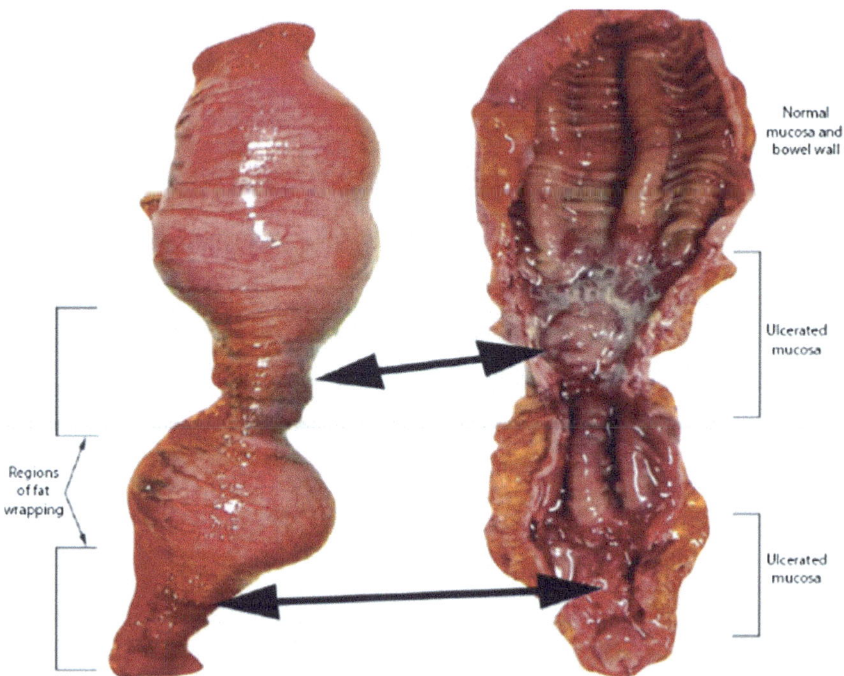

Fig. 40.1 The mesenteric transition zone in Crohn's disease. The zone at which the mesentery changes from normal to abnormal is shown. The start of mesenteric, mural and mucosal disease are demonstrated (Copyright J. Calvin Coffey)

Advanced Mesenteric Disease Predicts Increased Surgical Recurrence

In addition, we also found that the extent to which the mesentery wrapped around the circumference of the intestine (i.e. fat wrapping) predicted the likelihood of postoperative progression requiring reoperation (Fig. 40.2). The greater the amount of fat wrapping the greater the likelihood of a patient requiring re-operation (assuming they had undergone a conventional resection as their index procedure). This finding means that patients can be identified who are increased risk of progressive postoperative disease. This helps gastroenterologists in identifying those patients who might benefit from adjuvant treatments.

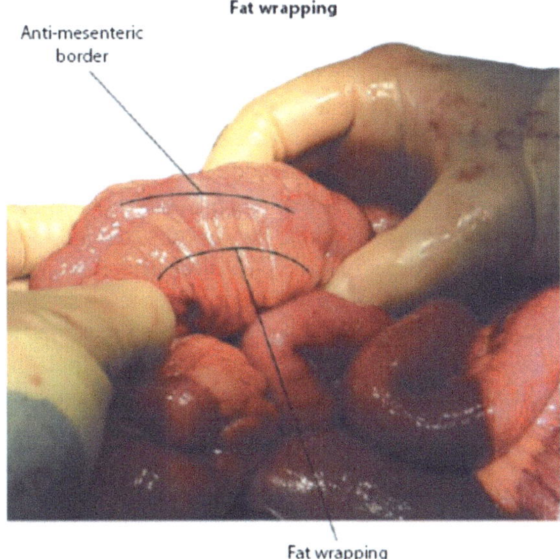

Fig. 40.2 Fat wrapping in Crohn's disease. The extent of fat wrapping on the bowel surface correlates with mucosal disease (Copyright J. Calvin Coffey)

Discussion

The data regarding removal of the mesentery in Crohn's disease can be considered as preliminary which means that it requires confirmation by other investigators and also further investigation. Early reports are emerging from other groups supporting the observation that removal of the mesentery is associated with reduced recurrence in Crohn's disease (Fig. 40.3). In addition, data is emerging that removing the mesorectum during proctectomy for Crohn's disease (i.e. where the rectum is being removed) is associated with improved outcomes.

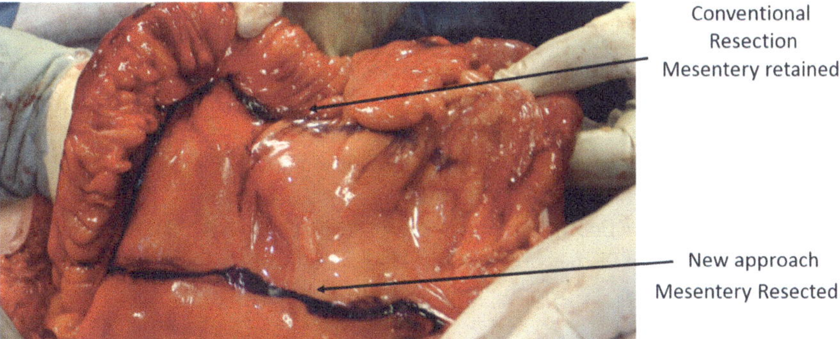

Fig. 40.3 Conventional versus mesenteric based small bowel resection for Crohn's. The purple lines demonstrate the traditional mesenteric sparing, 'non-oncological' Crohn's small bowel resection versus the novel mesenteric based excision

Over the past years several studies that focus on the link between adipocytes, lymphatics and intestinal inflammation have identified mechanistic cellular and molecular signaling pathways that could explain the benefits being seen with removal of the mesentery. Studies suggest variations in signaling pathways in Crohn's involved mesenteric adipose tissue. Supportive evidence has also emerged in relation to another operation for Crohn's disease that is also increasing in acceptance. This is the Kono-S procedure. At first sceptics criticized inclusion of the mesentery in resections for Crohn's disease. They said that data related to the Kono-S procedure (in which the mesentery is retained) indicates that the mesentery does not have to be removed. However, a closer look at the Kono-S will demonstrate that as part of the procedure, an intestinal anastomosis is created that is circumferentially free of mesentery (and any inputs the mesentery may be generate). As such, observations related to the Kono-S support those related to removal of the mesentery.

Others who criticize removal of the mesentery cite reduced recurrence rates in association with stricturoplasty (where the mesentery is retained). However, the fundamental aim of stricturoplasty is to widen the lumen of the intestine. By so doing the circumferential length of the intestine is increased. This means that the zone where the mesentery intersects the intestine is altered relative to the remainder of the circumference, thus altering the effects of mesenteric inputs. So, it can also be argued that the benefits seen with stricturoplasty may also support a role for altering mesenteric inputs.

The above findings signal that we are at an exciting time point in the surgical management of Crohn's disease. Some suggest that the recurrence rates associated with new procedures are such that perhaps we should be considering introducing these earlier in the treatment paradigm, rather than subjecting patients to many years of medical treatments which interrupt their lives considerably and ultimately which only serve to delay the inevitable requirement for surgery.

Conclusion

In conclusion, the mesentery may be safely included in resections in patients undergoing surgery for Crohn's disease. Early data indicate this is associated with reduced rates of progression to requiring reoperation. These findings need to be repeated by independent investigators and then compared with other conventional resection (in which the mesentery is retained) before inclusion of the mesentery becomes standard practice in the managing of Crohn's disease. The findings also point to the benefits of further investigation of disease features that are characteristic of Crohn's disease, i.e. fat wrapping and transmural disease.

Suggested Reading

1. Strong S, Steele SR, Boutrous M, Bordineau L, Chun J, Stewart DB, et al. Clinical practice guideline for the surgical management of Crohn's disease. Dis Colon Rectum. 2015;58 (11):1021–36.
2. Heald RJ, Husband EM, Ryall RD. The mesorectum in rectal cancer surgery–the clue to pelvic recurrence? Br J Surg. 1982;69(10):613–6.
3. Bunni J. The primacy of embryological, ontogenetic and specimen orientated (mesenteric) surgery as the most important tool in treating visceral (colorectal) cancer. Mesentery Peritoneum. 2017;1(3).
4. Brown SR, Fearnhead NS, Faiz OD, Abercrombie JF, Acheson AG, Arnott RG, et al. The association of coloproctology of great Britain and Ireland consensus guidelines in surgery for inflammatory bowel disease. Colorectal Dis. 2018;20(Suppl 8):3–117.
5. Coffey JC, O'Leary DP. The mesentery: structure, function, and role in disease. Lancet Gastroenterol Hepatol. 2016;1(3):238–47.
6. Buskens CJ, de Groof EJ, Bemelman WA, Wildenberg ME. The role of the mesentery in Crohn's disease. Lancet Gastroenterol Hepatol. 2017;2(4):245–6.
7. de Groof EJ, van der Meer JHM, Tanis PJ, de Bruyn JR, van Ruler O, D'Haens G, et al. Persistent mesorectal inflammatory activity is associated with complications after proctectomy in Crohn's disease. J Crohns Colitis. 2018.
8. Mege D, Michelassi F. Readmission after abdominal surgery for Crohn's disease: identification of high-risk patients. J Gastrointest Surg. 2018.
9. Connelly TM, Messaris E. Predictors of recurrence of Crohn's disease after ileocolectomy: a review. World J Gastroenterol. 2014;20(39):14393–406.
10. Fichera A, Lovadina S, Rubin M, Cimino F, Hurst RD, Michelassi F. Patterns and operative treatment of recurrent Crohn's disease: a prospective longitudinal study. Surgery. 2006;140 (4):649–54.
11. Bardhan KD, Simmonds N, Royston C, Dhar A, Edwards CM, Rotherham IBDDUG. A United Kingdom inflammatory bowel disease database: making the effort worthwhile. J Crohns Colitis. 2010;4(4):405–12.
12. Greenstein AJ, Lachman P, Sachar DB, Springhorn J, Heimann T, Janowitz HD, et al. Perforating and non-perforating indications for repeated operations in Crohn's disease: evidence for two clinical forms. Gut. 1988;29(5):588–92.
13. Maconi G, Colombo E, Sampietro GM, Lamboglia F, D'Inca R, Daperno M, et al. CARD15 gene variants and risk of reoperation in Crohn's disease patients. Am J Gastroenterol. 2009;104(10):2483–91.
14. Coffey CJ, Kiernan MG, Sahebally SM, Jarrar A, Burke JP, Kiely PA, et al. Inclusion of the mesentery in ileocolic resection for Crohn's disease is associated with reduced surgical recurrence. J Crohns Colitis. 2018;12(10):1139–50.
15. Louis E, Collard A, Oger AF, Degroote E, El Yafi FAN, Belaiche J. Behaviour of Crohn's disease according to the Vienna classification: changing pattern over the course of the disease. Gut. 2001;49(6):777–82.
16. Rivera ED, Coffey JC, Walsh D, Ehrenpreis ED. The mesentery, systemic inflammation, and Crohn's disease. Inflamm Bowel Dis. 2019;25(2):226–34.
17. Mao R, Kurada S, Gordon IO, Baker ME, Gandhi N, McDonald C, et al. The mesenteric fat and intestinal muscle interface: creeping fat influencing stricture formation in Crohn's disease. Inflamm Bowel Dis. 2018.
18. Peyrin-Biroulet L, Chamaillard M, Gonzalez F, Beclin E, Decourcelle C, Antunes L, et al. Mesenteric fat in Crohn's disease: a pathogenetic hallmark or an innocent bystander? Gut. 2007;56(4):577–83.
19. Wu Z, Tan J, Chi Y, Zhang F, Xu J, Song Y, et al. Mesenteric adipose tissue contributes to intestinal barrier integrity and protects against nonalcoholic fatty liver disease in mice. Am J Physiol Gastrointest Liver Physiol. 2018;315(5):G659–70.

20. Zuo L, Li Y, Zhu W, Shen B, Gong J, Guo Z, et al. Mesenteric adipocyte dysfunction in Crohn's disease is associated with hypoxia. Inflamm Bowel Dis. 2016;22(1):114–26.
21. Kono T, Fichera A, Maeda K, Sakai Y, Ohge H, Krane M, et al. Kono-S anastomosis for surgical prophylaxis of anastomotic recurrence in Crohn's disease: an international multicenter study. J Gastrointest Surg. 2016;20(4):783–90.
22. Fichera A, Zoccali M, Kono T. Antimesenteric functional end-to-end handsewn (Kono-S) anastomosis. J Gastrointest Surg. 2012;16(7):1412–6.
23. Shimada N, Ohge H, Kono T, Sugitani A, Yano R, Watadani Y, et al. Surgical recurrence at anastomotic site after bowel resection in Crohn's disease: comparison of Kono-S and end-to-end anastomosis. J Gastrointest Surg. 2018.
24. Michelassi F, Upadhyay GA. Side-to-side isoperistaltic strictureplasty in the treatment of extensive Crohn's disease. J Surg Res. 2004;117(1):71–8.
25. Bharadwaj S, Fleshner P, Shen B. Therapeutic armamentarium for stricturing Crohn's disease: medical versus endoscopic versus surgical approaches. Inflamm Bowel Dis. 2015;21 (9):2194–213.
26. Campbell L, Ambe R, Weaver J, Marcus SM, Cagir B. Comparison of conventional and nonconventional strictureplasties in Crohn's disease: a systematic review and meta-analysis. Dis Colon Rectum. 2012;55(6):714–26.
27. Connelly TM, Malik Z, Sehgal R, et al. Should surgical intervention become a primary treatment modality in Crohn's disease?—a review of the role of surgery and emerging surgical techniques. Mesentery Peritoneum. 2018;2(2).

Part VIII
Surgery for Individual Conditions

Surgical Management of Intestinal Volvulus

41

Ashley J. Williamson and John C. Alverdy

Definition and Clinical Features

Volvulus refers to the torsion or twisting of a viscera and mainly involves organs of the intestinal track. Organs may twist along two directions: organo-axial (longitudinally along the mesenteric axis) or mesentero-axial (perpendicular to the mesenteric axis). Organo-axial volvulus is the most common type in the adult population. Volvulus involves either the sigmoid colon or cecum in >95% of the cases that occur in the adult population. Rare cases of small intestinal volvulus may also occur. Gastric volvulus, a complication of large hiatal hernias generally seen in the elderly, is beyond the scope of the current discussion. Symptoms of cecal and sigmoid volvulus are consistent with large intestinal obstruction and include abdominal pain, constipation or obstipation, nausea, and occasionally emesis. Abdominal distension commonly is present. Physical examination often demonstrates marked abdominal distension, varying degrees of tenderness, and an empty rectal vault on digital rectal exam. The greatest concern in patients with intestinal volvulus is compromise of the blood supply to the organ. Compromised blood flow to the segment of affected bowel is related to the relative twisting and obstruction of the blood vessels within the mesentery. Diminished blood flow in affected blood vessels may lead to bowel ischemia, visceral perforation and death. In the United States, intestinal volvulus is more common in elderly patients, particularly those with medical co-morbidities, chronic constipation, and neurophysiological impairment. Risk factors for intestinal volvulus are shown in Table 41.1.

Imaging is the most appropriate method for diagnosing intestinal volvulus. Figures 41.1, 41.2 and 41.3 demonstrate imaging for cecal volvulus (Fig. 41.1) and sigmoid volvulus (Figs. 41.2 and 41.3).

A. J. Williamson · J. C. Alverdy (✉)
Department of Surgery, University of Chicago, 5841 S. Maryland, Chicago, IL 60637, USA
e-mail: jalverdy@surgery.bsd.uchicago.edu

E. D. Ehrenpreis et al. (eds.), *The Mesenteric Organ in Health and Disease*,
https://doi.org/10.1007/978-3-030-71963-0_41

Table 41.1 Risk factors for colonic volvulus

Risk factors
Chronic constipation
Elderly patient
Frequent use of laxatives
Previous laparotomy
Anatomic predisopositions

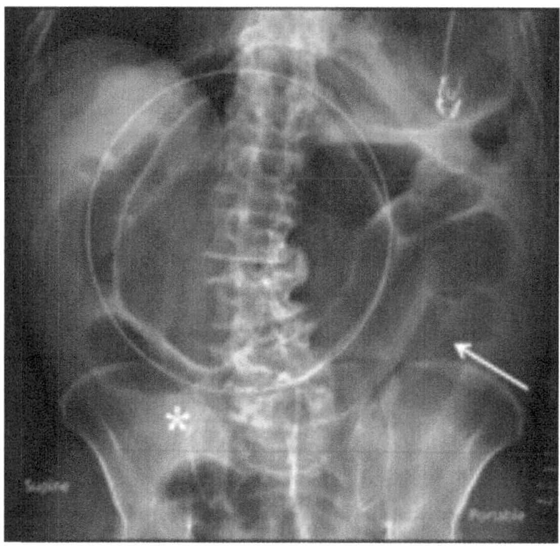

Fig. 41.1 Supine view of the abdomen in a patient with acute abdominal pain. There is a saccular dilated structure in the right mid abdomen (circle) corresponding to a dilated cecum. The cecum should normally be located in the right lower quadrant (asterisk). There is a partial small bowel obstruction. Dilated loops of small bowel are seen best in the left lower quadrant (arrow), secondary to the volvulus. With permission from Dr. Abraham Dachman, Dr. Scott Sorensen and Dr. Justin Ramirez

Description of Surgery

Initial intervention for intestinal volvulus is largely dictated by the location of the volvulus. In sigmoid volvulus, rigid or flexible endoscopy should be employed first in order to evaluate the extent of ischemia of the mucosa. In the absence of mucosal ischemia or colonic perforation, endoscopic detorsion is the initial treatment of choice for sigmoid volvulus and is successful in more than 90% of cases. First, rectal insufflation is performed and may result in detorsion. If the volvulus remains, the endoscope is advanced to the area where the spiraled appearing mucosa reaches a narrowed portion or apex. Gentle pressure is applied to this area, producing the desired effect of release of trapped air and detorsion. A soft tube is then passed

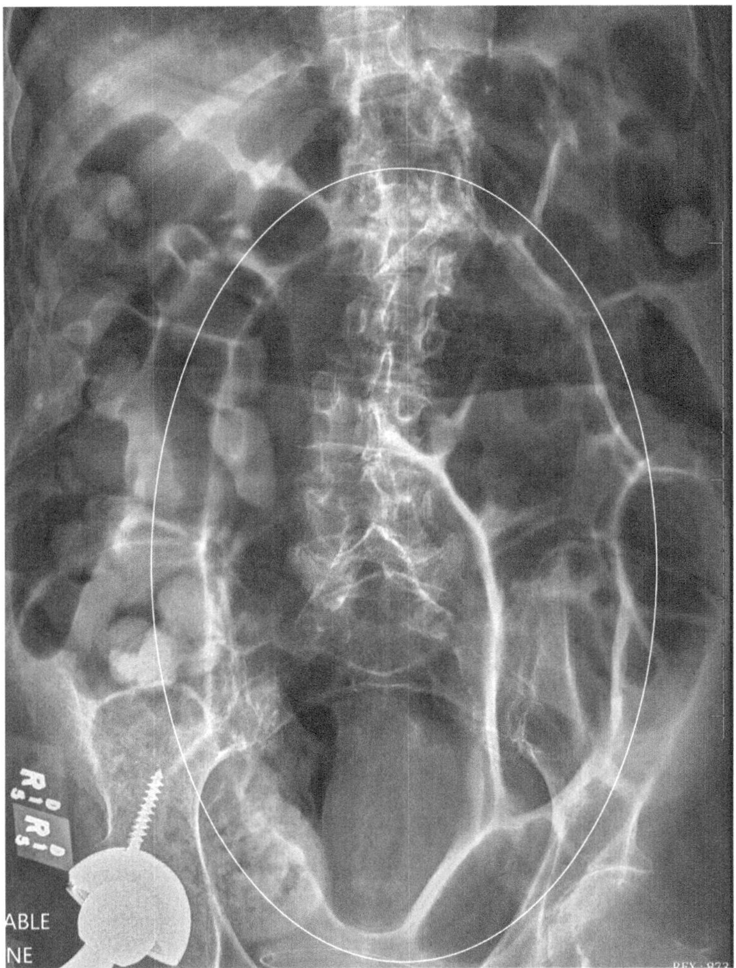

Fig. 41.2 Sigmoid volvulus demonstrated on plain abdominal film. Pathognomonic coffee bean sign outlined by white circle

beyond the site of torsion and left in position. Imaging is used to demonstrate locate the position of the tip of the tube and to confirm the efficacy of the procedure.

The tube is left in place after endoscopic decompression in order to facilitate bowel movement and preparation as indicated. Urgent surgical resection of the sigmoid colon should occur in patients with signs or symptoms of sigmoid ischemia or unsuccessful detorsion of the sigmoid volvulus.

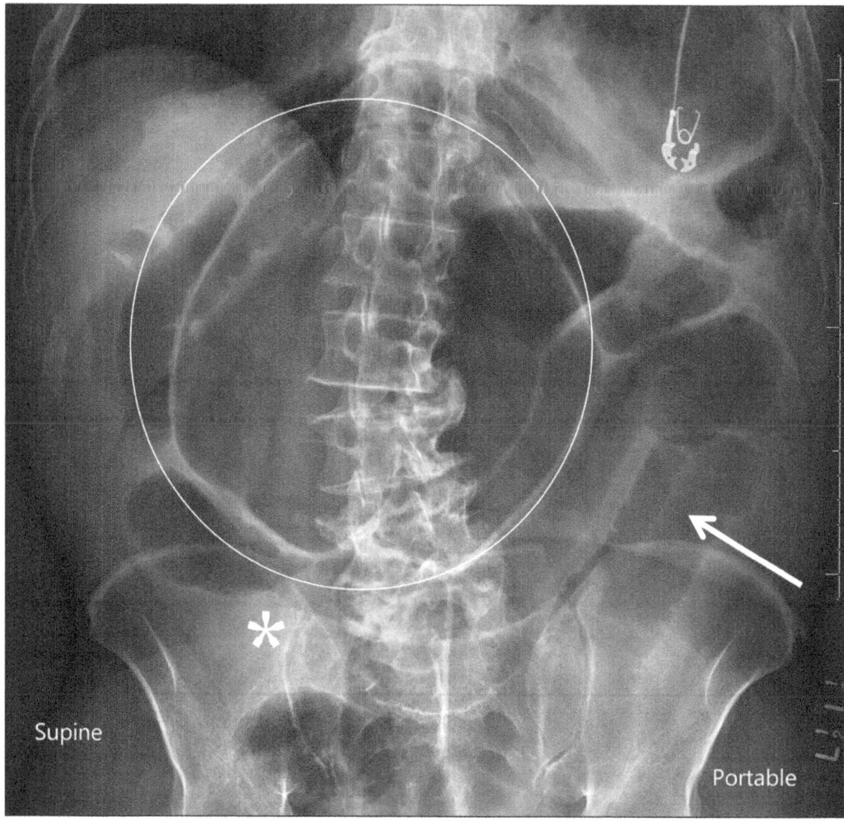

Fig. 41.3 Cecal volvulus demonstrated on plain film with volvulus outlined by white circle

Cecal volvulus is most often managed operatively. Endoscopic techniques produce a favorable result in only 12% of cases. Furthermore, the endoscopic manipulation of cecal volvulus carries the risk for injury to bowel that is already compromised.

Indications for Surgery

Cecal volvulus is an indication for urgent operative intervention. Signs and symptoms of ischemia including peritonitis on physical exam and hemodynamic instability are indications for surgery in patients with sigmoid volvulus. Laboratory and radiographic findings suggestive of the requirement for surgery rather than initial endoscopic decompression include leukocytosis, acute kidney injury, acidosis, free air, bowel wall edema, free fluid within the abdomen and pneumatosis of the bowel wall.

Contraindications

Endoscopic reduction should not be attempted in patients with signs of ischemia as indicated above. Operative contraindications include patients who are unfit to undergo general anesthesia, at which time a detailed goals of care conversation with the patient and family must take place. Patients who are not operative candidates may become candidates for endoscopic decompression.

Preparation of Patient

Upon initial concern for intestinal volvulus, patients should be urgently evaluated by a surgical team. Often, patients require rapid rehydration and adequate IV access should be obtained. Use of a nasogastric tube to decompress proximal dilation is often performed. Electrolyte abnormalities are corrected prior to intervention as release of torsion may result in the release of stored metabolites into the bloodstream. If there is concern for ischemic compromise of the bowel, antibiotic coverage should be chosen to cover gut flora. Accurate measurement of urine production during resuscitation and treatment is performed following insertion of an indwelling bladder catheter.

How the Procedures Are Performed

In cases of cecal volvulus and sigmoid volvulus with signs and symptoms of bowel compromise, the patient should be taken to the operating room for urgent bowel resection. For sigmoid volvulus, patients are most often placed in the supine position; however, a lithotomy position may be considered for ease of endoscopic evaluation. Perioperative antibiotic prophylaxis is indicated given the risk of infection. Most commonly, the procedure is performed as an open exploratory laparotomy as compared to laparoscopy given the severe dilation of the bowel and need to completely evaluate cause of intestinal volvulus (the lead point). The role of minimally invasive laparoscopy in the management is currently under evaluation. A midline incision is made with careful exposure of bowel, taking note of any murky fluid, blood, or air present in the abdomen upon entry. Areas of ischemia or perforation should be identified for segmental resection and bowel anastomosis to areas of well perfused tissue. In general, necrotic bowel should be resected with minimal manipulation and without detorsing in an effort to minimize the release of endotoxins from necrotic bowel after detorsion. The bowel shall be explored in its entirety in an effort to the extent of affected bowel or new diagnoses.

Operation for cecal volvulus involves the performance of laparotomy, bowel exposure and exploration, and identification of non-viable bowel. Segmental resection for the cecum is performed with the understanding that removal of the ileocecal valve may have some long-term outcomes on bowel function.

The most critical step in both operations occurs with the decision of whether to reconnect the bowel with a primary anastomosis with creation of a defunctionalized anastomosis with diverting loop, or an end ostomy. The most common operation described for sigmoid pathology is the *Hartmann's procedure* in which the sigmoid colon is resected, a rectal stump is left in place, and an end colostomy is matured at the skin to proximally divert the fecal stream. Diversion is favored in settings where the bowel is edematous and friable, there is a size mismatch between proximal and distal portions, or the patient is critically ill and requires a shorter operation. Defunctionalize anastomoses involve the creation of an anastomosis, with proximal loop colostomy diversion to eliminate passage of fecal material through the anastomosis while it heals. The decision on which surgical approach is best should be individualized for each patient and their clinical condition.

Specific operations for cecal volvulus involve either an ileocolonic anastomosis or ileostomy. Surgical decision making is the same as described above, with the understanding that ileostomies present a higher risk for dehydration and increased maintenance with liquid stool being present when compared to colostomy.

Typical Abnormal Findings

Typical findings including a massively dilated large bowel with potential small bowel dilatation. The bowel is often friable from edema, and rates of ischemia vary with location. Cecal volvulus is at greatest risk for ischemia at the time of diagnosis, with a reported rate of nonviable or gangrenous cecum occurring in up to 44% of patients in medical literature reviews.

Alternatives to Surgery

If a sigmoid volvulus is able to be successfully reduced, it is recommended that the patient been seen for follow up for an elective sigmoidectomy. These surgeries involve removal of sigmoid redundancy in its entirety to reduce risk of recurrent sigmoid volvulus. If the patients have megacolon (that may be primarily present or may occur secondary to chronic constipation), a sub-total colectomy is considered.

Operative detorsion, or detorsion with fixation of the sigmoid to the abdominal peritoneum (abdominal pexy) represents an alternative to other forms of operative intervention. These procedures are viewed less favorably given their increased recurrence rates of volvulus as compared to operative resections. Endoscopic fixation or pexy onto the abdominal wall in patients unfit to undergo general anesthesia or a major operation have been described but the long-term outcome of this procedure is not well known at this time.

When viable bowel is present in patients with cecal volvulus, operative detorsion, detorsion with cecopexy, and cecostomy are potential alternatives to cecal resection. Studies have evaluated cecopexy or detorsion alone in patients with cecal volvulus and have been shown the development of fewer wound infections, mortality rates of 10–13%, and recurrent volvulus in 12–13% of cases. Cecostomy decreases recurrent cecal volvulus to approximately 14% of patients but overall morbidity with this procedure is over 50%.

Outcomes

Largest trials report success rates of 60–95% of endoscopic detorsion for sigmoid volvulus. Recurrent sigmoid volvulus during admission is reported in approximately 4% of patients, while long term recurrence noted in 43–75% of patients.

Operative resection remains the gold standard for cecal volvulus. Abdominal and wound complications approach 28% in reviews of the procedure, whereas recurrent volvulus after cecal resection is an exceptionally rare event.

Complications

Hartmann's procedure and colonic resection with primary anastomosis remain the two most common surgical approaches for these patients, with Hartmann's procedures traditionally occurring in patients who are more seriously ill. Due to their performance in clinically compromised patients, Hartmann's procedures are not surprisingly found to have more postoperative complications and mortality when compared to primary anastomosis operations (8% vs. 5%). Anastomotic leak is reported in approximately 7% of cases of patients undergoing primary anastomosis. Of note, creation of a colostomy creates an increased socio-economic burden on patients. This includes the need for ongoing post-operative care and the purchase of ostomy supplies. While dehydration is uncommon in patients with a colostomy compared to those having an ileostomy, economic factors should be considered procedure selection.

The presence of ischemia during cecal volvulus is an independent risk factor for mortality when surgical resection is required. A cecal volvulus with evidence of ischemia has reported rates of mortality between 31 and 44% given that patients are often frail, elderly and present in emergent conditions.

Summary

While the Hartmann's procedure has been traditionally viewed as a procedure with high associated complications, more recent data suggest that it safe and effective in complex situations such as sigmoid volvulus. Rapid diagnosis and treatment is critical in all patients with suspected intestinal volvulus. As many patients are frail and elderly who present with volvulus, goals of care should be discussed with the patient and their caregivers.

Suggested Reading

1. https://www.fascrs.org/sites/default/files/downloads/publication/clinical_practice_guidelines_for_colon_volvulus.pdf.
2. Vogel JD, Feingold DL, Sterward DR, Turner JS, Boutros M, Chun J, Steele SR. Clinical practice guidelines for colon volvulus and acute colonic pseudo-obstruction. Dis Colon Rectum. 2016;59:589–600.
3. Gingold D, Murrell Z. Management of colonic volvulus. Clin Colon Rectal Surg. 2012;25:236–44.
4. Bruzzi M, Lefevre JH, Desaint B, et al. Management of acute sigmoid volvulus: short-and long-term results. Colorectal Dis. 2015;17:922–8.
5. Mabdida TE, Thomson SR. The management of cecal volvulus. Dis Colon Rectum. 2002;45:264–7.
6. Kuzu MA, Aslar AK, Soran A, Polat A, Topcu U, Hengirmen S. Emergent resection for acute sigmoid volvulus: results of 106 consecutive cases. Dis Colon Rectum. 2002;45:1085–90.
7. Perrot L, Fohlen A, Alves A, Lubrano J. Management of the colonic volvulus. J Visc Surg. 2016;153(3):183–92.

Embryologic Abnormalities of the Mesentery

42

Sara Gaines, Thomas Q. Xu, Richard A. Jacobson, and John C. Alverdy

Background

Starting as early as week five of gestation, the intestines of the midgut undergo three distinct events that are critical to normal intestinal fixation: (1) herniation of the midgut intestinal loop into the umbilical cord, (2) reduction of the herniated contents back into the abdomen, and (3) fixation of the intestines to the posterior body wall. During these three events, counterclockwise rotation around a vascular axis is completed followed by migration of the proximal and distal portions of the midgut to their normal adult anatomic positions to fix the intestines on a broad-based mesentery. As a result, the jejunoileal mesentery extends down obliquely from the ligament of Treitz to the ileocecal valve (Fig. 42.1). For additional information, see Chap. 2, Embryology of the Mesentery.

Malrotations and Indications for Surgery

Embryologic abnormalities of the mesentery that require surgical intervention are most commonly related to abnormal positioning of the bowel within the peritoneal cavity resulting from failure of any of the above 3 steps. These abnormalities are categorized under various forms of malrotation. Malrotation is thought to affect 1:500 live births with the majority becoming symptomatic within the first year of life. Classic malrotation presents as bilious emesis in infancy from duodenal obstruction or midgut volvulus, and it is usually associated with abnormal or absent intestinal fixation by mesenteric bands. The bands, known as Ladd bands, fix the

S. Gaines · J. C. Alverdy (✉)
Department of Surgery, University of Chicago, 5841 S. Maryland, Chicago, IL 60637, USA
e-mail: jalverdy@surgery.bsd.uchicago.edu

T. Q. Xu · R. A. Jacobson
Department of Surgery, Rush University Medical Center, Chicago, IL, USA

E. D. Ehrenpreis et al. (eds.), *The Mesenteric Organ in Health and Disease*,
https://doi.org/10.1007/978-3-030-71963-0_42

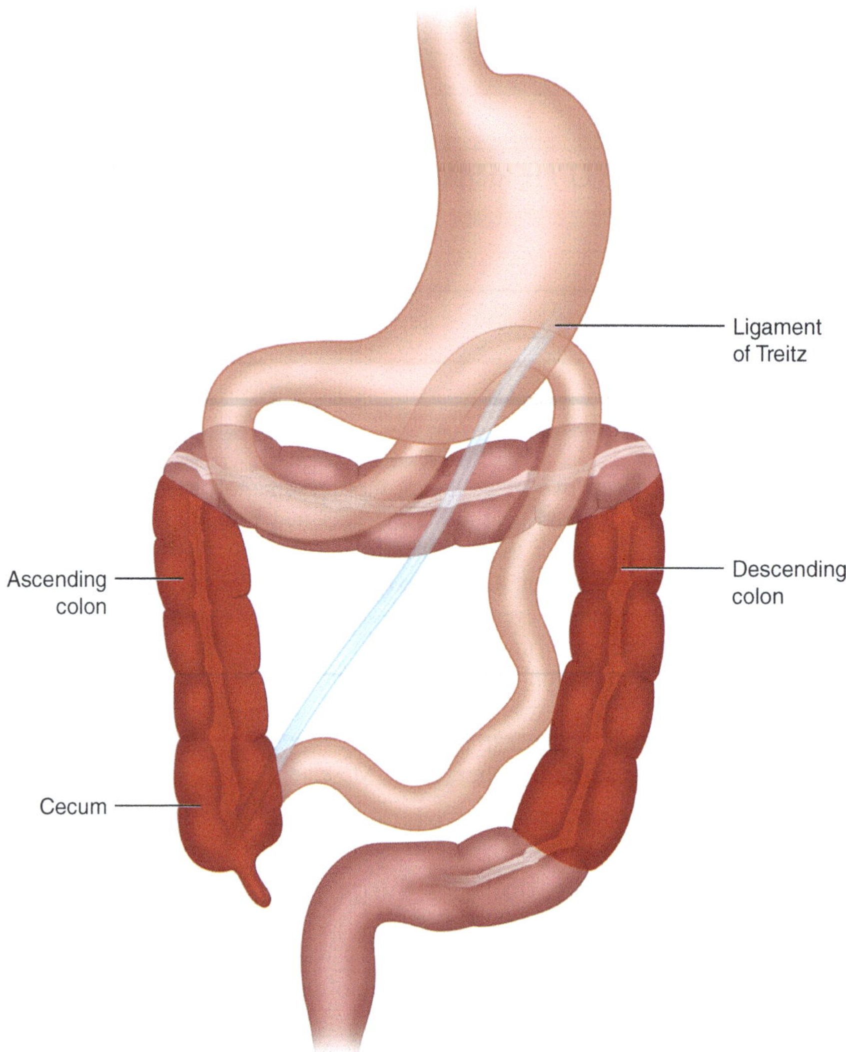

Fig. 42.1 Normal oblique fixation of the midgut mesentery at the ligament of Treitz to the cecum in the right lower quadrant

cecum to the posterior body which traps the duodenum leading to obstruction. The duodenojejunal attachment is abnormally short which makes the gut prone to twist counterclockwise around the superior mesenteric vessels leading a volvulus. An upper GI series is the standard of care for diagnosis and will show a variation in the normal positioning of the duodenojejunal junction (Figs. 42.2, 42.3 and 42.4). It has a sensitivity of 96% to detect malrotation and 79% sensitivity for midgut volvulus.

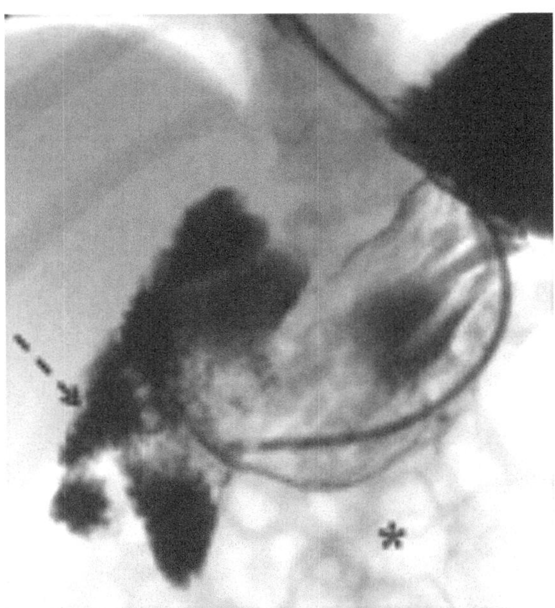

Fig. 42.2 This fluoroscopic image from a patient with malrotation is a single-contrast upper-GI exam. The exam shows dense barium within the stomach and proximal small bowel loops in the right upper abdomen (arrow) with a lack of the expected course of the duodenum to the ligament of Treitz in the left upper quadrant (asterisk). Note that, by convention, fluoroscopic images have the grey-scale inverted compared with regular X-ray

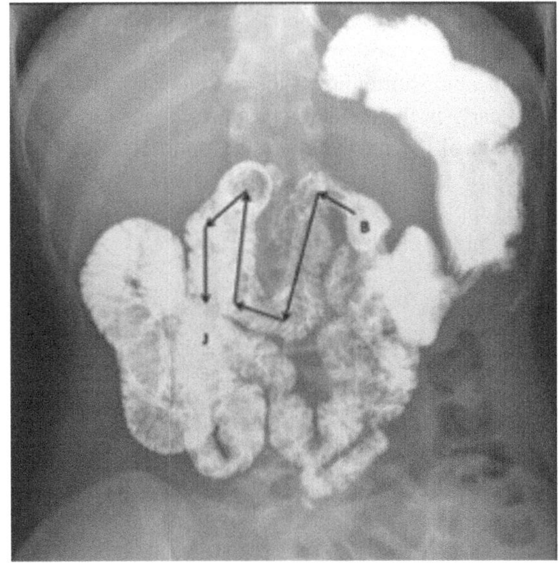

Fig. 42.3 Radiographs from a different patient during a small bowel examination again demonstrate the atypical course of the duodenum (arrows) with duodenojejunal junction positioned to the right of midline. With permission from Dr. Abraham Dachman and Dr. Justin Ramirez

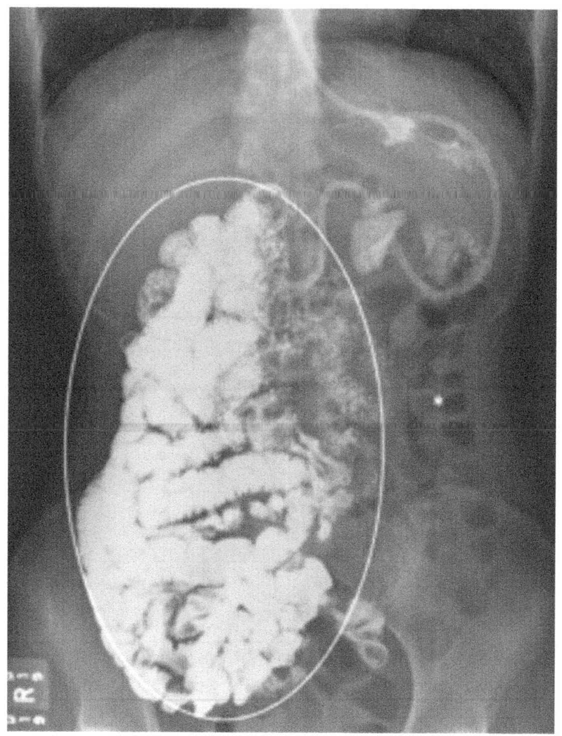

Fig. 42.4 A radiograph obtained later in the examination demonstrates progression of barium contrast into the remainder of the small bowel (oval) which is positioned entirely to the right of midline. Non-opacified, air-filled colon (asterisk) can be seen within the left abdomen. With permission from Dr. Abraham Dachman and Dr. Justin Ramirez

Mesocolic hernias may also occur when the right or left mesocolon fail to fix to the posterior body wall. These hernias carry a risk of small bowel obstruction, incarceration or strangulation. A less common cause of an embryologic mesenteric abnormality relates to an intrauterine vascular accident leading to atresia of the intestine and shortening of the mesentery. Intestinal atresia is caused by an ischemic event early in the developing gut. In a rare variant known as the "apple peel", there is atresia of the proximal jejunum and absence of the distal SMA and mesentery. The apple peel appearance results from a spiraling of the intestine around its vascular supply (Fig. 42.5). Infants born with this condition may suffer from necrotizing enterocolitis shortly after birth.

Surgeries for Malrotations and Mesocolic Hernias

Malrotation or hernias that cause obstruction are treated with an urgent laparotomy to reduce the chance of ischemic injury. Midgut volvulus that has progressed to complete infarction is rapidly lethal. Malrotation is treated with the Ladd procedure. The first step is to relieve the volvulus by rotating the bowel in a counterclockwise direction (Fig. 42.4). The peritoneal attachments between the cecum and the

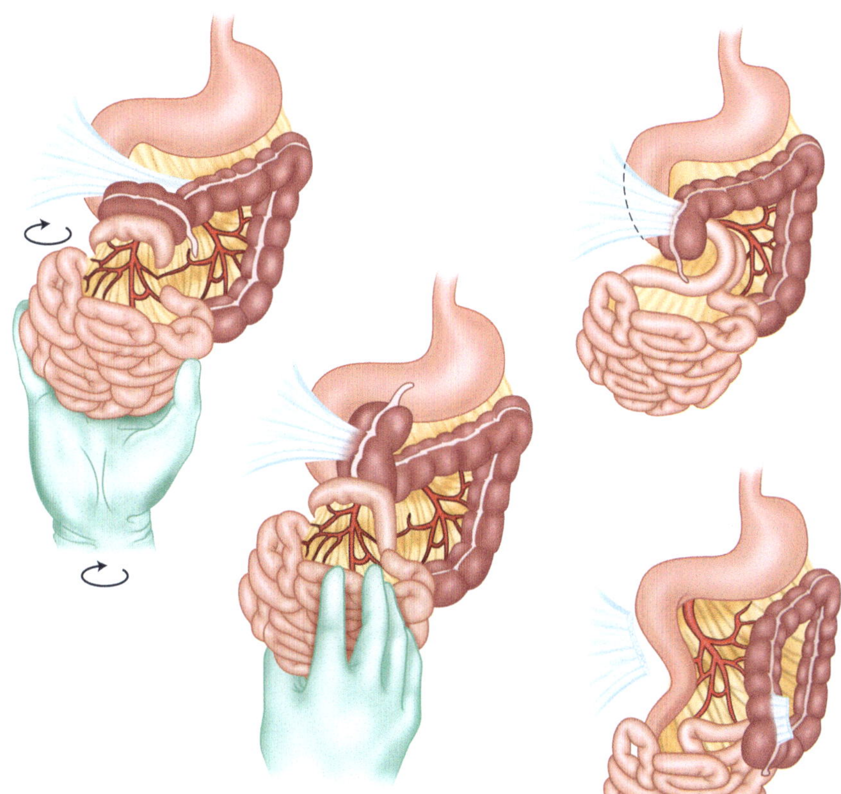

Fig. 42.5 Ladd procedure demonstrating counterclockwise detorsion of midgut volvulus

abdominal wall are then released. The duodenum is then completely mobilized (Kocher maneuver) to divide any remaining attachments. An appendectomy is performed prior to closing to avoid any potential confusion from acute appendicitis in the future.

Right mesocolic hernias are treated with division of the lateral peritoneal attachments of the cecum and right colon. Left mesocolic hernias are treated by mobilization of the inferior mesenteric vein to allow reduction of the small bowel from the hernia. The hernia is then closed to prevent recurrence.

Intestinal atresia that causes obstruction requires operative intervention to restore continuity. The goal is to preserve as much of the intestinal length as possible. This is achieved through short segmental bowel resection with end-end or end-oblique anastomosis. The apple peel jejunal atresia may require multiple serial anastomoses, and the ileocecal valve is preserved if possible to allow for toleration of enteral nutrition.

Complications

Complications include recurrence of the volvulus or duodenal obstruction. Small bowel obstruction due to adhesions is thought to occur in 1–10% of patients following the Ladd procedure.

Summary

Malrotation that causes an intestinal obstruction is a surgical emergency; a high index of suspicion is needed so as to not miss the diagnosis. Swift identification of malrotation prevents extent of intestinal necrosis and unnecessary bowel resection. Recurrence of volvulus or obstruction is rare with a technically complete Ladd procedure.

Suggested Reading

1. Soffers JHM, Hikspoors JPJM, Mekonen HK, Koehler E, Lamers WH. The growth pattern of the human intestine and its mesentery. BMC Dev Biol. 2015;15:31.
2. Berrocal T, Lamas M, Gutierrez J, Torres I, Prieto, del Hoyo ML. Congenital anomalies of the small intestine, colon, and rectum. RadioGraphics. 1999;19:1219–36.
3. Applegate KE, Anderson JM, Klatte EC. Intestinal malrotation in children: a problem-solving approach to the upper gastrointestinal series. Radiographics. 2006;26:5.
4. Mulholland MW, Lillemoe KD, Doherty GM, Maier RV, Simeone DM, Upchurch GR, editors. Greenfield's surgery. Philadelphia, PA: Lippincott Williams & Wilkins.
5. Sizemore AW, Rabbani KZ, Ladd A, Applegate KE. Diagnostic performance of the upper gastrointestinal eries I the evaluation of children with clinically suspected malrotation. Pediatr Radiol. 2008;38:518–28.

Mesenteric Hernia

43

Richard A. Jacobson, Robert C. Keskey, and John C. Alverdy

Background

Defects in the mesentery capable of permitting small bowel herniation can be congenital or acquired after surgery. Based on autopsy studies, congenital mesenteric defects exist in less than 2% of the American population. However, only a fraction of these become symptomatic over the course of a lifetime. Congenital hernias can occur through enlarged natural openings such as the foramen of Winslow, or due to abnormal rotation and fusion of the peritoneum creating paraduodenal fossae, pericecal folds, or spaces in the transverse or sigmoid mesocolon (Fig. 43.1). Hernias occur in the foramen of Winslow in 8%, paraduodenal (left and right) in 53%, transmesenteric in 8%, transomental in 1–4%, pericecal in 13%, intersigmoid in 6% and supravesical and pelvic in 6%. A large and growing portion of the population is at risk for mesenteric hernia through surgically acquired defects, especially those undergoing a Roux-en-Y gastric bypass procedure. Similar to congenital hernias, there are specific anatomic locations where internal hernias can occur following intestinal surgeries. In order to appreciate where these hernias can arise, it is important to understand intestinal anatomy following bypass surgery. In gastric bypass surgery, a gastric pouch is created and anastomosed to a distal limb of jejunum, referred to as a 'Roux' limb. The Roux limb is used to bypass 75–150 cm of small bowel before it is anastomosed to the biliopancreatic limb; the proximal jejunum carries biliary and pancreatic secretions (Fig. 43.2). The areas at risk for the development of mesenteric hernias are: (1) the space between the Roux limb mesentery and the transverse mesocolon (Petersen's space), (2) The mesenteric opening at the biliopancreatic

R. A. Jacobson
Department of Surgery, Rush University Medical Center, Chicago, IL, USA

R. C. Keskey · J. C. Alverdy (✉)
Department of Surgery, University of Chicago, 5841 S. Maryland, Chicago, IL 60637, USA
e-mail: jalverdy@surgery.bsd.uchicago.edu

E. D. Ehrenpreis et al. (eds.), *The Mesenteric Organ in Health and Disease*,
https://doi.org/10.1007/978-3-030-71963-0_43

limb, (3) The opening through the transverse mesentery when the bypass is retrocolic (Fig. 43.3). Any patient who has had a bowel resection is technically at risk, however this is particularly true following bariatric surgery since there is an ongoing loss of mesenteric fat that occurs over a long time period. When mesenteric hernias become symptomatic, they are more likely than to contain closed loop obstructions than external hernia. This increases the likelihood that mesenteric hernias will require operative intervention.

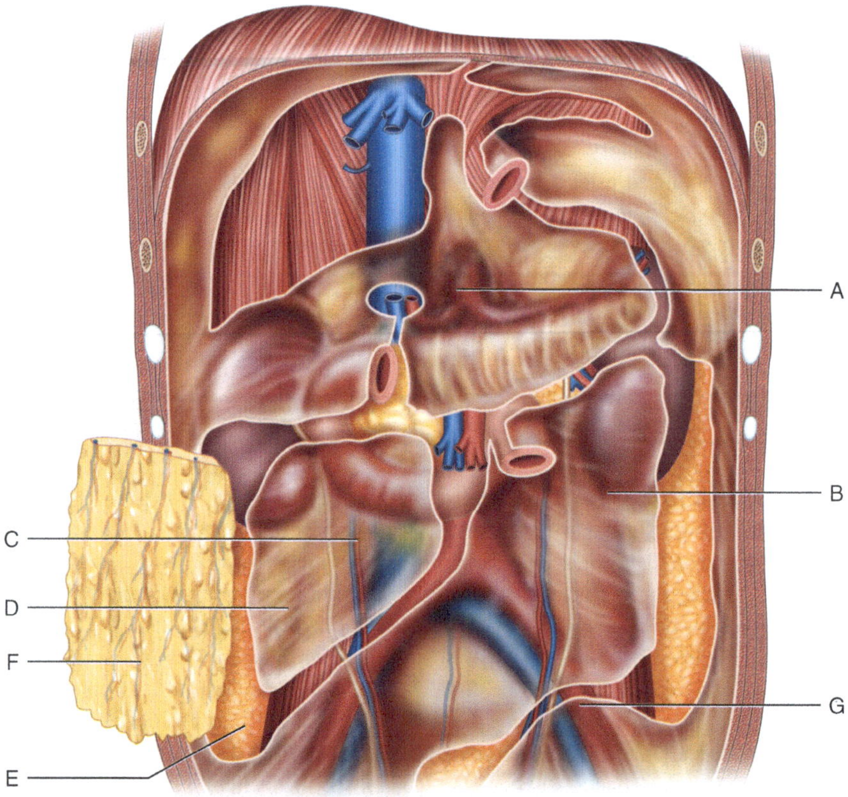

Fig. 43.1 Potential sites of internal hernias. A = foramen of Winslow, B = left paraduodenal, C = right paraduodenal, D = transmesenteric hernia, E = pericecal hernia, F = transomental hernia, G = intersignmoid hernia

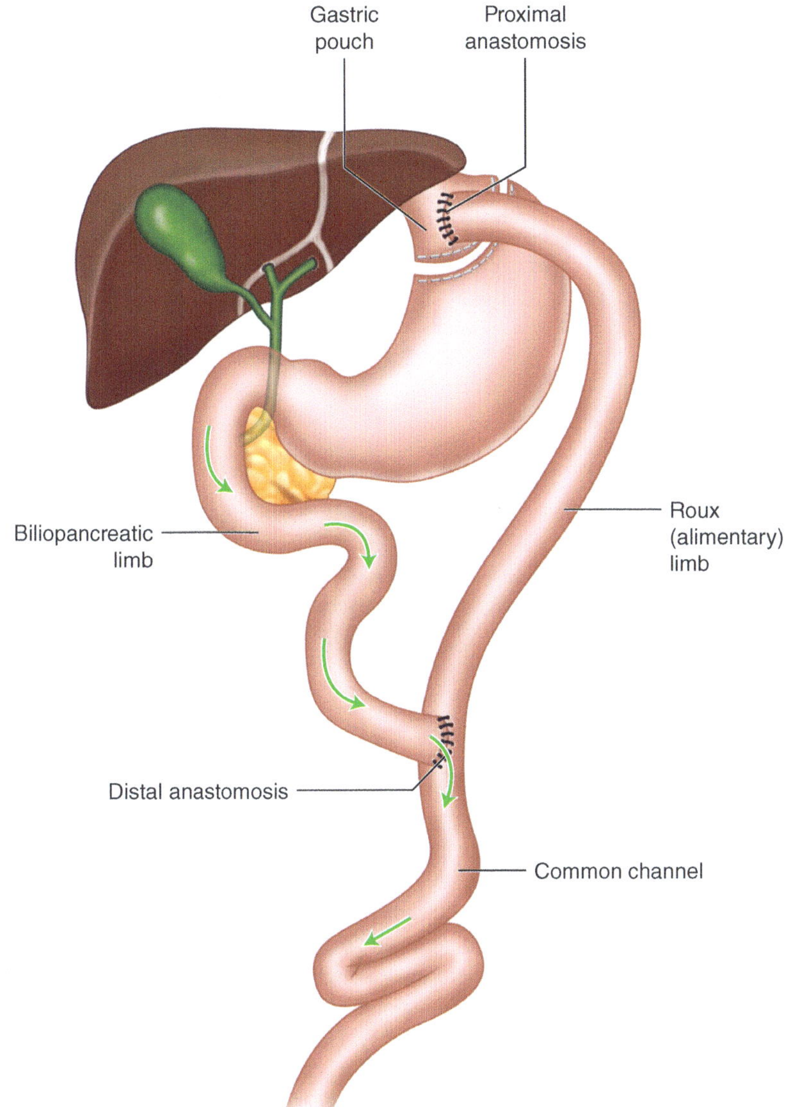

Fig. 43.2 Anatomy following Roux-en-y gastric bypass. The stomach is stapled to form a gastric pouch. A piece of jujenum is anastomosed to the neostomach and forms the "Roux" or alimentary limb. Of note, the roux limb can be placed overtop of the colon referred to as antecolic or through the transverse mesocolon referred to as retrocolic. The proximal jejunum draining bile acids and pancreatic enzymes is referred to as the 'biliopancreatic' limb and is connected to the Roux limb as shown. The intestine distal to this connection is referred to as the common channel

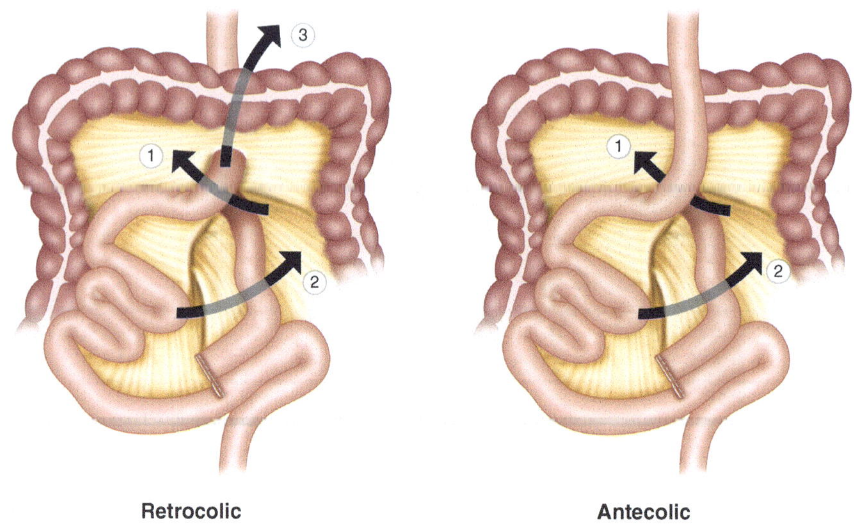

Fig. 43.3 Sites of potential internal hernias following bypass surgery. (1) Herniation of small bowel between the roux limb mesentery and the transverse mesocolon (Petersen's space), (2) The mesenteric opening at the biliopancreatic limb, (3) The opening through the transverse mesentery when the bypass is retrocolic

Clinical Presentation

Mesenteric hernias are considered internal hernias because they do not extrude outside of the abdominal cavity. As with any internal hernia, patients with mesenteric hernia present with a spectrum of clinical severity from asymptomatic abnormalities incidentally detected to acute mesenteric ischemia. Symptoms are typical of a bowel obstruction, including nausea, vomiting, abdominal pain and bloating, and may resolve if the hernia spontaneously reduces. Internal hernias are difficult to detect on physical exam as they lack an external bulge through the musculature of the abdominal wall. Laboratory testing may disclose contraction alkalosis after a prolonged obstruction and leukocytosis if the herniated bowel is ischemic. Computerized tomography (CT) with and without intravenous contrast is the imaging test of choice, as it can determine the location and type of hernia based on its relationship to mesenteric vessels and provide evidence of bowel ischemia.

Diagnosis

Patients with a surgical history who present with obstructive symptoms and abdominal distention on physical examination should be evaluated for adhesive small bowel obstruction (the most likely diagnosis) as well as internal hernia. Patients without a surgical history should be evaluated for endoluminal causes of obstruction such as gallstone ileus, bezoar, intussusception, inflammatory stricture or neoplasm, however internal hernia should be considered in the differential diagnosis as well [Beardsley Am J Surg]. In all the above cases, abdominal CT is warranted and necessary to make the diagnosis. CT findings associated with internal hernias include: dilated small bowel consistent with obstruction, swirling of mesenteric fat around the superior mesenteric artery known as the swirl sign, strangulation of the superior mesenteric vein, and engorged mesenteric vessels and edema. A CT scan of an internal hernia can be seen in Fig. 43.4a. A demonstrating the worrisome swirl sign associated with an internal hernia following bariatric surgery. Barium swallow can also be used to determine the presence and level of obstruction. If fluoroscopic imaging is performed in the presence of an internal hernia, clustering of loops of small bowel along with distended bowel proximal to the site of obstruction are present (Fig. 43.4b).

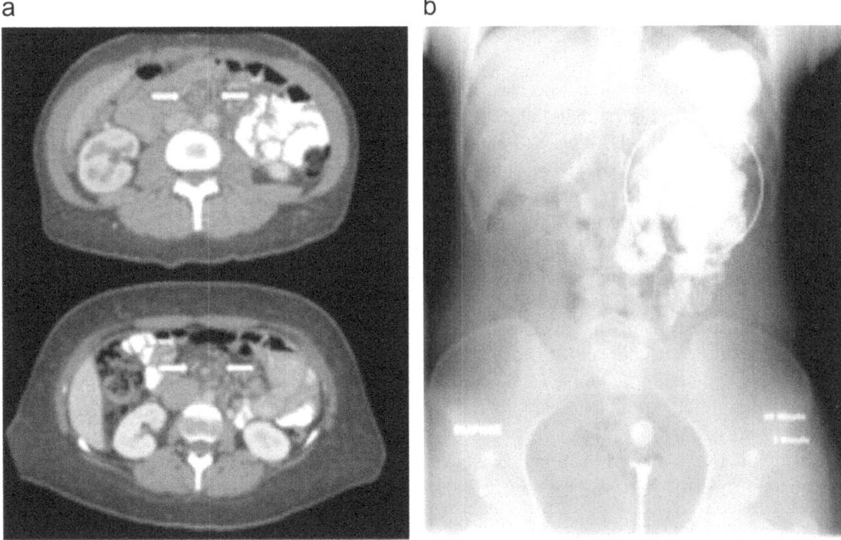

Fig. 43.4 Imaging of internal hernias. **a** Shows a CT scan demonstrating the presence of a mesenteric swirl sign concerning for the presence of an internal hernia. Jens Brøndum Frøkjær et al. The diagnostic performance and interrater agreement of seven CT findings in the diagnosis of internal hernia after gastric bypass operation. Abdominal Radiology 2018. 43 (12): 3220–3226. Reused with permission. © Springer Nature. **b** is a barium swallow showing an unusual clustering of jejunum in the left upper quadrant. Courtesy of Dr. Abraham Dachman and Dr. Justin Ramirez. Reused with permission

Management

Confirmed symptomatic mesenteric hernias should be surgically repaired. Patients with obstructive symptoms should be admitted to the hospital, given nasogastric decompression, resuscitated and medically optimized for surgery. Those with clinical, laboratory or radiographic evidence of bowel ischemia should proceed to the operating room expediently. Repair can be performed open or laparoscopically, depending on the history and specific condition of the patient.

Summary

Any patient with a previous Roux-en-Y reconstruction of the foregut who presents with a bowel obstruction should be suspected to have an internal hernia-CT scan is indicated. Inspection for swirling or spiraling or the mesenteric arteries about the axial rotation of the gut can be subtle signs of an internal hernia. Rapid surgical exploration is indicated when the diagnosis is suspected; this may result in a negative laparoscopy or laparotomy, which is an acceptable outcome.

Suggested Reading

1. Nobuyki, et al. CT of internal hernias. RSNA education exhibits. RadioGraphics. 2015;25(4). https://doi.org/10.1148/rg.254045035.
2. Shadhu K, Ramlagun D, Ping X. Para-duodenal hernia: a report of five cases and review of literature. BMC Surg. 2018;18:32. https://doi.org/10.1186/s12893-018-0365-8.
3. Jones TW. Paraduodenal hernia and hernias of the foramen of Winslow. In: Nyhus LM, Harkins HN, editors. Hernia. Philadelphia: JB Lippincott Co.; 1964. p. 577–601.
4. Eckhauser A, Torquati A, Youssef Y, Kaiser JL, Richards WO. Internal hernia: postoperative complication of Roux-en-Y gastric bypass surgery. Am Surg. 2006;72(7):581–5.
5. Doishita S, Takeshita T, Uchima Y, Kawasaki M, Shimono T, Yamashita A, Miki Y. Internal hernias in the era of multidetector CT: correlation of imaging and surgical findings. Radiographics. 2015;36(1):88–106.
6. Beardsley C, Furtado R, Mosse C, Gananadha S, Fergusson J, Jeans P, Beenen E. Small bowel obstruction in the virgin abdomen: the need for a mandatory laparotomy explored. Am J Surg. 2014;208(2):243–8.
7. Champion JK, Williams M. Small bowel obstruction and internal hernias after laparoscopic Roux-en-Y gastric bypass. Obes Surg. 2003;13(4):596–600.

Surgical Management of Bands and Adhesions

44

Ashley J. Williamson, Robert C. Keskey, and John C. Alverdy

Definition and Clinical Features

The intestine is fixed at two peritoneal reflections: the ligament of Treitz proximally and the ileocecal fold, or ligament of Treves distally. Normal embryologic rotation is required for the fixation of these points and orientation of the intestine. Embryologic rotation may be arrested resulting in malrotation most commonly diagnosed within the first year of life which is discussed in detail in another chapter. Rarely, malrotation is diagnosed in adult patients and adhesions between the cecum and duodenum (Ladd bands), or a narrow mesentery base are identified. By far the most common cause of intestinal obstruction in Western world is adhesions caused by previous surgery and scarring. Approximately 95% of patients who undergo abdominal surgery will develop adhesive disease which is a result of normal healing.

Unlike children, malrotation in adults may be asymptomatic or found incidentally on imaging studies. If symptomatic, both mesenteric bands and adhesions will present with symptoms of obstruction. Patients are likely to present with nausea, bilious emesis, dehydration, decreased flatus or bowel movement, and abdominal distension.

Description of Procedures

If malrotation in adults is suspected or diagnosed and the patient presents with an indolent course, an elective *Ladd procedure* (Fig. 44.1) may be performed with goal to broaden the base of the mesentery and reduce risk of volvulus. Obstruction from surgical adhesive disease is attempted to be managed conservatively with

A. J. Williamson · R. C. Keskey · J. C. Alverdy (✉)
Department of Surgery, University of Chicago, 5841 S. Maryland, Chicago, IL 60637, USA
e-mail: jalverdy@surgery.bsd.uchicago.edu

E. D. Ehrenpreis et al. (eds.), *The Mesenteric Organ in Health and Disease*,
https://doi.org/10.1007/978-3-030-71963-0_44

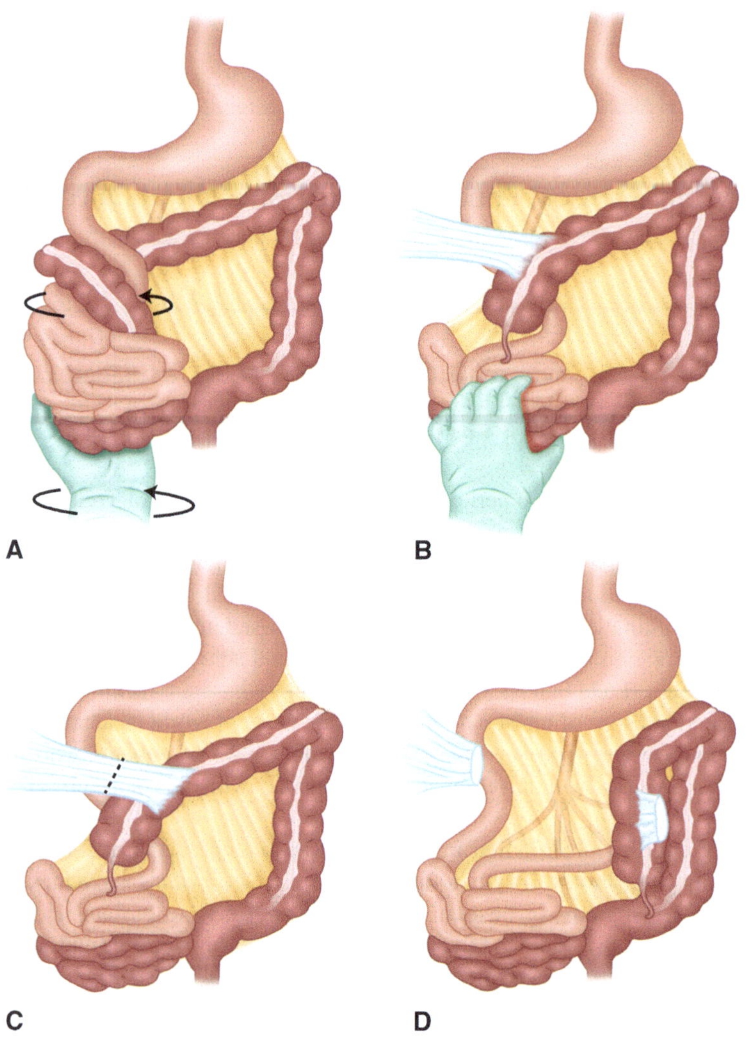

Fig. 44.1 Ladd's Procedure: The bowel is detorsed in a counterclockwise fashion (**a** and **b**). The Ladd bands are divided (**c**). The base of the mesentery is widened and the appendix is removed

nothing per mouth, nasogastric tube decompression, serial abdominal exams, and hydration prior to operating as any operation will further disease.

However, in patients with signs of bowel ischemia or hemodynamic instability: i.e. leukocytosis, acute kidney injury, peritonitis, or patients who do not improve with nasogastric tube decompression, urgent operative evaluation with exploratory laparoscopy or laparotomy is indicated based on surgeon comfort.

Indications for Surgery

The diagnosis of adult band or adhesive disease should be considered in patients who present with indolent symptoms of intermittent bilious emesis, food intolerance, failure to thrive and possibly chronic diarrhea (Table 44.1).

Patients with known abdominal surgical history presenting with small bowel obstruction without symptoms of ischemia are likely candidates for initial conservative management with nasogastric tube and rehydration. However, signs of compromised bowel or hemodynamic instability require urgent operative intervention. Further, patients whose abdominal tenderness does not improve or who are unable to progress their diets with nasogastric tube decompression warrant exploration in the operating room.

Specific indications for an elective adhesiolysis also include treatment of infertility and recurrent pregnancy loss in patients with prior pelvic surgery. New data is emerging for the use of adhesiolysis as treatment for chronic abdominal pain, however this body of literature is still evolving.

Contraindications

Patients who have never gone an operation and have no known surgical history require appropriate work up to indicate that scarring from adhesions is the cause of their bowel obstruction. Although certainly possible in the "virgin" abdomen, adhesive bowel obstruction is highly unlikely. Malignancy, hernias, and intestinal malrotation should all be looked for in these patients by using multi-modality diagnostic techniques and an operative evaluation to definitely rule out this diagnosis may also be warranted.

Laparoscopy has become more commonly practiced for abdominal exploration. Relative contraindications for use of laparoscopy in these situations include ascites and compromised cardiovascular status. Absolute contraindications include surgeon's inability to completely assess the abdomen using this approach. Laparotomy is utilized in these cases as an alternative.

Table 44.1 Indications for surgery in the setting of adhesive bowel obstruction

1. Signs of peritonitis: rebound, guarding, rigid abdomen, diffuse tenderness
2. Concern for bowel ischemia: hypo-enhancement of intestinal wall or free fluid on CT, pain out of proportion to physical exam, leukocytosis, hemodynamic instability
3. Failure of conservative management: persistently elevated NG tube output, persistent or worsening abdominal distension, persistent lack of bowel function (obstipation/constipation)

Preparation of the Patient

Nasogastric decompression, intravenous access and hydration, electrolyte repletion and a detailed medical and surgical history are imperative in patients presenting with obstruction secondary to suspected bands or adhesions. If the patients is stable, obtaining outside operative reports are helpful, if not essential, to understand and anticipate potential finding in the operating room.

How the Procedure is Performed

The Ladd procedure (Fig. 44.1) is performed for patients with suspected malrotation and may be performed using a laparoscopic or open method. The initial assessment for a volvulus is accomplished by eviscerating the bowel outside of the abdomen and assessing for any twist in the mesentery or signs of ischemia-induced bowel compromise. If a volvulus is present, the bowel is untwisted in a *counter-clockwise* fashion. Any necrotic bowel should undergo resection. The surgeon will perform a close examination of the bowel to look for any Ladd bands. Ladd bands are fibrinous bands that are present between the duodenum and the cecum. If present, are surgically divided. The duodenum is straightened for complete evaluation using a *Kocher maneuver*. The surgeon will also assess for any intermesenteric bands or fibrous bands between other bowel loops. These have the capacity to compress the blood supply in the mesentery. The bowel is investigated or "run" in its entirety and these bands are divided. An appendectomy is also performed at the time of this operation since the appendix is often in a different position than the right lower quadrant and leads to a delay in diagnosis of acute appendicitis and increased morbidity in patients with malrotation. The bowel is then placed in its appropriate anatomic location, as the anatomy allows, and the abdomen is closed.

Exploration in an acute setting or for adhesive disease involves the performance of either laparoscopic or laparotomy depending on the skill and preference of the operating surgeon. The abdomen should be explored in entirety looking for evidence of perforation. Other suggestions that a perforation is present include the occurrence of murky fluid or rush of air upon entry to the abdomen. The bowel should be examined in its entirety, with a thorough evaluation for any potential bands or adhesions. Often lysis of these bands will result in resolution of the symptoms. The viability of the bowel based on perfusion should be evaluated, with resection of any necrotic bowel after lysis of adhesions. Compromised bowel should be evaluated after lysis for resolution of perfusion. If it is unclear if the bowel is viable after adhesiolysis, a temporary abdominal closure device may be placed and a second look operation planned for 24–48 h from initial surgery. This technique is used to assess need for bowel resection prior to closure of the abdomen.

Typical Abnormal Findings M

Adult malrotation with volvulus may be visualized on a barium enema when there is evidence of complete obstruction at the level of the transverse colon and the oral contrast tapers to a "birds beak" appearance (Fig. 44.2a and b).

Further, malrotation may be identified on abdominal CT scan which may show the duodenum not crossing midline or the cecum no in the right lower quadrant.

Outcomes and Complications

Given the rarity of adult malrotation, there is often a delay in diagnosis and increased morbidity. For instance, the erroneous diagnosis of more common surgical diseases such as appendicitis and diverticulitis (that may not occur in the expected location) can lead to delayed treatment. These patients are more likely to experience advanced stages of adhesive small bowel disease prior to intervention.

It is estimated that 1% of all general surgery admissions and 3% of laparotomies are secondary to adhesive disease. Each operation including a lysis of adhesions operation predisposes the patient to more healing and adhesions. This must be considered when deciding before the performance of semi-elective or elective operations.

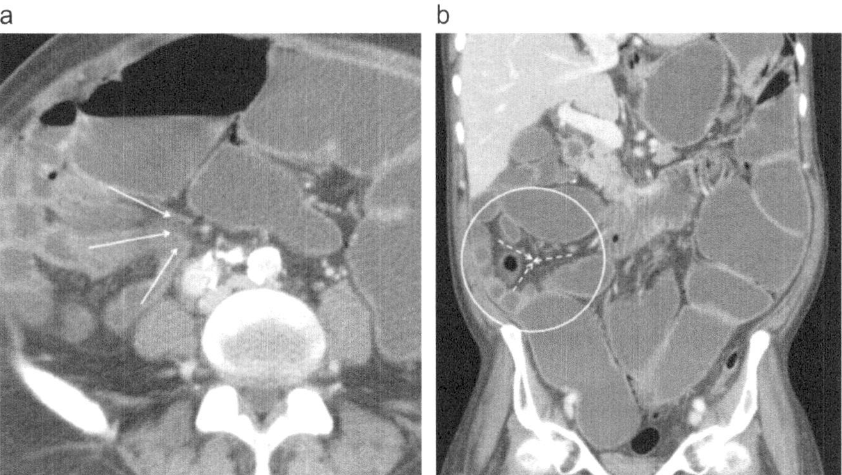

Fig. 44.2 Axial (**a**) and coronal (**b**) CT images demonstrate sharply angulated small bowel loops radiating to a single point (arrows) suggestive of tethering due to adhesive disease in this patient who has h ad a prior laparotomy. Note the adjacent, fluid-filled proximal small bowel loops indicative of small bowel obstruction. With permission from Dr. Abraham Dachman and Dr. Justin Ramirez)

In true cases of intestinal malrotation, morbidity rates remain high post-operatively in adults and approach 60%. This high morbidity rate is generally attributed to a delay in diagnosis. Most common complications of operations for small intestinal adhesions and obstructive small bowel diseases include ileus (or delayed return of bowel function), small bowel obstruction secondary to adhesive disease in the future (15%), recurrent volvulus, and rarely short gut syndrome when midgut resections lead to chronic malabsorption of nutrients, fluids and electrolytes.

Summary

Adhesiolysis in patients without prior surgery is rarely indicated unless the patient fails non-surgical management or imaging indicates a specific cause. It is essential that the past medical and surgical history of patients presenting with bowel obstruction is clearly outlined. A review of all prior operative reports should be obtained whenever possible. Prior to closing the abdomen, it is critical that the mesentery is flattened out and that all the bowel is in its proper orientation (**Suggested Reading**).

Suggested Reading

1. Durkin ET, Lund DP, Shaaban AF, et al. Age-related differences in diagnosis and morbidity of intestinal malrotation. J AM Col Surg. 2008;658.
2. Dietz DW, Walsh RM, Grundgest-Browniatowski S, et al. Intestinal malrotation: a rare but important cause of bowel obstruction in adults. Dis Colon Rectum. 2002;45:1381.
3. Davey AK, Maher PJ. Surgical adhesions: a timely update, a great challenge for the future. J Minim Invasive Gynecol. 2007;14:15.
4. Wilson MS. Practicalities and costs of adhesions. Colorectal Dis. 2007;9(Suppl):60.

Mesenteric Trauma

45

Evan G. Wong and Joseph V. Sakran

Epidemiology

Bowel and mesenteric injuries are relatively rare but can have devastating consequences. Following blunt abdominal trauma, these injuries can occur in isolation, seen in about 1–2% of patients and associated with other injuries in around 10–15% of cases. Mesenteric injuries are found in approximately 10% of laparotomies done for trauma. Small bowel mesenteric injuries are 5 to 7.5 times more common than colonic mesenteric injuries. Given the relatively low incidence, clinicians must ensure to suspect mesenteric injuries in the right clinical context.

A landmark multi-institutional trial through the Eastern Association for the Surgery of Trauma (EAST) demonstrated that a delay in diagnosis and intervention was associated with a significant increase in morbidity and mortality. Delays as short as 8 h were directly attributable to mortality. Further delays in surgical intervention beyond 24 h were associated with a three-fold increase in death.

Early diagnosis and intervention for bowel and mesenteric injuries is therefore crucial. The morbidity and mortality from these injuries are mainly attributable to hemorrhage and sepsis, the latter from the associated bowel perforations. However, early presentations can be subtle and diagnostic modalities notoriously lack sensitivity. Injuries that lead to laceration of the mesentery may result in a contained hematoma that could result in an atypical clinical presentation such as peritonitis. Therefore, the clinical care team must maintain a high index of suspicion for these potential injuries.

E. G. Wong · J. V. Sakran (✉)
Division of Acute Care Surgery and Adult Trauma Services, Department of Surgery, Johns Hopkins University, 1800 Orleans St, Sheikh Zayed Tower/Suite 6107A, Baltimore, MD 21287, USA
e-mail: Jsakran1@jhmi.edu

E. D. Ehrenpreis et al. (eds.), *The Mesenteric Organ in Health and Disease*,
https://doi.org/10.1007/978-3-030-71963-0_45

Pathophysiology

Mesenteric injuries from penetrating trauma typically result from direct contact of the object or missile's entry into or through the abdominal cavity. In circumstances where patients are injured with high energy weapons, the missile may result in a blast effect that pulverizes the surrounding tissue. Gunshot wounds are also more often associated with multiple intra-abdominal injuries. Although much attention is often given to the size and type of the implicated object, any penetrating injury to the abdomen could potentially injure the underlying structures, especially in the context of prior surgery where adhesions could appose mesentery and bowel to the abdominal wall.

Bowel and mesenteric injuries as a result of blunt abdominal trauma are most commonly associated with motor vehicle collisions and bicycle-related handlebar injuries. Crush injuries and falls from height are also well-described mechanisms. The remainder of bowel injuries can be seen in unusual circumstances such as contact sports or after the Heimlich maneuver.

These mechanisms lead to bowel and mesenteric injuries in a variety of ways, including rapid increases in intra-abdominal pressure, shearing forces from an acute acceleration or deceleration and compression against hard structures. For example, the use of a seatbelt in a motor vehicle collision is associated with bowel perforation and mesenteric injuries. The compression of the intra-abdominal structures by the seatbelt leads to an acute increase in pressure and blowout of the contained bowel as well as compression against structures such as the vertebrae. The rapid deceleration of a vehicle further contributes to mesenteric injuries, most notably at fixed points such as the Ligament of Treitz. These injuries can lead to mesenteric lacerations or hematomas, or partial thickness or full thickness bowel injuries.

Diagnosis

Clinical Exam

The signs and symptoms of bowel and mesenteric injuries can be subtle. In a review of nearly 3,000 patients who sustained blunt abdominal trauma, physical exam lacked both sensitivity (53%) and specificity (63%) for the diagnosis of bowel perforations.

While hemodynamic instability may be indicative of intra-abdominal hemorrhage, potentially due to a large mesenteric bleed, smaller lacerations or contained hematomas may not present in the same manner. Initial evaluation of patients with even bowel perforation may not present with the alteration in vital signs or physical exam that clinicians often expect. The development of fever, tachycardia and/or hypotension during the first 24 h following trauma should raise the suspicion of a missed bowel injury.

The presence of abdominal pain should alert the clinician to the possibility of a bowel or mesenteric injury. However, the absence of pain upon initial presentation does not exclude the diagnosis as a contained hematoma may not lead to peritoneal irritation and even frank bowel spillage may take hours to manifest as peritonitis. In addition, many trauma patients may be difficult to examine due to concomitant head injuries or the need for intubation for airway protection. Nevertheless, a key instrument in the diagnosis of mesenteric or bowel injuries is the monitoring of vital signs and serial abdominal examinations, preferably by the same provider. An increase in abdominal pain or development of peritonitis should be a warning for a missed intra-abdominal injury.

The seatbelt sign is a pattern of ecchymoses across the neck, chest and/or abdomen that mirror the position of the protective belt at the moment of the motor vehicle collision. Its presence has classically been associated with a pattern of injury, including blunt cerebrovascular injuries, sternal fractures, bowel and mesenteric injuries, pancreaticoduodenal injuries and Chance vertebral fractures. As such, the presence of a seatbelt sign should raise the suspicion of a bowel or mesenteric injury. However, it is far from pathognomonic—with a reported sensitivity of approximately 25%, and specificity of 85%. The presence of a seatbelt sign in the absence of abdominal pain or tenderness is associated with a low-risk of bowel and/or mesenteric injury.

Initial blood tests are often non-contributory, however an increasing leukocytosis and lactatemia, along with a concerning abdominal exam, raise the suspicion of missed internal injuries.

Late presentations of missed bowel and mesenteric injuries, weeks to months after the trauma, have been reported in the literature. These often arise from contained bowel perforations, which can present in a variety of ways such as intra-abdominal abscesses. Mesenteric avulsions, often called "bucket handle injuries" because of its characteristic resemblance to this object, may lead to localized bowel ischemia and eventual stricture. Thus, one possible late presentation of a missed mesenteric injury is evidence of a partial bowel obstruction.

Diagnostic Peritoneal Lavage (DPL) and Focused Assessment with Sonography in Trauma (FAST)

DPL has classically been described as a bedside diagnostic modality for intra-abdominal hemorrhage or bowel injury. The presence of red blood cells points towards hemorrhage whereas the presence of white blood cells, amylase, alkaline phosphatase or frank food particulate is associated with bowel injury. DPL has a high sensitivity but is associated with a significant false positive rate. For bowel injury in particular, DPL has been shown to be superior to Computed Tomography (CT). However, the use of DPL has significantly decreased over the years, and in many centers not performed anymore. This change has been in large part secondary to the increased use of ultrasound, specifically the FAST, and diagnostic laparoscopy. Therefore, DPL remains in the armamentarium of the trauma clinician but the authors do not recommend its routine use at this time.

FAST has largely supplanted DPL in the evaluation of blunt abdominal trauma. Less invasive, it provides a rapid point of care assessment of the presence of intra-abdominal free fluid. However, its sensitivity is extremely user-dependent and suboptimal for retroperitoneal injuries. In the evaluation for bowel and mesenteric injuries, FAST does not provide an etiology for the free fluid and is not able to distinguish physiologic from pathologic fluid. As such, the presence of free fluid on FAST in the right clinical context should prompt further evaluation for bowel or mesenteric injury. FAST therefore represents a bridge in the continuum of evaluation rather than a definitive modality.

CT

In the patient with a suspicion of bowel or mesenteric injury without absolute indications for immediate operative exploration such as hemodynamic instability, peritonitis, or evisceration, a CT scan of the abdomen/pelvis should be performed with intravenous (IV) contrast. Oral contrast is not necessary and often not feasible given the emergent nature of the imaging required.

CT scans notoriously lack sensitivity for bowel and mesenteric injuries, with a reported sensitivity of 45% and specificity of 95%. In a multi-institutional trial of over 275,000 patients, Fakhry et al. found a 13% false negative rate for the detection of small bowel injury with initial CT. Even with more modern CT technology, a 2019 update of this trial evaluating blunt small bowel perforation demonstrated that 4.2% of patients had a completely normal initial CT scan. There is a vast amount of literature describing more subtle radiological signs in an effort to increase its utility (Table 45.1). Overt signs of injury include an obvious mural defect, free air or contrast extravasation, either from the bowel lumen or

Table 45.1 CT signs of bowel or mesenteric injury

CT findings
Visible transmural defect
Extraluminal gas
Oral/intravenous contrast extravasation
Bowel thickening
Abnormal wall enhancement
Vascular thrombosis
Beading of mesenteric vessels
Mesenteric infiltration, fluid or hematoma
Intraperitoneal free fluid

Adapted from Bennett AE, Levenson RB, Dorfman JD. Multidetector CT Imaging of Bowel and Mesenteric Injury: Review of Key Signs. *Semin Ultrasound CT MRI* 2018; 39:363–373

vasculature. More subtle signs include the presence of intraperitoneal free fluid without solid organ injury, thickened bowel, abnormal mural enhancement or contained hematoma. The presence of a contained mesenteric hematoma should alert the clinician to the possibility of an underlying bowel injury and the risk of ischemia. Subtle but specific signs for mesenteric injury are vascular thrombosis, mesenteric fluid or infiltration or beading of mesenteric vessels. Importantly, the bowel and mesentery may even appear normal. Thus, a normal CT scan in a patient clinically at risk for bowel injury should still prompt further monitoring.

Associated injuries should raise the suspicion of bowel and mesenteric trauma (Table 45.2). Bladder injuries, perhaps as a correlate for other hollow viscera, are associated with bowel trauma. In the same pattern as the seatbelt syndrome, anterior wall hematomas, traumatic hernias, particularly of the anterior abdominal wall, and Chance fractures have all been found to be associated with bowel injuries.

Operative Assessment

The use of laparoscopy in trauma has been significantly increasing during the past few years. Its use has mirrored its increased utility during surgical training; however, proficiency with the technique is also an important limitation to the utility of laparoscopy for diagnostic purposes. The surgeon should be facile at manipulating

Table 45.2 Injuries associated with bowel and mesenteric injury

Associated injuries
Abdominal wall hematoma
Traumatic hernia (anterior, lateral, diaphragmatic)
Solid organ injuries (liver, spleen kidney)
Closed head trauma
Hemothorax
Pneumothorax
Rib fractures
Pulmonary contusion
Retroperitoneal hematoma
Pelvic fracture
Femur fracture
Pancreaticobiliary injuries
Bladder disruption
Chance fractures
Anorectal trauma

Adapted from: From Frick EJ, Pasquale M, Cipolle M. Small bowel and mesentery injuries in blunt trauma. *J Trauma.* 1999; 46(5):920–6 and Bennett AE, Levenson RB, Dorfman JD. Multidetector CT Imaging of Bowel and Mesenteric Injury: Review of Key Signs. *Semin Ultrasound CT MRI* 2018; 39:363–373

bowel in order to examine its entire length as well as its supplying mesentery. At any point where the surgeon is not entirely confident in the sensitivity of the exploration or suspicious of injury, the case should be converted to open laparotomy. The increased use of laparoscopy in trauma over the years has led to a decrease in negative trauma laparotomies in a select subset of patients.

One patient population that has particularly benefited from laparoscopy has been victims of anterior abdominal stab wounds. The use of laparoscopy in this population has led to a decrease in non-therapeutic laparotomies, hospital lengths of stay and costs. Equally applicable for victims of blunt trauma, laparoscopy should be considered in hemodynamically patients with a suspicion for bowel or mesenteric injury but without another overt indication for laparotomy. The gold standard diagnostic modality for bowel and mesenteric injury is an exploratory laparotomy. As such, any patient in whom the clinical suspicion remains high despite negative results from the previous modalities should undergo a laparotomy.

Grading

In order to provide a standard for defining bowel injuries, the American Association for the Surgery of Trauma (AAST) has developed an injury scale based on tissue loss and vascularity (Table 45.3). Specifically, for mesenteric injuries, a CT-based grading system has been developed that is based off of the previously described CT findings (Table 45.4). Finally, an intraoperative grading scale has also been developed for grading of mesenteric injuries (Table 45.5). These grading scales should be used to better communicate, study and manage both bowel and mesenteric injuries.

Table 45.3 AAST grading scale for small bowel injuries

Grade	Type	Description
1	Hematoma	Contusion or hematoma without devascularization
	Laceration	Partial thickness, no perforation
2	Laceration	Laceration <50% of circumference
3	Laceration	Laceration $\geq$ 50% of circumference without transection
4	Laceration	Transection of the small bowel
5	Laceration	Transection of the small bowel with segmental tissue loss
	Vascular	Devascularized segment

Adapted from the American Association for the Surgery of Trauma (AAST)https://www.aast.org/resources-detail/injury-scoring-scale#bowel

Table 45.4 Proposed CT-based mesenteric injury scale

Grade	Description
1	Contusion without bowel thickening or interloop collection
2	Hematoma <5 cm without bowel thickening or interloop collection
3	Hematoma >5 cm without bowel thickening or interloop collection
4	Contusion or hematoma (any size) with bowel thickening or interloop collection
5	Vascular or oral contrast extravasation, bowel transection or pneumoperitoneum

Adapted from McNutt MK, Chinapuvvula NR, Beckmann NM, Camp EA, Pommerening MJ, Laney RW, West OC, Gill BS, Kozar RA, Cotton BA, Wade CE, Adams PR, Holcomb JB. Early Surgical Intervention for Blunt Bowel Injury: The Bowel Injury Prediction Score (BIPS). *J Trauma Acute Care Surg* 2015; 78(1): 105–111

Table 45.5 Proposed intra-operative mesenteric injury scale

Grade	Hemorrhage	Extent	Bowel viability
1	Hematoma/contusion	Intact peritoneum	Viable
2	Exposed vessels	<1 cm peritoneal tear	Viable
3	Active bleeding	Gaping mesenteric fat	Viable
4	Bleed/thrombus	Defect through which daylight is visible	Questionable
5	Bleed/thrombus	Avulsion injury	Non-viable

Adapted from Bekker W, Hernandez MC, Zielinski MD,Kong VY, Laing GL, Bruce JL, Manchev V, Smith MTD, Clarke DL. Defining an intra-operative blunt mesenteric injury grading system and its use as a tool for surgical-decision making. *Injury, Int. J. Care Injured* 2019; 50 (2019) 27–32

Predictive Variables

In the 2019 multi-institutional EAST study on blunt small bowel perforation, factors associated with an injury were sex, abdominal tenderness, distention, peritonitis, bowel wall thickening, free fluid, and contrast extravasation.

McNutt et al. further developed a Bowel Injury Prediction Score (BIPS) composed of three variables: white blood cell count > 17.0×10^9 cells/L, abdominal tenderness and a CT grade of $\geq$ 4. After a review of 380 mesenteric injuries, the authors found a BIPS of 2 or more was associated with 19-fold increased odds of blunt bowel injury. These findings have not been prospectively validated, given the occasional subtle presentations of these injuries, and therefore further research in this area is warranted.

Management

Concern for Injury

In patients at risk of mesenteric and bowel injury but without absolute indications for operative exploration or clear signs of injury on imaging, observation with close serial abdominal exams in a monitored setting is warranted. Ideally, these exams should be performed by the same provider. In systems with multiple providers within a 24 h period, the incoming provider should examine the patient with the admitting provider. The patient should be kept fasting in preparation for a potential intervention. Antibiotics are not indicated and may even mask clinical signs.

Any evidence of worsening abdominal pain, tachycardia, hypotension, fever, tachypnea or desaturation should prompt a reassessment of the patient to determine if operative exploration would be indicated. Serial lab results can also be helpful, including an increase in white blood cell count, lactate or amylase. A decrease in hemoglobin, although nonspecific, can suggest an uncontained mesenteric laceration.

The optimal length of observation has not been well defined. Unnecessary lengthier admissions are labor-intensive and costly; however, the early discharge of a trauma patient with a bowel or mesenteric injury is unacceptable. Early evidence suggested that most bowel injuries manifested themselves in the first 24 h of admission. More recent literature suggests that an even shorter period of observation may be sufficient. For example, a review of 3,574 blunt abdominal trauma patients revealed that those with intra-abdominal injuries exhibited signs or symptoms and were diagnosed within 9 hours. All patients with an intra-abdominal injury requiring an intervention were diagnosed within the first 60 min of presentation. Currently, the authors advocate for an observation period of 24 h but this may change as more research is published.

High Suspicion for Injury

Patients with gunshot wounds to the abdomen are at high risk of bowel and mesenteric injury; unless the tracts are unequivocally tangential or isolated to the right upper quadrant in a stable patient, they should prompt urgent operative exploration.

As previously alluded to, patients with stab wounds to the anterior abdominal wall without peritonitis and who are hemodynamically stable can be investigated with multiple modalities. Local wound exploration is an option in the isolated stab wound but may be difficult in the combative trauma patient and puts health care providers at risk for needle stick injuries. Imaging with CT is fast becoming the initial investigative test of choice. Radiographic signs as noted in Table 45.1, paired with clinical data, helps with the decision of whether or not to take the patient to the operating room. Initial surgical management with laparoscopy may be considered depending upon the skillset of the surgeon, particularly in the hemodynamically stable patient with signs of intraperitoneal free fluid without solid organ injury.

In a review of 121 blunt abdominal trauma patients with bowel and mesenteric injuries, the use of laparoscopy was associated with a significant shorter length of stay and decreased rate of wound infections. Interestingly, conversion to laparotomy was only required in 8.5% of cases and there were no differences in complication rates between the laparoscopy and laparotomy groups.

Nevertheless, laparotomy remains the gold standard and is an acceptable initial management for any trauma patient with a high index of suspicion for bowel or mesenteric injury.

Intraoperative Finding

Once the decision is made to operate on the patient, the presence of a bowel or mesenteric injury can be confirmed and graded according to the previously described scales (Tables 45.3 and 45.5). Often, these injuries are encountered intra-operatively associated with other pathologies such as solid organ or vascular injuries that prompted the operative exploration in the first place.

Bowel wall contusions and hematomas (Grade 1 hematoma) should be explored to ensure that there is no underlying bowel injury. Partial thickness lacerations (Grade 1 laceration) and lacerations involving less than 50% of the bowel circumference (Grade 2) can be primarily repaired if closure does not lead to narrowing of the lumen. Single and two-layer closures appear to be equivocal and are at the surgeon's discretion. More severe injuries (Grades 3–5) with evidence of significant (>50%) disruption to the bowel wall, transection or devascularization require resection. Primary anastomosis is warranted unless the patient's physiology or injury burden require a damage control approach. In that case, the patient can be left in discontinuity with a plan for definitive management at a later time after physiologic resuscitation.

Early evidence suggested that hand sewn anastomosis techniques were associated with fewer complications as compared to stapled techniques in trauma patients, perhaps related to the underlying bowel edema. However, a recent AAST multi-institutional trial evaluating both techniques in 595 patients across 15 institutions showed similar anastomotic failure rates and operative times.

Mesenteric injuries present as contained hematomas, active bleeding and avulsion injuries with bowel ischemia. As for Grade 1 small bowel injuries, hematomas within the mesentery should be explored to exclude a concomitant bowel perforation. If the bowel is intact and viable, no further management is required. If a distal mesenteric vessel is visible or actively bleeding (Grade 2–3), we advocate simple suture ligation with close monitoring for bowel viability. With any avulsion injury ("bucket handle") or questionable bowel viability, the involved bowel segment should be resected. The patient's physiological status and associated injuries should guide the decision-making process for damage control surgery, which may require the bowel to be left in discontinuity and the abdomen left open with a temporary abdominal closure device in preparation for a second look in the coming days. If the mesenteric injury is isolated with clear demarcation in bowel viability in a stable patient, we advocate primary anastomosis.

Bowel viability can be clinically assessed by color, palpable pulses and intact peristalsis. Useful adjuncts include intra-operative Doppler of the distal mesentery and indocyanine green-enhanced fluorescence angiography. There are case reports of microvascular repairs of jejunal and ileal vessels in patients with extensive avulsion injuries; however, ligation and resection remain the mainstays of treatment.

Proximal Mesenteric Injury

Proximal mesenteric injuries are rare but devastating and represent some of the most challenging injuries faced by trauma surgeons. They are often associated with multiple other complex injuries, such as pancreaticoduodenal trauma, and patients are often in extremis. The management of superior mesenteric vessels often requires creative solutions as simple ligation can have dire consequences.

The AAST also describes an abdominal vascular injury scale. Specifically, for mesenteric injuries, Grades 1 and 2 refer to unnamed mesenteric vessels and the inferior mesenteric artery (IMA) or vein (IMV), respectively, all of which can be ligated with impunity provided there is adequate collateral flow.

Grades 3 and 4 refer to superior mesenteric vein (SMV) and artery (SMA) injuries, respectively. SMA injuries can be further divided according to Fullen's classification based on anatomic zone and ischemia grade (Table 45.6).

Table 45.6 Fullen's classification of superior mesenteric artery injuries

Anatomic classification		
Zone	SMA segment	
1	Trunk proximal to first major branch	
2	Trunk between pancreaticoduodenal and middle colic	
3	Trunk distal to middle colic	
4	Segmental branches	
Ischemic classification		
Grade	Category	Segments affected
1	Maximal	Jejunum, ileum or right colon
2	Moderate	Major segment, small bowel and/or right colon
3	Minimal	Minor segment, small bowel or right colon
4	None	None

Adapted from Fullen WD, Hunt J, Altemeier WA. The clinical spectrum of penetrating injury to the superior mesenteric arterial circulation. *J Trauma* 1972; Aug;12(8):656–64

Surgical options for the management of SMV injuries include ligation or primary repair. A retrospective study of 51 SMV injuries revealed a survival rate of 47%, with 59% of patients undergoing ligation, 31% primary repair and the remaining 10% exsanguinating before any operative intervention. Patients undergoing primary repair had higher survival rates (63%) as compared to ligation (40%); however, those undergoing primary repair also had less associated injuries implying a clear selection bias. Thus, the authors concluded that ligation is safe in patients with multiple other associated injuries.

The surgical management of SMA injuries includes ligation, primary repair or the use of interposition grafts with either saphenous vein or polytetrafluroethylene (PTFE) grafts. The specific management depends on extent and location of injury, associated injuries as well as the patient's physiological status.

A multi-institutional study of 250 patients with SMA injuries revealed an overall mortality of 39%. Fullen's classification of anatomic zones and ischemia grades as well as the AAST abdominal vascular trauma scale correlated with mortality. Patients were treated with ligation (72%), primary repair (22%), venous interposition (4%) and PTFE interposition (2%).

Patients treated with ligation had superior survival rates; however, a higher proportion of these patients had more distal injuries whereas those treated with primary repair were more likely to have proximal injuries. Complications included bowel ischemia or infarction, short bowel syndrome, repair failure, thrombosis and pseudoaneurysm.

Future Directions

Mesenteric trauma represents a particular challenge as the diagnosis can be difficult based solely on clinical data and imaging, and even short delays in diagnosis are associated with significant morbidity and mortality. Laparoscopy has an increasing role in the diagnosis of mesenteric injuries, particularly in stable victims of anterior abdominal stab wounds. The surgical management of these injuries centers around the extent of hemorrhage and bowel viability. The management of proximal mesenteric injuries is particularly challenging and is associated with significant morbidity and mortality.

Given the need to detect mesenteric injuries early and to avoid non-therapeutic laparotomies in those without them, much attention is currently being paid to further clinical diagnostic methods. There is currently an ongoing multi-institutional effort that is performing a prospective observational trial that aims to validate the previously described BIPS method of detection of bowel and mesenteric injuries. This study may be a useful adjunct to the armamentarium of clinicians in the management of mesenteric injuries.

Suggested Reading

1. Fullen WD, Hunt J, Altemeier WA. The clinical spectrum of penetrating injury to the superior mesenteric arterial circulation. J Trauma. 1972, Aug;12(8):656–64.
2. Frick EJ, Pasquale M, Cipolle M. Small bowel and mesentery injuries in blunt trauma. J Trauma. 1999;46(5):920–6.
3. Fakhry SM, Brownstein M, Watts DD, Baker CC, Oller D. Relatively short diagnostic delays (< 8 Hours) produce morbidity and mortality in blunt small bowel injury: an analysis of time to operative intervention in 198 patients from a multicenter experience. J Trauma. 2000;48:408–15.
4. Asensio JA, Britt LD, Borzotta A, Peitzman A, Miller FB, Mackersie RC, Pasquale MD, Pachter HL, Hoyt DB, Rodriguez JL, Falcone R, Davis K, Anderson JT, Ali J, Chand L. Multiinstitutional experience with the management of superior mesenteric, artery injuries. J Am Coll Surg. 2001;193(4):354–66.
5. Fakhry SM, Watts DD, Luchette FA. Current diagnostic approaches lack sensitivity in the diagnosis of perforated blunt small injury: analysis from 275,557 trauma admissions from the east multi-institutional HVI trial. J Trauma. 2003;54:295–306.
6. Asensio JA, Petrone P, Garcia-Nuñez L, Healy M, Martin M, Kuncir E. Superior mesenteric venous injuries: to ligate or to repair remains the question. J Trauma. 2007;62(3):668–75.
7. Joseph DK, Kunac A, Kinler RL, Staff I, Butler KL. Diagnosing blunt hollow viscus injury: is computed tomography the answer? Am J Surg. 2013;205:414–8.
8. Jones EL, Stovall RT, Jones TS, BEnsard DD, Burlew CC, Johnsons JL, Jurkovich GJ, Barnett CC, Peracci FM, Biffl WL, Moore EE. Intra-abdominal injury following blunt trauma becomes clinically apparent within 9 hours. J Trauma Acute Care Surg. 2014 April;76 (4):1020–3.
9. McNutt MK, Chinapuvvula NR, Beckmann NM, Camp EA, Pommerening MJ, Laney RW, West OC, Gill BS, Kozar RA, Cotton BA, Wade CE, Adams PR, Holcomb JB. Early surgical intervention for blunt bowel injury: the bowel injury prediction score (BIPS). J Trauma Acute Care Surg. 2015;78(1):105–11.
10. Lin HF, Chen YD, Lin KL, Wu MC, Wu CY, Chen SC. Laparoscopy decreases the laparotomy rate for hemodynamically stable patients with blunt hollow viscus and mesenteric injuries. Am J Surg. 2015;210:326–33.
11. Bennett AE, Levenson RB, Dorfman JD. Multidetector CT imaging of bowel and mesenteric injury: review of key signs. Semin Ultrasound CT MRI. 2018;39:363–73.
12. Bekker W, Hernandez MC, Zielinski MD, Kong VY, Laing GL, Bruce JL, Manchev V, Smith MTD, Clarke DL. Defining an intra-operative blunt mesenteric injury grading system and its use as a tool for surgical-decision making. Injury, Int J Care Injured. 2019;50 (2019):27–32.
13. American Association for the Surgery of Trauma (AAST). Injury Scoring Scales. Available at: https://www.aast.org/Library/TraumaTools/InjuryScoringScales.aspx.
14. Fakhry SM, Allawi A, Ferguson PL, Michetti CP, Newcomb AB, Liu C, Brownstein MR. Blunt small bowel perforation (SBP): an eastern association for the surgery of trauma multicenter update 15 years later. J Trauma. 2019;86(4):642–50.

Mesenteric Cysts

46

Paul T. Hernandez and Nicole M. Saur

Cystic Tumors of the Mesentery

Cystic tumors of the mesentery are a diverse group of rare tumors that occur across all ages and genders. Given the rarity of these tumors, with the cumulative incidence of approximately 1 in 100,000 to 1 in 250,000 hospitalized patients, they are often lumped under the catchall term "mesenteric cysts". Though frequently treated as a single entity, mesenteric cysts can be divided into six distinct groups based on their histopathologic features or cell-type of origin, with each group of tumors having a distinct epidemiology and biologic behavior. These groups include: (1) cysts of lymphatic origin including lymphangioma, (2) cysts of mesothelial origin including simple mesothelial cysts and multi cystic mesothelioma, (3) enteric duplication cysts, (4) urogenital cysts, (5) cysts derived from germ cells including mature cystic teratoma, and (6) post-traumatic/post-infectious pseudocysts. The management of these lesions varies from observation to emergent surgery depending on patient factors and the presence of and severity of symptoms.

Description of Condition

Mesenteric Lymphangioma

Lymphangioma are benign, congenital tumors characterized by the proliferation of thin-walled lymphatic channels. It is hypothesized that lymphangiomas are formed when primitive lymph vessels fail to connect with the remainder of the lympho-

P. T. Hernandez · N. M. Saur (✉)
Department of Surgery, Division of Colon and Rectal Surgery, University of Pennsylvania, Perelman School of Medicine, 800 Walnut St. 20th Floor, Philadelphia, PA 19106, USA
e-mail: nicole.saur@uphs.upenn.edu

E. D. Ehrenpreis et al. (eds.), *The Mesenteric Organ in Health and Disease*,
https://doi.org/10.1007/978-3-030-71963-0_46

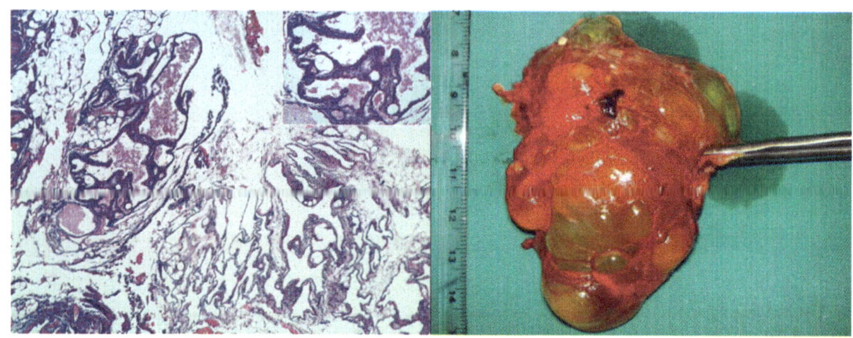

Fig. 46.1 Left: Numerous dilated lymphatic spaces, sometimes filled with lymph (EE 5X). Inset: focally lymphatic spaces are covered by plump endothelial cells (EE 20X). Right: Thin walled, multiseptated cyst filled with serous fluid. Aprea, Giovanni, Francesco Guida, Alfonso Canfora, Antonio Ferronetti, Antonio Giugliano, Melania Ciciriello, Antonio Savanelli, and Bruno Amato. "Mesenteric Cystic Lymphangioma in Adult: A Case Series and Review of the Literature." *BMC Surgery* 13, no. Suppl 1 (2013): A4. https://doi.org/10.1186/1471-2482-13-S1-A4. Reused with permission © Springer Nature

vascular system during embryonic development. Grossly, these tumors appear as thin-walled, multiseptated cysts that contain serous to chylous lymphatic fluid. On pathology, these tumors are characterized by dilated lymphatic channels and large, irregular lymphatic spaces lined by bland epithelial cells and smooth muscle. Among the cystic tumors of the mesentery, lymphangiomas are the most common. The most common sites for lymphangioma are the head, neck, and axilla; approximately 9% of lymphangioma occur in the peritoneal cavity. Less than 1% occur in the mesentery, with the most common site being the mesentery of the small bowel. Mesenteric lymphangioma may also involve the wall of the adjacent bowel. Almost 90% of intraabdominal lymphangioma are diagnosed by 2 years of age. Malignant transformation of mesenteric lymphangioma is exceedingly rare, but has been reported in the literature (Fig. 46.1).

Simple Mesothelial Cysts

It is hypothesized that simple mesothelial cysts form when the mesenteric leaves fail to fuse during development. The resulting tumor is a thin-walled unilocular cyst filled with serous fluid. On histology, these tumors have a fibrous wall that is devoid of a defined muscular layer and are lined by an attenuated layer of mesothelium. While generally considered benign, there have been isolated reports of malignant transformation in the literature. Mesothelial cysts have been described in all portions of the mesentery from the ligament of Treitz to the mesorectum, but the most commonly reported site is the ileal mesentery (Fig. 46.2).

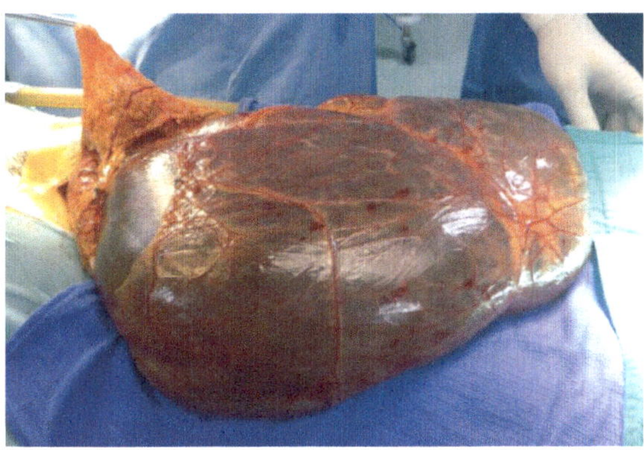

Fig. 46.2 Large thin-walled, unilocular cyst. Ousadden, Abdelmalek, Hicham Elbouhaddouti, Karim Hassani Ibnmajdoub, Taoufiq Harmouch, Khalid Mazaz, and Khalid AitTaleb. "A Giant Peritoneal Simple Mesothelial Cyst: A Case Report." *Journal of Medical Case Reports* 5, no. 1 (December 2011). https://doi.org/10.1186/1752-1947-5-361

Multicystic Mesothelioma

Multicystic mesothelioma of the peritoneum is a rare tumor that arises from the epithelial and mesenchymal elements of mesothelial tissue. In contrast to malignant peritoneal mesothelioma, which is an aggressive and universally fatal disease, multicystic mesothelioma is considered benign and has an extremely low potential for malignant transformation. The incidence of the disease is ~0.15 per 100,000 cases annually. Unlike malignant mesothelioma which typically occurs in men and has a strong association with asbestos exposure, multicystic mesothelioma has been reported in both sexes but typically affects the pelvic peritoneum of premenopausal women though involvement of the small bowel mesentery is possible. Involvement of the mesocolon of the sigmoid and transverse colon are also well described. Grossly, the tumor is characterized by its multicystic "bunch of grapes" appearance with individual cysts that range in size from several millimeters to 20 cm and are filled with thin serous or gelatinous fluid. Microscopically, the cysts are lined by a thin layer of flattened cuboidal or hobnail mesothelial cells and are separated by fibrous tissue. Multicystic mesothelioma has an association with previous surgery, endometriosis and familial Mediterranean fever. Given the significant differences in the biological and clinical behavior between malignant peritoneal mesothelioma and multicystic mesothelioma, some hypothesize that multicystic mesothelioma is a hyperplastic rather than neoplastic process. Nonetheless, multicystic mesothelioma has a high propensity for local recurrence with 25–50% recurrence rates after surgical resection, and while very rare, malignant transformation has been described.

Enteric Duplication Cysts

These developmental anomalies are cystic or tubular bowel-like structures situated adjacent to the luminal GI tract. Like the adjacent bowel, duplication cysts are lined by intestinal epithelium and have a well-defined outer layer of smooth muscle complete with myenteric neural elements. Heterotopic gastric mucosa can also be present in up to 50% of cases, especially in duplications of the ileum and the colon.

There are two morphologic types of enteric duplication cysts, cystic-type and tubular-type. Cystic-type duplications, which make up the majority, are spherical in shape and have no communication with the lumen of the adjacent bowel, though they typically share a common wall and a blood supply with the adjacent segment of intestine. Tubular duplications, which account for approximately 20% of the cases, are often long and tubular and do typically communicate with the adjacent bowel. While duplication cysts can occur anywhere along the GI tract from the mouth to the anus, they are most frequently encountered in the jejunum and ileum on the mesenteric side of the lumen.

There have been many proposed embryologic mechanisms for the formation of enteric duplication cysts, and no mechanism has gained universal acceptance. The split notochord hypothesis, which proposes an abnormal separation of the notochord from the gut endoderm resulting in enteric duplication or diverticula, has gained the most favor as this hypothesis also explains the 15% association of developmental vertebral abnormalities with enteric duplications. As they are congenital tumors, the majority become symptomatic or are discovered incidentally before 2 years of age.

Malignancy is rare, but can arise in enteric duplications. They occur most commonly in colonic duplications as up to 67% of malignancies diagnosed in duplication cysts occur in this location. Malignant transformation can also occur in small bowel duplications, and is more common among duplications with heterotopic gastric mucosa.

Urogenital Cysts of the Mesentery

Urogenital mesenteric cysts are rare congenital tumors thought to arise from vestigial remnants of the embryonic urogenital apparatus. On histology, these cysts are lined by cuboidal to columnar epithelium. They may also demonstrate some elements of distinct pronephric, mesonephric, metanephric, or Mullerian differentiation including various primitive renal elements and fallopian tube-like ciliated epithelium. Grossly, the cysts are typically thin-walled and contain a mix of serous fluid and adipose tissue (Fig. 46.3).

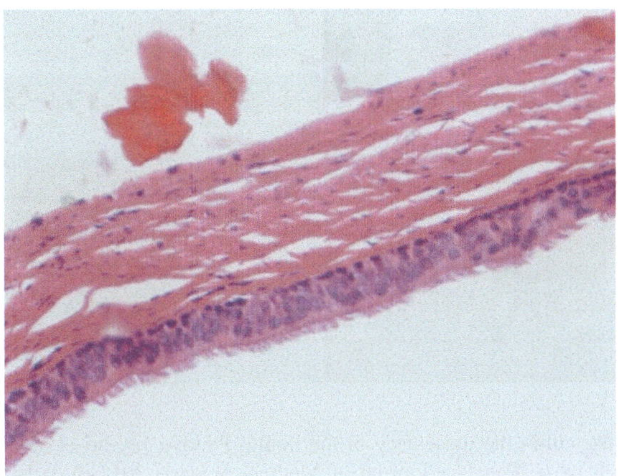

Fig. 46.3 Cyst with fibromuscular wall covered by ciliated epithelium, of Mullerian type. Mokhtari M, Kumar PV. Cytologic findings of urogenital mesenteric cyst. Arch Iran Med. 2013; 16(7):436–438.

Mature Cystic Teratoma

Mature cystic teratomas, otherwise known as dermoid cysts, are germ cell tumors that arise from totipotent primordial germ cells. These tumors have a cyst wall lined by flattened cuboidal epithelium and contain a mix of tissues derived from all three germ cell layers including tissue of ectodermal (i.e. skin, hair follicles, sebaceous glands, teeth), mesodermal (i.e. muscle, urinary), and endoderm origin (i.e. lung, gastrointestinal) as well as liquid sebaceous material. Macroscopically, these tumors are thick walled, septated cysts that contain a thick, fatty, gelatinous material as well as teeth and hair.

Teratomas can occur anywhere along the pathway of germ cell migration, most commonly in the midline, from the cranium to the sacrococcygeal region. The most common location is the sacrococcygeal region and the gonads, but there are numerous reports of these tumors occurring in the small bowel mesentery, making mature cystic teratomas an important entity in the differential diagnosis of a mesenteric cyst.

While the vast majority of these tumors occur in adult females, isolated cases have been reported in male patients. Mature cystic teratomas are considered benign, but malignant transformation occurs in 0.2 to 2% of these tumors with increased risk of malignancy in patients 45 years and older and those with rapidly growing tumors or tumors greater than 10 cm in diameter (Fig. 46.4).

Mesenteric Pseudocyst

Similar to the more common pancreatic pseudocysts, mesenteric pseudocysts are fluid collections encased fibrous tissue without an epithelial lining. Mesenteric pseudocysts are not associated with pancreatitis and are thought to occur secondary

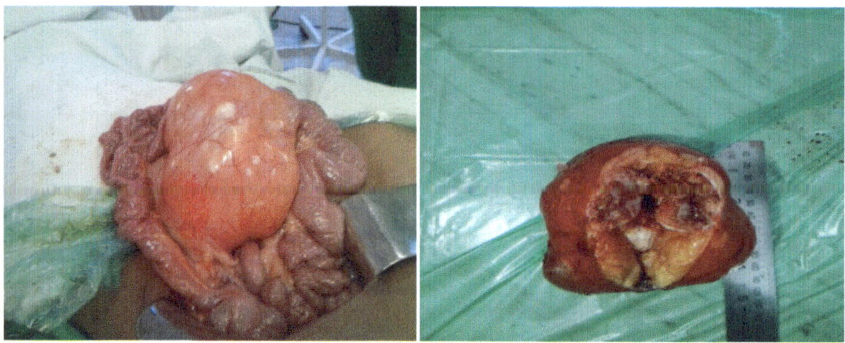

Fig. 46.4 Teratoma within the mesentery of the ileum. Pandya Jayshri et al. Mature mesenteric teratoma in an adult male. Panacca Journal of Medical Science 2015; 5(3):161–163. Copyright © 2015 by author(s) and Panacea J Med Sci. This is an Open Access article distributed under the terms of the Creative Commons Attribution 4.0 International License (creativecommons.org)

to trauma, infections, or prior surgery when a hematoma or abscess is formed in the mesentery and not completely absorbed. As a result, the persistent cyst is often filled with hemorrhagic contents and cellular debris.

Mesenteric pseudocysts occur most commonly in patients in their 40 s and are reported more commonly in females than in males (Fig. 46.5a and b).

Presentation and Diagnosis

Cystic tumors of the mesentery have a wide spectrum of presentations. Approximately 40% of mesenteric cysts are discovered incidentally on physical exam, on imaging, or during abdominal surgery for other indications. Of the remaining 60% of patients, the majority present with chronic, non-specific symptoms including mild abdominal pain, abdominal bloating, or abdominal distension. Only 5 to 10% of patients will ultimately present with acute complications of their cysts.

Diagnostic Studies in Uncomplicated Mesenteric Cysts

The majority of cystic tumors of the mesentery are discovered incidentally on abdominal imaging, physical exam, or in the setting on non-specific abdominal symptoms. Ultrasound (US) and commuted tomography (CT) are the most commonly utilized imaging modalities in the diagnosis of mesenteric cysts. While less commonly employed, magnetic resonance (MR) imaging can sometimes more accurately describe the relation between the mass and the adjacent organs and soft tissues or provide additional information regarding the nature of the cyst contents.

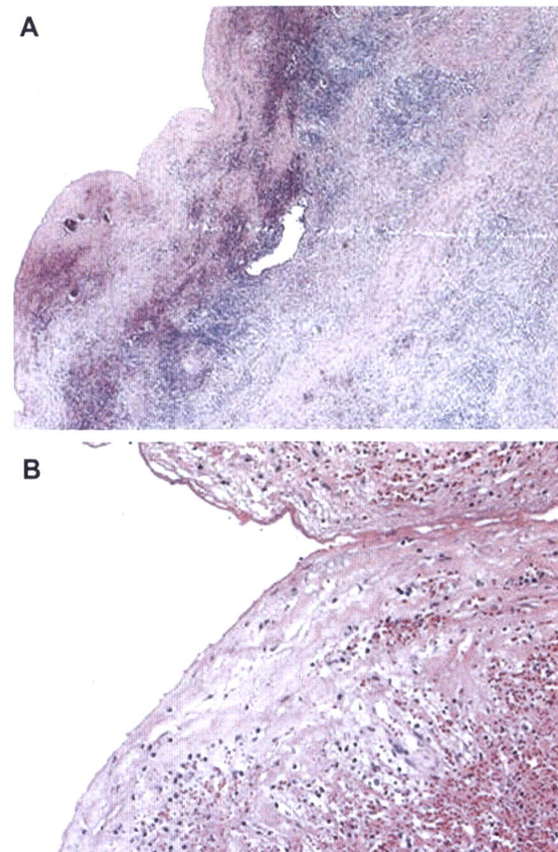

Fig. 46.5 a and b Histopathology of the pseudocyst wall showing dense fibrous tissue with no epithelial lining. Geng, Jiun-Hung, Chun-Hsiung Huang, Wen-Jeng Wu, Shu-Pin Huang, and Yi-Ting Chen. "Huge Retroperitoneal Nonpancreatic Pseudocyst." *Urological Science* 23, no. 2 (June 2012): 61–63. https://doi.org/10.1016/j.urols.2012.03.002. Reused with permission © Elsevier

Physical Exam

When discovered on physical exam, a mesenteric cyst may demonstrate a triad of findings first described by Paul Jules Tillaux in the late nineteenth century. Tillaux's triad includes (1) a fluctuant abdominal mass near the umbilicus that (2) is mobile only from the patient's right to left or left to right and (3) is dull to percussion but encircled by zone of tympani representing the surrounding bowel.

Imaging

Each imaging modality for the evaluation of mesenteric cyst has its own benefits and limitations. US has the benefits of being both widely available and cost effective. Unlike CT, US does not emit ionizing radiation making it the study of choice in children. US, however, is operator dependent and does not demonstrate the relationship of the mesenteric cyst to adjacent organs as well as CT and MR.

While all varieties of mesenteric cysts share the common characteristics of being well-defined, hypoechoic/hypoattenuating masses, subtle differences in the radiologic appearances of the different types of mesenteric cysts can facilitate the making a pre-operative diagnosis.

Lymphangioma

By US, lymphangiomas appear as thin-walled, cystic, multiseptated/lobulated mases that may contain internal echoes, sedimentation or fluid–fluid levels caused by cellular debris.

On CT, mesenteric lymphangioma appear as cystic masses with Hounsfield units (HU) ranging from water attenuation for cysts with predominantly serous contents to fat attenuation for cysts with predominantly chylous contents. When performed with intravenous contrast, the cyst wall may show enhancement. The fine septa, readily appreciable by ultrasound, are usually less apparent on CT though there are occasionally calcifications that form along the septa and revealing their presence.

For mesenteric lymphangioma, MR can provide additional information that can further clarify the diagnosis. These tumors are hyperintense on T2-weighted images, and their internal septae are much more apparent when compared to CT. Typically, the cyst contents demonstrate high signal intensity on T1-weighted images, and a comparison of in-phase and opposed-phase T1-weighted images can demonstrate the fact that these tumors contain both a significant amount of fat and water as expected in lymphatic fluid. This technique allows lymphatic cysts to be differentiated from cystic masses that contain purely serous fluid and those that are purely fat containing.

Mesothelial Cysts

By US, mesothelial cysts are similarly thin-walled and anechoic, but they typically lack the septations and internal debris seen in lymphangiomas. These cysts often demonstrate strong posterior acoustic enhancement. CT and MR demonstrate a cystic mesenteric mass without a discernable wall or septations. The cyst fluid is also different from that of lymphangiomas in that it has Hounsfield units by CT closer to water than fat, and the fluid has low signal intensity on T1-weighted images as compared to the high signal intensity in lymphangiomas.

Multicystic Mesothelioma

With a similar appearance to lymphangioma, multicystic mesothelioma appear as a hypoechoic, multiloculated mass on US. CT commonly demonstrates a cystic mass with homogeneous fluid attenuation and no calcifications. As with lymphangioma, CT may or may not demonstrate the multilocular nature of the mass. On MR, these cystic lesions show low signal intensity on T1-weighted images and high signal intensity on T2-weighted images without any solid components or enhancement.

Enteric Duplication Cysts

Ultrasonography of enteric duplication cysts classically demonstrates a double-layered wall with an inner layer of hyperechoic mucosa and an outer layer of hypoechoic muscle often referred to as the "gut signature". When present in a cyst the "gut signature" has a high specificity for an enteric duplication cyst, but in many cases inflammation may obscure this classic sign (Fig. 46.6).

By CT, enteric duplications are flat, spherical or tubular tumors situated adjacent to a loop of bowel. They can displace but show no evidence of infiltrating other organs. As with the adjacent bowel, the wall of duplications generally demonstrates mucosal enhancement with intravenous contrast. If distended with fluid at the time of CT, cyst contents have Hounsfield units in the 0–20 range. MR imaging similarly demonstrates a well-circumscribed, homogeneous, hypointense mass on T1-weighted images. On T2-weighted images of the distended cyst contents are hyperintense, though there may be variability in the signal due to associated hemorrhage or mucous secretion. As with CT, gadolinium-enhanced MRI shows enhancement of the cyst wall.

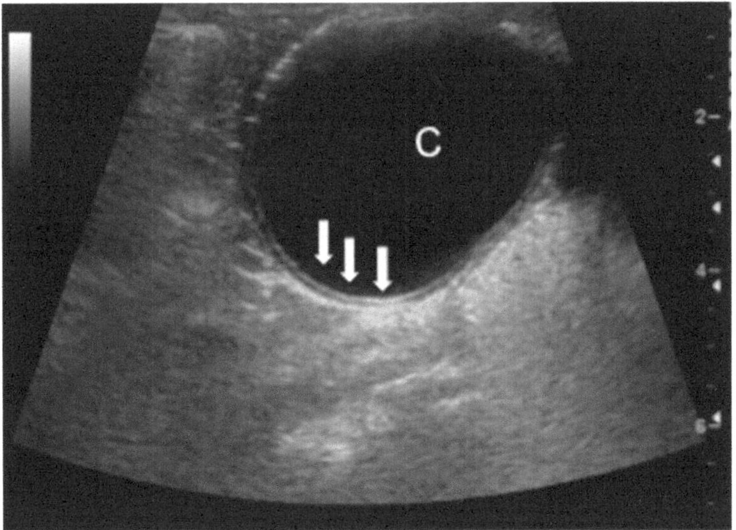

Fig. 46.6 Portions of the cyst wall show the characteristic double-layered appearance with echogenic mucosa internally (*arrows*) and hypoechoic muscle externally. Di Serafino, Marco, Carmela Mercogliano, and Gianfranco Vallone. "Ultrasound Evaluation of the Enteric Duplication Cyst: The Gut Signature." *Journal of Ultrasound* 19, no. 2 (June 2016): 131–33. https://doi.org/10.1007/s40477-015-0188-8. Reused with permission © Springer Nature

Barium studies, though less-frequently used, can show unique findings. Most commonly these demonstrate a submucosal mass causing extrinsic indentation of the adjacent bowel, but if the duplication communicates with the lumen, a spherical or tubular structure filled with barium can be seen adjacent to the GI tract. This is a finding with high specificity for this entity.

Tc 99 pertechnate scans can demonstrate the presence ectopic gastric mucosa in enteric duplications with a sensitivity of up to 75%.

Mature Cystic Teratoma

Of the cystic tumors of the mesentery, mature cystic teratomas have the most characteristic ultrasound appearance. Identification of a "dermoid plug" or echogenic mass of teeth, hair, or fat floating within the cyst is one specific finding. The presence of a "dermoid mesh", a mesh-like network of hyperechoic lines and dots with in the cyst due to small hairs floating in the cystic fluid, and a "fat-fluid level" or the interface of layering serous fluid and sebum allow the diagnosis of a dermoid cyst by ultrasound with a positive predictive value close to 100% (Fig. 46.7).

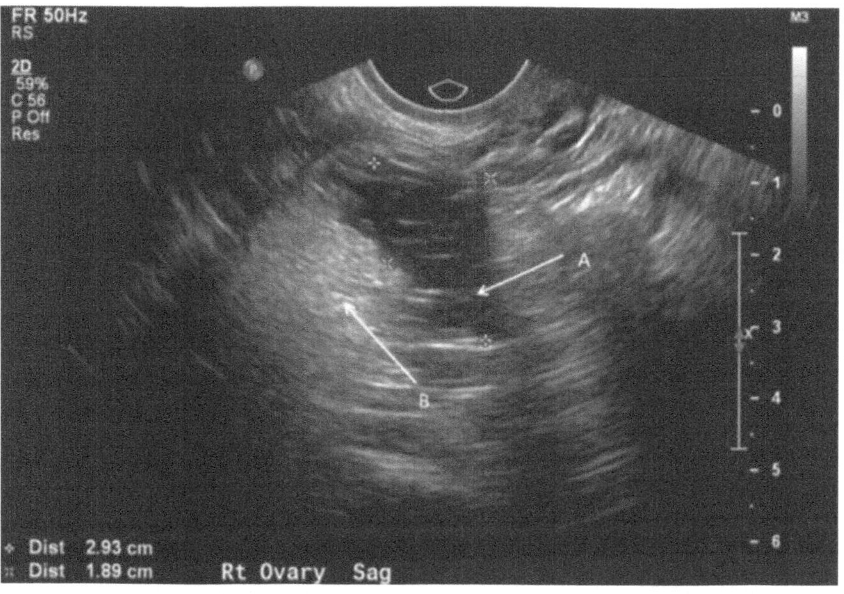

Fig. 46.7 Dermoid mesh (**a**) and dermoid plug (**b**). Kite, Lauren, and Talat Uppal. "Ultrasound of Ovarian Dermoids - Sonographic Findings of a Dermoid Cyst in a 41-Year-Old Woman with an Elevated Serum HCG." *Australasian Journal of Ultrasound in Medicine* 14, no. 3 (August 2011): 19–21. https://doi.org/10.1002/j.2205-0140.2011.tb00119.x. Reused with permission © John Wiley and Sons

Various imaging findings on CT have high specificity as well. CT demonstrates a thick-walled cystic tumor with central fat attenuation and scattered areas of calcification often representing teeth. Tufts of hair are also frequently identified on CT, but the fat-fluid level seen on ultrasound is less frequently appreciated.

By MR, the sebaceous components of mature cystic teratoma have high signal intensity on T1-weighted imaging though this signal is lost on fat-suppressed T1-weighted sequences reflecting the macroscopic fat within the cyst.

Mesenteric Pseudocyst

By US, mesenteric pseudocysts are unilocular or multilocular cystic masses with a hypoattenuating, fluid-filled center and varying amounts of intralesional debris. The wall of these cysts is often very thick and lacks the "gut signature" seen in enteric duplications. CT and MR show a thick-walled cyst with varying degrees of internal septation and debris.

Endoscopy and Endoscopic Ultrasound

Traditional endoscopy is not considered a diagnostic modality of choice for cystic tumors of the mesentery as only those cysts affecting the colon and very proximal small bowel are accessible with traditional endoscopic techniques. Furthermore, given the majority of cystic tumors of the mesentery, save communicating duplication cysts, do not communicate with the lumen of the bowel, endoscopy generally demonstrates little more than evidence of extrinsic compression or a submucosal tumor. Still, mesenteric cysts have been detected incidentally on screening endoscopies and when endoscopy is chosen as a diagnostic modality to investigate vague abdominal symptoms. Adding endoscopic ultrasound to traditional endoscopy, however, has been shown to be effective in clarifying the diagnosis as this modality provides the same information as traditional ultrasound often with higher resolution.

Acute Complications of Mesenteric Cysts

Obstruction, Volvulus and Intussusception

Mesenteric cysts can result in bowel obstruction due to mass effect on the adjacent bowel, intussusception, or intestinal volvulus. In the setting of intestinal volvulus, patients may present with signs of intestinal ischemia including severe abdominal pain, hemodynamic instability, leukocytosis, lactic acidosis, and peritonitis, in addition to the bilious/feculent emesis and obstipation seen in uncomplicated obstruction. Often, patients presenting with a mesenteric cyst causing obstruction or volvulus will describe a history of intermittent obstructions and/or abdominal pains in the past. CT in an obstructed patient demonstrates dilated loops of small bowel often with air fluid levels and a transition point at the site of compression by the cystic tumor. Contrast enhanced CT in patients with intestinal volvulus classically demonstrates whirling of the small bowel and mesenteric vessels around the superior mesenteric artery with evidence of a closed-loop bowel obstruction. Frequently, the mesenteric cyst is seen lodged in the pelvis where is acts as a point of fixation, allowing volvulus to occur.

Ureteral Obstruction

Mesenteric cysts of any etiology can cause external compression of the ureter resulting in hydronephrosis. Unilateral ureteral compression at the sacral promontory is most commonly reported. Frequently these cysts are large enough to cause symptoms of abdominal distension and pain though some patients may be asymptomatic. In patients with normal renal function, unilateral ureteral obstruction does not significantly elevate markers of renal insufficiency. Conventional urography demonstrates ureteral obstruction and CT urography demonstrates ureteral obstruction at the site of the compression by the cystic tumor (Fig. 46.8).

Cyst Rupture

Cyst rupture typically presents on a spectrum from focal, mild abdominal pain to severe generalized abdominal pain and peritonitis often depending on the size and etiology of the cyst. Ruptured, infected cysts and ruptured, hemorrhagic cysts produce the most severe symptoms. CT findings typically demonstrate free fluid in the dependent portions of the abdomen. There may or may not be evidence of a residual cyst depending on the presence of septations/loculations prior to cyst rupture.

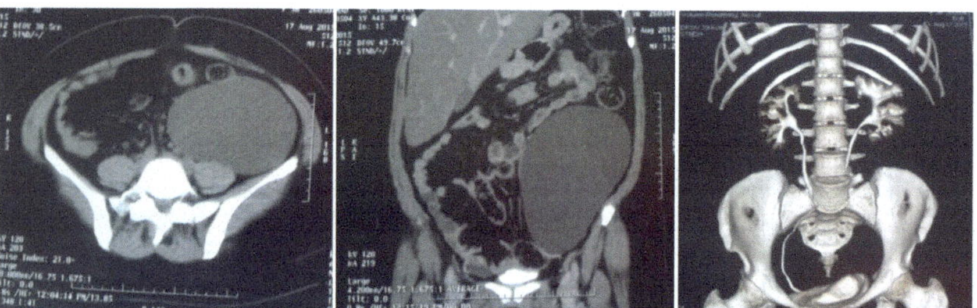

Fig. 46.8 Mesothelial cyst arising from the jejunal mesentery causing L sided ureteral compression and L sided hydronephrosis. El-Agwany, Ahmed Mohammed Samy. "Huge Mesenteric Cyst: Pelvic Cysts Differential Diagnoses Dilemma." *The Egyptian Journal of Radiology and Nuclear Medicine* 47, no. 2 (June 2016): 373–76. https://doi.org/10.1016/j.ejrnm.2016.01.008 Reused with permission © Elsevier

Cyst Hemorrhage

Internal hemorrhage in an isolated cyst often results in severe, localized abdominal pain due to internal pressurization of the cyst causing stretching and inflammation of the overlying peritoneum. Significant bleeding leading to hemorrhagic shock is very rare, but has been described, generally in the setting hemorrhage that results in cyst rupture. Ultrasound and CT often demonstrate cystic masses with evidence of layering hyperechoic or high attenuating blood products as well as evidence of inflammation of the adjacent mesentery in the form of hyperechoic fat or high attenuating fat.

Infected Mesenteric Cyst

All mesenteric cysts have the potential for spontaneous infection of the cyst contents, but infected mesenteric pseudocysts are most commonly reported in the literature. The most commonly reported infectious agents include *Staphlococcus aureus* and *Escheria coli* though *Streptococcus pneumoniae*, *Salmonella enteritis*, and *Mycobacterium tuberculosis* have also been implicated.

Infected non-ruptured mesenteric cysts most commonly present with abdominal pain, fevers, focal tenderness to palpation, and leukocytosis. In the case of a non-ruptured, infected mesenteric cyst, CT frequently demonstrates the cystic mass with rim enhancement in the setting of intravenous contrast.

When infected mesenteric cysts rupture, the presentation is more acute. These patients typically present with marker of a systemic inflammatory response including tachycardia, significant leukocytosis, high fevers, and diffuse peritonitis. Some patients may even demonstrate end organ dysfunction and shock.

Peptic Ulceration

Mucosal ulceration resulting in gastrointestinal hemorrhage and perforation has also been described in the setting of enteric duplication cysts of the small or large bowel that contain ectopic, acid-secreting gastric mucosa.

Patients presenting with perforation typically demonstrate the signs and symptoms classically associated with a perforated viscus including the acute onset of severe abdominal pain, peritonitis, tachycardia and leukocytosis. CT in this setting usually demonstrates some degree of free air and free fluid within the abdomen, and the duplication cyst may be apparent.

The presentation of bleeding from mucosal ulceration due to heterotopic gastric mucosa is often variable depending on the location of the cyst. Lesions associated with the proximal small bowel often present with anemia and a history of melena, while distal lesions, especially those in the distal colon and rectum, commonly present with bright red rectal bleeding. For lesions in the colon and the very proximal small bowel, traditional endoscopic techniques may be helpful in identifying and temporarily treating the site of and may demonstrate the connection between the duplication cyst and the normal bowel. Bleeding caused by duplications in the jejunum and ileum are more challenging to localize. As these lesions are typically out of the reach of traditional endoscopy, double balloon enteroscopy or capsule endoscopy may be helpful. As with other etiologies of gastrointestinal bleeding, CT angiography, mesenteric angiography and Tc^{99} tagged red blood cell nuclear scan can identify active bleeding caused by duplication cysts. If an enteric duplication cyst or with ectopic gastric mucosa is suspected, a 99^{m} Tc pertechnetate scan or "Meckel's scan" can be performed to demonstrate the presence of ectopic gastric mucosa with a positive predictive value approaching 100%. This study will not, however, differentiate between duplication cysts and the more common Meckel's diverticulum.

Management

Indications for Surgery

Given the rarity of cystic tumors of the mesentery there is no definitive evidence-based consensus on their management. The most common approach to these tumors, regardless of etiology, involves urgent surgical management in all patients presenting with acute complications, including elective resection in patients whose cysts are symptomatic and in good surgical candidates whose cysts are discovered incidentally.

The rationale given by most authors regarding the decision for surgery in the asymptomatic patient is first to prevent these tumors from becoming symptomatic or from causing acute complications, and second to prevent the rare event of malignant transformation. Surgery is generally recommended for all types of mesenteric cysts regardless of the presumed pre-operative diagnosis.

Contraindications to Surgery

In patients presenting with bowel obstruction, mesenteric ischemia, sepsis, or peritonitis secondary to complications of a mesenteric cyst, urgent surgical management is essential and without contraindication apart from patient refusal. Among symptomatic patients and patients whose cysts are discovered incidentally, surgery should be recommended to those who are deemed low-risk surgical candidates. Observation and serial imaging is reasonable in asymptomatic and mildly symptomatic patients deemed moderate to high-risk, but given the rarity of these tumors there is no consensus on the optimal interval of, duration of, or best imaging modality for surveillance.

Surgical Options

For good surgical candidates, complete cyst excision is the recommended surgical procedure for all varieties of mesenteric cysts as aspiration/drainage or marsupialization results in significant rates of recurrence, precludes pathologic examination of the entire cyst, and does not eliminate the risk of malignant transformation.

Both open and laparoscopic approaches to resection of mesenteric cysts have been described. In most cases, mesenteric cysts can be carefully enucleated without injuring the vessels of the mesentery, and this is particularly true in simple mesothelial cysts, benign cystic mesothelioma, mature cystic teratoma, and urogenital cysts. To facilitate laparoscopic enucleation, many authors describe cyst drainage prior to beginning dissection.

If the mesenteric vessels cannot be preserved, or if the tumor involves the wall of the adjacent bowel as commonly seen in duplications and lymphangioma, a segmental bowel resection with primary anastomosis should be performed.

While not common in clinical practice, some authors have advocated for the use of hyperthermic intraperitoneal chemotherapy at the time of surgical resection of benign multicystic mesothelioma given its relatively high recurrence rates of 25–50% after surgical excision. While this has technique has been demonstrated to be safe, there is currently no data to support that decreases the rates of recurrence as compared to surgery alone.

For symptomatic patients who are poor surgical candidates, percutaneous drainage of mesothelial cysts may provide temporary relief, and endoscopic internal drainage of colonic duplications has also been described.

Summary

Mesenteric cysts are rare entities that can be discovered incidentally, present symptomatically, or present acutely. If the presentation is subacute, imaging can be utilized to classify the cyst type and plan for treatment. Cysts are generally excised in fit patients to prevent the development of cyst-related symptoms or malignant transformation. If the presentation is acute, the patient should be resuscitated and taken to the operating room emergently and, if possible, the cyst should be resected at that time.

Suggested Reading

1. Stoupis C, Ros PR, Abbitt PL, Burton SS, Gauger J. Bubbles in the belly: imaging of cystic mesenteric or omental masses. RadioGraphics. 1994, July; 14(4):729–37. https://doi.org/10.1148/radiographics.14.4.7938764.
2. de Perrot M, Bründler M-A, Tötsch M, Mentha G, Morel P. Mesenteric cysts. Dig Surg. 2000;17(4):323–28. https://doi.org/10.1159/000018872.
3. Shamiyeh A, Rieger R, Schrenk P, Wayand W. Role of laparoscopic surgery in treatment of mesenteric cysts. Surg Endos. 1999, September;13(9):937–39. https://doi.org/10.1007/s004649901140.
4. Prakash A, Agrawal A, Gupta RahulK, Sanghvi B, Parelkar S. Early management of mesenteric cyst prevents catastrophes: a single centre analysis of 17 cases. Afr J Paed Surg. 2010;7(3):140. https://doi.org/10.4103/0189-6725.70411.
5. Tan JJY, Tan KK, Chew SP. Mesenteric cysts: an institution experience over 14 years and review of literature. World J Surg. 2009, September;33(9):1961–65. https://doi.org/10.1007/s00268-009-0133-0.

Mesenteric Abscess

47

Sameh Hany Emile

Mesenteric abscess is a term that describes the collection of pus or infected material between any of the several folds of peritoneum that envelop the intestine and connect them to the posterior abdominal wall.

Description of the Condition

The pathogenesis of mesenteric abscess entails a wide spectrum of conditions. Based on its etiology, mesenteric abscess can be broadly classified into primary and secondary abscesses.

Primary abscess of the small bowel mesentery is quite rare. Dudely and MacLaren attributed primary mesenteric abscess to diffuse suppurative mesenteric adenitis in which the enlarged mesenteric lymph nodes develop small foci of necrosis and suppuration. This condition may present with generalized peritonitis or may have subtle course and is incidentally detected during laparotomy. Patients with mesenteric abscess may also present with colicky abdominal pain that gradually becomes more frequent and severe. Diarrhea associated with passage of mucous with stool is another possible presentation of primary mesenteric abscess due to irritation of the intestinal walls which results in increased intestinal peristalsis. The majority of patients have toxemia manifested with high grade fever, tachycardia, and dehydration.

S. H. Emile (✉)
Department of Colorectal Surgery, Mansoura University Hospitals,
60 Elgomhouria St, Mansoura 35516, Dakahlia, Egypt
e-mail: Sameh200@hotmail.com

E. D. Ehrenpreis et al. (eds.), *The Mesenteric Organ in Health and Disease*,
https://doi.org/10.1007/978-3-030-71963-0_47

Nonetheless, Fraenkel questioned the presence of primary mesenteric abscess and considered it a misnomer. He reported finding foreign bodies as fishbone and rolled-up tomato skin in two cases of mesenteric abscess which suggested that the inflammatory process was secondary to the presence of foreign body, and was not an idiopathic process.

However, mesenteric abscess can also be secondary to a variety of causes including inflammatory bowel diseases, particularly Crohn's disease, Meckel's diverticulitis, tuberculosis, infected mesenteric cysts, complicated mesenteric adenocarcinoma, and parasitic infestations.

Patients with Crohn's disease can spontaneously develop abdominal or pelvic abscess. The transmural inflammation of Crohn's disease eventually results in microperforations of the intestinal segment. Contained perforations lead to the formation of abscess cavity, mostly at the ileocecal region. However, Crohn's disease-related abscesses have also been reported in the abdominal wall, psoas and gluteus muscles, subphrenic space, and in the mesentery of the bowel.

Mesenteric abscess causing extrinsic compression of the bowel can actually be common in Crohn's disease. As Hokama and colleagues reported, the abscess usually presents as a round, egg-sized, tender mass in either quadrants of the abdomen accompanied by fever and leukocytosis in patients with Crohn's disease.

Teeples and Tabibian described a case of Crohn's disease-related mesenteric abscess that presented with symptoms of bowel obstruction including progressive abdominal distention, nausea, and feculent vomiting. Weight loss is another notable manifestation of the disease. In contrast to the usual acute presentation, patients with mesenteric abscess may present without fever, tachycardia, or other signs of toxemia. Interestingly, some patients may have normal abdominal examination apart from mild infraumbilical tenderness. The definitive diagnosis of mesenteric abscess in patients with this subtle clinical picture can be only made by CT or MR enterography which can clearly detect the air-fluid filled abscess cavity within the mesentery, displacing the intestine.

Although uncommon, mesenteric abscess may be a consequence of inflamed or ruptured Meckel's diverticulum as has been reported in a few reports. In a report by Duncan D, an otherwise healthy man complained of vomiting and central abdominal pain that became localized to the right of the umbilicus and was misdiagnosed as having acute appendicitis. However, exploration revealed a 5-cm mesenteric abscess with an inflamed Meckel's diverticulum adherent to its posterior surface. En-bloc resection of the inflamed diverticulum and the affected mesentery was conducted and was followed by uneventful recovery. An explanation of this highly unusual condition was made that prolonged fixation of the diverticulum to the mesentery, probably caused by congenital adhesions, allowed stagnation of the intestinal contents and formation of an enetrolith that became lodged inside the diverticulum and initiated the acute inflammatory process.

In another report by Wasike and Saidi in 2006, a young man presented with symptoms and signs suggestive of acute appendicitis associated with elevated leucocyte count and fever. High resolution spiral abdominal CT scan revealed an abscess cavity within the small bowel mesentery of about 2.5 inches in size. On exploratory laparotomy a perforated Meckel's diverticulum was found adherent to the mesentery and walled off with gangrenous ileal loops which were resected along with the perforated diverticulum. This report highlighted the important fact that complicated Meckel's diverticulum may disguise as mesenteric abscess, and the differentiation between the two entities is usually difficult, even after performing abdominal CT scan.

Mesenteric abscess may also be secondary to tuberculosis. The incidence of extrapulmonary tuberculosis has increased considerably with the growing numbers of patients with human immune deficiency virus. While abdominal tuberculosis mainly affects the terminal ileum and the cecum, cases of isolated mesenteric tuberculosis have also been reported.

Many patients with mesenteric tuberculosis have asymptomatic pulmonary tuberculosis. Swallowing the infected sputum leads to spread of mycobacterial infection to the small bowel and subsequently to the mesenteric lymph nodes. As the disease progresses, the mesenteric lymph nodes undergo caseation, forming caseous granulomas that eventually develop into cold abscess.

Clinical presentation of mesenteric abscess secondary to tuberculosis may include classical symptoms of tuberculosis such as loss of weight, loss of appetite, night fever and night sweats. In addition, nonspecific symptoms as abdominal pain and nausea may also be present. What makes the diagnosis even more difficult is that many of these patients have no history of contact with tuberculosis patients, have no pulmonary symptoms suggestive of tuberculosis, and have negative sputum to acid-fast bacilli as Anand and Sushmita highlighted in their report. Vijaya et al. found fibrocavitation in the lung in chest X-ray helpful in leading the diagnosis whereas abdominal CT scanning may reveal an ileocecal or mesenteric mass. Nonetheless, in some patients the diagnosis can be only made after laparotomy.

A more aggressive presentation of mesenteric tuberculosis can occur in the setting of disseminated tuberculosis associating HIV infection as once described by Pandit et al. The immune deficiency makes the patient more susceptible to acquire pulmonary and extrapulmonary tuberculosis which follows a more aggrieve course and disseminated pattern than the usual one. Abdominal CT or MRI scanning may show multiseptated peripherally enhancing hypodense lesion in the mesentery, suggestive of mesenteric abscess. CT-guided fine needle aspiration cytology (FNAC) from the abdominal lesion may help conclude the diagnosis of mesenteric tuberculous abscess after staining for acid-fast bacilli and doing PCR of the aspirated pus.

In addition to mycobacterium tuberculosis, patients with HIV can develop mesenteric abscess secondary to mycobacterium avium infection which was once reported by Mohar and colleagues. Mycobacterium avium is one of the most serious opportunistic infections that commonly affects the lungs, yet localized mycobacterial lymphadenopathy or lymphadenitis can also occur. Intra-abdominal affection by mycobacterium avium infection is very rare, however, once diagnosed it could indicate a disseminated pattern of infection that warrants a combination of at least two antibiotics including clarithromycin or azithromycin in addition to ethambutol and/or rifabutin.

Mesenteric abscess may represent a secondary infection of a mesenteric cyst as Kim et al. described in one report. Mesenteric cysts are collection of fluids between the two layers. They can be categorized as true cysts with epithelial lining and pseudocysts without an epithelial lining. True mesenteric cysts include chylolymphatic cyst which arise due to abnormal lymphatic development or degeneration of lymphatics and lymph nodes; enetrogenous cyst due to sequestration of the developing bowel into the developing mesentery; teratomatous or dermoid cyst; and hydatid cyst. Pseudocysts also include traumatic cysts; cold abscess; and degenerating tumors.

Non-complicated mesenteric cysts are usually asymptomatic or may present with Tillaux's triad which involves a fluctuant swelling near umbilicus that moves freely across but not along attachment of mesentery and is dull on percussion with a zone of resonance around and a band of resonance across. Spontaneous infection of mesenteric cysts is very rare and in most cases Escherichia coli is the causative organism. Patients who develop infection of mesenteric cysts rapidly develop acute abdomen with high grade fever and leukocytosis. Diagnostic modalities include abdominal ultrasonography, CT scanning with contrast, and magnetic resonance imaging.

One of the rarest presentations of mesenteric abscess reported in the literature is primary mesenteric adenocarcinoma complicated with infection and abscess in the mesocolon. In a case report by Luo and colleagues, a middle-aged male presented with acute intestinal obstruction without specific findings in the abdominal CT scan. Laparotomy revealed a 6 × 5 × 4 cm abscess in the mesocolon below the splenic flexure. Marked hypermeia, edema, and induration of the left colon mesentery was noted with significant tissue adhesions. Intraoperative frozen section analysis revealed a primary mesenteric adenocarcinoma. Despite being a rare diagnosis, the presence of a mesenteric neoplasm in the setting of mesenteric abscess should be excluded by histopathologic diagnosis, particularly in high-risk patients.

Another rare cause of mesenteric abscess is parasitic infestation, namely by thread worms or Enterobius vermicularis. Beeson and Woodruff assumed that extra-intestinal deposition of ova of thread worms may result in mesenteric abscess in children. An 11 year old child was operated on for acute appendicitis and during the procedure a nodular patch of thickened tissue was found in the omentum covering the appendix. Pathologic examination of the diseased tissue revealed that Entrobius ova were the cause of the inflammatory process.

Diagnosis of Mesenteric Abscess

Being an uncommon entity, the diagnosis of mesenteric abscess is usually difficult and tends to be delayed. Patients with mesenteric abscess may have a suggestive history of underlying diseases such as Crohn's disease or tuberculosis. The clinical presentation may vary according to the underlying pathology. Some patients present with typical symptoms and signs of acute inflammatory process in the abdomen manifesting as acute abdominal pain and tenderness, high grade fever, toxemia, and elevated leucocyte count and C-reactive protein. In contrast, other patients may present with less pronounced, nonspecific symptoms such as loss of weight and appetite, low grade fever, abdominal mass, and diarrhea.

Imaging studies play a crucial role in the diagnosis of mesenteric abscess; plain abdominal X-ray may give a clue by showing air and fluid inside the abscess cavity. Abdominal ultrasonography may demonstrate a well-defined cystic lesion inside the bowel mesentery, with uniform low level internal echoes; however; it may not be reliable in several cases in which the gaseous distention of the bowel can obscure the examined field.

The investigation of choice for the diagnosis of mesenteric abscess, or intraabdominal abscess in general, is the contrast-enhanced CT scanning of the abdomen. Compared with ultrasound, CT scan is more accurate and sensitive, less operator dependent, and provides better anatomic details, especially for retroperitoneal structures. The multidetector CT scan with three-dimension volume rendering has proved to be more precise in evaluating neoplastic, infectious, and inflammatory processes that affect the mesentery. Mesenteric abscess varies in appearance in CT scan. While acute abscesses tend to have thin wall, chronic abscesses have much thicker walls. The abscess may also appear as unilocular, multilocular or septate cavity with infiltration of surrounding fat. Large mesenteric abscesses can produce a mass effect, displacing the mesenteric vessels and neighboring structures. In patients with Crohn's disease, multiple necrotic mesenteric lymph nodes appear in CT scan as suppurative non-caseating abscesses, which is known as disseminated aseptic abscesses.

CT-guided FANC from the abscess cavity may confer further useful data for establishing the diagnosis of mesenteric abscess. The aspirate is subjected to thorough chemical and microbiological examination in attempt to decipher the underlying cause of the abscess. Staining for acid-fast bacilli and PCR of the aspirated contents may help reach the diagnosis of tuberculous mesenteric abscess. However, in some cases the examination of the aspirate is negative to bacteria and fungi, indicative of a condition named "sterile abscess syndrome".

None of the reports of mesenteric abscess in the literatures used MRI for diagnosis, perhaps because of its limited availability and high costs. However, MRI is a useful tool for diagnosis of intra-abdominal abscess in general. Intra-abdominal abscess, including mesenteric abscess, are detected as well-defined fluid collections with rim enhancement and decreased signal intensity on gadolinium-enhanced, T1-weighted images.

Adjunct measures that may play a complementary role in identifying the etiology of mesenteric abscess include barium studies and colonoscopy, particularly in suspected Crohn's disease. Hokama and colleagues reported a mesenteric abscess close to the sigmoid colon as detected by CT scanning in a patient with Crohn's disease. Barium enema revealed irregular narrowing of the sigmoid lumen secondary to the inflammatory change caused by the extraintestinal lesion. Colonoscopy showed an elevated mass with spontaneous drainage of whitish pus from the inflamed mass.

Management of Mesenteric Abscess

The treatment of mesenteric abscess is mainly dependent on the underlying etiology, hence accurate diagnosis is imperative for decision making. Some patients respond well to medical treatment while others may need further intervention, including percutaneous drainage and laparotomy.

Treatment of primary mesenteric abscess is by open or closed percutaneous drainage and broad-spectrum antibiotics. However, this approach was refuted by other investigators who assumed that the use of large bore drains would lead to further complications and that primary closure of the mesentery after drainage of the abscess without leaving a tube inside the abscess cavity could be a better alternative.

Mesenteric abscess secondary to Crohn's disease almost always represents a contained microperforation. Hemodynamically stable patients would greatly benefit of drainage of the abscess which allows resolution of the acute inflammation associated with the septic process and enables limited, one-stage resection at a later date.

CT-guided percutaneous drainage has replaced open drainage for most Crohn's disease –related mesenteric abscess. The approach of percutaneous drainage relies on the location of the abscess; transgluteal, transabdominal, and perineal approaches of drainage have been described in the literature. The size of the drainage catheter depends on the viscosity of the material aspirated, but in all cases the catheter should be large enough to ensure successful drainage. Improvement in the general condition within two days indicates successful drainage; however; persistent fever warrants repeating the CT scan to confirm that the catheter is in the proper position. Culture of the aspirated material would reveal the most appropriate antibiotic for the organism isolated, antibiotics are usually administered intravenously for less than seven days or until the fever subsides. Percutaneous drainage is, however, not possible in multilocular abscess and those associated with fistula, thus in these situations, open drainage by laparotomy would be necessary.

Following success of percutaneous drainage, which is defined as "resolution of symptoms with collapse of the abscess cavity and avoidance of early surgery", patients are placed on mesalazine therapy. Some authors advocate the use of Anti-tumor necrosis factor (TNF)-α after percutaneous drainage of mesenteric abscess, nevertheless, the safety of (TNF)-α after percutaneous drainage has not been well assessed.

Although it is unknown whether elective surgery is indicated in all patients with Crohn's disease or not, elective resection of the affected bowel loops is usually indicated after full resolution of symptoms in order to avoid leaving a focus for future abscess recurrence. There is no consensus on the timing of elective surgery after successful drainage, most authors recommend waiting for 6–8 weeks. However, Poritz and colleagues were able to perform elective one-stage resection in more than 80% of patients after only one week of successful drainage. The timing of surgery essentially depends on the nutritional status of the patients and prior use of steroid therapy, hence should be carefully tailored to each patient in light of clinical and radiologic parameters.

Mesenteric abscess secondary to tuberculosis usually responds well to a full course of anti-tuberculosis therapy for six to twelve months. Surgery is not warranted in tuberculous abscess; however; since preoperative diagnosis is often difficult, many of these patients undergo surgical exploration and drainage of the mesenteric abscess. In patients with associated HIV infection, antiretroviral therapy along with antifungal treatment should be also administered to hasten the resolution of mesenteric abscess.

If properly diagnosed, infected mesenteric cysts may be treated with CT-guided percutaneous drainage with intravenous antibiotics according to the results of culture of the aspirated pus. Unfortunately, many of these patients present with acute abdomen that warrants surgical exploration, in such case surgical treatment involves either resection of the affected bowel segment with primary anastomosis, or complete enucleation of the infected cyst taking care to avoid rupture of the cyst wall and compromising the bowel vascularity.

In the case of perforated Meckel's diverticulum causing mesenteric abscess, the optimal treatment is by surgical intervention and en bloc resection of the affected ileal loops together with the perforated Meckel's diverticulum with primary anastomosis of the bowel ends. Mesenteric abscess on top of mesenteric adenocarcinoma is an indication of radical resection of the affected bowel, but since most adenocarcinomas are located at the root of the mesentery and tend to infiltrate major mesenteric blood, radical resection of the tumor is not usually feasible, leaving adjuvant chemotherapy and radiotherapy as the last resort.

Data on the use of laparoscopy in diagnosis and treatment of mesenteric abscess are quite sparse. In a single report, Terashita et al. documented the case of an elderly woman that presented with acute abdomen and fever with mild ascites and thickening of the ileum in the abdominal CT scan. The authors used laparoscopic-assisted surgery and found an abscess in the mesentery of the ileum that was perforated causing generalized peritonitis, however, no perforations were detected in the ileum. The authors made a diagnosis of ruptured idiopathic mesenteric abscess that warranted resection of the affected ileal loops.

Indications, Contraindications, and Challenges of Surgery for Mesenteric Abscess

Indications for surgical intervention for mesenteric abscess include failure of percutaneous drainage of Crohn's disease-related abscess, elective resection of the affected bowel segment within 6–8 weeks after successful drainage, rupture of abscess with generalized peritonitis, infected mesenteric cyst not amenable to drainage, small bowel obstruction caused by adhesions between the abscess and the bowel loops, abscess secondary to perforated Meckel's diverticulum, and abscess on top of underlying mesenteric adenocarcinoma.

Relative contraindications for surgery for mesenteric abscess include unilocular singular abscess that can be adequately treated with percutaneous drainage and tuberculous mesenteric abscess with confirmed diagnosis by staining to acid-fast bacilli or PCR.

It is worthy to highlight the technical challenges of surgery for mesenteric abscess that may predispose to serious consequences. The acute inflammation renders the affected mesentery very friable and vulnerable to injury and bleeding. The dissection used for enucleation of the abscess may induce injury of the mesenteric blood vessels, with subsequent hemorrhage and bowel ischemia that may end with devitalization of a considerable segment of the intestine. In several occasions the mesenteric abscess is associated with dense adhesions with the intestine that are liable to injury during adhesolysis. Hence, gentle handling of the bowel loops and the mesentery and meticulous dissection technique are crucial to avoid inadvertent injuries. Finally, owing to the associated peritonitis and the effect of surgical manipulations, abdominal adhesions and small bowel obstruction may eventually ensue and add further morbidity.

Complications of Mesenteric Abscess

Complications of mesenteric abscess are also dependent on the underlying etiology. Crohn's disease-related mesenteric abscess may be associated with significant stricture of the neighboring bowel segment, may perforate freely causing generalized peritonitis, or may cause fistula with a nearby organ.

Tuberculous mesenteric abscess may be a single manifestation of diffuse abdominal tuberculosis affecting the peritoneum and the intestine. Patients with immune deficiency are also liable to disseminated military tuberculosis that may affect other organs in the body.

Small bowel obstruction can be a possible complication of mesenteric abscess associated with significant tissue adhesions. Adhesive bands between the wall of the abscess and the intestine may compress the underlying bowel and necessitates emergency surgery and adhesolysis.

Compromise of the bowel vascularity is another possible consequence of mesenteric abscess due compression or thrombophlebitis of the mesenteric blood vessels which may lead to gangrene of the bowel loop warranting emergency resection.

Peritonitis that ensues rupture of mesenteric abscess can result in marked toxemia, septicemia, and eventually septic shock. Adequate resuscitation with intravenous fluids and antibiotics is crucial in patients with peritonitis before surgical exploration to optimize their outcome.

Outcome of Mesenteric Abscess

Crohn's disease-related mesenteric abscess is liable to recurrence after successful drainage since the focus of disease is left in situ. Therefore, elective resection of the affected bowel segment is indicated.

After complete course of anti-tuberculosis treatment, the related mesenteric abscess tends to subside completely, nonetheless, chances of recurrence are still possible in case the primary disease has relapsed.

The outcome of mesenteric abscess caused by infected cyst or perforated Meckel's diverticulum is usually good after proper surgical intervention with no recorded recurrences since the primary pathology was eradicated.

Although very rare, mesenteric abscess on top of mesenteric adenocarcinoma has a bad prognosis owing to the difficulty of radical resection of the primary tumor and possibility of intraperitoneal metastatic disease that precludes curative resection of the tumor.

Summary

Mesenteric abscess is an uncommon condition that can be a primary disease or secondary to a variety of conditions. The most common type of secondary mesenteric abscess is the Crohn's disease-related abscess. Other causes include tuberculosis, infected mesenteric cysts, and perforated Meckel's diverticulum. Diagnosis of mesenteric abscess is mainly made by abdominal CT scanning with CT-guided FNAC to shed light on the underlying etiology. Treatment of mesenteric abscess depends mainly on its etiology. Treatment options include medical treatment with antibiotics, anti-tuberculosis drugs, and antifungal medications, and CT-guided percutaneous drainage. Surgical treatments include drainage of the abscess, enucleation of infected mesenteric cyst, and resection of the affected bowel segment. Complications of mesenteric abscess involve free rupture and peritonitis, fistula with nearby organ, disseminated intraperitoneal infection, bowel obstruction, and bowel gangrene, in addition to the possible complications of surgery for mesenteric abscess.

Colonic Abscess

Definition and Incidence

Colonic abscess, also known as pericolic abscess, is a localized walled-off collection of pus and bacteria within or adjacent to the colonic wall. Relevant to mesenteric abscess, colonic abscess is formed as a sequel to a small perforation (microperforation) at the mesenteric side of the colonic wall, then the infection is contained by pericolic fat and the mesentery. Therefore, the mesentery forms part of the colonic abscess wall. Although there are no documented statistics on the incidence of colonic abscess; it is considered an uncommon condition with an increased incidence in patients older than 45 years.

Causes

A colonic abscess almost always indicates an underlying colonic perforation. While large, uncontained perforations usually result in localized or generalized peritonitis, smaller and contained perforations can lead to the formation of pericolic abscesses.

The common causes of colonic perforation include acute diverticulitis, inflammatory bowel diseases (IBD), namely ulcerative colitis, and advanced colon cancer. Less commonly, acute or recurrent volvulus of the sigmoid colon can be associated with colonic perforation and abscess owing to the increased tension and compromised blood supply of the colonic wall.

Since complicated diverticular disease is by far the most common cause of colonic abscess, both share the same risk factors. Patients who are more amenable to develop complicated diverticulitis with pericolic abscess are elderly, those with chronic constipation due to low dietary fiber and high intake of fat and red meat, those with connective tissue diseases, and those with increased body mass index. The pathogenesis of colonic abscess on top of acute diverticulitis entails increased intraluminal pressure of the colon, associated with acute inflammation and focal necrosis followed by an erosion of the diverticular wall which results in abscess formation.

A multitude of bacteria can be involved in the formation of colonic abscess, most of which are considered natural flora of the colon such as Enterococcus, Escherichia coli, and Bacteroides. In addition, Klebsiella, Pseudomonas aeruginosa, and Staphylococcus aureus can be also isolated from colonic abscesses.

Diagnosis

Most colonic abscesses are secondary to complicated diverticular disease of the colon. Therefore, most patients with colonic abscess report a long history of chronic constipation with or without recurrent episodes of acute pain in the left lower quadrant of the abdomen implying recurrent attacks of uncomplicated diverticulitis. Although most attacks of acute diverticulitis tend to be uncomplicated, up to 20% of patients may present with pericolic or intra-abdominal abscess in CT scanning. Patients with colonic abscess secondary to IBD or advanced colon cancer may report a history of altered bowel habits, bloody diarrhea, recurrent abdominal pain, and weight loss prior to their presentation with an abscess.

On clinical examination, patients present with obvious physical signs of localized peritonitis in the left iliac fossa that include tenderness, rebound tenderness, and rigidity. A tender swelling may be felt in the left lower quadrant of the abdomen. Many patients present with high fever and chills when a significant amount of pus has been formed and entrapped within the abscess cavity. It has been noted that the severity of symptoms is proportional to the extent of the disease and size of the abscessA rare presentation of diverticular colonic abscess was reported in a 72-year-old woman who presented with chronic refractory diarrhea and upon assessment with CT scanning a large local air-fluid level within the culdesac area was found. Afterwards, exploratory laparotomy revealed a large pelvic abscess between the sigmoid colon and uterus.

Initial laboratory findings in patients with acute colonic abscess include leucocytosis, although this finding may be absent in elderly, debilitated, or immunocompromised patients. Additional findings may include low serum hemoglobin, especially in patients with underlying IBD or colon cancer, altered liver function tests, and elevated C reactive protein. Blood culture may be indicated in severely ill patients with marked toxemia and usually reveals persistent bacteremia with more than one bacterial species isolated. Urine analysis may detect red or white blood cells owing to the acute inflammation adjacent to the ureter or the bladder, or due to colovaginal fistula.

The diagnostic modality of choice to detect and properly assess colonic abscesses is abdominal CT scanning with contrast. CT scan can be useful in the detection of the abscess, measurement of its size, identification of the underlying cause, and assessment of the colonic wall and nearby structures. Inflammation of the colonic wall with symmetric thickening and enhancement is usually observed and diverticulae are recognized as outpouching of the colonic wall. Free air or fluid in the peritoneal cavity indicates free perforation of the abscess wall and free air in the bladder may imply a colovesical fistula.

CT scanning can also classify abscess into local (pericolic), pelvic, or distantDifferent cut-off levels have been used to define small and large colonic abscess in CT scan. Mora Lopez et al. proposed a cut-off of 4 cm whereas Sallinen et al suggested a cut-off of 6 cm to differentiate small from large colonic abscess as a complication of diverticulitis.

Treatment

Once the diagnosis of colonic abscess has been established, prompt treatment is indicated to avoid the consequences of abscess rupture and peritonitis or fistulation with a nearby organ, particularly the urinary bladder.

Treatment usually commences with non-operative management including antibiotics and/or percutaneous drainage. While antibiotics are not routinely indicated in in acute uncomplicated diverticulitis, aggressive intravenous antibiotic treatment has been recommended by an expert panel as a first-line therapy when pericolic extraluminal gas is present. Antibiotic regimens may involve a single broad spectrum agent such as Meropenem, Imipenem-Cilastatin, and Piperacillin-tazobactam or combined regimens such as cefepime+ metronidazole, ciprofloxacin + metronidazole, or levofloxacin + metronidazole. The recommended treatment duration is up to seven days or longer if adequate source control was not achieved.

Antibiotic treatment is usually sufficient in small colonic abscess, measuring less than 3 cm in size. However, larger abscesses may require further treatment with percutaneous drainage. Percutaneous drainage is also indicated in patients with small abscesses that did not improve with antibiotic treatment. Contingent upon local resources, drainage is performed under ultrasound or CT guidance using the Seldinger technique. CT guided drainage may be preferred when the abscess is small or obscured by nearby structures. The route of drainage differs according to the location of the collection. Access through the gluteal region, vagina, or rectum through a safe window into the collection is attempted using a sterile technique then a drain is placed inside the abscess cavity.

A systematic review found that failure of non-operative treatment of diverticular colonic abscess was recorded in 20% of patients and one-quarter of patients who initially had a successful non-operative treatment experienced recurrent attacks on long follow-up. Overall, drainage was associated with higher success and less recurrence than antibiotic treatment only.

Failure of non-operative treatment indicates emergency surgery which entails resection of the affected colon segment with creation of end-colostomy (Hartmann's procedure). Delayed closure of colostomy is planned within six months, after resolution of the acute inflammatory state. However, the difficulties encountered during the reversal of Hartmann's procedure may preclude closure of colostomy in many patients. Therefore, a single stage operation of resection and primary anastomosis has been suggested as an alternative to Hartmann's procedure. Nonetheless, acute surgery for colonic abscess should be considered as the last resort as it has a substantially increased risk of mortality as compared to non-operative management (12.1% versus 1.1%), according to the collective evidence.

Conclusions

Colonic abscess is an uncommon condition that is secondary to complicated acute diverticulitis of the colon in the majority of cases. Patients with colonic abscess present with acute pain and clinical signs of peritonitis in the left iliac fossa. Patients with large colonic abscesses may have high fever and chills. Diagnosis is made by CT scanning and non-operative treatment with antibiotics and/or percutaneous drainage is usually sufficient in most patients. Emergency surgery is indicated when non-operative management fails to improve the patients, yet it is associated with a higher risk of morbidity and mortality.

Suggested Reading

1. Dudley HA, Maclaren IF. Primary mesenteric abscess. Lancet. 1956 Dec 8;271(6954):1182–4.
2. Fraenkel GJ. Primary mesenteric abscess. Lancet. 1956 Dec, 271(6954):1309.
3. Hokama A1, Kinjo F, Tomiyama R, Kishimoto K, Maeda K, Miyagi S, Nakama M, Saito A. Mesenteric abscess in Crohn's disease. Gastrointest Endosc. 2005 Aug;62(2):306;discussion 306.
4. Teeples TJ1, Tabibian JH. Images in clinical medicine. Mesenteric abscess in Crohn's disease. N Engl J Med. 2013 Feb 14;368(7):663. https://doi.org/10.1056/NEJMicm1110648.
5. Duncan D. Meckel's diverticulum with mesenteric abscess. BJS. 1956 May;43(182):664.
6. Wasike R, Saidi H. Perforated Meckel's diverticulitis presenting as a mesenteric abscess: case report. East Afr Med J. 2006;83(10):580–4.
7. Anand P, Sushmita C. Isolated mesenteric tuberculosis presented as abdominal mass. Nat J Med Res. 2012;2(3):389–90.
8. Richards RJ. Management of abdominal and pelvic abscess in Crohn's disease. World J Gastro Endosc. 2011;3(11):209–12.
9. Vijaya D, Lakshmi Kanth TK, Khadri SIS, Suresh Chandr HC, Malini A. Isolated tuberculous mesenteric abscess—a case report. Ind J Tub. 2000;47:101.
10. Pandit V1, Valsalan R, Seshadri S, Bahuleyan S. Disseminated tuberculosis presenting as mesenteric and cerebral abscess in HIV infection: casereport. Braz J Infect Dis. 2009 Oct;13(5):383–6. https://doi.org/10.1590/S1413-86702009000500014.
11. Mohar SM, Saeed S, Ramcharan A, Depaz H. Small bowel obstruction due to mesenteric abscess caused by Mycobacterium avium complex in an HIV patient: a case report and literature review. J Surg Case Rep. 2017;2017(7):rjx129. Published 2017 Jul 12. https://doi.org/10.1093/jscr/rjx129.
12. Kim EJ, Lee SH, Ahn BK, Baek SU. Acute abdomen caused by an infected mesenteric cyst in the ascending colon: a case report. J Korean Soc Coloproctol. 2011;27(3):153–6.
13. Luo Y, Ou M, Zhang Y, Fu Z, Wang C. Primary mesenteric adenocarcinoma covered by abscess of the mesocolon and intestinal obstruction: a case report. Mol Clin Oncol. 2017;6(4):593–6.
14. Beeson BB, Woodruff AW. Mesenteric abscess caused by threadworm infection. Trans R Soc Trop Med Hyg. 1971;65(4):433.
15. Montgomery RS, Wilson SE. Intraabdominal abscesses: image-guided diagnosis and therapy. Clin Infect Dis. 1996, Jul;23(1):28–36.
16. Terashita Y, Haruki N, Mori Y, Harata K, Naito A, Sato A. A case of a ruptured idiopathic mesenteric abscess successfully treated by laparoscopic-assisted surgery. Nihon Rinsho Geka Gakkai Zasshi (J Japan Surg Assoc). 2012;73(4):1008–12.

17. Gregersen R, Mortensen LQ, Burcharth J, Pommergaard HC, Rosenberg J. Treatment of patients with acute colonic diverticulitis complicated by abscess formation: A systematic review. Int J Surg. 2016 Nov;35:201–208. https://doi.org/10.1016/j.ijsu.2016.10.006.
18. Solomkin JS, Mazuski JE, Bradley JS, Rodvold KA, Goldstein EJ, Baron EJ, O'Neill PJ, Chow AW, Dellinger EP, Eachempati SR, Gorbach S, Hilfiker M, May AK, Nathens AB, Sawyer RG, Bartlett JG. Diagnosis and management of complicated intra-abdominal infection in adults and children: guidelines by the Surgical Infection Society and the Infectious Diseases Society of America. Surg Infect (Larchmt). 2010 Feb;11(1):79–109. https://doi.org/10.1089/sur.2009.9930.
19. Andersen JC, Bundgaard L, Elbrønd H, Laurberg S, Walker LR, Støvring J. Danish national guidelines for treatment of diverticular disease. Dan Med J. 2012;59(5).
20. Ambrosetti P. Acute left-sided colonic diverticulitis: clinical expressions, therapeutic insights, and role of computed tomography. Clin Exp Gastroenterol. 2016 Aug 18;9:249–57. https://doi.org/10.2147/CEG.S110428.
21. Daryani NE, Keramati MR, Habibollahi P, Pashaei MR, Ansarinejad N, Ajdarkosh H. Colonic diverticular abscess presenting as chronic diarrhea: a case report. Cases J. 2009 Dec 23;2:9389. https://doi.org/10.1186/1757-1626-2-9389.
22. Ghoulam EM. Diverticulitis work-up. Medscape. https://emedicine.medscape.com/article/173388-workup.
23. Neff CC, van Sonnenberg E. CT of diverticulitis. Diagnosis and treatment. Radiol Clin N Am. 1989;27:743–52.
24. Mora Lopez L, Serra Pla S, Serra-Aracil X, Ballesteros E, Navarro S. Application of a modified Neff classification to patients with uncomplicated diverticulitis. Color Dis. 2013;15:1442–7.
25. Sallinen VJ, Leppäniemi AK, Mentula PJ. Staging of acute diverticulitis based on clinical, radiologic, and physiologic parameters. J Trauma Acute Care Surg. 2015;78:543–51.
26. Emile SH, Elfeki H, Sakr A, Shalaby M. Management of acute uncomplicated diverticulitis without antibiotics: a systematic review, meta-analysis, and meta-regression of predictors of treatment failure. Tech Coloproctol. 2018 Jul;22(7):499–509. https://doi.org/10.1007/s10151-018-1817-y.
27. Sartelli M, Moore FA, Ansaloni L, Di Saverio S, Coccolini F, Griffiths EA, et al. A proposal for a CT driven classification of left colon acute diverticulitis. World J Emerg Surg. 2015;10:3.
28. Lué A, Laredo V, Lanas A (2016). "Medical Treatment of Diverticular Disease: Antibiotics". J. Clin. Gastroenterol. 50 Suppl 1: S57–9. https://doi.org/10.1097/MCG.0000000000000593.
29. Sartelli M, Viale P, Catena F, et al. 2013 WSES guidelines for management of intra-abdominal infections. World J Emerg Surg. 2013 Jan 8;8(1):3. https://doi.org/10.1186/1749-7922-8-3.
30. Gregersen R, Mortensen LQ, Burcharth J, Pommergaard HC, Rosenberg J. Treatment of patients with acute colonic diverticulitis complicated by abscess formation: A systematic review. Int J Surg. 2016 Nov;35:201–208. doi: 10.1016/j.ijsu.2016.10.006. Epub 2016 Oct 11. PMID: 27741423.
31. Bossert FR, Parsons LC, Tsaltas T (2015). "Laparoscopic Diverticular Abscess With Drainage". J Minim Invasive Gynecol. 22(6S): S149. https://doi.org/10.1016/j.jmig.2015.08.541.
32. Bauer VP. Emergency management of diverticulitis. Clin Colon Rectal Surg. 2009 Aug;22(3):161–8. https://doi.org/10.1055/s-0029-1236160.

Mesenteric Neoplasms

48

Rishabh Sehgal, J. Calvin Coffey, and Deborah S. Keller

Introduction

Traditional teaching has perceived the mesentery as a fragmented structure with complex relationships to surrounding intraabdominal structures. This concept has been indoctrinated into mainstream surgical and anatomical literature for over a century. Recent research and formal appraisal of the mesentery has defined it as a unique organ, continuous from the duodeno-jejunal flexure to the mesorectum. Histological and electron-microscopic studies have expanded on the details of the mesenteric organ, demonstrating it consists of a surface mesothelium and underlying connective tissue throughout. Fibrous septae divide adipocyte lobules within the body of the mesocolon. Where in contact with the retroperitoneum, the mesentery is the mesocolon, and two mesothelial layers are separated by a distinct layer of connective tissue—Toldt's fascia—separates both. Lymphatic channels are evident both in mesenteric connective tissue and Toldt's fascia. The new model of the mesentery organ provides us with an opportunity to understand disease states that involve the mesentery.

The rich network of lymphatics and close relationship of the mesentery to other intra-abdominal structure, mean the mesentery is prone to a myriad of disease states. When pathological process originate from the mesentery itself, and spread to

R. Sehgal
Department of Colorectal Surgery, University Hospital Limerick, Limerick, Ireland

J. C. Coffey
School of Medicine and Department of Surgery, University of Limerick Hospital Group, Limerick, Ireland

D. S. Keller (✉)
Division of Colon and Rectal Surgery, NewYork-Presbyterian, Columbia University Medical Center, Herbert Irving Pavilion, 8th Floor, 161 Fort Washington Avenue, New York, NY 10032, USA
e-mail: debby_keller@hotmail.com

E. D. Ehrenpreis et al. (eds.), *The Mesenteric Organ in Health and Disease*,
https://doi.org/10.1007/978-3-030-71963-0_48

involve adjacent structures, this is known as a primary mesenteropathy. In contrast secondary mesenteropathies are diseases where the primary abnormality originates outside the mesentery and indirectly involves the mesentery. Neoplasms of the mesentery are rare and usually diagnosed incidentally on radiological imaging. They encompass a heterogenous group of lesions ranging from benign cysts to aggressive malignancies. They can arise from distinct cellular components of the mesenteric organ, including the peritoneal surface, connective tissue, adipocytes, lymph nodes, and blood vessels, with a wide spectrum of surgical management options. In this chapter, we describe the major types of mesenteric neoplasms, indications and contraindications for surgery, and the surgical management for mesenteric neoplasms.

Description of Mesenteric Neoplasms

Tumors of the mesentery are rare neoplasms first described in the early twentieth century. Since the initial description of mesenteric tumors, experience has accumulated in treating these lesions, and a more complete picture has emerged of diseases that manifest as mesenteric masses (e.g. lymphoma, desmoid tumors, Castleman's disease). Although uncommon, they are encountered in all age groups, so an increased awareness of these tumors is important in the differential diagnosis of abdominal masses and to aid in recognition and proper management (Table 48.1).

Anatomically, the mesentery spans the entire length of the intestine and consists of fibrofatty, fanlike structure containing arterial, venous, lymphatic, and neural structures in contiguity. The mesentery of the small bowel mesentery and portions of the colon mesentery are mobile, while the ascending and descending colon mesenteries are normally fixed against the retroperitoneum. The greater and lesser omentum are also included in the classification of mesentery, and tumors of these structures are generally deemed as mesenteric. According to this model, the vast majority of mesenteric tumors originate in the small bowel or omentum.

Mesenteric tumors can be either cystic or solid. Cystic mesenteric masses affect males and females equally with up to 60% involving the small bowel mesentery. The estimated incidence of cystic mesenteric masses is approximately 1/100,000 in

Table 48.1 Mesentericopathies

Mesenteropathies		
Primary	Secondary	Indeterminate
Crohn's disease	Mesenteric adenitis	Adhesions
Ischaemia	Appendicitis	Diabetes mellitus
Mal or nonrotation	Phlegmon	Atherosclerosis
Volvulus	Metastatic disease	Metabolic syndrome
Internal herniation	Desmoid tumor	Irritable bowel syndrome
Cysts	Peritoneal carcinomatosis	
Schwannoma	Castleman's disease	
Sclerosing mesenteritis	Lymphoma	
Epiploic appendagitis	Mesenteric cavitation syndrome	

the adult population and 1/20,000 in the paediatric population. Up to 50% of cystic mesenteric masses are cystic lymphangiomas. Lymphomas are the most common solid primary mesenteric neoplasms. They affect 1 in 200,000–350,000 population, with the B-cell subtype being most prevalent. Carcinoid tumors are slow growing solid neuroendocrine neoplasms involving the gastrointestinal tract in 90% of cases. Carcinoid tumors affecting the mesentery are mostly metastases. Mid-gut carcinoid tumors involve the mesentery in 40 to 80% of reported cases. Intra-abdominal desmoid tumors account for 8% of all desmoid tumors and usually involve the small bowel mesentery. They are most commonly associated with Familial adenomatous polyposis (FAP) (in 70% of cases) in which their incidence ranges from 4–32%. The remaining types of mesenteric tumors are rare and sporadically mentioned in literature. In this work, we will describe the most common types of mesenteric neoplasms, with attention to their diagnostic algorithm, surgical options, indications and contraindications for surgery, and common complications encountered.

Lymphoma

Solid primary tumours of the mesentery are rare. Lymphoma is the most common malignancy involving the mesentery. The incidence ranges between 1 in 200,000 to 1 in 350,000 population and the majority present after 60 years of age. Non-Hodgkin's lymphoma accounts for the majority of cases, with up to 70% being large B-cell neoplastic subtype. The mesentery can be the first site in which lymphoma arises. This is referred to as primary lymphoma. Mesenteric lymphoma may also involve the small bowel by means of direct extension or displacement due to the mass effect. They most commonly present with abdominal pain and a palpable abdominal mass. They can remain asymptomatic (even when the tumor burden is high) or they can present with systemic symptoms such as fevers, rigors, weight loss and night sweats. The latter symptoms are suggestive of advanced disease. They rarely cause bowel obstruction, perforation or gastrointestinal bleeding. Rarely, mesenteric lymphomas occur in association with immune thrombocytopenia and dermatitis herpetiformis.

Early in the course of the disease, involved mesenteric nodes are small, soft and discrete. As disease progresses, enlarged nodes develop and often coalesce and tend to form a conglomerate mass. Ultrasound scanning helps in confirming the clinical suspicion and allows for localization and biopsy of lesions. Computed tomography (CT) has a greater sensitivity and provides more detailed spatial resolution and ideally should be performed after ultrasound to further stage the disease. Bulky retroperitoneal adenopathy can often accompany primary mesenteric disease. CT may demonstrate a large, lobulated and "cakelike" heterogeneous mass displacing small bowel and containing areas of necrosis (low attenuation); or, ill-defined infiltration of mesenteric fat. Pathognomonic features of mesenteric lymphoma (on CT) include a round well-defined homogenous mass encasing mesenteric vessels

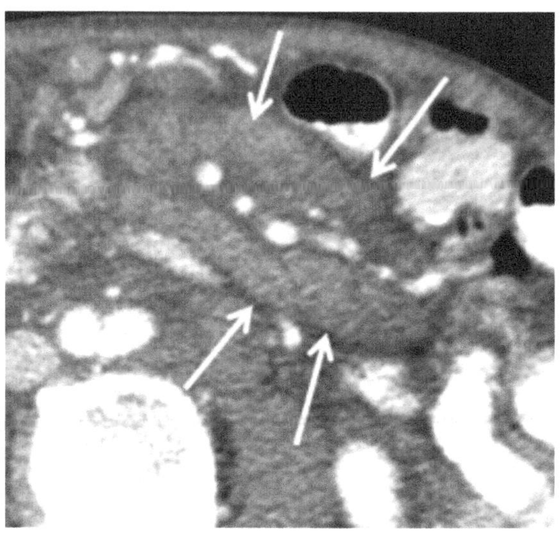

Fig. 48.1 CT image mesenteric lymphoma. Mesenteric lymphoma. A large mass is seen representing confluent mesenteric adenopathy (arrows) on both sides of the mesenteric vessels. The radiographic term for this finding is the "sandwich sign."

giving rise to the so-called 'sandwich or hamburger sign'. In this sign, mesenteric adipose tissue and tubular vascular structures represent the "filling," whilst the homogeneous soft tissue masses represent the "sandwich bun" (Fig. 48.1).

Differential diagnosis includes carcinoma, sarcoma, carcinoid tumor, tuberculosis, Whipple disease, lymphadenopathy associated with acquired immunodeficiency syndrome, and inflammatory bowel disease. Although diagnosis of mesenteric lymphoma can be obtained on radiologic findings alone, a tissue sample is required for definitive diagnosis and treatment planning. Tissue may be obtained laparoscopically or via open surgical means. Whilst the application of laparoscopic methods to investigate and treat mesenteric masses represents an evolving area of surgical care, the open approach may be safer, given proximity of masses to mesenteric vessels. In patients whose disease must be distinguished only from that requiring resection (i.e., lymphoma, mesenteric lipodystrophy), laparoscopy offers considerable benefits as it avoids a full laparotomy incision.

Mesenteric lymphoma is treated by cytotoxic chemotherapy, with CHOP (cyclophosphamide, doxorubicin, vincristine, and prednisone) regimen, plus the monoclonal antibody rituximab the gold standard for the management of diffuse large B cell lymphoma. Although some cases are diagnosed following resection of an uncharacterized mesenteric mass, surgical treatment is best used as a diagnostic tool when the diagnosis is probable but uncertain. Moreover, surgery is reserved for the management of any complications that may arise due to the presence of the lymphoma. Rare cases of surgical resection combined with chemotherapy in patients with bulky mesenteric diffuse large B cell lymphoma have been reported.

Desmoid Tumors

Desmoid tumors are clonal fibroblastic proliferations that arise from deep soft tissue and are characterized by infiltrative growth and a tendency toward local recurrence but an inability to metastasize. Desmoids can manifest as flat, fibrous, sheet-like lesions or well defined and discrete masses (Fig. 48.2a–d). Desmoids account for less than 3% of all soft-tissue sarcomas and 0.03% of all malignancies. The etiology is unknown, however they are associated with familial adenomatous polyposis (FAP), especially in patients with a mutation at the 3′ end of the 1440 codon of

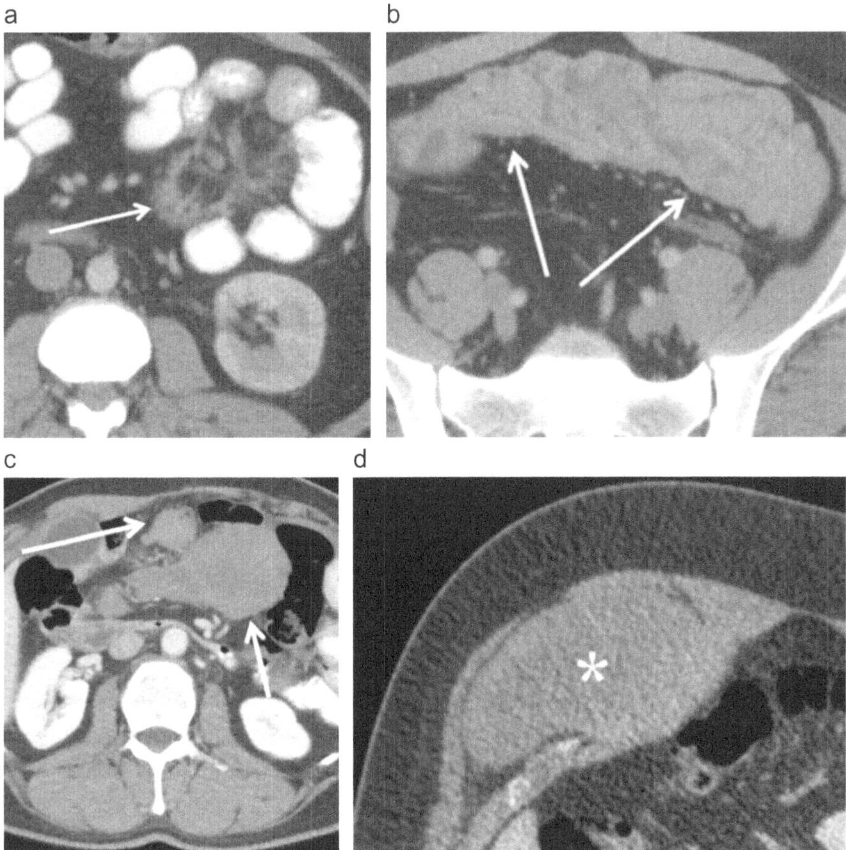

Fig. 48.2 Contrast-enhanced axial CT images from a patient with previously undiagnosed Gardener syndrome. **a**: shows soft tissue attenuation in a stellate pattern within the mesenteric fat that is intimately associated with small bowel loops (arrow). **b**: shows innumerable polyps throughout the imaged colon. **c**: Desmoids are benign, inflammatory fibroblastic tumors. They appear as well-circumscribed enhancing masses in the mesentery (arrows). **d**: Desmoids may also present as masses of the anterior abdominal wall (asterix)

the APC gene, and trauma, developing within 5 years after abdominal surgery. Their development following trauma is thought to arise from prolonged acute and chronic inflammatory. They have an equal gender distribution and affecting adults in their 30 and 40s. Desmoids affect the extremities, trunk, and abdominal cavity in the majority of cases. Although only 5% of sporadic desmoids are intra-abdominal, 80% of patients with FAP-associated desmoids develop intra abdominal disease involving the small bowel mesentery. Patients with FAP and desmoid disease tend to present with multiple but smaller lesions, whereas sporadic cases tend to present with a solitary large lesion.

Mesenteric desmoids can be asymptomatic and diagnosed incidentally. They may cause symptoms due to external pressure effect on the adjacent bowel, or by impinging on the intestinal blood supply. Lesions that can extend retroperitoneally leading to additional complications such as hydronephrosis. CT and MRI are the main radiologic modalities used for investigating such mesenteric masses. Radiologically, mesenteric desmoids are represented by a soft-tissue mass, with displacement of bowel loops and surrounding vasculature, with or without intestinal involvement and serosal changes. The soft-tissue component is a dominant feature that can mimic other solid tumors, including gastrointestinal stromal tumor (GIST), lymphoma, carcinoids, or fibrosarcomas.

In the case of mesenteric desmoid tumors, medical management with Sulindac, Raloxifene, and chemotherapeutic agents such as Methotrexate, Vinorelbine, Doxil, and Adriamycin are used to slow growth and ameliorate obstructive symptoms. Radiotherapy is not generally recommended for retroperitoneal or intra-abdominal desmoids, due to their proximity to key intra-abdominal structures. Surgery is indicated for treatment of complications such as bowel obstruction, enterocutaneous fistula, and ureteric obstruction. For resectable lesions, wide local excision with macroscopic negative margins is the standard. The resultant mesenteric defect is usually closed to avoid internal hernias. Smaller desmoids at a distance from major vascular trunks, can be safely resected with minimal complications. Contraindications to surgery are location at the root of the small bowel mesentery or near other critical structures; in these cases proximity to critical small bowel blood supply may render the mass unresectable. Palliative options include enteroenteric or enterocolic bypass. These can be performed either laparoscopically or via an open approach. Small bowel and multi-visceral transplant have been described as surgical options. Although Chatzipetrou et al. reported more than 80% enteral independence and one recurrence in patients who underwent intestinal transplantation for desmoid tumors other transplant groups have reported very poor outcomes for transplantation with respect to complicated intra-abdominal desmoids.

Mesenteric Gastrointestinal Stromal Tumors

Gastrointestinal stromal tumors (GISTs) are mesenchymal tumors of the gastrointestinal tract arising from the 'pacemaker cells' that regulate peristalsis, i.e. the interstitial cells of Cajal. GISTs account for less than 5% of all gastrointestinal malignancies and most commonly affect adults in between ages 40 and 70 years. Mazur and Clark were first to describe GIST as gastrointestinal tract nonepithelial neoplasm that did not possess the ultrastructural characteristics of smooth muscle cells, and immunohistochemical features of Schwann cells. Immunohistochemical staining demonstrates expression of CD-117 (95%) and CD-34 (70%) protein and mutations of C-kit and PDGFRA genes are frequent.

Primary GIST mostly occur as solitary lesions involving the stomach (60–70%) followed by small bowel (20–25%). Less frequent sites include the colon, rectum, esophagus and appendix (Fig. 48.3a and b). There are also sporadic reports of GIST arising in the omentum, mesentery and retroperitoneum. They can be asymptomatic, present as a palpable abdominal mass or as acute abdomen due to rupture. Ultrasound and CT are the main diagnostic modalities used to asses stage. CT scan features include well-defined lobular mass with heterogenous contrast enhancement and areas of hypodensity.

Surgery remains the mainstay of treatment for resectable GISTs. Favourable prognostic features include size less than five cm, complete resection (R0), absence of perforation and low histological grade. Conversely size greater than five centimetres, tumor necrosis, mitosis greater than 5/50 high power filed, positive

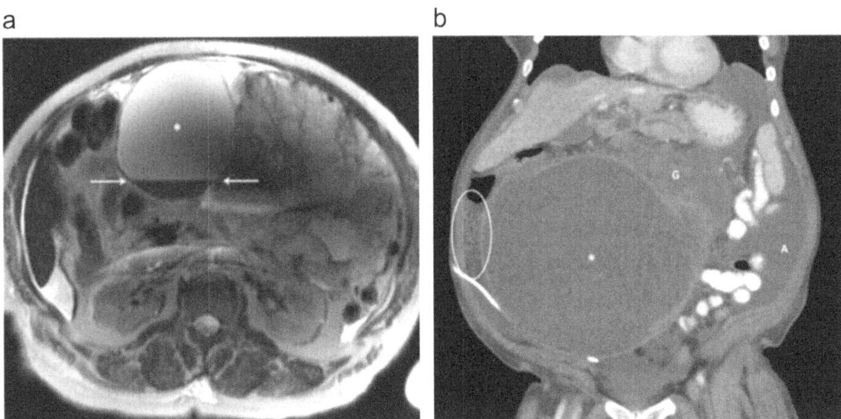

Fig. 48.3 **a** and **b**: Mesenteric GIST. Coronal contrast-enhanced computed tomography and axial T2-weighted magnetic resonance imaging demonstrate an ill-defined, heterogeneously enhancing GIST (G) with a large cystic component (asterisk). Additional heterogeneous soft tissue along the periphery of the lesion (oval) is suggestive of peritoneal sarcomatosis and associated ascites (A). Layering T2-hypointense signal (arrows) is indicative of necrosis with layering blood products. With permission from Dr. Abraham Dachman and Dr. Justin Ramirez

resection margins, and distant metastasis are poor prognostic indices. The tyrosine kinase inhibitor, imanitib mesylate, has been utilized in the adjuvant setting in patients who present with very large sized tumors that are at high risk of disease recurrence and metastatic spread. Overall the malignant potential of GIST varies, with benign tumours having a 5-year survival of 95% in contrast to the malignant type where it is 21%.

Mesenteric Carcinoid Tumors

Carcinoid tumors are rare neuroendocrine tumors that arise within the gastrointestinal tract in 90% of cases. Gastrointestinal (GIT) tract carcinoid tumors are classified according to embryologic origin as foregut, midgut and hindgut tumors. 46–64% of GIT carcinoid tumors arise in the midgut and most midgut carcinoid tumors originate in the terminal ileum, beyond the ligament of Treitz. Primary mesenteric carcinoid tumors are rare, and while the mesentery is involved in 40 −80% of cases, the majority of carcinoid tumors in the mesentery are metastases. As mid-gut carcinoid tumors secrete serotonin, patients can develop carcinoid syndrome with flushing and diarrhoea (especially though not exclusively, when the tumor has spread to the liver). Furthermore, carcinoid tumors can induce an intense desmoplastic reaction leading to mesenteric scarring and shortening. This can sometimes lead to intestinal obstruction, segmental portal hypertension, and chronic mesenteric ischaemia.

Microscopically, typical carcinoid tumors have one of five growth patterns: insular, trabecular, glandular, undifferentiated, or mixed. Most mid-gut carcinoids have a mixed insular and glandular growth pattern, and express have synaptophysin, chromogranin A, cytokeratins and neuron-specific enolase.

Before a diagnosis of primary mesenteric carcinoid can be made, involvement of other sites that are more commonly affected sites must be out-ruled by performing CT of thorax, abdomen and pelvis, colonoscopy, small bowel series and scintigraphy. Abdominal CT and an octreotide scanning may demonstrate a mesenteric tumor and/or liver metastases. CT scan demonstrates a solitary mass with varying degrees of fibrosis, calcification, focal or diffuse neurovascular bundle invasion. A colonoscopy, small bowel series and surgical exploration of the abdomen can confirm the diagnosis of mesenteric carcinoid.

Surgical excision is a mainstay of treatment for carcinoid tumors. For tumors smaller than 2 cm without lymph node involvement, local segmental resection is adequate. Tumors measuring more than 2 cm with regional mesenteric metastasis and lymphadenopathy are associated with 80–90% incidence of metastasis. Therefore, wide excision of the bowel and associated mesentery with lymph node dissection is recommended. Moreover surgical debulking of local or metastatic tumors is recommended in order to relieve symptoms and prolong survival.

Mesenteric Liposarcoma

Sarcomas are a heterogeneous group of malignant tumors which arise from mesenchymal tissues. Liposarcoma is one of the most common histological variants seen in adults. They can be subdivided into several histological subtypes including well-differentiated, pleomorphic, round-cell, myxoid and dedifferentiated type. The well-differentiated subtype are most common, have the highest fat content and are also referred to as atypical lipoma. Myxoid lesions are the second most common and are of intermediate-grade. They are associated with balanced chromosomal translocation t(12,16). Pleomorphic and round-cell lesions are considered high-grade and metastasize frequently. Approximately 75% of well-differentiated liposarcomas develop in the deep soft tissue of the limbs, followed by 20% in the retroperitoneum and inguinal area. Primary mesenteric liposarcoma is a rare entity, with less than 30 reported cases in the literature.

Mesenteric liposarcoma more commonly affect males between ages 50 and 70 years. They carry low metastatic potential and are locally aggressive. They are most commonly deep seated solitary masses located close to the root of the mesentery. The majority are asymptomatic or present with a mobile abdominal mass. However they can present with non-specific symptoms such as anorexia, early satiety, and abdominal distension, and occasionally can cause subacute intestinal obstruction or volvulus.

CT and MRI scan are the radiological investigations of choice. They provide detail regarding tumor characteristics, size and relationship to adjacent structures. Typical CT findings favouring a liposarcoma include inhomogeneous attenuation, with evidence of significant amounts of soft-tissue within the fatty mass; infiltration or poor margination; CT houndsfield units greater than the patients normal fats; and contrast enhancement. Enhancement on CT changes according to the degree of histological grade. It is reported that well-differentiated liposarcomas are hyperintense on T2-weighted MRI, and that they demonstrate faint or no enhancement. CT and MRI appearances of well differentiated liposarcoma, especially lipoma-like tumors resemble those of fat, but those findings may be differentiated from other types of tumors by their largely lipomatous appearance.

Surgery with wide local resection with adequate margins remains the mainstay of treatment of localized mesenteric liposarcoma. Complete surgical resection at the time of primary presentation is likely to improve long-term survival, as well as distant recurrence-free survival, and positive surgical margins are the main predictors of local relapse. The role of radiotherapy and adjuvant chemotherapy for high-risk patients remains unclear. Interestingly Ishiguro et al. demonstrated significant tumor shrinkage by pre-operative chemotherapy using doxorubicin, cisplatin and ifosfamide and that this facilitated a R0 resection. Trabectedin has emerged as a favourable option for patients with advanced soft tissue sarcomas and has been approved in the European Union as a second-line option for advanced previously-treated soft tissue sarcomas.

Prognosis depends on the histological subtype. In 1979, Evans reported that the median survival for patients with the well-differentiated, myxoid, dedifferentiated and pleomorphic types was 119, 113, 59 and 24 months, respectively.

Castleman's Disease

Castleman's disease is a rare benign lymphoproliferative disorders that affects adults in the third and fourth decade of life. It was first described by the American physician and pathologist Dr. Benjamin Castleman in 1956 when he described a group of patients with large thymoma-like masses in the anterior mediastinum. The aetiology is unknown, however it has been shown associated with the human immunodeficiency virus (HIV) and human herpes virus 8 (HHV8). Moreover it has been linked to several malignancies such as Kaposi's sarcoma, non-Hodgkin's lymphoma, Hodgkin's lymphoma and POEMS syndrome (polyneuropathy, organomegaly, endocrinopathy, M-protein, skin manifestation). Although Castleman's disease usually involves the mediastinum in 70% of cases, reports of extrathoracic lesions involving the neck, axilla, shoulders, pelvis, pancreas, nasopharynx and retroperitoneum have been described. Castleman's disease involving the mesentery is a rare.

Castleman's disease can be classified into two histological types, hyaline type and plasma cell type. The hyaline type is typified by the presence of small hyaline follicles and intrafollicular capillary proliferation. This type accounts for more than 90% of Castleman's disease cases, usually presents with localized disease (unicenteric) and is asymptomatic in the majority. The plasma cell type is less common and usually presents with disseminated lymphadenopathy (i.e. multicenteric), pyrexia, diaphoresis, weight loss, anaemia, elevated C-reactive protein (CRP), erythrocyte sedimentation rate (ESR) and hypergammaglobulinemia. However the hyaline and plasma cell type can exhibit considerable and histologic overlap, thus a so-called mixed variant of Castleman's disease is occasionally seen.

Radiological investigations are nonspecific and not diagnostic. Abdominal ultrasound can detect homogenous, hypoechoic mass like lesions that point towards Castleman's disease. The classic CT appearance of hyaline Castleman's disease is that of a solitary enlarged lymph node or localized nodal masses demonstrating homogeneous, intense contract enhancement. Three patterns of disease have been described; these include a solitary non-invasive mass (most common: 50% of cases), a dominant infiltrative mass with associated lymphadenopathy (40% of cases), and matted lymphadenopathy without a dominant mass (10% of cases). On magnetic resonance imaging (MRI), lesions in hyaline vascular Castleman's disease classically exhibit heterogeneous T1 and T2 hyperintensity compared with skeletal muscle. Prominent flow voids may be seen, identifying feeding vessels. MRI enables assessment of the extent of disease as well as its relationship to adjacent structures (although evaluation of calcification is limited.

Due to lack of specific radiographic features for Castleman's disease CT or ultrasound-guided fine-needle aspiration of the mass lesion should be performed to obtain an accurate tissue diagnosis. Solitary lesions or unicentric type Castleman's disease virtually always behave in a benign manner. Indications for surgery include the presence of focal lesions suited to curative resections, or multicentric lesions, for palliative resection or attempt for curative resection with multimodal medical and chemotherapy. Surgical excision of focal mesenteric lesions is recommended and associated with excellent prognosis. Recurrence of localised disease is rare after complete surgical resection, but has been reported in patients after incomplete surgical removal. Limited success with radiotherapy (30–45 Gy) and chemotherapy has been reported. Multicentric Castleman's disease carries a poor prognosis as the disease is systemic. Management is controversial for multicentric lesions and is largely considered multimodal encompassing surgery, chemotherapy, steroid therapy, antiviral medication, or the use of antiproliferative regimens. In most cases, diagnosis of mesenteric Castleman's disease is confirmed only after resection and histopathological examination of the specimen.

Cystic Lesions

Mesenteric cysts are benign intra-abdominal lesions first described in 1507 by the Florentine anatomist Benevenni while performing an autopsy in an 8-year old boy. This was followed by the first description of a chylous cyst by von Rotitansky in 1842. In 1852, Gairdner described an omental cyst. Tillaux described the first successful surgical treatment of a mesenteric cyst in 1880. He was succeeded by Pean in 1883, who described the first marsupialization technique for mesenteric cysts. However, it was not until 1993 that Mackenzie reported the first successful laparoscopic resection of a mesenteric cyst.

Mesenteric cysts have an incidence ranging from 1 in 100,000 in the adult population and 1 in 20,000 in the pediatric population. They affect females more than males the fourth decade of life. The majority of mesenteric cysts are single however can be uni- or multi-locular. They are asymptomatic in approximately 45% of cases and diagnosed incidentally during routine physical examination, during abdominal surgery, or routine imaging. However they can present with a variety of nonspecific symptoms such as abdominal pain, distension, palpable mass, nausea, vomiting, change of bowel habit and weight loss. Up to one third of patients present acutely due to complications from the cyst. These include intestinal obstruction, volvulus, haemorrhage, infection or cyst rupture.

Mesenteric cysts range between 2 to 35 cm in size and the majority (50–67%) of reported cases are in the small bowel mesentery. This is followed by 24–37% located in the mesocolon and the remainder within the retroperitoneum. The majority are solitary lesions however uni- and multilocular varieties exist. Most are comprised of a single layer of columnar or cuboidal epithelial cells. The cyst fluid can be chylous, serous, or haemorrhagic and associated with small bowel mesentery, mesocolon and trauma respectively. Mesenteric cysts can be classified based on histological features into six subgroups: (1) cysts of lymphatic origin (simple lymphatic cyst and lymphangioma); (2) cysts of mesothelial origin (simple mesothelial cyst, benign cystic mesothelioma and malignant cystic mesothelioma); (3) cysts of enteric origin (enteric cyst and enteric duplication cyst); (4) cysts of urogenital origin; (5) mature cystic teratoma (dermoidcysts); and (6) pseudocysts (infectious and traumatic cysts).

Ultrasound and CT are the main imaging modalities utilized characterizing mesenteric cysts. Ultrasound has a higher sensitivity in characterizing the internal nature of the cyst and can differentiate between septations, debris, and fluid levels. CT will provide greater detail regarding location of the cyst and its relation to surrounding structures. Mesenteric cysts carry a 3% overall malignant potential. Majority of cancers developing out of mesenteric cysts are sarcomas with a few cases describing adenocarcinomas. Poor prognosticators on radiological imagining include rapid growing cysts that possess a solid component within the cyst.

Surgery with complete enucleation is the recommended treatment of choice for mesenteric cysts. Older techniques such as simple aspiration and marsupialization are associated with greater recurrence and infection rates. Depending on the association of the cyst to surrounding structures, an en-bloc resection incorporating bowel may be necessary. While performing surgery either open or via the minimally invasive approach it is recommended to abide to fundamental oncological principles to prevent recurrence. In the setting of managing mesenteric cysts conservatively no clear consensus exists regarding follow-up. Several authors have recommended frequent imaging for cysts that are bigger in size, possess a solid component within the cyst and have a faster rate of growth as such would be indicative of increased malignant potential and the risk of developing complications.

The prognosis for mesenteric cysts is generally good as the majority are benign and recurrence rates are low with a R0 resection. Recurrence is greater in those cysts located in the retroperitoneum as they are more technically demanding to excise due to their close proximity to great vessels, ureter and other organs.

Conclusions

With the new understanding of the mesentery as a unique continuous organ, we can review the disease states that involve the mesentery. While rare overall, the basic understanding of the primary lesions of the mesentery, their distinguishing features, diagnosis and management options, is key to optimize management of mesenteric neoplasms.

Suggested Reading

1. Coffey JC, O'Leary DP. The mesentery: structure, function, and role in disease. The Lancet Gastroen Hepatol. 2016;1(3):238–47.
2. Coffey JC, Sehgal R, Walsh D. Pathology of the mesentery. Mesenteric Principles of Gastrointestinal Surgery: Basic and Applied Science: CRC Press; 2017. p. 85–109.
3. Dufay C, Abdelli A, Le Pennec V, Chiche L. Mesenteric tumors: diagnosis and treatment. J Visc Surg. 2012;149(4):e239–51.
4. Mueller PR, Ferrucci JT Jr, Harbin WP, Kirkpatrick RH, Simeone JF, Wittenberg J. Appearance of lymphomatous involvement of the mesentery by ultrasonography and body computed tomography: the "sandwich sign" . Radiology. 1980;134(2):467–73.
5. Salemis NS, Gourgiotis S, Tsiambas E, Karagkiouzis G, Nakos G, Karathanasis V. Diffuse large B cell lymphoma of the mesentery: an unusual presentation and review of the literature. J Gastroint Cancer. 2009;40(3–4):79–82.
6. Kasper B, Strobel P, Hohenberger P. Desmoid tumors: clinical features and treatment options for advanced disease. Oncologist. 2011;16(5):682–93.
7. Bonvalot S, Desai A, Coppola S, Le Péchoux C, Terrier P, Dômont J, et al. The treatment of desmoid tumors: a stepwise clinical approach. Annals Oncol. 2012;23(suppl_10):x158−66.
8. Mazur MT, Clark HB. Gastric stromal tumors. Reappraisal of histogenesis. Am J Surg Pathol. 1983;7(6):507–19.
9. Schwameis K, Fochtmann A, Schwameis M, Asari R, Schur S, Köstler W, et al. Surgical treatment of GIST—an institutional experience of a high-volume center. Int J Surg. 2013;11 (9):801–6.
10. Dematteo RP, Ballman KV, Antonescu CR, Maki RG, Pisters PW, Demetri GD, et al. Adjuvant imatinib mesylate after resection of localised, primary gastrointestinal stromal tumour: a randomised, double-blind, placebo-controlled trial. Lancet (London, England). 2009;373(9669):1097–104.
11. Modlin IM, Kidd M, Latich I, Zikusoka MN, Shapiro MD. Current status of gastrointestinal carcinoids. Gastroenterology. 2005;128(6):1717–51.
12. Jain SK, Mitra A, Kaza RC, Malagi S. Primary mesenteric liposarcoma: an unusual presentation of a rare condition. J Gastrointest Oncol. 2012;3(2):147–50.
13. Moyana TN. Primary mesenteric liposarcoma. Am J Gastroent. 1988;83(1):89–92.
14. Soft tissue and visceral sarcomas: ESMO Clinical Practice Guidelines for diagnosis, treatment and follow-up. Annals Oncol Official J Eur Soc Med Oncol. 2012;23 Suppl 7:vii92–9.
15. Chronowski GM, Ha CS, Wilder RB, Cabanillas F, Manning J, Cox JD. Treatment of unicentric and multicentric Castleman disease and the role of radiotherapy. Cancer. 2001;92 (3):670–6.
16. de Perrot M, Brundler M, Totsch M, Mentha G, Morel P. Mesenteric cysts. Toward less confusion? Dig Surg. 2000;17(4):323–8.

Mesenteric Artery Thrombosis and Embolism

49

Salvatore Parascandola and Vincent Obias

Description of This Condition

Mesenteric vascular occlusive disease is a rare but potentially devastating condition, characterized by intestinal hypoperfusion of the small intestine caused by complete or partial obstruction of the mesenteric arteries. Mesenteric ischemia accounts for roughly 1 in every 1000 hospital admissions in this country, with mortality rates ranging as high as 50–75%. Untreated, acute mesenteric ischemia carries a mortality of close to 100%. The most common causes of acute mesenteric ischemia (AMI) are arterial embolism (40 to 50%) and arterial thrombosis (25 to 30%).

Anatomy and Pathophysiology

Perfusion to the gastrointestinal system is provided by three main mesenteric arteries: the celiac artery (CA), the superior mesenteric artery (SMA), and the inferior mesenteric artery (IMA) (Fig. 49.1). The CA, arising from the infradiaphragmatic suprarenal abdominal aorta, provides circulation to the foregut (esophagus to duodenum), hepatobiliary system, and the spleen. The SMA, arising just distal on the suprarenal aorta to the CA, supplies the midgut (jejunum to mid-transverse colon). The IMA, which originates from the left lateral portion of the infrarenal aorta, supplies the hindgut (mid-transverse colon to rectum). The superior and inferior pancreaticoduodenal arteries provide collateral perfusion between the CA and SMA, while the marginal artery of Drummond, arc of Riolan,

S. Parascandola (✉) · V. Obias
Department of Colorectal Surgery, George Washington University Medical Faculty Associates, Washington, DC, US
e-mail: salvatore.parascandola@gmail.com

E. D. Ehrenpreis et al. (eds.), *The Mesenteric Organ in Health and Disease*,
https://doi.org/10.1007/978-3-030-71963-0_49

and retroperitoneal meandering arteries form collateral networks between the SMA and IMA. Due to this redundancy in circulation, *chronic* reduction of flow in one or two mesenteric vessels is typically tolerated, provided sufficient temporal compensatory perfusion by the uninvolved branches. *Acute* occlusion a main mesenteric trunk, by contrast, can produce profound ischemia and necrosis in the absence of collateral perfusion. Prompt recognition and treatment before the onset of irreversible ischemia are critical.

In embolic occlusion, emboli lodge at branch points distal to the middle colic artery and early jejunal branches, sparing the SMA origin (Fig. 49.2). Risk factors for those with embolic occlusion of the mesenteric arterial circulation include atrial fibrillation, recent myocardial infarction, mitral valve disease, mural thrombus, or left ventricular aneurysm. In the absence of well-formed collateral perfusion, embolization results in a pattern of ischemia that compromises the majority of small bowel and ascending colon while sparing the proximal jejunum and distal transverse colon (Fig. 49.3a).

Thrombosis of the SMA is usually associated with systemic, pre-existing atherosclerosis. Many of these patients have a history of chronic mesenteric ischemia (CMI) including 'food fear', post-prandial pain, and weight loss. SMA thrombosis is also seen in the setting of vasculitis, mesenteric dissection, or mycotic aneurysm. Thrombosis typically occurs at the origins of visceral arteries, leading to continuous intestinal involvement in the distribution of either the CA or SMA, without sparing the proximal jejunum or colon (Fig. 49.3b).

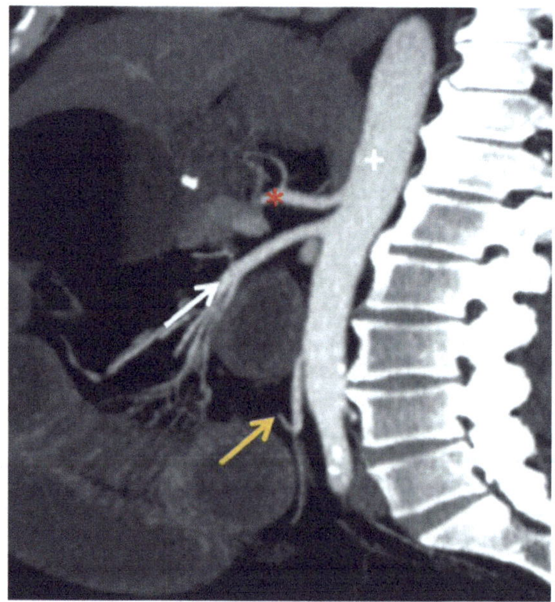

Fig. 49.1 Sagittal CT demonstrates the three major arteries that supply the bowel, celiac axis (asterisk), superior mesenteric artery (white arrow) and inferior mesenteric artery (orange arrow). All are visceral branches of the abdominal aorta (+)

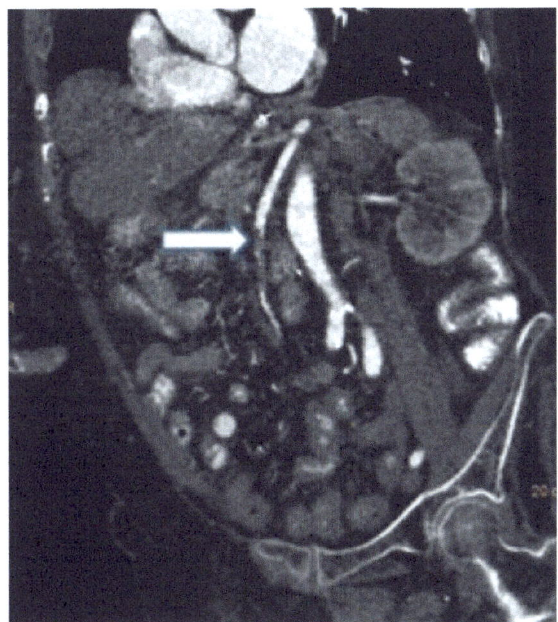

Fig. 49.2 Sagittal CTA scan of a patient with acute mesenteric ischemia secondary to embolic occlusion of the SMA (arrow)

Presentation

Patients with acute mesenteric artery occlusion may present with significant abdominal pain that is out of proportion to a benign abdominal exam. In these patients, high level of suspicion is key to early diagnosis. Nausea, vomiting, and diarrhea may follow the onset of abdominal pain. Fever, guarding, and abdominal tenderness are late findings, typically associated with bowel infarction. Patients may be tachycardic and demonstrate melena or heme positive stools. History of arrhythmia, prosthetic heart values, recent myocardial infection, or prior embolization to other arterial trees should raise the suspicion of mesenteric emboli in patients with acute onset of abdominal pain.

Thrombotic mesenteric occlusion may also present with sudden onset of severe midabdominal pain that is out of proportion to the physical findings. However, unlike patients with acute embolic occlusion, these patients often have a history consistent with chronic mesenteric ischemia, which manifests as postprandial abdominal pain leading to food avoidance and significant weight loss. These patients may also demonstrate other sequelae of diffuse atherosclerotic disease e.g. coronary or peripheral artery disease.

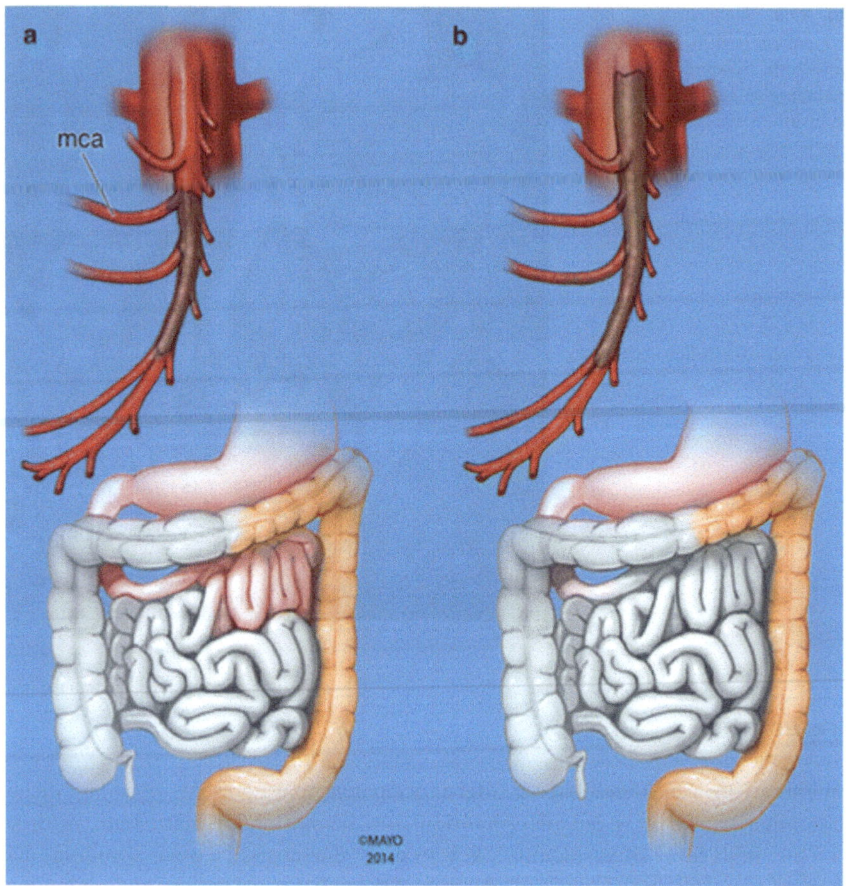

Fig. 49.3 **a** Embolic occlusion of the superior mesenteric artery with jejunal-sparing ischemia. **b** Mesenteric arterial thrombosis with continuous intestinal involvement in the distribution of the SMA, without sparing the proximal jejunum or colon. Reused with permission. Copyright © Springer Nature

Laboratory Findings

Although laboratory findings are often insensitive and nonspecific for the diagnosis of mesenteric artery embolism or thrombosis, though they can help substantiate clinical suspicion. The majority of patients will have an abnormally elevated leukocyte count. Metabolic acidosis with elevated lactate develops as a result of anaerobic metabolism. While an elevated lactate can be seen in the setting of decreased oral intake and dehydration, a lactate level of >2 mmol/l is associated with irreversible intestinal ischemia. Additionally, elevated D-dimer is associated with intestinal ischemia, reflecting ongoing clot formation and degradation by fibrinolysis. Elevated serum amylase and aspartate aminotransferase may also be seen.

Imaging Workup

Plain abdominal radiographs are typically the first test ordered in a patient with acute abdominal pain. While they are helpful in ruling out other causes of abdominal pain, they have a limited role in diagnosing acute mesenteric ischemia. Plain films may demonstrate ileus and vascular calcifications, suggesting ischemia. Such findings as bowel wall edema (thumbprinting), pneumatosis, pneumobilia, or pneumoperitoneum may indicate infarction. Importantly, a negative radiograph does not exclude acute mesenteric ischemia, as up to 25% of abdominal radiographs will be normal.

Today, multi-detector CTA represents the most useful and rapid tool for the fast and accurate diagnosis of AMI. In the presence of irreversible ischemia, CTA will demonstrate intestinal dilatation and thickness, reduction or absence of visceral enhancement, pneumatosis intestinalis, portal venous gas, or free intraperitoneal air (Fig. 49.4a and b). Biphasic CTA is the diagnostic modality of choice and includes the following steps: (1) Pre-contrast scans to detect hyper attenuating vascular calcification, intravascular thrombus, and intramural hemorrhage, (2) arterial and venous phases to demonstrate thrombus in the mesenteric arteries and veins, presence of embolism or infarction of other organs, and abnormal enhancement of the bowel wall, and (3) multi-planar reconstructions (MPR) to assess the origin of the mesenteric arteries. Three-dimensional reconstructions generated from CTA datasets can also provide valuable guidance for preoperative planning (Fig. 49.5). For the diagnosis of AMI, a sensitivity of 93%, specificity of 100%, and positive

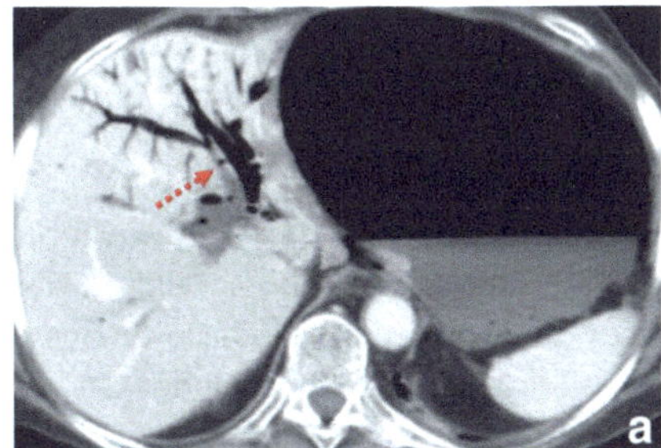

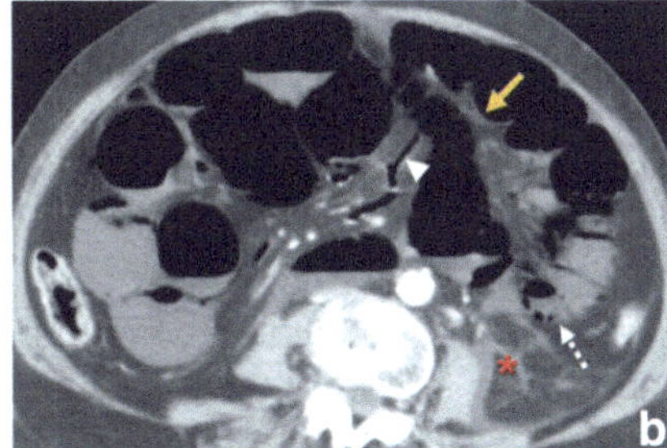

Fig. 49.4 CTA of the abdomen in a patient with embolism to the superior mesenteric artery demonstrating **a** pronounced intrahepatic portal venous gas (red arrow) and **b** dilated, gas-filled loops of bowel and transmural necrosis (orange arrow). Pneumatosis (white arrow), mesenteric venous gas (white arrowhead), and fat stranding (asterisk) are seen

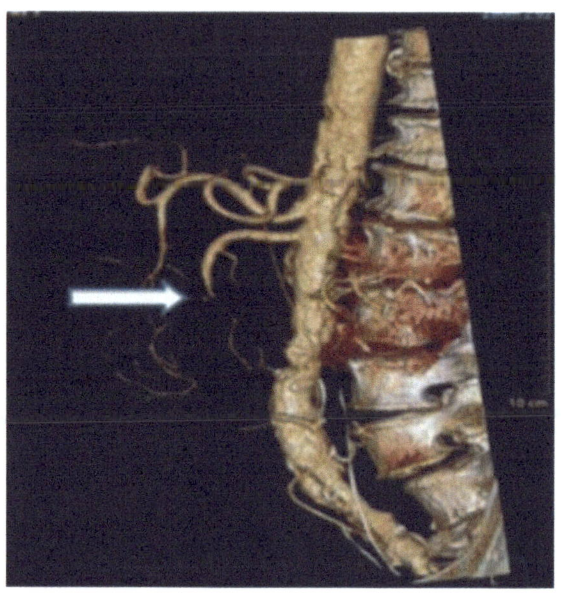

Fig. 49.5 3D CTA reconstruction demonstrating mid occlusion of SMA (arrow)

and negative predictive values of 100 and 94%, respectively, have been achieved with this imaging modality.

Catheter-based angiography is reserved for cases where the diagnosis is not clear, endovascular interventions are contemplated, or further details of SMA anatomy are required for SMA revascularization. Multiple views are needed for the sufficient evaluation of the SMA and CA. Lateral views are used to assess the origin of the CA and the SMA, while anteroposterior views allow for the evaluation of the circulation of the distal CA and SMA. Selective catheterization of the SMA may reveal a 'meniscus sign' at the branch points where emboli lodge.

Gadolinium-enhanced magnetic resonance angiography (MRA) avoids the risk of allergic reaction and nephrotoxicity associated with iodinated CT imaging. While MRA is helpful in diagnosing chronic mesenteric ischemia, in the setting of acute ischemia, MRA and post processing are too time consuming for these critically ill patients. Given the consequences of a missed or delayed diagnosis, CTA should be considered over MRA even in the setting of renal insufficiency.

Duplex ultrasonography has little value in the diagnosis of AMI. However, it is a valuable noninvasive adjunct for follow up after surgical or endovascular revascularization and can be used to assess patency and recurrence of disease. Endoscopy, colonoscopy, or barium radiography offer no useful information in the diagnostic evaluation of acute mesenteric ischemia. Furthermore, if mesenteric ischemia is considered, a barium enema is contraindicated as intraperitoneal extravasation of barium can occur in the setting of intestinal perforation.

Treatment

Patients with acute embolic or thrombotic mesenteric ischemia warrant prompt fluid resuscitation, correction of electrolyte imbalances, and systemic anticoagulation with heparin to prevent further thrombus propagation. Empiric antibiotics should be administered with regimens that cover gram-negative enteric flora and anaerobes. Patients often present with organ dysfunction and require close hemodynamic monitoring. While these patients may be hypotensive, vasopressors can exacerbate ongoing intestinally ischemia, thus these patients should undergo fluid resuscitation with isotonic crystalloid solution before initiation of vasopressors.

The goals of surgical therapy for acute mesenteric ischemia are (1) restoration of arterial perfusion and (2) resection of non-viable bowel. Prompt laparotomy should be done in any patient with peritonitis or evidence of threatened intestinal viability. In embolic SMA disease, laparotomy may reveal intestinal ischemia from the mid-jejunum to the ascending or transverse colon. Frankly necrotic or perforated bowel should be resected and left in discontinuity prior to revascularization in order to limit gross contamination. The SMA may be exposed in one of three approaches. First, it may be exposed caudal to the inferior border of the by entering the lesser sac (Fig. 49.6a). Second, the transverse colon may be lifted superiorly and a horizontal incision made in the mesentery, exposing the SMA at the root of the small bowel mesentery (Fig. 49.6b). Finally, the SMA can be exposed by dividing the ligament of Treitz and peritoneal attachments and laterally mobilizing the fourth portion of the duodenum laterally (Fig. 49.6c).

After systemic heparinization, proximal and distal control of the vessel is achieved with vascular clamps, and a transverse arteriotomy is made sharply. A balloon thromboembolectomy can then be performed with a 3-Fr or 4-Fr catheter through the arteriotomy (Fig. 49.7). In the event the embolus has lodged more distally at the jejunal and ileal branches, thromboembolectomy is best accomplished with a smaller catheter (i.e., 2 Fr or 3 Fr), given the fragility of this vasculature. This maneuver is repeated until the balloon is free of thrombus upon withdrawal and pulsatile flow is restored. Following successful thromboembolectomy, the arteriotomy can be closed primarily using simple interrupted 5–0 or 6–0 monofilament sutures or patched using cryopreserved vein. For repair of diminutive branch vessels, vein patch angioplasty is preferred.

In patients with acute mesenteric thrombosis, ischemia typically involves a heavily atherosclerotic proximal CA or SMA. In these patients, vascular bypass is required for successful revascularization and typically requires single vessel reconstruction in order to bypass the in-situ thrombosis. Autogenous or synthetic materials may be used for the graft. Synthetic bypass grafts of 6–8 mm Dacron or polytetrafluoroethylene (PTFE) are most often used given their availability, ease of handling, and size match. In the setting of gross abdominal contamination and risk of subsequent graft infection, saphenous vein and thigh femoral vein may be used. Rifampin-soaked Dacron is another option, as vein harvest may be too time consuming in critically ill patients. Several options exist for graft orientation.

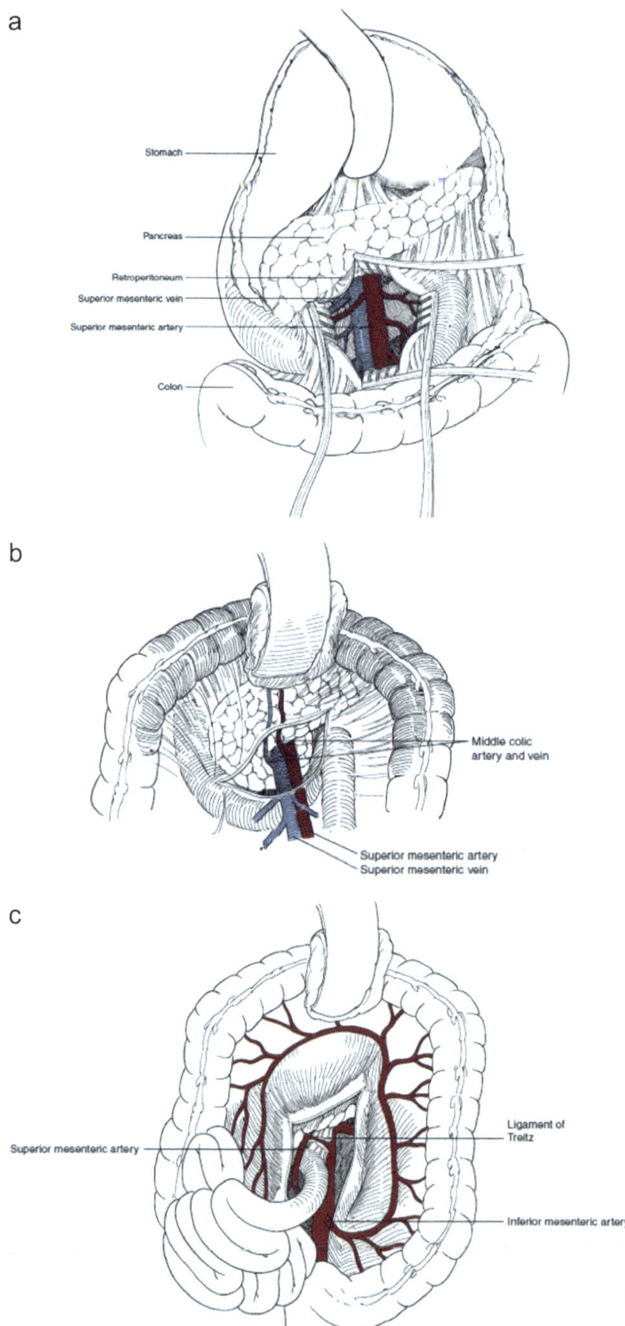

Fig. 49.6 Exposure of the superior mesenteric artery through **a**. longitudinal midline incision in the retroperitoneal tissue immediately inferior to the border of the pancreas. **b**. At the base of the transverse mesocolon through a horizontal incision in the mesentery. **c**. Above the superior margin of the duodenum by completely mobilizing the fourth portion of the duodenum after incising the ligament of Treitz and the other peritoneal attachments. Reused with permission. Copyright © Elsevier

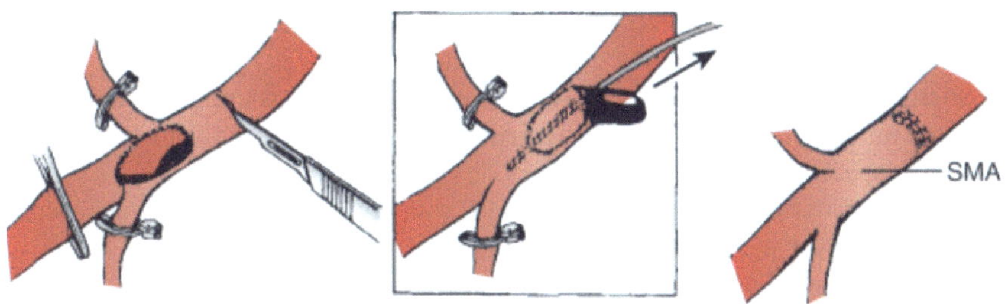

Fig. 49.7 Balloon thromboembolectomy through transverse arteriotomy in the superior mesenteric artery. Reused with permission. Copyright © Elsevier

Retrograde approaches are most often used. For retrograde bypasses, the proximal anastomosis can be placed on the right common iliac artery (most common), infrarenal aorta, or left common iliac artery and is performed in an end-to-side fashion. The distal anastomosis on the SMA can be performed in an end-to-side or end-to-end fashion. The graft is typically tunneled so that it forms a curve or "C loop" between the proximal and distal anastomosis as it traverses caudal to cephalad (Fig. 49.8).

In the cases of prohibitive anatomy, antegrade bypass from the supraceliac aorta may be considered. The supraceliac aorta is exposed by mobilizing the left triangular ligament of the liver and incising the median arcuate ligament and crus of the diaphragm. Once the patient is systemically heparinized, the aorta is occluded. An

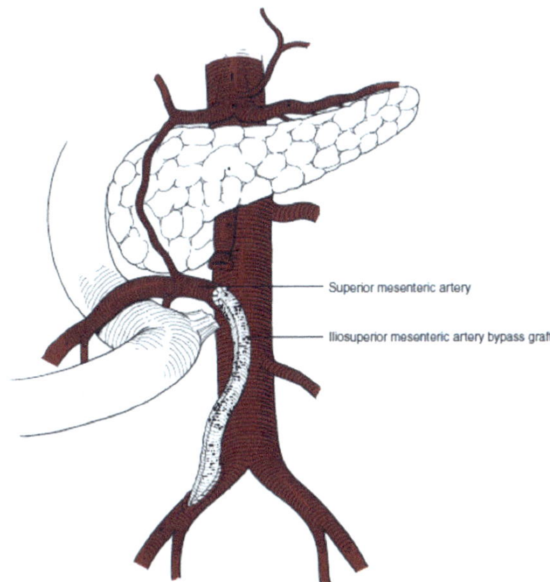

Fig. 49.8 Completed retrograde bypass from the proximal right common iliac artery to the superior mesenteric artery. Reused with permission. Copyright © Elsevier

arteriotomy is made in the aorta and anastomosis performed using 3–0 nonabsorbable, monofilament suture, and 5–0 sutures with felt pledgets. A limited endarterectomy may be necessary. The distal anastomosis to the celiac artery and superior mesenteric artery is constructed using an end-to-end and end-to-side manner, respectively (Fig. 49.9).

After revascularization by thromboembolectomy or bypass, an assessment of the viability of bowel must be performed and includes running the entire length of bowel, assessing the color and peristaltic activity, and palpating the mesenteric arcades. Adjuncts include Doppler interrogation of the mesenteric arcade, fluorescein injection with Wood's lamp inspection, or perfusion fluorometry. Non-viable bowel must be resected. In cases of questionable viability, a second-look procedure in 24–48 h should be considered. Often the full extent of bowel viability is not immediately obvious following initial vascularization. Additional non-viable bowel resections should be performed at that time.

In patients with acute mesenteric thrombosis without evidence of bowel ischemia, successful cases of endovascular treatments for AMI have been reported. Major drawbacks to endovascular approaches include inability to inspect the viability of bowel both before and after revascularization and the prolonged time that

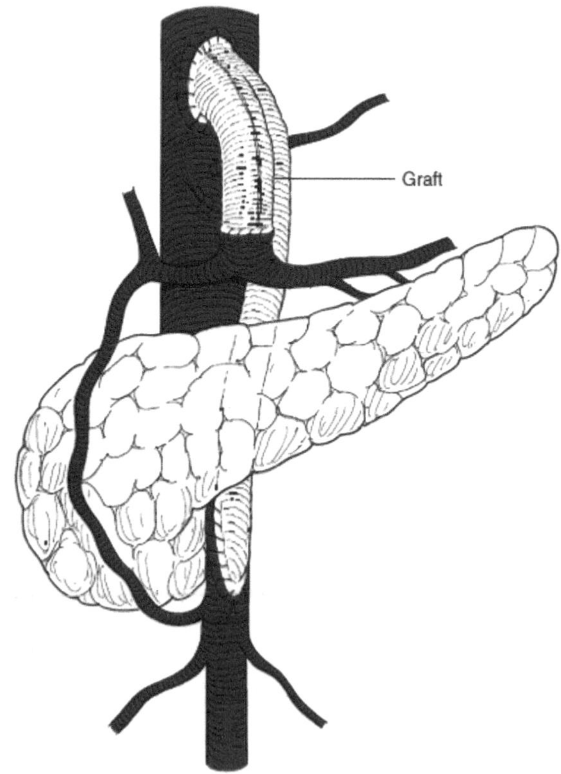

Fig. 49.9 Completed antegrade bypass from the supraceliac aorta to the celiac axis and the superior mesenteric artery. Reused with permission. Copyright © Elsevier

may be required to achieve successful thrombolysis. Given these limitations, purely endovascular approaches may be used in select groups of patients. Percutaneous mechanical thrombectomy, aspiration thrombectomy, and intra-arterial thrombolysis, with or without adjuvant angioplasty and stenting, have been increasingly described in recent literature. To perform an endovascular revascularization, vascular access can be gained through a retrograde femoral approach or antegrade brachial approach. Once an introducer sheath is placed, a lateral and anteroposterior aortogram is performed to identify the origin of the CA and SMA. Next, systemic heparin is given and guidewire placed across the lesion. Once wire placement has crossed the lesion, the catheter is advanced and its position confirmed by angiogram. Thrombectomy, aspiration, or balloon angioplasty can then be completed and a postangioplasty angiogram performed. If the completion angiogram demonstrates incomplete angioplasty, placement of mesenteric stent is warranted. Involvement of the orifice or proximal vessel should be treated with balloon-expandable stent (Fig. 49.10a–d). To reduce vasospasm, an intra-arterial infusion of nitroglycerine or papaverine can be used. Intra-arterial delivery of thrombolytic agents into the thrombus is another potentially useful treatment modality in acute thrombotic mesenteric ischemia.

Hybrid approaches to acute mesenteric ischemia offer the ability to examine the abdominal viscera and control sepsis while employing less invasive mesenteric revascularization techniques. After laparotomy, the SMA is exposed and a puncture made in the vessel, though which retrograde guidewire access is gained. The SMA is clamped to avoid distal embolization. The SMA lesion is then crossed and the guidewire passed through brachial or femoral access, creating "through-and-through" access. If thrombectomy is required, an arteriotomy in the SMA is made and Fogarty balloon passed over the wire and into the aorta. Thrombectomy is performed and completion and restoration of flow assessed. If thrombectomy is not required, no arteriotomy is necessary. Using the through-and-through wire, a stent can be placed in either an antegrade or retrograde fashion (Fig. 49.11a–d).

Complications after endovascular repair include access site complications (hematoma, pseudoaneurysm, access artery thrombosis, retroperitoneal hematomas), distal embolization during wire and catheter manipulation, and recurrent stenosis requiring repeat angioplasty and stenting. A baseline duplex of the treated vessel should be obtained before discharge, every 6 months for 1 year, and annually thereafter. Cumulative stent patency rate over 3 years is reported between 44 and 88%.

Outcomes

Postoperatively, all patients who undergo revascularization for acute mesenteric ischemia require close monitoring. The postoperative course for these patients is frequently complicated by the development of multiple organ dysfunction, which likely accounts for prolonged hospitalization and is a leading cause of death. Postoperative care often involves intensive care unit admission, prolonged

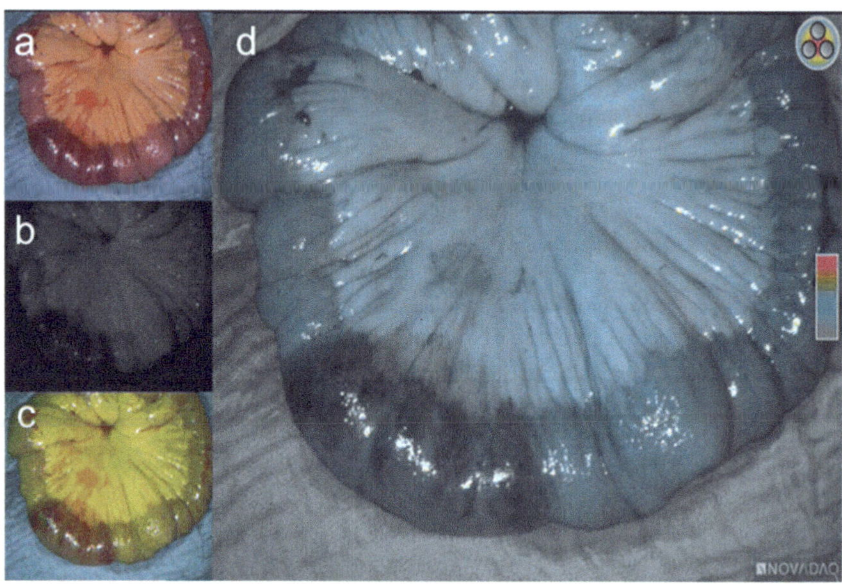

Fig. 49.10 A malperfused area of the small intestine **a** macroscopically. **b**In the SPY mode. **c** In the ICG fluorescence mode. **d** In the Pinpoint semi-quantitative assessment mode. Reused with permission. Copyright © Springer Nature

endotracheal intubation, and parenteral nutrition. Coagulopathy and thrombocytopenia may be present and can be managed with platelets, plasma, or both. Depending on the extent of necrosis and/or contamination, broad-spectrum antibiotic therapy is typically required. In patients who underwent endovascular revascularization, any new abdominal complaints should immediately raise suspicion for missed intestinal non-viability.

The aggregate mortality rates in patients with acute mesenteric ischemia have been historically reported as 54 and 74% in patients with acute embolism and thrombosis, respectively. Risk factors for increased mortality include age greater than 60 years, intestinal non-viability requiring resection, colon involvement, and duration of symptoms before treatment.

Recently, there has been increasing utilization of endovascular techniques over traditional laparotomy as the initial management in patients with AMI. Beaulieu and colleagues conducted a review using the National Inpatient Sample database which identified a significant increase in endovascular techniques from 2005 through 2009. Their study reported that mortality, hospital length of stay, need for bowel resection,

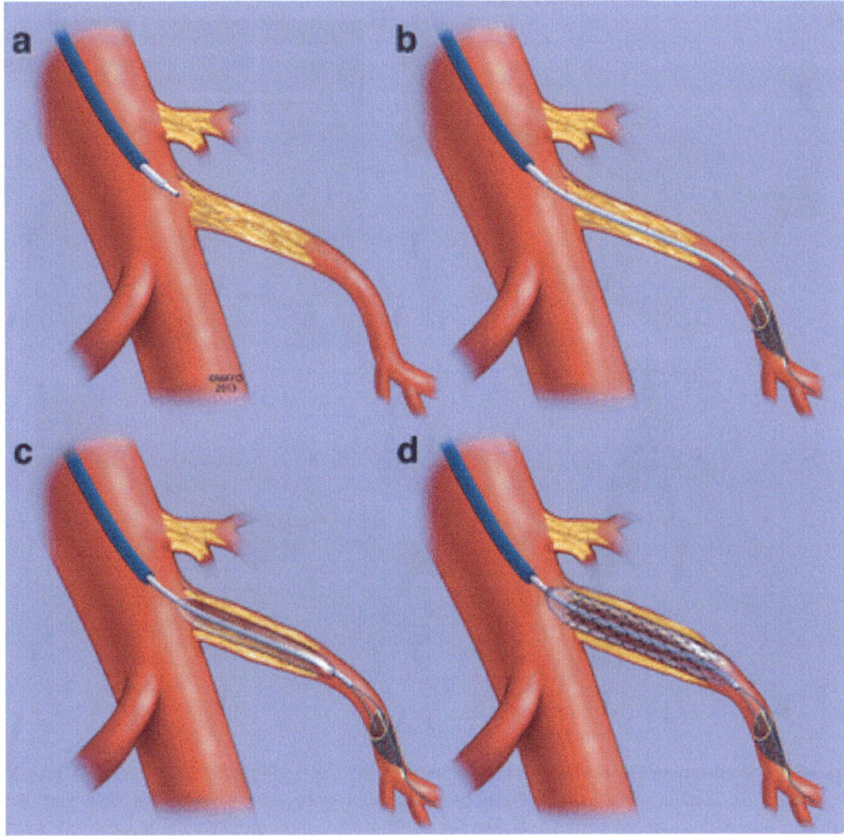

Fig. 49.11 Endovascular balloon angioplasty and stenting in acute mesenteric thrombosis. **a** Introducer sheath placed at the lesion. **b** Guidewire catheter placement across the lesion. **c** Balloon angioplasty. **d** Stent placement. Reused with permission. Copyright © Springer Nature

and need for postoperative parenteral nutrition were reduced in those patients receiving endovascular therapy compared to traditional open surgical revascularization. A large study by Arthurs and colleagues compared traditional open surgery versus endovascular revascularization in 70 patients with acute embolism or thrombosis. In those who underwent endovascular therapy, technical success rate was seen in 78% of patients. Endovascular therapy resulted in lower rates of acute renal failure and pulmonary failure. They noted an overall mortality rate of 39% for patients undergoing endovascular therapy vs 50% for traditional therapy. Notably, as abdominal exploration was reserved for peritonitis or clinical deterioration, 31% of patients who underwent endovascular revascularization did not require laparotomy.

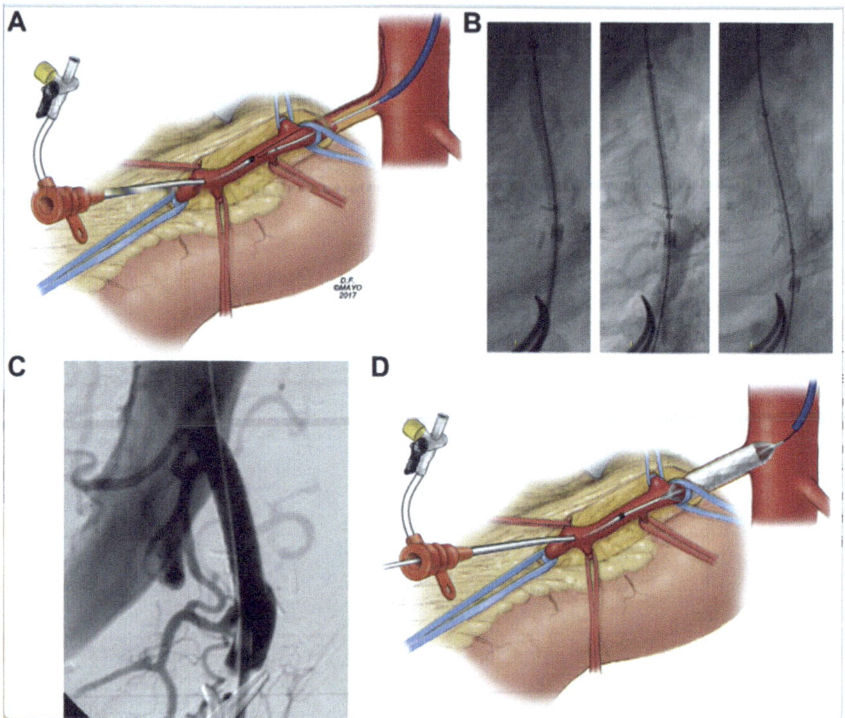

Fig. 49.12 Retrograde Open Mesenteric Stenting (ROMS). **a** A guidewire is exteriorized through the retrograde sheath, establishing through-and-through access between the access site and superior mesenteric artery (SMA). **b** Deployment of the balloon-expandable covered stent. **c** Completion angiography without residual stenosis. **d** During catheter manipulations, the distal SMA and its branches are controlled to avoid distal embolization. Reused with permission. Copyright © Elsevier

Though an uncommon disease, the physiologic consequences of acute mesenteric embolism and thrombosis are devastating. Early diagnosis and resuscitation is critical for survival. In addition to prompt recognition, patient-specific and hybrid management approaches may lead to a reduction in the historically high mortality rates seen in this disease process.

Suggested Readings

1. Acosta S, Björck M. Modern treatment of acute mesenteric ischaemia. Br J Surg. 2014;101 (1):e100–8. https://doi.org/10.1002/bjs.9330.
2. Arthurs ZM, Titus J, Bannazadeh M, et al. A comparison of endovascular revascularization with traditional therapy for the treatment of acute mesenteric ischemia. J Vasc Surg. 2011;201:698–705.

3. Bala M, Kashuk J, Moore EE, et al. Acute mesenteric ischemia: guidelines of the world society of emergency surgery. World J Emerg Surg. 2017;12:38. Published 2017 Aug 7. https://doi.org/10.1186/s13017-017-0150-5.
4. Branco BC, Montero-Baker MF, Aziz H, Taylor Z, Mills JL. Endovascular therapy for acute mesenteric ischemia: an NSQIP analysis. Am Surg. 2015;81(11):1170–6.
5. Beaulieu RJ, Arnaoutakis KD, Abularrage CJ, et al. Comparison of open and endovascular treatment of acute mesenteric ischemia. J Vasc Surg. 2014;59:159–64.
6. Coleman DM, Rectenwald JE, Upchurch GR. Renal and Splachnic Vascular Disease. Mulholland, MW, Lillemoe, KD, Doherty, GM, Upchurch, GR, Alam, HB, Pawlik, TM. Greenfield's surgery: scientific principles and practice. Philadelphia: Lippincott Williams & Wilkins; 2017:1648–68.
7. Holscher C, Reifsnyder T. Acute mesenteric ischemia. In: Cameron, J. L., & Cameron, A. M. Current Surgical Therapy. Philadelphia: Elsevier; 2020.pp. 1057–1061.
8. Leone M, Bechis C, Baumstarck K, et al. Outcome of acute mesenteric ischemia in the intensive care unit: a retrospective, multicenter study of 780 cases [published correction appears in Intensive Care Med. 2015 May;41(5):966–8]. Intensive Care Med. 2015;41 (4):667–76. https://doi.org/10.1007/s00134-015-3690-8.
9. Lin PH, Poi M, Matos J, Kougias P, Bechara C, Chen C. Arterial Disease. In: Brunicardi F, Andersen DK, Billiar TR, Dunn DL, Hunter JG, Matthews JB, Pollock RE. eds. Schwartz's Principles of Surgery, 10e New York, NY: McGraw-Hill; 2015.
10. Lo R, Schermerhorn M. Mesenteric Arterial Disease: Epidemiology, Pathophysiology, and Clinical Evaluation In: idawy, A, Perler, B. Rutherford's Vascular Surgery and Endovascular Therapy, 8e Philadelphia: Elsevier; 1725–1734.
11. Myers S. Acute Embolic and Thrombotic Mesenteric Ischemia. In: Cameron, J. L., & Cameron, A. M. Current Surgical Therapy. Philadelphia: Elsevier; 2020. 718–720.
12. Savlania A, Tripathi RK. Acute mesenteric ischemia: current multidisciplinary approach. J Cardiovasc Surg (Torino). 2017;58(2):339–50. https://doi.org/10.23736/S0021-9509.16.09751-2.
13. Scali ST, Ayo D, Giles KA, et al. Outcomes of antegrade and retrograde open mesenteric bypass for acute mesenteric ischemia. J Vasc Surg. 2019;69(1):129–40. https://doi.org/10.1016/j.jvs.2018.04.063.
14. Karampinis I, Keese M, Jakob J, et al. Indocyanine Green Tissue Angiography CanReduce Extended Bowel Resections in Acute Mesenteric Ischemia. J Gastrointest Surg.2018;22 (12):2117–2124. https://doi.org/10.1007/s11605-018-3855-1.

Part IX
Future

Future Research on the Role of the Mesentery in Health and Disease

50

Eli D. Ehrenpreis, Steven D. Wexner, John C. Alverdy, and David H. Kruchko

Introduction

Throughout this book, we have emphasized the recent recognition of the mesentery as an independently functioning organ. The designation of the mesentery as a "new" organ in the body encourages reinterpretation of the mesentery's role in health and disease. Redefinition of the mesentery from apportioned connective tissue aggregates within the abdominal cavity to a fully integrated, operational system allows us to investigate this "new" organ from basic and clinical research standpoints. Future studies will provide a deeper understanding of the role that the mesentery plays in healthy and diseased states. In turn, new investigative breakthroughs will translate into preventive and therapeutic advances. In this chapter, the authors review selected fields of potential research on the mesentery related to their specific areas of interest. These areas include genetics, clinical pharmacology, imaging, mesenteric pathologic disorders, and surgical approaches to the mesentery.

E. D. Ehrenpreis (✉) · D. H. Kruchko
Department of Medicine, Advocate Lutheran General Hospital, 1775 Dempster St, 6 South, Park Ridge, IL 60068, USA
e-mail: e2bioconsultants@gmail.com

S. D. Wexner
Department of Colon and Rectal Surgery, Cleveland Clinic Florida, Weston, IL 33331, USA

J. C. Alverdy
Department of Surgery, University of Chicago, Chicago, IL 60637, USA

E. D. Ehrenpreis et al. (eds.), *The Mesenteric Organ in Health and Disease*,
https://doi.org/10.1007/978-3-030-71963-0_50

Genetic Investigations and Mesenteric Disease

Improved understanding of mesenteric diseases will require additional research involving comprehensive genetic profiling. Type I autoimmune pancreatitis (AIP) is a disease of the mesentery that has been well studied in Japanese populations. AIP research has identified strong genetic associations that will ultimately lead to improved and more personalized future treatment options. However, less is known about the genetic profile of other autoimmune mesenteric diseases, such as mesenteric panniculitis. In order to elucidate key genetic components contributing to these diseases, recently developed variations in next-generation sequencing will be utilized. Techniques that could prove useful include those that specifically analyze genetic expression (such as methylation sequencing). Understanding the immunologic structure and function of the mesentery may be accomplished using immune-repertoire sequencing.

Clinical Pharmacology

While mesenteric-based strategies are being developed in the surgical treatment of the disease, medical therapeutics have lagged behind. Since we now understand that the mesentery is an important component of health and disease states, mesenteric-directed pharmacotherapy may hold the key to enhancing treatments for diseases as diverse as adult onset diabetes, autoimmune disorders and the systemic inflammatory response. For example, biologic therapies that interact directly with cells within the mesentery that are responsible for the sampling and reaction to ingested food-borne peptides could prevent or treat conditions such as Crohn's disease and food allergies.

At present, the clinical pharmacokinetic properties of the mesentery are unknown. There are a variety of reasons for this knowledge void, the principle one being that it is not possible to sample drug levels within the mesentery in vivo. This limits the ability to optimize small molecule pharmaceuticals that concentrate within the mesentery. Use of noninvasive methods to measure mesenteric drug levels are anticipated for the future. These methods in turn hold the promise of improving therapy for primary mesenteric diseases such as mesenteric panniculitis, and the development of drugs that can be used to treat or prevent conditions in which the mesentery plays an important role in their pathophysiology.

Mesenteric Panniculitis

Mesenteric panniculitis can be incidentally found by abdominal imaging studies. These patients have undergone imaging studies for one of two reasons; they have unexplained abdominal pain or they have imaging performed as part of a diagnostic evaluation in the setting of a known neoplastic disease. Future studies are anticipated that will better differentiate mesenteric panniculitis from mesenteric neoplasms, perhaps resulting in fewer patients requiring biopsies of lesions for a definitive diagnosis. Furthermore, better longitudinal data when more patients have been studied may allow for stratification of treatment and prognosis for individual patients. Determining the specific components of immune dysregulation that are pathognomonic of mesenteric panniculitis will allow for the optimization of treatments for the disease. In addition, a better understanding of psychosocial factors associated with rare diseases and specific occurrences related to the symptoms and uncertain nature of mesenteric panniculitis at present has not been studied.

New and Future Developments in Mesenteric Surgery

Dr. Coffey has pointed out that a great deal of time is directed at accessing and mobilizing portions of the mesentery prior to their actual resection. New surgical instruments and techniques are anticipated to improve the processes of peritonotomy, mesofascial separation, and mesenterectomy. Some of these may include the use of gas and hydrodissection. He also anticipated the possibility that mesofascial separation may be performed with the assistance of a radiologist via radiologic cannulation of the mesofascial interface. This technique may eventually be combined with magnetic-based mobilization of the intestine, to allow for less invasive forms of mesosigmoidectomy. New understanding the the anatomy of the mesorectal region of the mesentery, in particular a clear demonstration of the circum-mesorectal plane separating the entirety of the mesorectum from adjacent structures may be helpful in maximizing performance of mesosigmoidectomy.

Inflammatory Bowel Disease

As has been discussed and debated in the chapters of this textbook, the role of the mesentery in patients with Crohn's disease remains unclear. Some surgeons feel that the mesentery suffers inflammation as a result of intestinal involvement. However, other surgeons feel that the mesentery in and of itself may harbor abnormalities related to Crohn's disease. Future research will hopefully help us elucidate the role of the mesentery in remission and recurrence of Crohn's disease and based upon that knowledge guide both medical therapy and surgical resections. In addition, pouchitis may well be related to mesenteric factors in either the small bowel mesentery or in any retained presacral mesorectum on which the pouch is resting. Our methods of pouch construction as well as preventative and therapeutic measures for pouchitis might be contingent upon a better understanding of any role of the mesorectum in ulcerative colitis and ulcerative proctosigmoiditis. It is theoretically possible that an inflamed mesentery may subsequently lead to inflammation of either the colon or the reconstructed pouch. Alternatively, it is possible that inflammation within the mesentery or mesorectum prevent or make more difficult the medical or surgical management of Crohn's disease and mucosal ulcerative colitis.

Radiologic Advances

Mesenteric and Bowel Injuries

Delay in the diagnosis of bowel and mesenteric injuries is relatively common and increases the likelihood of morbidity and mortality. Use of multidetector computed tomography (CT) has resulted in the identification of specific imaging findings of bowel and mesenteric injury from penetrating and blunt trauma. Because these findings may be subtle and occur in the setting of other complex injuries, future work in improving the visualization of bowel and mesenteric injuries as well as education of radiologists in the identification of these findings is anticipated to produce timely diagnoses and improved outcomes in these patients.

Fluorescence Lymphangiography

Keller et al have used fluorescence lymphangiography to demonstrate a watershed area in the ileocolic region. Fluorescence lymphangiography (also called lymphoscintigraphy) using intravenous indocyanine green enhances the visualization of lymphatic channels in real-time. By highlighting sentinel lymph nodes, this technique holds promise for increasing the precision of mesenteric dissection during bowel resection for colon cancer.

Use of Next Generation Technology to Advance Our Understanding of the Role of the Mesentery in Human Disease

Access to the mesentery via endoscopic fine needle aspiration, laparoscopy and direct image-guided puncture are making mesenteric tissue harvest a non-morbid practical advance in diagnosing disorders of the mesentery. Advances in immunohistochemistry, metabolomics, use of mass spectrometry, 16 s rRNA analysis of microbial communities, CyTOF analysis of lymphocyte phenotypes, flow cytometry, etc. can now be applied to procured mesenteric tissues. Use of techniques such as single cell RNA-seq can now be applied to individual cell types (i.e.fat cells, lymphocytes, macrophages, etc.) to be able to transcriptionally analyze specific subsets of cells. In addition, dual RNA seq can be used to determine the transcriptome of both microbes and host cells to define the "interactome" that characterizes specific mesenteric diseases. The future lies in properly applying these techniques to advance our understanding of the mesentery as an independently function organ in human disease.

Suggested Reading

Medical and Pharmacology

1. Li Y, Zhu W, Zuo L, Shen B. The role of the mesentery in crohn's disease: the contributions of nerves, vessels, lymphatics, and fat to the pathogenesis and disease course. Inflamm Bowel Dis. 2016 Jun;22(6):1483–95.
2. Rivera ED, Coffey JC, Walsh D, Ehrenpreis ED. The mesentery, systemic inflammation, and crohn's disease. Inflamm Bowel Dis. 2019 Jan 10;25(2):226–34.
3. Ehrenpreis ED, Roginsky G, Gore RM. Clinical significance of mesenteric panniculitis-like abnormalities on abdominal computerized tomography in patients with malignant neoplasms. World J Gastroenterol. 2016;22(48):10601–8.
4. Argikar AA, Argikar UA. The mesentery: an ADME perspective on a 'new' organ. Drug Metab Rev. 2018 Aug;50(3):398–405.

Surgery

5. Keller DS, Joshi HM, Rodriguez-Justo M, Walsh D, Coffey JC, Chand M. Using fluorescence lymphangiography to define the ileocolic mesentery: proof of concept for the watershed area using real-time imaging. Tech Coloproctol. 2017 Sep;21(9):757–60.
6. Coffey JC. Future directions. In: Mesenteric Principles of Gastrointestinal Surgery: Basic and Applied Science. 1st Edition. John Calvin Coffey, Rishabh Sehgal, Dara Walsh. CRC Press, Taylor & Francis Group, Boca Raton, Florida, 2017.
7. Rosenthal RJ, Bashankaev B, Wexner SD. Laparoscopic management of inflammatory bowel disease. Dig Dis. 2009;27(4):560–4.

Radiology

8. Bates DD, Wasserman M, Malek A, Gorantla V, Anderson SW, Soto JA, LeBedis CA. Multidetector CT of surgically proven blunt bowel and mesenteric injury. Radiographics. 2017 Mar–Apr;37(2):613–25
9. Landry BA, Patlas MN, Faidi S, Coates A, Nicolaou S. Are we missing traumatic bowel and mesenteric injuries? Can Assoc Radiol J. 2016 Nov;67(4):420–5.
10. Li Y, Zhu W, Zuo L, Shen B. The role of the mesentery in crohn's disease: the contributions of nerves, vessels, lymphatics, and fat to the pathogenesis and disease course. Inflamm Bowel Dis. 2016 Jun;22(6):1483–95.

Next Generation Technology

12. Adey AC. Integration of single-cell genomics sets. Cell. 2019;177(7):1677–9.
13. Westermann AJ, Gorski SA, Vogel J. Dual RNA-seq of pathogen and host. Nat Rev Microbiol. 2012 Sep;10(9):618–30.

Index

E. D. Ehrenpreis et al. (eds.), *The Mesenteric Organ in Health and Disease*,
https://doi.org/10.1007/978-3-030-71963-0

W

The manufacturer's authorised representative in the EU is Springer Nature Customer Service Centre GmbH, Europaplatz 3, 69115 Heidelberg, Germany. If you have any concerns regarding our products, please contact ProductSafety@springernature.com

Printed and bound by CPI Group (UK) Ltd, Croydon, CR0 4YY
17/07/2026
02170250-0003